HARMONISATION OF TECHNICAL STANDARDS IN THE EC

HARMONISATION OF TECHNICAL STANDARDS IN THE EC

Second Edition

Sebastian Farr, LLB,
Partner, Simmons & Simmons,
London and Brussels

JOHN WILEY & SONS

Chichester · New York · Brisbane · Toronto · Singapore

First edition published 1992
This edition published in the United Kingdom by John Wiley & Sons Ltd,
Baffins Lane, Chichester,
West Sussex PO19 1UD, England

National 01243 779777
International (+44) 1243 779777

Published in North America by John Wiley & Sons,. Inc
7222 Commerce Center Drive
Colorado Springs
CO 80919, USA

Other Wiley Editorial Offices

John Wiley & Sons Inc., 605 Third Avenue,
New York, NY 10158-0012, USA

Jacaranda Wiley Ltd, 33 Park Road, Milton,
Queensland 4064, Australia

John Wiley & Sons (Canada) Ltd, 22 Worcester Road,
Rexdale, Ontario M9W 1L1, Canada

John Wiley & Sons (SEA) Pte Ltd, 37 Jalan Pemimpin #05-04,
Block B, Union Industrial Building, Singapore 2057

British Library in Cataloguing Publication Data

A catalogue record for this book is available from the British Library

ISBN 0-0471-95926-X

Typeset in 10/12 Baskerville by Servis Filmsetting Ltd, Manchester
Printed and bound in Great Britain by Bookcraft (Bath) Ltd

This book is printed on acid-free paper responsibly manufactured from sustainable forestation, for which at least two trees are planted for each one used for paper production.

Contents

Preface to the Second Edition

Since the First Edition of *Harmonisation of Technical Standards in the EC* was produced in November 1992, much has happened in the field of standards. At an EC level, we have seen the introduction of 4 more "new approach" Directives, as well as a proposal for a further one. In addition, the Low Voltage Directive, so important to a number of sectors of industry, has been deemed to be a new approach Directive. All of those Directives have been amended by Directive 93/68 which harmonised the CE marking and the amendments made by that Directive are fully reflected in Annex 6 to the Second Edition.

Furthermore, many of the Directives dealt with in the First Edition have now entered or are now entering into force and this has resulted in an even greater degree of interest from sectors of industry affected by the new approach Directives.

At an international level, we have seen the signing of the GATT/WTO Agreement on Technical Barriers to Trade in April 1994.

The way in which the process of standardisation under the EC's new approach and global approach might impinge upon other aspects of the EC Treaty (principally the competition rules and the rules on free movement of goods) has also had an airing with the Commission's investigation into the ETSI policy for standardisation. That is dealt with in a new section in the book.

There has also been an enormous increase in interest on the part of industry in the European Community's approach to standardisation. This is not only because it is necessary to ensure that products comply with the essential requirements of the Directives, but also in terms of a marketing tool. As a marketing tool it is very useful for industry to advertise that its products already comply with the requirements which may not yet be mandatory. I have also heard stories of time being booked with various testing houses for the purpose of testing *competitors'* products (presumably with a hope that they fail the tests).

In compiling the Second Edition to the book, I have had tremendous support from my publishers, which in turn led me to deliver the manuscript (almost) on time.

I would like to thank Monique Blondin and Cathy Allen who had the unenviable task of typing the manuscript and making the necessary amendments. Caroline Harris researched the annexes.

The law is stated as at 1 December 1995 and if there are any mistakes, they are mine.

I would very much welcome constructive criticism by readers of the book and my contact address, telephone and fax numbers are set out below.

Finally, my wife Pat deserves a great deal of thanks for encouraging me to produce a Second Edition of the book. She probably knows more about the Directives referred to in this work than I do.

Simmons & Simmons
118 Rue d'Arlon
1040 Brussels
Belgium

Telephone No: (32 2) 280 16.70
Fax No: (32 2) 230 04.82

Sebastian Farr
Brussels
1 December 1995

Preface to the First Edition

The European Community's approach to the harmonisation of technical standards is of vital importance in its quest to secure a truly single European market within which goods may circulate freely. Despite the importance of this topic very little has been written on it. I decided to write the book in response to an overwhelming desire for information on the subject from manufacturers and their professional advisers, trade associations and Chambers of Commerce both within the EC and in America and the Far East.

I have endeavoured to describe the framework of the EC's "new approach" to the harmonisation of technical standards and the global approach to conformity assessment. In so doing, I have attempted to describe briefly the EC's old approach, which was characterised by years of negotiation over highly complicated Directives, some of which were never adopted. After outlining the role of the European standards organisations, I have summarised the existing new approach Directives, and the full texts of these are set out in the Annexes.

In writing the book I have assumed a basic knowledge in the reader of the workings of the EC institutions. Where necessary I have made reference in the footnotes to other works on particular topics (for example on the detailed rules concerning Articles 30–36 of the EEC Treaty on the free movement of goods).

Overall, the book is intended to provide the reader with an easy guide through this complex field, but inevitably individual readers will have questions relating to areas going beyond the scope of the book.

The law is stated as at 30 September 1992.

I would like to thank my publishers for the opportunity to produce this book and for providing the correct amount of encouragement when it was most needed.

Finally, I would like to thank my wife, Pat, who typed the manuscript, had the patience to bear with me and to whom the book is dedicated.

Sebastian Farr
Brussels
30 September 1992

Tables

TREATIES

EC DIRECTIVES

DECISIONS

PROPOSED LEGISLATION

CASES

Contents of Chapter 1
Introduction

Chapter 1
Introduction

1. The concept of a single European market

The concept of a single European market in which goods must be allowed to flow freely across Member States' borders is as old as the date of signature of the EEC's Treaty of Rome itself (25 March 1957). Article 30 of the EEC Treaty, for example, prohibits quantitative restrictions and measures having an equivalent effect which hinder trade between Member States. Furthermore, Articles 85 *et seq* prohibit anti-competitive behaviour which has as its object or effect the prevention, restriction or distortion of competition insofar as it affects trade between Member States. These articles of the Treaty also cover abuses of monopoly power and require organisations to which states have granted special or exclusive rights to conform with the competition rules, save where this would prevent them carrying out their tasks. **1.1**

However, the rules set out in the EEC Treaty, have still not resulted in a single European market. The aim of creating a frontier-free European Community was given greater impetus by the publication of the EC's Commission's White Paper in 1985 entitled "Completing the Internal Market" (Com (85) 310 final June 1985). This identified three areas essential for the completion of the internal market (abolishing physical, fiscal and technical barriers). It laid down an ambitious 300 point legislative programme necessary for abolishing the three types of barriers mentioned. In fact, the EC Member States had already taken a step towards a common system of standardisation, regulation and certification with the adoption in 1983 of Directive 83/189 (OJ 1983 L109/8 as amended). This Directive (which is known as the Technical Standards Directive) obliges Member States to notify all draft standards and technical regulations to the Commission and other Member States for prior examination, thus enhancing transparency in this area.[1] **1.2**

In July 1984 the Member States adopted a statement in which they accepted that their policies on health, safety and consumer protection were equivalent. Having established this, future EC legislation could concentrate on establishing the essential requirements for safety, health and consumer **1.3**

[1] Dir 83/189 is dealt with in Chap 2 and the consolidated text is set out in Annex 1. It entered into force on 1 January 1985.

protection in order to complete the single European market. The identification in EC legislation of "essential requirements" is the basis of the EC's "new approach" to the harmonisation of technical standards and has greatly speeded up the process of abolishing technical barriers to trade within the EC.

1.4 The new approach to the harmonisation of legislation was officially confirmed in May 1985 by the EC's Council of Ministers which adopted a Resolution on the New Approach to Technical Harmonisation and Standardisation.[2] This Resolution was supplemented in December 1989 by the adoption of a Council Resolution on a Global Approach to Conformity Assessment. The latter Resolution aimed at defining the EC's policy regarding conformity assessment (*i.e.* the methods by which manufacturers must prove that their products conform to the relevant legislation).[3]

1.5 Finally, the EC's approach to the harmonisation of technical standards and the adoption of a global approach to conformity assessment was taken further forward with the adoption in October 1990 of a Commission Green Paper on the Development of European Standardisation: Action for Faster Technological Convergence in Europe (OJ 1991 C120/1). This consultation document invited comments from all interested parties and constituted the most wide-ranging review to date of European standardisation. Following receipt of comments from interested parties, including third countries, the EC Commission published a second communication in December 1991 entitled Standardisation in the European Economy which was a follow-up to the Green Paper, outlining the responses which had been received and the Commission's recommendations for follow-up action.[4]

2. The "old approach" to technical harmonisation

1.6 The need for a new approach and the adoption of the EC's policy as outlined above can perhaps best be seen by a brief analysis of the EC's "old approach" to technical harmonisation. The old approach of the 1960s and 1970s was characterised by "maximilistic" harmonisation. The EC legislation set out in exhaustive detail the performance objectives and design specifications for the products covered. The need to produce detailed draft legislation on the basis of preparatory work carried out by experts created a number of difficulties which can be summarised as follows:[5]

---

[2] Council Resolution of 7 May 1985 on the new approach to technical harmonisation and standardisation (OJ 1985 C136/1). The terms of the Resolution are described in Chap 3, *post*.

[3] Council Resolution on the global approach to conformity assessment of 21 December 1989 (OJ 1990 C10/1). See Chap 5.

[4] Standardisation in the European Economy (Follow up to the Commission Green Paper of October 1990), 16 December 1991 COM (91) 521 final.

[5] For a detailed assessment of the problems associated with the "old approach" to harmonisation see: CEPS Standards Programme : Paper No 1 (CEPS Working Party Report No 5 – March 1992).

(i) Time-consuming and cumbersome procedures

The old approach necessitated time-consuming preparation of detailed technical legislative proposals. This in turn meant that only a small number of proposals for technical legislation could be addressed at any one time. **1.7**

(ii) Unanimity

The basic legislative rule of Article 100 of the EEC Treaty requiring unanimity within the Council of Ministers before legislation could be adopted, meant that individual States could easily block measures they did not find acceptable. The time taken to agree compromise proposals meant that the rate of adoption of EC legislation was unacceptably low. Furthermore, a number of harmonisation proposals were dropped when it was clear that no agreement would be reached within the EC Council of Ministers. **1.8**

(iii) Too much uniformity

The maximilistic approach which sought to achieve total harmonisation was at odds with the necessity for product flexibility to take into account varying regional and national considerations. It also hampered innovation by product manufacturers. **1.9**

(iv) Failure to draw upon European standardisation work

Separate from the work being carried out by the EC Commission, various voluntary European standards bodies were producing European-wide product specifications. These bodies, which were able to draw upon considerable expertise from national standards bodies, were not sufficiently involved in the Commission's work. **1.10**

(v) Proliferation of new national regulations

Although the EC's General Programme for the Elimination of Technical Barriers to Trade Caused by Disparities Among National Legislation[6] (the old approach) contained a "gentlemen's agreement" preventing national legislation in areas where EC legislation was planned, the agreement was of little value. Other technical barriers were created by the implementation of new national regulations in areas *outside* the ambit of the Programme. **1.11**

---

[6] Adopted by the EC Council of Ministers in May 1969. The General Programme embodied the EC's "old approach" to technical harmonisation. It was supplemented by a further programme in 1973 concerning technical barriers in the foodstuffs and industrial product sectors.

Furthermore, Member States were slow in providing information to the EC Commission on new national measures in areas covered by the Programme. As a result, even where detailed EC legislation was adopted and replaced national technical barriers, new national technical barriers were erected in other sectors. It was to increase the flow of information on national rules that the Council of Ministers adopted Directive 83/189 (OJ 1983 L109/8, as amended).

(vi) Non-recognition of testing and certification processes

1.12 Member States did not respond to the invitation contained in the harmonising Directives to accept other Member States' tests and national approvals. Futhermore, no provision was made for the mutual recognition of test certificates, so the Member States continued to apply their own certification procedures to imports, thus duplicating certification costs. These difficulties also acted as a significant disincentive to export to other Member States.

(vii) The position of third country producers

1.13 The interests of third country producers were seldom considered when implementing the old approach to technical harmonisation.

(viii) Failure to implement by the Member States

1.14 Member States were notoriously slow to implement EC Directives into national legislation, and in some instances failed completely to do so. The result was that technical barriers often survived the adoption of EC legislation, the only available remedy being for the EC Commission to take action under Article 169 of the EEC Treaty against recalcitrant Member States for "failure to fulfil a Treaty obligation". But even that procedure is time-consuming and does not guarantee that a Directive will be speedily implemented into national legislation.

1.15 Although some progress was made under the old approach, it was clear that the problems outlined above meant that the EC's desire to create a single European market for the free flow of, *inter alia*, goods could not be achieved without a change in policy. An example of the difficulties created by the old approach can be seen from the fact that the EC Commission originally envisaged the adoption of 124 Directives in the 18 month period up to the end of 1970. In the end the Council of Ministers succeeded in adopting only one measure within the deadline. Although this legislative record subsequently improved, the defects in the old approach were manifest.

3. Technical standards and conformity assessment as barriers to trade

Before looking at the EC's new approach to technical harmonisation and conformity assessment, it is important to summarise the way in which technical regulations, industry standards and conformity assessment procedures can constitute barriers to trade. **1.16**

(i) Technical regulations

Technical regulations lay down specifications regarding the form, manufacture and the performance of products, services and processes. They are set out in laws adopted by governments and enforced by the appropriate public authorities. Products which do not comply with the regulations are generally illegal and may not be sold on the national market. Furthermore, sanctions are generally attached to the marketing of products which do not satisfy the technical regulations applicable to them. The traditional justification for the adoption of technical regulations include the safeguarding of a number of public policy objectives including health and safety, protection of the environment and the regulation of equity and fairness in commercial transactions. Technical regulations are generally supplemented by legislation covering product liability and misleading advertising. **1.17**

In a number of respects, technical regulations for products provide certainty for producers. They help engender consumer confidence in the products marketed and they also may provide manufacturers with a comprehensive set of technical criteria against which to manufacture and market their products. However, these benefits can be offset by a number of disadvantages. Technical regulations may be inflexible and excessively detailed. They might also overlap or duplicate other technical regulations which offer equally satisfactory safeguards for consumers. However, one of the most important negative effects of national technical regulations is the possible hindrances to international trade. Where technical regulations apply to a product, other Member States' products may be excluded from the export market unless they conform to local specifications. This gives rise to import prohibitions or at least disincentives. The only solution for the exporting producer is to adapt the product to meet the local requirements of the importing country. One example given by the UK Department of Trade and Industry is of the manufacturer of woodworking and metal sawing equipment for the professional market who was obliged to produce up to 8 different models of each machine to meet differing national standards.[7] **1.18**

---

[7] See "Keeping your product on the market – new common rules" the Department of Trade and Industry, December 1991.

1.19 The barriers to trade caused by differences in national technical product regulations are principally the increase in production costs which arise in the following ways:

(a) Product development

1.20 Even before a manufacturer can commence production, he must gather detailed and accurate information regarding the scope and content of regulations which are applicable in each of his potential export markets. This is both time-consuming and expensive in terms of fees paid to professional advisers.

(b) Direct costs

1.21 Different national product rules may oblige manufacturers to manufacture different models for different export markets (as with the manufacturer of woodworking equipment noted above). This feature reduces the benefit of economies of scale and adds to the average unit cost of production.

(c) Distribution and stock costs

1.22 A manufacturer will have to maintain separate stocks for each model it manufactures. This in turn means that capital is tied up without yielding any return.

The cumulative effect of the barriers to trade created by different technical regulations mean that unit cost is increased, prices to the consumer are increased and this in turn may lead to a fall-off in demand for the manufacturer's product.

(ii) Industry technical standards

1.23 The essential feature of a standard is that it is of a voluntary nature.[8] In this respect standards are to be distinguished from technical regulations which have been considered above. Furthermore, standards can be distinguished from manufacturers' specifications which are drawn up by individual manufacturers and cover only their products. However, it is not unusual for the specifications of the market leader to become the industry standard. Industry technical standards produce considerable benefits both to producers and to consumers. They generally result in the reduction of average production costs for a number of reasons including: exploitation of economies of scale, use of interchangable parts, simplification of production and construction processes, and reduction in product liability insurance as insurers may be prepared to demand lower premiums for products which conform to recognised industry standards.

1.24 In addition, industry standards benefit consumers through lower sale prices which can be passed on to the consumer by virtue of the reduction in

[8] The International Organisation for Standardisation (ISO) Guide 2 (1991) states that a standard is "... a document, established by consensus and approved by a recognised body, that provides, for common and repeated use, rules, procedures or charactertistics for activities or their results, aimed at the achievement of the optimum degree of cover in a given context."

production costs. The CEPS Standards Programme, Paper No 1[9] identifies three different types of industry standards:

> quality standards, which enable users to ascertain the fitness of a product for a particular purpose (including its reliability and health and safety features);
> compatibility standards, which permit different products or components to be used together (for example computers and printers);
> standards which reduce the variety of products, which are intended to prevent any unnecessary duplication and diversity of models.

Furthermore, it is possible to classify industry technical standards in terms of design or performance standards. Design standards lay down the requirements of the physical attributes of a given product, and performance standards relate to the use to which a product may be put. Normally, performance standards concern health and safety requirements. **1.25**

Prior to the EC's programme of harmonising technical standards, national standards organisations were producing industry standards which duplicated one another and therefore resulted in differing (but equally acceptable) standards. For the exporter this presented obvious problems of having to satisfy different national standards in different market places. Products which were mass produced in, for example, the United Kingdom, might have encountered difficulties when being marketed in other Member States. This in turn meant that the free movement of goods, and indeed competition, was hindered because national standards generally favour national products. It is therefore clear that industry standards can become significant obstacles to exporters wishing to penetrate foreign markets. The combined effects of the negative consequences of a proliferation of national standards result in increased production costs, lack of economies of scale and disincentives to export to foreign markets. This in turn results in competition which is less keen. The end result is generally that consumers have to pay more for their products. **1.26**

(iii) Conformity assessment

Certification and conformity assessment is the process by which the manufacturer proves that his products satisfy the relevant regulations and standards. In some cases, self-certification may be possible, but in other cases the intervention of an independent testing house may be required by national legislation. Once a manufacturer has adapted his product in order to meet the importing country's rules, he must then submit his product for testing and certification to the appropriate authorities in the importing country. But similar conformity assessment procedures may have had to have been fulfilled in his own country if the product is to be sold on the **1.27**

[9] CEPS Standards Programme: Paper No 1 (March 1992) "The European Community without technical barriers" – Report No 5 at 7.

domestic market. In other words, a product may well have to go through a number of conformity assessment procedures in different Member States. This will usually involve the supply to the importing country's authorities of the necessary documentation and product models for examination. This entails time delays and expense including the fees charged by the testing houses. These cost barriers to trade are characterised by:

1.28 higher production costs which are passed on to the consumers (and which may lead to a falling off of demand);
a reduction in external competition, allowing manufacturers in the importing countries to raise their prices in the knowledge that their overseas competitors' products will not reach the market for some time and when they do, they will be saddled with higher costs.

4. Concept of mutual recognition in eliminating barriers to trade

(i) Introduction

1.29 The question of eliminating technical barriers to trade was considered by the European Court of Justice (ECJ) in 1979 when it pronounced upon the concept of "mutual recognition" in the context of Articles 30 to 36 of the EC Treaty (on the free movement of goods) in the *Cassis de Dijon* case.[10] The ECJ held (at 662) that in the absence of harmonised EC rules:

> ". . . obstacles to movement within the Community resulting from disparities of . . . national laws . . . must be accepted insofar as those provisions may be recognised as being necessary in order to satisfy mandatory requirements relating in particular to the effectiveness of fiscal supervision, the protection of public health, the fairness of commercial transactions and the defence of the consumer . . ."

1.30 In other words, with the limited exceptions mentioned, Member States must accept the principle of equivalence. Thus, a product which is lawfully produced and marketed in one Member State must be admitted into another. In this way, Member States are obliged to accord "mutual recognition" to each other's national rules.[11] The importance of this concept was specified in the Commission's 1985 White Paper (Com (85) 310 final, June 1985) where it is stated:

---

[10] Case 102/78 *Rewe Zentrale* v *Bundesmonopolverwaltung für Branntwein* [1979] ECR 649.
[11] A full examination of the rules contained in Articles 30–36 of the EEC Treaty goes beyond the scope of this book. Reference should be made to Gormley, *Prohibiting Restrictions on Trade within the EEC* (1985) and Oliver, *Free Movement of Goods* (1995).

> "... the general principle should be approved that, if a product is lawfully manufactured and marketed in one Member State, there is no reason why that should not be sold freely throughout the Community. Indeed, the objectives of national legislation, such as the protection of human health and life and of the environment, are more often than not identical. It follows that the rules and controls developed to achieve those objectives, although they may take different forms, essentiually come down to the same thing, and so should normally be accorded recognition in all Member States...."

The White Paper recognised that in view of the Council of Ministers' acceptance[12] of the essential equivalence of the objectives of national legislation, mutual recognition could be an effective strategy for bringing about a common market in a trading sense. This strategy is supported in particular by Articles 30 to 36 of the EEC Treaty, which prohibit national measures which would have excessively and unjustifiably restrictive effects on the free movement of goods. **1.31**

The White Paper stressed that in cases where harmonisation of regulations and standards is not considered essential either from a health, safety or an industrial point of view, immediate and full recognition of differing quality standards and other national regulations must be the rule. In particular, the prohibition of imports into one Member State could not be based upon the sole argument that an imported product has been manufactured according to different specifications from those used in the importing country. Similarly, the importer should not be required to submit the imported products to additional technical tests or to certification procedures in the importing state (White Paper, sections 77, 78). Finally, the Commission stated that it would use all its powers available under the EEC Treaty, particularly Articles 30 to 36, to reinforce the principle of mutual recognition. **1.32**

Whilst the mutual recognition of national standards and conformity assessment procedures is an important tool in seeking to achieve a European Community without technical barriers to trade, it might prove insufficient for establishing an expanding European market based upon open competition. Therefore, the approach of the EC Commission in proposing legislation to the Council of Ministers has been to decide what it is necessary to harmonise on the one hand, and what may be left to mutual recognition of national regulations and standards on the other. In other words, on the occasion of each harmonisation initiative, the Commission must determine whether the existing national regulations are excessive in relation to health and safety requirements, and therefore constitute unjustified barriers to trade according to Articles 30 to 36 of the EEC Treaty (White Paper, section 65). **1.33**

---

[12] For the Council of Ministers' July 1984 "Conclusions on Standardisation" see Chap 3, *post* and OJ 1985 C136/2.

(ii) Limitations on the concept of mutual recognition as a method of eliminating barriers to trade

1.34 There are two important limitations on the concept of mutual recognition as a tool for eliminating barriers to trade:

> The principle of mutual recognition is inapplicable if a Member State can point to "mandatory requirements" concerning, *inter alia*, fiscal supervision, the protection of public health, the fairness of commercial transactions and the defence of the consumer.[13]
> Secondly, it may be difficult for an exporter in Member State "A" to know whether the technical regulations in Member State "B" are in fact equivalent to those in Member State "A". If the goods are seized by the customs authorities on import in Member State "B" on the grounds that the goods do not meet national requirements, it may be necessary for the exporter to take proceedings in Court, which is both time consuming and expensive.

The existence of harmonised standards under EC Directives avoids these difficulties.

[13] These instances in which Member States may decline the principle of mutual recognition were specifically referred to in the passage quoted from Case 102/78 (see above). See also Art 36 of the EEC Treaty which spells out similar circumstances in which Member States may refuse to apply the principle of the free movement of goods, laid down in Art 30 of the EEC Treaty.

Contents of Chapter 2

The Technical Standards Directive

Chapter 2
The Technical Standards Directive

1. Introduction

As indicated above (Chap 1, 1.6 *et seq*) one of the major difficulties with the old approach to the harmonisation of technical standards was that the EC could not effectively prevent Member States from adopting new product regulations giving rise to technical trade barriers. The adoption in March 1983 of Council Directive 83/189 (OJ 1983 L109/8) was an attempt to tackle this difficulty by imposing information and notification requirements when Member States were preparing the adoption of new technical legislation. The operation of the Directive is examined below. Its scope was widened in 1988 to include agricultural products, foodstuffs, medicinal products and cosmetics, and again in 1994 so as to cover *de facto* technical regulations. **2.1**

2. Application of Directive 83/189

Article 1 of the Directive begins by defining a number of terms for the purposes of the Directive, and these definitions establish the scope of the Directive. **2.2**

The term *technical specification* is defined as a specification contained in a document laying down the characteristics required of the product (such as levels of quality, performance, safety or dimensions), including the requirements applicable to the product as regards the name under which the product is sold, terminology, symbols, testing and test methods, packaging, marketing or labelling and the production methods and procedures for agricultural products as defined in Article 38(1) of the EEC Treaty and for products intended for human and animal consumption and for medicinal products as defined in Directive 65/65 (JO 1965, 369). **2.3**

The amended directive also refers to any "other requirement", being a requirement, other than a technical specification, imposed on a product for the purpose of protecting, in particular, consumers or the environmment, and which affects its life cycle after it has been placed on the market, such as conditions of use, recycling, re-use or disposal, where such conditions can significantly influence the composition or nature of the product or its marketing. **2.4**

2.5 A *standard* is defined as a technical specification approved by a recognised standardising body for repeated or continuous application, with which compliance is not compulsory, and which is one of the following: an international standard, a European standard or a national standard.[1]

2.6 A *standards programme* is a work programme of a recognised standardisation body listing subjects on which standardisation work is being carried out.

2.7 A *draft standard* is a document containing the text of the technical specifications concerning a given subject, which is being considered for adoption in accordance with the national standards procedure, as that document stands after the preparatory work and is circulated for public comment or scrutiny.

2.8 *Technical regulation* covers technical specifications, including the relevant administrative provisions, the observance of which is compulsory, *de jure* or *de facto*, in the case of marketing or use in a Member State or a major part thereof, except those provided for in Article 10 of the Directive, prohibiting the manufacture, importation, marketing or use of a product. *De facto* technical regulations include: Member States' measures which refer either to technical specifications or other requirements or to professional codes or codes of practice which in turn refer to technical specifications and compliance with which confers a presumption of conformity with the obligations imposed by the laws; voluntary agreements to which a public authority is a contracting party and which provide for compliance with technical specifications or other requirements, excluding public procurement tenders' specifications; technical specifications or other requirements which are linked to fiscal or financial measures affecting the consumption of products by encouraging compliance with such technical specifications or requirements.

2.9 *Draft technical regulation* means the text of a technical specification including the administrative provisions formulated with the aim of enacting it or of ultimately having it enacted as a technical regulation, the text being at a preparatory stage at which substantial amendments can still be made.

2.10 Finally, the Directive applies to *products* which are defined as any individually manufactured products and any agricultural products.

---

[1] This definition of a "technical standard" accords with the 1979 GATT Agreement on Technical Barriers to Trade. See also now the Uruguay Round Agreement on Technical Barriers to Trade ((1994) Cm 52564). A wider definition of the term "standard" is accepted by the International Organisation for Standardisation (ISO) and also the United Nations Economic Commission for Europe (UNECE). That definition is that a standard is "... a technical specification or other document available to the public, drawn up with the co-operation and consensus or general approval of all interests affected by it based upon the consolidated results of science; technology and experience, aimed at the promotion of optimum community benefits and approved by a body recognised on the national, regional or international level". As pointed out by Nicholas and Repoussard (*Common Standards for Enterprises*), the principal reason for the difference is that the GATT and Dir 83/189 are concerned with standards in terms of barriers to trade, whereas the ISO and UNECE use their definition for the purpose of drawing up standards, hence the broader definition.

The directive does not apply to measures which Member States consider necessary under the Treaty for the protection of persons, in particular workers, when products are used, provided that such measures do not affect the products. **2.11**

3. Requirement on Member States to provide information

Article 2 of the Directive states that the EC Commission and the standards institutions in Lists I and II of the Annex (the various national standards institutes, as well as CEN, CENELEC and ETSI) shall be informed of the new subjects for which the national or standards bodies have decided to prepare or amend a standard, unless it is an identical or equivalent transposition of an international or European standard. This provision of information requirement under the Directive means that each Member State must inform the EC Commission and the National Standards Institutes as well as CEN, CENELEC and ETSI of its forthcoming standards programme. When supplying this information the Member States must indicate whether the proposed standards will represent the transposition of existing international or European standards and/or whether they will incorporate any national divergences or amendments, or whether the programme constitutes a new national standard or amendment thereto.[2] **2.12**

If one or more of the standards institutions wishes to be involved in activities planned by other standards institutions or wish a European standard or other document leading to uniform technical specifications to be drawn up, it must so inform the Commission and the other standards institutions. **2.13**

In addition to receiving the standards programmes, at least every four months the standards institutions and the EC Commission should receive all the draft standards, except where such standards merely transpose the full text of an international or European standard. Again, when a draft is communicated it must be indicated whether the standard will be the transposition of a European or international standard incorporating national divergences, a new national standard or an amendment thereto. **2.14**

[2] The Commission's notice concerning non-respect for Council Directive 83/189 states that national technical standards and regulations adopted in breach of the provisions of Directive 83/189 are unenforceable against third parties, and the Commission would expect national courts to refuse to enforce them. It also points out that in the case where a national draft regulation might be justified under Article 36 EC Treaty but might nonetheless create barriers to intracommunity trade, the Commission can oblige the Member State to suspend the adoption of its technical regulation for a period of twelve months to enable the Commission to initiate EC legislation on the subject. The statement also underlines the absolute obligation to notify national standards (OJ [1986] C245/4).

4. The Standing Committee

2.15 Article 5 of the Directive establishes a Standing Committee which consists of representatives appointed by the Member States and chaired by a representative of the Commission. The Committee can and does call upon the assistance of experts or advisers in carrying out its tasks. It meets at least twice a year with representatives of the national standards institutions, CEN/CENELEC and ETSI. In addition, the EC Commission submits to it a report on the implementation and application of proposals aimed at eliminating existing foreseeable barriers to trade. The Committee draws up an opinion and may propose that the Commission:

2.16 request that the European standards institutions draw up a European standard within a time limit;
ensures that, in order to avoid the risk of barriers to trade, initially the Member States concerned decide amongst themselves on appropriate measures;
takes all appropriate measures;
identifies the areas where harmonisation appears necessary and, in appropriate cases, undertakes harmonisation in given sectors.

2.17 In certain circumstances, the Commission is obliged to consult the Committee, and may consult the Committee in other circumstances (for example, on any preliminary draft technical regulation received by the Commission).

5. Standstill provision – Article 7

2.18 Article 7 lays down a standstill provision which obliges Member States to ensure their standards institutions do not draw up or introduce standards in a given field whilst the European standards requested by the EC Commission from the European standards institutions are being drawn up. The standstill lapses unless a European standard has been introduced within six months following expiry of the time limit fixed for its adoption. The standstill does not apply to the work of standards institutions undertaken at the request of the public authorities to draw up technical specifications or a standard for specific products for the purposes of enacting a technical regulation for such products. The Member States are obliged to notify the EC Commission of such requests and state the grounds for their enactment.

6. Draft technical regulations

2.19 Article 8 of the Directive obliges Member States to immediately communicate to the EC Commission any draft technical regulations, except where such technical regulations merely transpose the full text of an international or European standard in which case information regarding the relevant standard will suffice. Member States must also state why the enactment of the technical regulation is necessary. Furthermore, the text of the basic legislative or regulatory provisions enacting the technical regulations must be supplied to the Commission where this is necessary to enable the Commission to assess the implications of the draft technical regulation.

The Commission will immediately notify the other Member States of any **2.20** draft it has received and may also refer the draft to the Standing Committee.

The other Member States may comment on the draft technical regulation which has been notified.

Article 9 of the Directive provides for the postponement of the adoption **2.21** of a draft technical regulation in certain circumstances. Those circumstances and the periods of postponement are as follows:

– 3 months from the date of notification of the draft technical regulation to the Commission; **2.22**

– 4 months in the case of a draft technical regulation in the form of a voluntary agreement (as defined in Article 1(9), second indent of Directive 83/189) ;

– 6 months in the case of other draft technical regulations;

– 12 months if, within 3 months of the notification, the Commission announces its intention to propose or adopt a Directive, Regulation or Decision on the matter, or if the Commission states that the technical regulation concerns a subject which is covered by an existing proposal for a Directive, Regulation or Decision.

If the Council adopts a "common position" during the standstill period, the **2.23** 12–month period referred to above is extended to 18 months.

The standstill period does not apply to technical specifications or other **2.24** requirements which are linked to fiscal or financial measures affecting the consumption of products by encouraging compliance with such measures, nor other requirements linked to social security systems.

Articles 8 and 9 of the Directive also do not apply to certain matters linked **2.25** to Article 10 of the amended Directive (for example national rules prohibiting manufacture in so far as they do not affect the free movement of goods).

Furthermore, the standstill provisions do not apply in "emergency situa- **2.26** tions" described in Article 9 (7).

2.27 Rules governing national regulations under the EC's technical standards Directive (83/189)

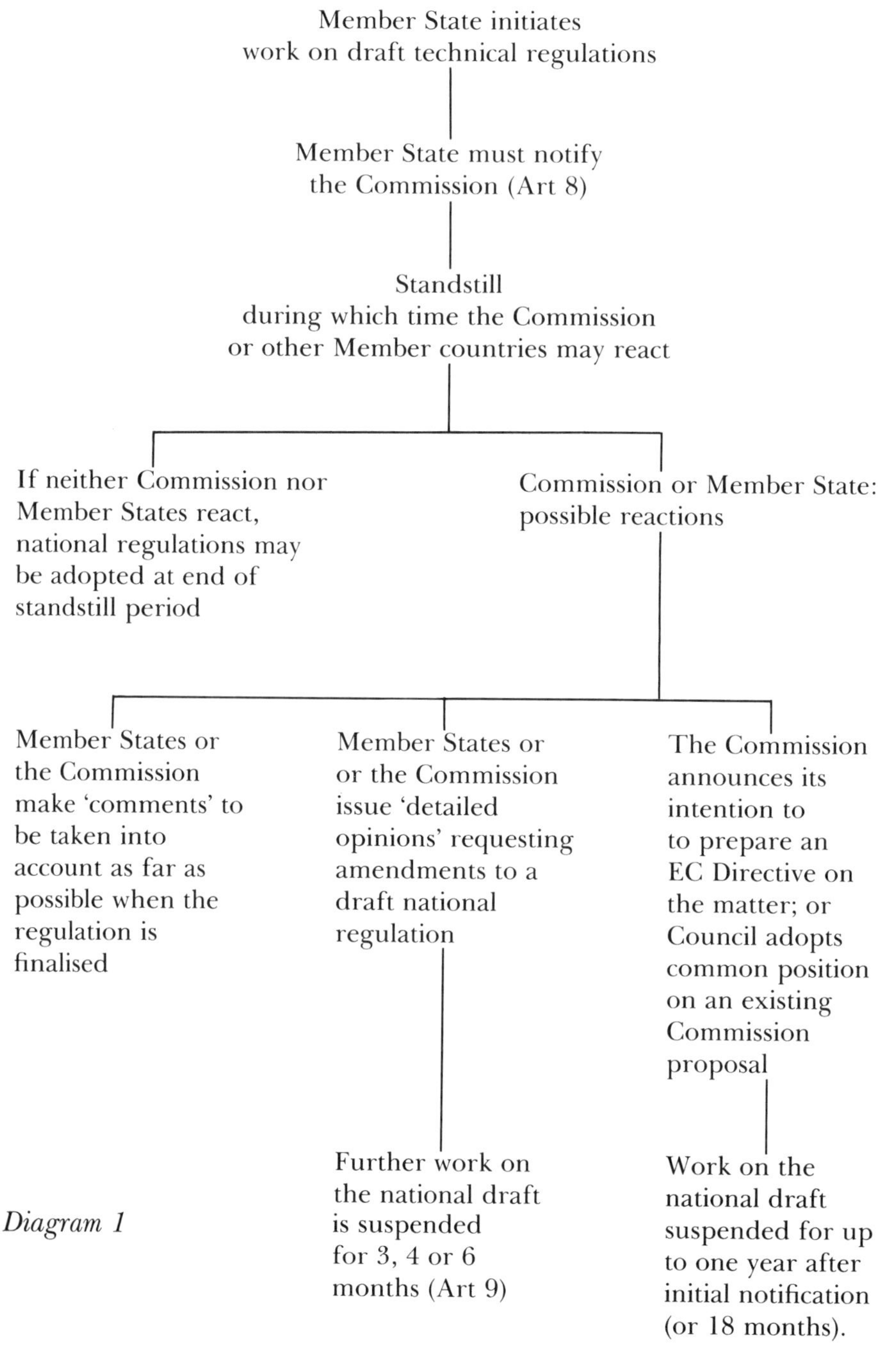

Diagram 1

Contents of Chapter 3

EC's New Approach to Technical Harmonisation and Standards

Chapter 3
EC's New Approach to Technical Harmonisation and Standards

1. Introduction

The starting point of the EC's "new approach" to harmonisation of technical standards is contained in the Council Resolution of 7 May 1985: "A new approach to technical harmonisation of standards" (OJ 1985 C136/1). This Resolution is stated as being an extension of the Council's conclusions on standardisation approved on 16 July 1984 and these are reproduced in Annex 1 to the Resolution of May 1985. Those conclusions stated that the Council believed that standardisation goes a long way to ensuring that industrial products can be marketed freely and also towards creating a standard technical environment for undertakings in all countries, which improves competitiveness not only on the Community market but also on external markets, especially in new technology. The conclusions set down a number of principles for a European standardisation policy. These include: **3.1**

– agreement by the Member States to ensure the mututal recognition of the results of tests in the establishment, where necessary, of harmonised rules as regards the operation of certification bodies; **3.2**
– an extension of the EC practice in matters of technical harmonisation of entrusting the task of defining the technical characteristics of products to standards, preferably European, but if necessary national, where the conditions necessary for this purpose, particularly as regards health protection and safety, are fulfilled;
– a rapid strengthening of the capacity to standardise, preferably at a European level, with a view to facilitating harmonisation of legislation by the EC and also to further industrial development.

The Council Resolution of 7 May 1985 built upon the conclusions of July 1984 by approving certain guidelines set out in Annex 2 to that Resolution, along with a list of main elements to be embodied in the new approach Directives. The Resolution also recognised that a new approach to technical harmonisation would have to be accompanied by a policy on conformity assessment.[1] **3.3**

[1] Conformity assessment is dealt with in Chap 5, *post*.

2. Guidelines for a new approach to technical harmonisation and standards

3.4 The Council Resolution sets out guidelines for the new approach, beginning with the *four fundamental principles* upon which the new approach is based. These principles are:

(a) that legislative harmonisation should be limited to the adoption in the form of Directives of the *essential safety requirements* (or other requirements in the general interest) with which products put on the market must conform, and which should therefore enjoy free movement through the Community;[2]

(b) that the task of drawing up the technical specifications needed for the production and placing on the market of products conforming to the essential health and safety requirements established by the Directives is entrusted to the European standards organisations;[3]

(c) that the technical specifications are not mandatory and maintain their status as voluntary standards;

(d) that national authorities are obliged to recognise that products manufactured in conformity with harmonised standards (or in the absence of European harmonised standards, national standards) *are presumed to conform to the essential requirements* established by the Directives. (In other words, the producer has the choice of manufacturing in conformity with the standards, or manufacturing in conformity with some other standards. In the latter event, however, he would have an obligation to prove that his products conformed to the essential requirements of the Directives. Conformity with the standards referable to the particular Directives is therefore an easier way of complying with the "essential requirements" of applicable Directives. The new approach therefore gives manufacturers greater flexibility. The technical standards produced by the European standards organisations under the Directives only confer one (privileged) method of conformity to the applicable "essential requirements").

3.5 The guidelines recognise that in order for the new approach based on the above-mentioned principles to operate effectively, it is necessary that on the one hand the standards offer a guarantee of quality with regard to the essential requirements under the Directives; furthermore, the public authorities must keep intact their responsibility for the protection of the safety (or other health requirements) on their territory.

---

[2] This principle refers to the adoption of Directives under Art 100 of the EEC Treaty. Since the adoption of the Single European Act, measures necessary to achieve the EC's internal market are adopted under Art 100A, which provides for majority voting in the Council of Ministers. This has also greatly facilitated the speed at which progress has been made in this area.

[3] The structure and role of the European standards organisations and the procedure for the drawing up of standards under the new approach Directives, is discussed in Chap 4, *post*.

The quality of the harmonised standards referred to in the Directives is to be ensured by standardisation mandates given to the European standards organisations, conferred by the EC Commission, the execution of which conforms to the general guidelines which have been agreed between the EC Commission and the European standards organisations.[4] Where national standards are (provisionally) used, their quality is to be verified by procedure at Community level managed by the EC Commission, assisted by a Standing Committee composed of officials from national administrations. 3.6

The guidelines for the new approach to technical harmonisation are designed to halt the proliferation of excessively technical separate Directives for each product (which characterised the old approach to harmonisation). The scope of the Directives according to the general reference to standards formula should encompass wide product categories and types of risk. The aim of the new approach is therefore to complete the complex task of harmonising technical legislation, and on the other hand promote the development and application of European standards. These conditions will not only facilitate the free movement of goods but also improve the competitiveness of European industry. 3.7

3. Main principles of the new approach Directives

The Council Resolution sets out the principals and main elements which should be present in each of the new approach Directives. To that extent the Resolution has provided a "framework" to which the new approach Directives should conform. The Resolution emphasises that the traditional principles justifying the use of legislating by means of Directives in this area are present and include the fact that Member States have the responsibility of ensuring safety on their territory of persons, animals and goods, or the respect of other essential protection requirements in the general interest such as health, consumer or environmental protection. National provisions ensuring such protection must be harmonised in order to ensure the free movement of goods, without lowering existing and justified levels of protection in the Member States. CEN, CENELEC and ETSI (depending on the products covered by the Directive) are the competent bodies to adopt European harmonised standards within the scope of the Directives, in accordance with the guidelines which the Commission, after consultation with the Member States, has signed with these bodies. 3.8

[4] These are: CEN, CENELEC and ETSI. See Chap 4, *post*.

4. Main elements of the new approach Directives: Commission Guidelines

3.9 The Council Resolution sets out the main elements constituting the various new approach Directives and this framework can be seen in the Directives and proposals themselves set out in Annexes 5(i) to 5(xii) to this book. In 1994 the Commission published the "Guide to the Implementation of Community Harmonisation Directives based on the new Approach and the Global Approach" (first edition). This guide is intended as a practical manual for those working with Community technical regulations. It has a modular format consisting of a set of sheets dealing with the main elements of the new approach directives. This format means that the guide can be added to and updated. The guide is a valuable tool to understanding the workings of the new approach directives. Its intention is to assist the various Directorates-General within the Commission to co-ordinate more closely, as well as assisting the vast number of public authorities in their supervisory tasks, as well as the notified bodies, who must co-operate with each other and co-ordinate their works so as to ensure the uniform application of the conformity assessment procedures. Where appropriate, reference to the Guidelines is made in paragraphs 3.10. to 3.34, as appropriate.

(i) Scope of the Directives

3.10 The Directives define the range of products covered as well as the nature of the hazards it is intended to avert. The aim is to avoid a proliferation of Directives on specific products, but rather define the scope in such a way that a consistent approach to the necessary action is ensured. The terms of the Directives do not preclude the possibility of several Directives being adopted on one and the same product according to the various types of hazard associated with that product. The example given is the mechnical safety of a machine on the one hand, and pollution by that machine on the other.

(ii) Placing the product on the market and putting into service

3.11 Products covered by the new approach Directives may be placed on the market only if they do not endanger the safety of persons, domestic animals or goods when properly installed and maintained and used for the purposes for which they are intended.

As the Directives are intended to provide for total harmonisation as a general rule, any product placed on the market falling within the scope of the Directives must be in conformity with the essential requirements of these Directives. The Resolution does not, however, exclude the possibility of optional harmonisation being appropriate in a particular case.

The object of the new approach Directives is to leave to the manufacturers the choice of means of attestation of conformity, and therefore eliminate the possibility of Member States setting up any system of control prior to placing on the market (except where such control is required by specific Directives). However, national authorities must be allowed to exercise control on the market by way of spot checks and must fulfil their responsibilities for the placing of products on the market. **3.12**

In certain cases, in particular concerning the protection of workers and consumers, the conditions set out in the general clause concerning placing products on the market may be strenthened to include the foreseeable use of the product. **3.13**

Certain definitions of "placing on the market" and "putting into service" are explained in the Commission Guidelines. **3.14**

Placing on the market is the initial action making available on the Community market, for payment or free of charge, a product covered by the Directive in question, with a view to distribution and/or use in the Community. Placing on the market means the time when the product passes from the manufacturing stage within the EC to the stage of distribution and/or use on the EC market. As the placing on the market refers only to the first time that the product is made available on the EC market for distribution or use in the Community, the Directives apply only to new or "reconditioned" products manufactured in the EC and to new (or reconditioned) or used products imported from a third country. The concept of "making available" a product on the Community market means: **3.15**

– The transfer of the products, in terms of ownership or the physical handover of the product by the manufacturer or his authorised representative or the importer to the person responsible for distributing the product on the Community market, or the passing of the product to the final consumer or user in a commercial transaction (whether for payment or free of charge). This is so whether the transfer is made on the basis of a sale, loan, hire, lease, gift or any other type of commercial transfer. The product should comply with the Directive at the moment of transfer;
– The offer of transfer (in cases where the manufacturer or his authorised representative or the importer makes a product available in his own commercial distribution chain with a view to direct transfer to the final consumer or user). The product must comply with the Directive from this moment onwards.

Certain transactions are not considered as "placing on the market". According to the Guidelines these include: **3.16**

– The transfer of the products from a manufacturer in a third country to his authorised representative in the EC whom the manufacturer has made responsible for completing the necessary procedures required to ensure that

the product conforms to the Directive in order to place it on the Community market;
– The import into the Community with a view to re-export (for example under inward processing arrangements) ;
– The transfer of a product manufactured within the EC with a view to export to a third country;
– Display of a product at trade fairs and exhibitions.
– The products must comply with the Directives from the date set by the Directives for their implementation in their entirety. The Directives will not apply to products placed on the market before that date (or put into service if the Directive contains such a provision). Furthermore, most Directives provide for a transitional period during which products which comply with the national rules in force on a particular date may still be manufactured and placed on the market or put into service during the transitional period.

3.17 "Placing on the market" refers to each individual product covered by the Directive in question and which exists physically and in finished form, regardless of when and where it was manufactured, and whether it was produced individually or as part of a batch. "Putting into service" means first use within the EC by the end user of the product covered by a new approach Directive.

(iii) Essential safety requirements

3.18 The Directives describe the essential safety requirements with which all products covered by the Directives must conform.

3.19 The essential requirements are to be worded precisely enough in order to create, on transposition into national law, legally binding obligations which can be enforced. They must also be formulated so as to enable the certification bodies immediately to certify products as being in conformity, having regard to those requirements in the absence of any standards.

(iv) Free movement of goods clause

3.20 The new approach Directives contain free movement of goods clauses. These lay down an obligation on the Member States to accept, under the conditions of the Directive, the free movement of products which satisfy the essential requirements of the Directives.

(v) Means of proof of conformity and effects

3.21 The Directives oblige Member Sates to presume that products which are accompanied by one of the means of attestation (described below) are in conformity with the essential safety requirements of the Directive, and therefore satisfy the requirements for placing that product on the market. The means of attestation declares that the products are in conformity with the harmonised standards adopted by the European standardisation body appointed under the Directive or (as a transitional measure and insofar as harmonised standards do not exist in the field covered by such standards) applicable national standards. Member States are obliged to notify the Commission of the text of those national standards which they consider meet the essential requirements of the Directives. The EC Commission ensures that these national standards are published in the *Official Journal*.

The new approach Directives also generally require that where a **3.22** manufacturer has not applied any standard (because no European or national standard exists), then conformity with the Directive must be accepted if the products are accompanied by certificates and marks of conformity issued by a third party or the results of tests carried out by a third party.

(vi) Standing Committee

A Standing Committee is established, chaired by a representative of the **3.23** Commission and consisting of representatives appointed by the Member States who may avail themselves of the help of experts or advisers. The Committee is to be convened by its chairman either on its own initiative or at the request of a Member State. Any question regarding the implementation of a new approach Directive may be submitted to the Committee. The tasks of the Committee are concerned with the implementation of the Directives. The object of the consultation of the Committee prior to the publication of the references to national standards is more to provide for a forum for the discussion of the objections which the Commission or Member States may formulate, than to carry out a systematic examination of the entire content of the standards.

(vii) Management of the list of standards

Where a Member State or the Commission considers that harmonised **3.24** standards do not satisfy the essential requirements of the Directives, the Commission or a Member State may bring this to the attention of the Standing Committee, setting out reasons for its view. The Committee will give an opinion as a matter of urgency.

Once the Member States have notified the Commission of the national **3.25** standards which they believe satisfy the essential requirements of the Directives, the Commission consults with the Standing Committee which will tender an opinion. Following this, the Commission notifies the Member State whether the national standard in question should or should not enjoy the presumption of conformity and, if so, be subject to national publication.

If the Commission or a Member State takes the view that the national **3.26** standard no longer fulfils the conditions or presumption of conformity to the safety requirements, the Commission consults with the Standing Committee which issues an opinion. In the light of that opinion, the Commission notifies the Member State whether or not the standard in question should continue to enjoy the presumption of conformity, and if not, that it should be withdrawn from publication.

(viii) Safeguard clause

3.27 The new approach Directives generally provide that where a Member State finds that a product might compromise the safety of individuals, domestic animals or property, it shall take all appropriate measures to withdraw or prohibit the placing on the market of the product or at least restrict its free movement, even if the product is accompanied by one of the means of attestation.

3.28 Where a Member State takes such action it must inform the Commission of the measures it has taken and of the reasons why the product no longer conforms with the terms of the Directive.

3.29 The Commission will consult the Member States and also the Standing Committee. If the Commission decides that the Member State's action is justified, it will inform the Member State in question and also notify other Member States that they are also obliged to prevent the product in question from being placed on the market.

3.30 If the failure of the product to comply with the essential requirements results from a defect in the harmonised standard or in the national standards, then the Commission shall notify the Member States and shall inform the European standards body concerned, and if necessary give it a new or revised mandate to modify the harmonised standard. In the case of national standards, it shall ask the Member State concerned to withdraw the standard.

(ix) Means of attestation of conformity

3.31 The means of attestation which manufacturers may use to take advantage of the presumption of conformity[5] with a new approach Directive are as follows:

- certificates and marks of conformity issued by a third party;
- results of tests carried out by a third party;
- declaration of conformity issued by a manufacturer (or his agent based in the Community);
- any other means of attestation set out in the Directive.

3.32 The choice between these different means of attestation may be limited or even removed, according to the nature of the product and the hazards covered by the particular Directive in question.[6]

---

[5] Details of the conformity assessment procedure are set out in Chap 5, *post*.
[6] See for example the difference in the conformity assessment possibilities under Dir 90/385 (implantable medical devices) and Dir 91/263 (telecommunications equipment).

National bodies which are authorised to issue marks or certificates of conformity must be notified by each Member State to the EC Commission and to other Member States. This requirement has given rise to the phrase *notified bodies* which is used to described these organisations. They must carry out their duties according to recognised international practices and principles and in accordance with ISO guides. The responsibility for control of the national bodies lies with the Member States. As regards manufacturers' declarations of conformity, the national authorities have the right to require the manufacturer or his importer to furnish data relating to the tests carried out concerning health and safety, if they have good grounds for believing that a product does not offer the degree of safety required in every respect. Refusal or inability on the part of a manufacturer or importer to supply these data amounts to sufficient reason to doubt the presumption of conformity laid down in the Directives. **3.33**

Having thus outlined the framework of the new approach Directives, the Council Resolution sets out the criteria for choosing the priority areas in which the new approach could initially be applied. The new approach calls for the "essential requirements" to be harmonised and made mandatory by Directives based on Article 100 (and now Art 100A) of the EEC Treaty. In view of this, the "general reference to standards" approach will be appropriate only where it is genuinely possible distinguish between "essential requirements" and "manufacturing specifications". Thus, in all areas in which the essential requirements in the public interest are such that a large number of manufacturing specifications have to be included if the public authorities are to fulfil their responsibility for protecting their citizens, the conditions for the "general reference to standards" will not be fulfilled. One of the main purposes of the new approach is to make it possible to settle, by adopting a single Directive, all the problems concerning regulations for a large number of products, without the need for frequent amendments or adaptations of that Directive. **3.34**

5. Summary of the new approach to technical harmonisation

In conjunction with the concept of "mutual recognition" (see Chap 1, 1.29 *et seq*) the new approach based upon the above mentioned Council Resolution permits the elimination of technical barriers to intra-Community trade caused by product regulations. **3.35**

The main features of the new approach may be summarised as: **3.36**

(i) the prevention of new technical barriers to trade; and
(ii) reference in the new approach Directives to European standards to be drawn up by the European standards organisations; and
(iii) mutual recognition of national standards and technical regulations in the absence of harmonisation at an EC level.

3.37 The new approach abandons the need for maximalistic harmonisation of product specifications, requiring instead only the laying down of *essential requirements* which must be satisfied by all products sold in the EC. The emphasis is therefore on performance criteria (instead of design specifications). This in turn produces the following benefits:

- it should be easier to reach agreement within the Council of Ministers as to the "essential requirements" and therefore it will be easier to adopt EC legislation;
- the EC Commission is no longer required to draw up detailed technical specifications; instead these are left to the European standards organisations who are far better qualified to carry out this work;
- the legislation can have broad application and therefore apply to a wide range of product categories.[7]

3.38 Once the new approach Directives are proposed by the EC Commission and then adopted by the Council of Ministers, the Commission issues mandates to the European standards bodies inviting them to prepare European standards satisfying the essential requirements of the EC Directives.

3.39 However, the new approach does have some weaknesses. In the first place, the essential requirements enumerated in the Directives are often stated in very general terms, and the European standards organisations do not have clear directions to carry out the technical task of drawing up the standards themselves. Thus, in the case of the construction products Directive (Dir 89/106, OJ 1989 L40/12) the Commission had to draw up interpretation documents providing guidelines for the European standards bodies. Delays in drawing up these interpretation documents held up the standardisation work under this Directive.

3.40 Secondly, the European standards bodies are now obliged to draw up thousands of high quality standards, and this in turn has placed a tremendous strain upon the resources of these organisations.

[7] According to CEPS Working Party Report No 5 (March 1992) at 35, Dir 89/392 on machine safety covers some 55,000 types of industrial machines.

Contents of Chapter 4

The European Standards Organisations

Chapter 4
The European Standards Organisations

1. Introduction

The essential feature of the EC's new approach to harmonisation of technical standards, as noted in Chapter 3, is that it aims to establish certain essential health and safety requirements in the various new approach Directives. These fairly abstract requirements are given meaning by the reference to European standards which fill out the technical details which products must satisfy. These standards are written by the European standards organisations, of which there are principally three: Comité Européen de Normalisation (CEN), Comité Européen de Normalisation Electrotechnique (CENELEC) and the European Telecommunications Standards Institute (ETSI). In this chapter, the role and working methods of these organisations are outlined, as is the process of standardisation and the types of standards and norms which are adopted by these organisations. **4.1**

2. The European standards organisations

(i) Comité Européen de Normalisation (CEN)

CEN is the European standards organisation which is responsible for planning, drafting and adopting technical standards in a variety of fields (except for electrotechnology and telecommunications).[1] CEN's main sectors of activity in the context of European technical legislation includes: mechanical engineering, building and civil engineering, health technology, the environment, health and safety in the work place, gas and energy, consumer goods, sports and leisure. It also works in other sectors including information technology, biotechnology and biology, quality, certification and testing, transport and packaging, food and chemistry. **4.2**

CEN is a non-profit making international association established under Belgian law. It was established in 1961 and shares premises with its sister **4.3**

[1] The preparation of standards in these fields is entrusted to the other two other European standards bodies, CENELEC (the European Committee for Electrotechnical Standardisation) and ETSI (the European Telecommunication Standards Institute).

organisation, CENELEC. The two organisations are sometimes referred to as the Joint European Standards Institution.

4.4 Membership of CEN consists of 18 national standards institutions of the Member States of the EEC and the European Free Trade Association (EFTA). In addition, there are a number of CEN affiliated bodies. A list of members and affiliates is given in Annex 2.

4.5 The structure of CEN is set out in diagram 2 opposite. CEN consists of a General Assembly and an Administrative Board, along with a technical management system organised around the major areas of CEN's activities (known as Sector Technical Boards), as well as standards drafting bodies (the Technical Committees) and their working groups. CEN's central secretariat serves the General Assembly, the Administrative Board, the Technical Board, its Sector Technical Boards and Programming Committees. In addition, CEN has four Associated Standardising Bodies (ASBs). CEN has delegated the task of drafting standards in specialised areas to these bodies. The ASBs are: the European Aerospace Industry Association (AECMA), the European Committee for Iron and Steel Standardisation (ECISS), the Western European Edifact Board (WE/EB) and the European Workshop for Open Systems (EWOS).

4.6 Requests to participate in the work of CEN should be addressed to the CEN Central Secretariat or to the National Standards Body which is a member of CEN for the country concerned.

4.7 Diagram 3 on p 38 indicates the procedural process involved in the preparation of standards with CEN. Where it is not possible to use an existing document (for example an ISO standard) then an Experts Committee is set up under the framework of the Technical Committee. These Committees draft the European standards. Once agreement on a draft standard is reached in the Technical Committee, a process of consultation which is meant to ensure the general acceptability of the proposed standard begins. The process includes a six month public enquiry and adoption of the standard through a vote by each national member of CEN.

4.8 Although the resulting standard is ratified in English, French and German (the three official languages of CEN), adoption of the standard is only completed once it has been transposed into national standards in each of the 18 member countries of CEN.[2]

(ii) Comité Européen de Normalisation Electrotechnique (CENELEC)

4.9 As with its sister organisation, CEN, CENELEC is a non-profit making international organisation established under Belgian law. It was set up in

[2] The structure of the European standards bodies and the procedure for the adoption of standards was one of the major issues addressed in the Commission's Green Paper of 3 October 1990 (published in OJ 1991 C120/1). The contents of the Green Paper and the follow-up to the Green Paper are discussed in more detail below.

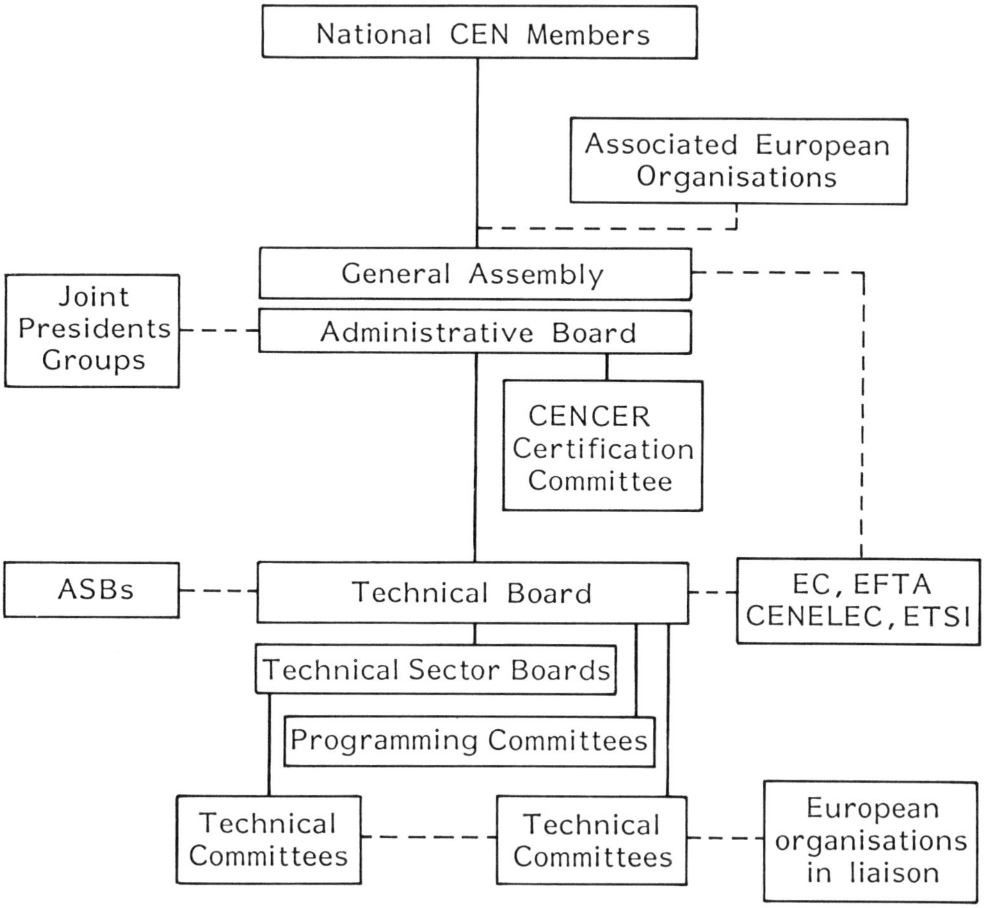

Diagram 2
Source: CEN

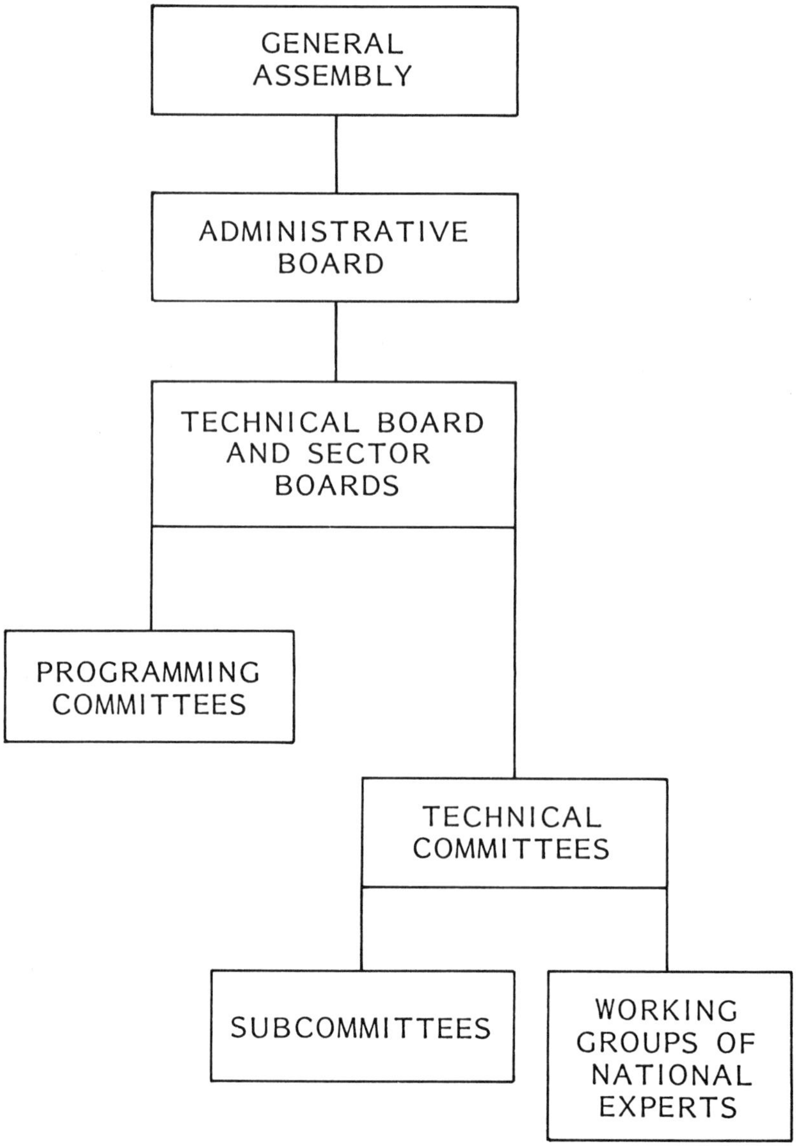

Diagram 3
Source: CEN

1973[3] and its members comprise the National Electrotechnical Committees of 18 countries in Europe.[4] The main purpose of CENELEC is to establish a single set of harmonised electrotechnical standards for Europe in order to facilitate the achievement of a single European market for goods and services. These European standards are prepared following specific requests from the Commission and from the EFTA. Where appropriate, the new standards are prepared in collaboration with CEN and also with the European Telecommunication Standards Institute (ETSI). Furthermore, matters concerning the use of standards in the certification of electrotechnical products are also considered.

Examples of priority areas for CENELEC work are those directly or indirectly related to EC Directives including:[5] **4.10**

- the scope of EC Directive 73/23 (the low voltage Directive) which covers all equipment, both domestic and industrial, designed for use with rated voltage between 50v and 1000v AC and 75v and 1500 v DC;[6]
- the scope of other specific EC Directives for electrical equipment; such as electromedical equipment, electromagnetic compatibility[7], equipment for use in potentially explosive atmospheres, metrology of measuring instruments including electronics, electrical safety of machines, building materials, power engineering, railway and signalisation equipment and broadcast receivers;
- accepted EC or EFTA mandates for standards for electrical equipment (either to prepare initial standards required at European level or to provide standardised solutions where the free movement of goods and services must be guaranteed);
- accepted EC or EFTA mandates for information technology equipment (ITE) including telecommunications equipment (in collaboration with CEN and ETSI).

Diagram 4 on p 40 shows the structure of CENELEC. The representatives of the national standards bodies meet in the forum of the General Assembly, but the actual drafting of the standards is carried out by the Technical Committees. CENELEC strives for consensus although a weighted majority vote is sufficient to adopt a European Norm. **4.11**

(iii) European Telecommunication Standards Institute (ETSI)

ETSI is in essence a trade association established to develop standards to meet the needs of its members, which are the national telecommunications **4.12**

---

[3] By the merger of CENEL and CENELCOM, which were the European counterparts of the International Electrotecnical Commission (IEC).

[4] A list of members is set out in Annex 3.

[5] See for example the CENELEC Annual Catalogue.

[6] Dir 73/23 (OJ 1973 L77/29). The low voltage Directive was adopted before the implementation of the "new approach" to technical harmonisation. However it has been "deemed" to be a new approach Directive since 1993.

[7] Dir 89/336 (OJ 1989 L139/19).

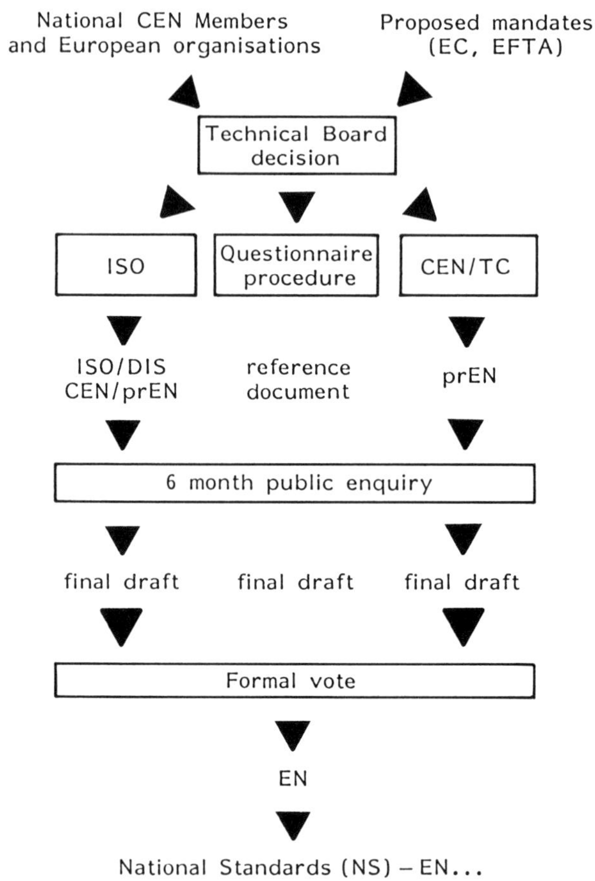
National CEN Members
and European organisations
Proposed mandates
(EC, EFTA)
Technical Board
decision
ISO
Questionnaire
procedure
CEN/TC
ISO/DIS
CEN/prEN
reference
document
prEN
6 month public enquiry
final draft
final draft
final draft
Formal vote
EN
National Standards (NS) – EN...

Diagram 4

administrations and the major suppliers and purchasers of telecommunications equipment. ETSI was established in March 1988. Its main objective is to speed up the development of European standards for the telecommunications sector. These are referred to as ETSs. A large part of the work of ETSI concerns the provision of standards necessary for the operation of the Integrated Digital Services Network (ISDN) and broad band telecommunications systems. There are over 1,000 technical experts participating in ETSI Technical Committees, partly because ETSI has committed itself to producing only "single option" standards, in view of the vital importance of compatibility considerations in the IDSN field. The experts work in project teams and the draft ETSs are discussed by the Technical Committees and are voted on in the General Assembly.

ETSI differs from CEN and CENELEC both in respect of membership **4.13**
and also voting procedures. Membership of ETSI consists of a system of individual membership for companies operating in the telecommunications field (including third country companies). It has about 290 members including the national telecommunications administrations. Some 21 countries are represented in ETSI, namely the EC Member States, the EFTA States and Cyprus, Malta and Turkey.

Because of the commitment to produce "single option" standards, **4.14**
agreement on a standard could prove difficult to achieve. For this reason the voting procedures are somewhat more relaxed in order to improve the chances of adoption. The various members from each country decide how the country should use its vote. A proposal put up for the vote must achieve 71% of a weighted vote in order to be adopted. Thus the only way in which a proposed ETS can be defeated is where a number of countries whose combined vote is greater than 30% vote against the proposal.

(iv) Other institutional bodies and associated standards bodies

There are a variety of other bodies which play a part in the standardisation **4.15**
process. Some of these have already been identified (see 4.2 *et seq*). Others include:

(a) The Information Technology Steering Committee (ITSTC)

The ITSTC was formed by CEN, CENELEC and ETSI. Its role is to set out **4.16**
the preparation of the European standards necessary to promote the "Open Systems Interconnection" (OSI) model within Europe to allow for the interconnection of information systems produced by different manufacturers.

(b) European Workshop for Open Systems Standardisation (EWOS)

EWOS produces (jointly with ETSI) functional standards for the telecom- **4.17**
munications sector.

(c) European Organisation for Testing and Certification (EOTC)

4.18 This organisation arises out of a Memorandum of Understanding signed between the EC, the EFTA countries, CEN and CENELEC in April 1990. The EOTC is aimed at promoting European certification systems. It also encourages the mutual recognition of test reports.

3. Adoption of standards by the European standards organisations

4.19 The standards organisations are entrusted with the task of drawing up the detailed standards by reference to which manufacturers may be presumed to satisfy the "essential requirements" of the various EC Directives. The work is assigned to them by the EC Commission by means of individual "standardisation mandates" which, after consultation with the standardisation body concerned, establish the scope of the work, lay down any further guidelines and fix the timetable by which the standards should be adopted. The Commission finances the standardisation work (with contributions from the EFTA countries whose national standardisation bodies are also members of the European standards organisations).

4.20 Between 1986 and 1990 about 30 standardisation mandates related to EC legislation were given to CEN and CENELEC in respect of approximately 800 European standards for completion by 1993. The demand for new standards work led to the doubling of the CEN/CENELEC technical committees and working groups. For example, between December 1987 and December 1989 the number of technical committees rose from 122 to 239 (source: Commission Green Paper). In 1986 there were only 220 European standards in the course of development in CEN, but by 1989 this had risen to 950. A useful summary of the standardisation work attributed to CEN, CENELEC and ETSI relating to the internal market, information technology and telecommunications is set out in the Annex to the Commission Communication entitled Standardisation in the European Economy (Follow up to the Commission Green Paper of October 1990, COM (91) 521 final, 16 December 1991). This disclosed work concerning the drafting of about 1,950 European standards broken down amongst the new approach Directives and other work, and also stating the amount financed by the Commission. This table is reproduced on p 43.

4.19

New Approach Directives	*Number of Standards*		*Amount Financed (million ECU)*	
Pressure vessels	42		1.15	
Toys	7		0.37	
Construction products	484		3.33	
Machines	184		3.79	
Personal protective equipment	102		3.17	
Medical devices	42		2.06	
Gas appliances	54		2.91	
Electromagnetic compatibility	23		0.42	
	___	938	_____	27.21
Other Work				
Information Technology	257		13.37	
Telecommunications	30		5.57	
Public Procurement	216		3.91	
Eurocodes	27		4.69	
Steel	129		3.80	
Advanced ceramics	42		0.59	
Aerospace	300		1.29	
	___	1001	_____	33.21
		_____		_____
Total		1939		60.42

Source: EC Commission Document COM (91) 521 final

Since the information in the chart above was compiled, things have moved on considerably. In presenting the CEN annual report to the General Assembly in October 1995, the Secretary General stated that in 1995 CEN would have produced more than 700 European Standards (double the number for 1992). This means that the total number is more than 2200 published by CEN alone. Of these, some 400 Standards now exist in support of the New Approach Directives. By way of example, approximately 100 Standards for personal protective equipment (Directive 89/686) have been published in the Official Journal. 4.21

Clearly the work of the European standards organisations is central to the success of the new approach Directives (and indeed, in other areas where industry is demanding standards even if that sector is not covered by EC legislation). In October 1990 the EC Commission published its Green Paper, prompted by a concern that the single European market may not become a reality if the development of standards was not accelerated. The Green Paper summarised the current status of standardisation within the EC and laid down a number of recommendations and proposals for amendment of the existing European standardisation system.[8] The Green Paper was widely circulated to interested parties and was supplemented by 4.22

[8] It should be noted also that the European standards organisations also draw up standards in sectors not yet covered by EC Directives.

a further Communication from the Commission in December 1991.[9] Both the Green Paper and the Follow up paper are discussed at 4.34 *et seq.* However, before doing so it is important briefly to summarise the types of European standards and pre-standards which the European standards organisations draw up.

4.23 Examples of the problems posed by the need to accelerate the standardisation process can be seen by reference to a number of Directives. For example, it was sometime before the toys directive (Dir 88/378, OJ 1988 L187/1) had referenced to it the standards which were requested of the European standards organisations under mandate from the Commission. Similarly, the simple pressure vessels Directive (Dir 87/404, OJ 1987 L220/48) was amended in order to prolong its transitional period. In the same way, the electromagnetic compatibility Directive (Dir 89/336, OJ 1989 L139/19) was also subject to amendment caused by virtue of the fact that the relevant standards were not yet available.

4. European standards

4.24 CEN and CENELEC produce European standards in three forms. Where possible, they draw upon ISO/IEC international standards as the basic document, and once this international standard is selected as the basic document for the work of CEN/CENELEC, all national committees belonging to CEN and CENELEC are obliged to stop any national work on the same subject until a decision has been taken on the introduction of the proposed standard at a national level. By way of example, the European standard for electrical safety of computer equipment (EN60–950) is based upon the international standard IEC950. (Note that, in turn, the European Electrical Safety of Computers Standard EN60950 is adopted in England by the British Standards Institute as BS7002 and implemented by the German Association of Electrical Engineers as VDE0805/5.90.)

4.25 The chosen international standard may be adopted without any change being made to its text, or alternatively it may be modified as necessary so as to meet the requirements of the European markets. In some cases of course, there may not be an international standard upon which to draw, in which case the European standards organisations must draft a standard from scratch. The various standards produced by CEN, CENELEC and ETSI are as follows:

(i) European standards (ENs)

4.26 ENs are implemented at a national level by being granted the status of a national standard and by the withdrawal of any conflicting national standard. The implementation may be done by the issue of an endorsement

[9] Standardisation in the European Economy (Follow up to the Commission Green Paper of October 1990) COM (91) 521 final.

notice declaring that the EN is the national standard together with, if appropriate, the publication of the agreed and voted upon common modifications. If the full EN text is published at a national level, no alteration to the content or layout of the EN is permitted in the resulting national standard.

National standards which conflict with the EN must be withdrawn after **4.27** the date fixed by the Technical Board for the withdrawal of these national standards.

ENs are mandatory in the sense that once an EN is adopted, each national **4.28** committee member of CEN and CENELEC is obliged to adopt it as a national standard and withdraw any conflicting national standards or deviations. However, ENs, being standards, retain the essential feature of standards, namely that they are voluntary. As previously pointed out, even where an EN, once adopted, is assigned to a particular EC Directive, manufacturers have the choice of either manufacturing in accordance with that EN (in which case they benefit from the presumption of conformity for that particular Directive) or manufacturing to some other standard (or indeed no standard) but in which case they must prove conformity.

(ii) Harmonisation Documents (HDs)

Harmonisation documents must be implemented at a national level either **4.29** by issuing a national standard, or at least by the public annoucement of the HD number and title, and as with ENs, no conflicting national standard may continue to exist after the date fixed by the Technical Board for withdrawal of these requirements. HDs are basically European standards which temporarily permit national deviations amongst the various participating countries.

(iii) European pre-norms (ENVs)

Draft European standards which may or may not be adopted in the future **4.30** are called ENVs (European pre-norms).

ENVs are generally established as prospective standards for provisional **4.31** application in technical fields where innovation is particularly high or at least where there is an urgent need for standards. National standards bodies are obliged to make ENVs available at a national level in an appropriate form and to announce their existence in the same way as they do with ENs and HDs. The difference with ENVs is that any conflicting national standards may be kept in force until such time as the ENV is converted into an EN.

Because membership of CEN and CENELEC is essentially limited to **4.32** European countries[10], it is important to note that although CEN and

[10] A full list of the National standards bodies which are members of CEN and CENELEC set out at Annexes 2 and 3 respectively.

CENELEC will generally refer to an international ISO or IEC standard as a basis for preparing ENs, Third Country standards institutes are not represented within CEN and CENELEC. Therefore, to the extent that the ENs depart from ISO or IEC documents, third countries have little or no say in the preparation of EN standards. It is therefore important for third country manufacturers to closely follow the development of standardisation within CEN and CENELEC. Third country goods which are designed only to satisfy international standards may find themselves disadvantaged on the European market because they do not conform to the deviations which CEN and CENELEC have allowed for in producing the ENs.[11]

(iv) European Telecommunication Standards (ETSs)

4.33 ETSI produces standards called ETSs. In addition, ETSI draws up Interim European Telecommunication Standards (I-ETSs) and ETSI Technical Reports (ETRs). I-ETSs are provisional and generally last for three years. After two years members are asked for their comments, and the I-ETR might then be converted into an ETS, or have its period of operation extended, or be withdrawn. ETRs are simply guidance on matters which are connected with but beyond the scope of an ETS or an I-ETS.

5. Commission's Green Paper on development of European standardisation

4.34 The purpose of the October 1990 Green Paper[12] published by the Commission was to establish a consultation document which could be circulated to all interested parties drawing to the attention of producers and users of industrial products in both the private and public sector the strategic significance of European standardisation in order to achieve completion of the internal market. A subsidiary purpose of the Green Paper was to make suggestions for the acceleration of the delivery of European standards, with particular reference to those required for the implementation of EC product legislation in the form of new approach Directives.[13] A third objective of the Green Paper was to stimulate debate

[11] The position of third country manufacturers is considered in Chap 8, *post*.
[12] Green Paper (OJ 1991 C20/1) and Follow Up Paper, COM (91) 521 final.
[13] A major criticism of the Green Paper was that it concentrated too much on standardisation in the framework of the EC Directives, and did not stress the importance of standardisation in other sectors not covered by EC Directives – see CEPS Working Document No 65 ("Fixing European Standards: Moving Beyond the Green Paper" – CEPS Standards Programme: Paper No 3, para 4.2).

on how to ensure the long term dynamism and stability in European standardisation so that the acceleration of standardisation could be sustained at a pace which would satisfy the needs of industry during the next ten years.

The Green Paper examines a number of issues concerning the organisational structure, financing, policies and practices of the European standards bodies, and also those of national standards authorities and tries to assess the changes which might be necessary to make standardisation serve the European market more effectively. The Green Paper is divided into two parts. Part 1 sets out the challenges and problems facing European standardisation, with reference to the structure and operation of the European standardisation bodies (CEN, CENELEC and ETSI). Part 2 puts forward possible solutions to the challenges facing European standardisation in the 1990s and addresses the role of European industry and other interested parties in the standardisation process. **4.35**

The importance of the Green Paper (and the Follow Up Paper) cannot be over-emphasised. The question of European standardisation is fundamental to the opening up of the European market. In the Green Paper, the Commission made a number of recommendations which may be summarised as follows: **4.36**

(a) The Commission called upon European industry to give European standardisation a much higher priority in its strategy for the internal market. The Commission took the view that without greater involvement of industry and standardisation work, and the commitment of more money and expertise to that process, the objectives of the Commission and the European standardised bodies may not be met.
(b) The standardisation bodies were asked to take further steps to improve their efficiency and to consider restructuring the European standardisation system to permit sectorial autonomy in standards making whilst ensuring co-ordination through new European level structures (a European Standardisation Council and Board) which would lay down the strategic direction of European standardisation.
(c) The other recommendations included greater direct participation of interested parties in European standardisation work, the creation of self standing European standards (as opposed to the adoption at a national level of standards created by the European standards bodies), and a long-term policy for the financing of European standardisation bodies which would allow for future Community funding of standardisation to reduce from its present high levels over the next few years.

A full summary of the Commission's recommendations was given at Section V of the Paper. The most important aspects of the Commission's recommendations set out therein are as follows: **4.37**

(i) European industry and interested parties

4.38 European standardisation should be given a much higher priority in company strategy for the single European market. Greater involvement of industry is needed in terms of:

– direct funding of standardisation bodies (both national and European);
– proposing priorities for standardisation work;
– releasing experts for standardisation work (especially by the establishment of project teams).

(ii) European and national standardisation bodies

(a) Efficiency

4.39 The Commission urged procedural changes in order to speed up the delivery of European standards. The Green Paper suggested changes such as:

– the use of "drafting secretariats" or "project teams" to accelerate drafting of common working documents;
– the active encouragement of more sectorial Associated Standardisation Bodies by CEN/CENELEC;
– the use of new communications technology to accelerate discussion on working documents;
– more systematic use of majority voting to agree draft European standards, particularly for standards requested under the various EC new approach Directives;
– shorter public enquiry periods (two months for concensus drafts, no more than four months for majority vote drafts);
– a maximum two month period for response to comments;
– direct applicability of adopted European standards, without waiting for national transposition.

(b) Co-ordination and structure of European and national standardisation organisations

4.40 The Green Paper urged the establishment by the end of 1991 of a European standardisation system comprising:

– a European Standardisation Council made up of persons reflecting the views of major economic interest groups, the EC Commission and EFTA secretariat and the Presidents of the European standardisation bodies, with responsibility for the strategic direction of European standardisation;
– a European Standardisation Board, made up of representatives of the European standardisation bodies responsible for the management and co-ordination of the European standardisation system on behalf of the Council;
– European standardisation bodies, recognised by the Council as exclusively competent in their particular area (for example machinery, etc);
– national standardisation bodies having exclusive responsiblity for carrying

out certain tasks for the European standardisation bodies at national level (for example, public enquiry and voting procedures).

(c) Membership and international co-operation

The Green Paper recommended that there be no enlargement of full membership of the European standardisation bodies for the moment, but went on to suggest that associate membership (participation without voting rights) should be granted to non-member European countries. 4.41

(d) Accountability, financing and information

A number of recommendations were put forward in the Green Paper concerning participation in the work and management of European level standards bodies and that this work and management be opened up to interested parties. In addition, part of the revenue from the sale of European standards would be directed to the European standards bodies. (At the moment the European standards are sold by the national standards authorities in each Member State.) 4.42

(e) Status for the European standard

The Green Paper suggested that European standards should exist in their own right (in other words, the requirement that they be transposed into national standards before they are used should be abolished). 4.43

A single set of European standards (possibly bearing the initials "ES") would be established, removing the current distinction between ENs (produced by CEN and CENELEC) and ETSs (which are produced by ETSI).

(f) Testing and Certification

The European standardisation bodies should define their relationship with the European Organisation for Testing and Certification (EOTC). 4.44

(iii) Member States' authorities

The new European standardisation system would be formally recognised in EC law and the terms of co-operation with the public authorities would be laid down in a Council Decision. 4.45

A further Council Decision on a multi-annual action programme would be necessary, and this would commit the EC to the principle of funding the development of European standardisation for a fixed period (1991–1995) within the limits set annually by the European Community budget. 4.46

Official Community funding of mandated European standardisation to be changed to reimbursement of incurred expenses. 4.47

General active promotion of European standardisation of European Member States. 4.48

6. The Follow up to the Green Paper

4.49 The Commission's Follow up to the Green Paper (COM (91) 521 final, 16 December 1991) summarised the responses to the Green Paper and made recommendations for the development of European standardisation in the 1990s. Although the Commission received overwhelming support for its objective (the establishment of a more efficient and market responsive machinery for European standardisation) there were a number of its detailed recommendations which were rejected by industry and standards bodies. For example, although the importance of standardisation for European industry was universally accepted, most commentators felt that it was unreasonable to ask for a greater financial contribution from industry, as opposed to government. Specific issues may be summarised as follows:

(i) Standards Organisations

4.50 The need for the three standardisation bodies to improve their efficiency was widely recognised and support was given to the Commission's recommendations that the bodies should use project teams and associated standardisation bodies to help them in their work.

4.51 However, the original suggestion that there could be more European standardisation organisations in addition to CEN, CENELEC and ETSI was strongly rejected. Similarly, there was considerable opposition to the creation of new bureaucratic layers to supervise or direct the activities of existing bodies (such as the creation of a European Standardisation Council). However, the three European standardisation organisations accepted that there were political arguments in favour of establishing a common consultative body (to be called the *European Standardisation Forum*) in which economic interests could be directly represented and to reinforce their co-operation within the CEN/CENELEC/ETSI Joint Presidents' Group. The precise character, tasks and composition of the forum is still open for discussion.

(ii) Status of the European Standard

4.52 The Commission's proposal that European standards should exist in their own right was amongst the most controversial of all of the recommendations. It was strongly supported by a number of industrial sectors (for example new technologies) and by some of the national standards organisations. However, a majority view was that presentation of European standards at national level is a major factor in their market acceptance, at least for the next few years. Both CEN and CENELEC accepted the need for greater visibility for the European standard at national level and much faster transposition into national standards.

(iii) Council Resolution of 9th July 1992.

The Commission's follow-up to the White Paper resulted in a Council Resolution of 9th July 1992 entitled "the Role of European Standardisation in the European Economy".[14] 4.53

The Resolution reviews the main steps taken with regard to standardisation, and stresses the need for high quality of European standards and for their systematic transposition into national standards. It also encourages the European Standards Organisation jointly to continue and increase their consultations on a regular basis with all the parties concerned including small and medium-sized undertakings.

7. Standards and intellectual property rights – a conflict

In the December 1991 Follow up to the Commission's Green Paper (referred to in section 4(6) above), the Commission stated that it would welcome the development by the European standards bodies of clear principles concerning the inclusion of intellectual property rights and standards, based upon the practice of the international standardisation organisations. The Follow up also stated that ". . . in view of the importance and complexity of the issue for intellectual property rights, standardisation, competition and trade policies . . . " the Commission intended to produce a separate Communication on the subject. It produced such a Communication on 27 October 1992 entitled "Intellectual Property Rights and Standardisation".[15] The Communication is a non-binding document which simply and clearly sets out the Commission policy on the interrelationship between intellectual property rights, the standardisation process and the EC rules on competition law and the free movement of goods. The Communication summarises the essential conflict between the interests of the owners of intellectual property rights and the standards bodies and others interested in the standardisation process.[16] That conflict arises out of the fact that intellectual property rights (such as patents, copyrights etc.) and the standardisation process have clearly different objectives but coexist in the same industrial and commercial environment. The process of standardisation aims at disseminating technology in the public interest, whilst the rules concerning the protection of intellectual property rights aim to secure monopoly or near-monopoly rights for their owners. In order for these two conflicting elements to successfully coexist and for the 4.54

[14] OJ 1992 C173/1
[15] COM (92) 445 Final
[16] See in particular Section 1 of COM(92)445 Final

standardisation process to move forward effectively, clear solutions must exist in order to resolve the potential conflicts between the principles of intellectual property rights and the aims of the standardisation process. The Communication also recognises that if the standardisation process is implemented without sufficient regard to the rights of the owners of intellectual property, then the incentive to produce new products and processes will be lost.

4.55 The aim of the Communication is not to regulate standard making directly by legislative proposal.[17] However, the Communication sets out a number of principles which it believes should form the basis of any internal rules which standards bodies may wish to elaborate. The Paper emphasises that standards bodies remain free to structure their membership rules and their internal organisational procedures as they wish. However, various types of behaviour either on the part of the standards bodies or alternatively on the part of owners of intellectual property rights could bring them into conflict with the competition and free movement of goods provisions of the EC Treaty, as well as with national legislation and even international conventions.[18] An example of such a conflict, which occured in the case of ETSI, is examined in paragraph [4.61] below.

4.56 Section 2 of the Communication summarises the principles and objectives of standardisation and distinguishes between two types of standard, namely those which result from a formal, consensus-building procedure managed by a standardisation body and a standard which arises spontaneously as a result of the degree of market penetration of a particular product or technical process (the latter are called "*de facto*" or "proprietary standards"). It goes on to summarise the principles used in national and international standards bodies and refers in particular to sub-paragraphs (b) and (c) of Annex A of the ISO document entitled "Reference to Patented Items IEC/ISO Directives – Part 2 Methodology for the Development of International Standards". These state as follows

> – ". . . if the proposal is accepted on technical grounds, the originator shall ask any known patent holder for a statement that he would be willing to negotiate licences under patent and like rights with applicants throughout the world on reasonable terms and conditions. A record of the patent holder's statement should be placed in the files of the ISO Central Secretariat or the IEC Central Office, as appropriate, and shall be referred to in the relevant international standard. If the patent holder does not provide such a statement, the Technical Committee shall not proceed with the inclusion of the patented item unless the respective Council gives permission.
> – should it be revealed after publication of the International Standard that licences under a patent and like rights cannot be obtained under reasonable terms and conditions, the International Standard shall be referred back to the Technical Committee for further consideration."

[17] See Section 1.1.3 of COM(92)445 Final
[18] For example the GATT/WTO Agreement on Technical Barriers to Trade

Reference to these principles are reflected in the closing section of the Communication (Section 6.2 "General Principles") which summarises the principles on which standardisation should take place in a manner which recognises the partnership which exists between standard makers and holders of intellectual property rights. Those general principles list responsibilities both on the standard making bodies and on the intellectual property right holders as follows: **4.57**

European Standards-making bodies should ensure that: **4.58**

(i) All persons wishing to use European standards must be given access to those standards;
(ii) Standards are available for use on fair, reasonable and non-discriminatory terms, regardless of whether the users participate in the work of the standard-making body or not but taking into account the circumstances of the use;
(iii) Users are able to use the above standards to manufacture in conformity with the standards in the Community and to import into the Community goods legitimately manufactured in third countries in conformity with those standards;
(iv) Best efforts are made to identify holders of any intellectual property rights
 – by conducting searches
 – by publication of adequate information and where appropriate by holding public enquries
 before adopting a standard, work on a particular solution only continuing if all known intellectual property rights can be licensed for use in the standard.
(v) Fair conditions are provided to the holders of intellectual property rights, especially with regard to the time limits for identifying IPRs and to agreeing to their use, and in respect of arbitration mechanisms as to royalty rates;

Intellectual property rightholders should: **4.59**

(vi) Use best efforts to identify in a timely manner any IPR which they hold which is relevant to a standard being developed and to confirm or refuse permission for its incorporation in that standard properly;
(vii) Offer fair, reasonable and non-discriminatory monetary or non-monetary terms for the licensee to use any IPR;
(viii) Regard agreement to the incorporation of an IPR in a standard as irrevocable unless the exceptional circumstances justify withdrawal of licences once the standard is adopted;

(Commission Communication on Intellectual Property Rights and Standardisation, Section 6.2).

8. Rules on Competition Law and Free Movement of Goods

4.60 The Communication states that the Commission may find itself obliged to consider whether Articles 30–36 (on the free movement of goods), 59 and 66 (freedom to provide services and to establish oneself) and Articles 85 and 86 (rules governing restrictive agreements and abuses of dominant positions) are applicable in certain cases. One example of the application of the Competition Rules to activities of a standardisation body can be seen by reference to the investigation into the ETSI interim IPR policy.[1]

4.61 In February 1994 ETSI notified the ETSI interim intellectual property rights policy ("IPR policy") and also the ETSI intellectual property rights undertaking ("the Undertaking") as approved by the ETSI General Assembly in March 1993, for the purposes of obtaining negative clearance or an exemption under Article 85 of the EC Treaty (which prohibits restrictive agreements). Subsequently, the members of ETSI voted to abandon the Undertaking and the Commission's examination of the notification under Article 85 proceeded purely on the basis of the IPR policy (which had itself been somewhat revised).

4.62 In drawing up the IPR policy and indeed the undertaking, ETSI's initial approach was one of disregarding intellectual property rights and subjecting members of ETSI to a regime of compulsory licensing. ETSI believed that the public interest, as well as the precise nature of the standards needed to make telecommunications networks inter-work required such a strategy. (It is to be noted that this approach conflicted with that set out in the October 1992 Communication from the Commission, paragraph 5.1.15, which states:)

> "Although it could be argued that consumers would benefit in the short term if intellectual property rights were compulsorily licensed to serve as the basis of standards, in the long term, investment in research and development in the standardised industrial sectors would dry up within the Community. Non-Community entities with extensive research activities would be encouraged to keep their technology out of Community markets, while low-cost manufacturing centres outside the Community would benefit from cheap licences to use Community technology."

4.63 The ETSI Policy and Undertaking adopted in March 1993 foresaw a system in which ETSI members would agree in advance to allow those IPRs deemed "essential" for an ETSI standard, to be included in the standard unless the IPR owner had identified any IPR it wished to withhold. Apart from establishing this licensing by default obligation (which differed from the practice of other standard making bodies where the IPR holders must explicitly agree to have their technology included in a standard), the ETSI IPR Undertaking set forth certain obligations regarding the terms of the licence to be granted to other members.

These arrangements gave rise to a complaint which was lodged in June 1993 by the Computer Business Equipment Manufacturers Association ("CBEMA"), most of whose members were also members of ETSI. The complaint alleged infringements of both Articles 85 and 86 of the EC Treaty. The issues raised by the CBEMA complaint were never decided on formally by the Commission because the Undertaking was abandoned and the Policy was amended, as stated above. **4.64**

The European Commission has indicated[19] that it intends to adopt a favourable view under Article 85 of the EC Treaty towards the ETSI interim IPR Policy, now that that Policy has been slightly amended and the Undertaking abandoned. The Policy, on which the Commission is likely to issue a favourable ruling, contains the following provisions: **4.65**

(i) each member of ETSI will use its reasonable endeavours to inform ETSI in a timely manner of essential IPRs it becomes aware of;

(ii) where a member notifies ETSI that it is not prepared to licence an IPR in respect of a standard, ETSI shall review the requirement for that standard and satisfy itself that a viable alternative technology is available for the standard which is not blocked by that IPR and which satisfies ETSI's requirements. Where, in the opinion of ETSI's Technical Assembly, no such viable alternative technology exists, work on the standard shall cease and ETSI shall request the member to reconsider its opposition to use of the IPR. If a member will not reconsider its position, it will inform the Director of ETSI and provide a written explanation of its reasons for refusing to licence that IPR;

(iii) any violation of the Policy by a member shall be deemed to be a breach by that member of its obligations to ETSI and the ETSI General Assembly shall have the authority to decide what action should be taken against that member;

(iv) IPR holders, whether members of ETSI or third parties, should be adequately and fairly rewarded for the use of their IPRs in the implementation of standards;

(v) where an essential IPR relating to a particular standard is brought to the attention of ETSI, ETSI's Director shall request the owner to give an undertaking in writing within three months that it is prepared to grant irrevocable licences on fair, reasonable and non-discriminatory terms and conditions to manufacture, sell, lease or otherwise dispose of equipment so manufactured.

Consideration of the ETSI Policy illustrates the potential conflict between the process of standardisation the rights of the IPR holders as summarised above.[20] **4.66**

Reference was made above to the complaints lodged by CBEMA based upon Articles 85 and 86 of the EC Treaty resulting from the initial ETSI **4.67**

---

[19] See OJ (1995) C76/5

[20] For further consideration of such issues, see Good "1992 and Product Standards: A Conflict with IPRs?" [1991]11 EIPR 398

approach of compulsory licensing. The Commission's Notice on the ETSI Policy was published shortly before the judgment of the ECJ in the *Magill* case on 6 April 1995.[21] In that case, the ECJ held that the refusal by an IPR holder to grant a licence of its IPR rights was not, generally, an abuse of a dominant position in breach of Article 86 but may be "in exceptional circumstances". On the facts of that case, the ECJ held that the refusal without justification to provide basic information by relying on national copyright law was prohibited by Article 86 as it prevented the appearance of a new product on the market, which the IPR holder did not offer and for which there was a potential consumer demand. In the field of standardisation, similar conflicts between the rights of IPR holders and the need for standards will no doubt be further refined.

[21] Cases C241/91 and C242/91 RTE and ITP v. Commission.

Contents of Chapter 5

Global Approach to Testing, Certification and Conformity Assessment

Chapter 5
Global Approach to Testing, Certification and Conformity Assessment

1. Introduction

The Council Resolution of 7th May 1985[1] refers to the fact that methods of assessing product conformity with the essential requirements of the new approach Directives would be mentioned in each such Directive. This ignored the fact that many Member States were reluctant to accept test results and certificates issued by other Member States. This reluctance to accept the conformity assessment of other Member States resulted in technical barriers to trade for two reasons: **5.1**

(i) Even where EC harmonised rules for particular products existed, the evidence of conformity provided to the manufacturer by the relevant authorities in the exporting Member State could be rejected by the authorities in the importing Member State. The *presumption of conformity* set out in the new approach Directives, would be of little value if the authorities in the importing Member State refused to recognise the certificates of the authorities in the exporting Member State.

(ii) Refusal to accept test certificates in this way would undermine the concept of "mutual recognition". As previously noted (1.29 *et seq*), under the concept of mutual recognition each Member State recognises that other Member States' rules on health and safety offer the same level of protection as the rules in the importing Member State.[2] However, the concept of mutual recognition does not prevent the authorities in the importing Member State from questioning the ability of the exporting Member State to monitor adequately conformity with its own technical regulations. The importing state might therefore impose additional conformity assessment checks, causing unnecessary duplication, delay and costs.

However, until July 1989 when the Commission forwarded a proposal[3], **5.2**

---

[1] Council Resolution of 7 May 1985 on a new approach to technical harmonisation (OJ 1985 C136/1), see Chap 3.

[2] Indeed, this is specifically referred to in Annex 1 to the Council Resolution of 7 May 1985.

[3] COM (89) 209 Proposal for a Council Decision concerning the Modules for the various phases of the conformity assessment procedures which are intended to be used in the technical harmonisation Directives (OJ 1989 C231/3). See now Council Decision 93/465 (OJ 1993 L220/1) and Annex 5.

no comprehensive plan for abolishing technical barriers caused by conformity assessment differences existed.

5.3 Following receipt of this Commission proposal, the Council of Ministers adopted a Resolution on 21 December 1989 on a global approach to conformity assessment.[4]

2. Content of Council Resolution on a global approach to conformity assessment

5.4 The Council Resolution specifically refers to its Resolution of 7 May 1985 on the new approach to technical harmonisation, and also to the Commission's proposal of July 1989 on conformity assessment Modules mentioned above. Taking into account the approach set out in the Commission's proposal, the Council Resolution adopts a number of *guiding principles* for a European policy of conformity assessment. The guiding principles may be summarised as follows:

(a) A consistent approach in EC legislation must be ensured by devising *Modules* for the various phases of conformity assessment procedures and by laying down criteria for the use of those procedures, for the designation and notification of bodies under those procedures and for use of the "CE mark" (this is the mark which goods will be entitled to bear if they satisfy the essential requirements of the Directives applicable to them).[5]

(b) Generalised use of the European standards relating to quality assurance (EN2900) and to the requirements to be fulfilled by the "notified bodies" concerned (EN45000)[6], the setting up of accreditation systems and the use of techniques of intercomparison should be promoted in all EC Member States.

(c) The promotion of mutual recognition agreements on certification and testing between bodies operating in the non-regulatory sphere are essential for the completion of the internal market. The establishment of a flexible, unbureaucratic testing and certification organisation at a European level with a basic role of promoting such agreements and of providing a prime forum within which to frame them should significantly contribute to the furtherance of that objective.

(d) Possible differences in levels of development in the EC and in industrial sectors with regard to quality infrastructure (especially calibration and metrology systems, testing laboratories, certification and inspection

[4] Council Resolution of 21 December 1989 on a global approach to conformity assessment (OJ 1990 C10/1).

[5] As to the CE Mark, generally see Chap 6, *post*.

[6] As to EN29000 and EN45000, see 5.5 *et seq*, below.

bodies, and accreditation systems) such as are likely to have an adverse effect on the operation of the single European market should be studied with a view to the preparation of a programme of Community measures, possibly including budgetary measures as soon as possible.

(e) In its relations with Third Countries the EC will endeavour to promote international trade in regulated products, in particular by concluding mutual recognition agreements on the basis of Article 113 of the EEC Treaty in accordance with EC law and the EC's international obligations[7], whilst ensuring in the latter case that:

– the competence of third country bodies remain on a par with that required of their EC counterparts;
– the mutual recognition arrangements are confined to reports, certificates and marks drawn up and issued directly by the bodies designated in the agreements;
– in cases where the EC wishes to have its own bodies recognised, the agreements establish a balanced situation with regard to the advantages derived by the parties in all matters relating to conformity assessment for the products concerned.

3. Observations on the Council Resolution on a global approach to conformity assessment

A number of observations must be made on the Council Resolution on a global approach to conformity assessment. 5.5

(i) Accreditation

The above mentioned guidelines include reference to a generalised use of the European standards relating to quality assurance systems (EN29000) and to the requirements to be fulfilled by the *notified bodies* concerned (EN45000). This is a reference to the possible lack of confidence by Member States in the testing and certification by the notified bodies of other Member States. Reference to the application of unified ground rules (laid down in the EN45000 series of European norms) which govern the competence of testing and certification bodies is an attempt to overcome this potential lack of confidence. In other words, national notified bodies responsible for testing compliance of products with the essential requirements of the new approach Directives should satisfy the EN45000 criteria 5.6

[7] Art 113 of the EEC Treaty provides *inter alia*, that where agreements with third countries need to be negotiated, the Commission shall make recommendations to the Council of Ministers which in turn will authorise the Commission to open the necessary negotiations. As to negotiations with third countries concerning the mutual recognition of conformity assessment, see Chap 8, *post*.

before they are accredited to take on the conformity assessment. Laboratories and certification bodies which monitor the application of national rules should also comply with EN45000.

5.7 In the case where testing and certification is carried out by the manufacturer ("self certification"), as opposed to by third parties[8], it is necessary to have unified rules for approving quality management systems. In such a case, the Global Approach Resolution suggests the use of the EN29000 series of norms.

(ii) Conformity assessment procedures

5.8 The Council Resolution of 7 May 1985 sets out a number of different means of evaluating conformity.[9] This initial statement of policy was taken one step forward in the Commission's proposal on a Global Approach (COM (89) 209 final, OJ 1989 C231/3) and was embodied in the Council Resolution of December 1989 as being the first guiding principle for a European policy on conformity assessment. The Commission's proposal resulted in Council Decision 90/683 (OJ 1990 L380/13) and its replacement, Decision 93/465 (OJ 1933 L220/23) which is examined in more detail below. The modular approach for assessing conformity to the EC Directives provides for Modules A to H, each Module differing from the others with regard to the body responsible for testing and the stage in the manufacturing process at which such testing occurs. The various Modules range in complexity from self-certification to establishing approved quality control systems at each of the design, production and testing phases of manufacture under the EN29000 series of manufacturing quality assurance standards.

5.9 When adopting new approach Directives, the EC Council of Ministers is able to specify the appropriate Module in each Directive, taking into account the safety risks involved, the products covered and also the production processes involved. Different Modules will therefore generally be specified in each new approach Directive.

(iii) The CE marking

5.10 Products which satisfy the essential requirements of the new approach Directives will be entitled to have affixed to them the CE marking (see Chap 6, *post*).

---

[8] Self-certification is one method of proving conformity, and is expressly referred to in the Council's Resolution of 7 May 1985, and provided for in Council Dec 93/465 concerning Modules for the various phases of the conformity assessment procedures to be used in the technical harmonisation Directives (as to which see section 4 below).

(iv) Conformity assessment and industry standards

The global approach set out in the Council Resolution is not confined to areas covered by EC product Directives. It also proposes structures for co-operation and testing and certification in the sphere of voluntary standards. To this end, in April 1990, a Memorandum of Understanding was signed by the EC Commission, EFTA, CEN and CENELEC which laid down principles for a new European Organisation for Testing and Certification to deal with these questions.[10] **5.11**

Finally, although the global approach to conformity assessment is aimed at reducing bureaucracy and simplifying the task of assessment, it has not been without its critics. For example, various consumer groups have argued that the decision to allow manufacturers self-certification of compliance with Directive 88/378 (the toys Directive) does not provide sufficient safeguards for consumers. **5.12**

4. Modules for various phases of the conformity assessment procedures – Council Decision 93/465

The Council Resolution on the global approach, and the Commission's proposal (COM (89) 209 final, OJ 1989 C231/3) resulted in the adoption of Council Decision 90/683 of 13 December 1990 (OJ 1990 L380/13). This was replaced by Decision 93/465. The text of this latter Decision is set out in Annex 5. The Decision contains only one main article. This states that the procedures for conformity assessment to be used in the technical harmonisation Directives will be chosen from among the Modules listed in the Annex to the Decision and the guidelines contained therein. In essence, the Decision provides a "menu" from which the Council of Ministers may choose when adopting a new approach Directive on a proposal from the Commission. The Decision states that the procedures for conformity assessment may only depart from the Modules when the specific circumstances of a particular sector or Directive so warrant. Such departures must be limited in extent and must be explicitly justified in the relevant Directive. The Decision also lays down the rules for affixing the CE conformity marking concerning the design, manufacture, placing on the market and entry into service or use of industrial products. The Commission will report periodically on the functioning of Council Decision 93/465 and on whether conformity assessment procedures are working satisfactorily or need to be modified. **5.13**

[10] The role of ETOC is discussed in Chap 4.

5.14 The Annex to the Decision is divided into three parts. Part 1 contains general guidelines, Part 2 sets out in detail the various Modules for conformity assessment and Part 3 consists of a diagram entitled "Conformity assessment procedures in Community Legislation".

(i) General guidelines

5.15 The main objective of a conformity assessment procedure is to allow the public authorities to ensure that products placed on the market conform to the requirements as expressed in the provisions of the new approach Directives, in particular with regard to the health and safety of users and consumers. Conformity assessment can be divided into a number of Modules which relate to the design phase of products and also to the production phase. The general guidelines set out in the Annex to the Decision state that, as a general rule, products should be subject to both phases before being able to be placed on the market if the results are positive.

5.16 However, the general guidelines allow for specific Directives to provide for different arrangements. There are a number of Modules which cover the two phases in a variety of ways. The new approach Directives must set the range of possible choices which can be considered by the Council of Ministers to give the public authorities the high level of safety they seek for a given product or product sector.

5.17 Thus, as indicated above, the Directives will, in setting the range of possible Modules for a given product or product sector, attempt to leave as wide a choice to the manufacturer as is consistent with ensuring compliance with the essential requirements of the new approach Directives. Furthermore, the new approach Directives will set out the criteria governing the conditions in which the manufacturer shall choose the most appropriate Modules for his production from the modules laid down in each Directive. One of the other general guidelines states that for the purposes of operating the Modules, the Member States shall notify the bodies ("the notified bodies") under their jurisdiction which they have chosen from the technically competent bodies complying with the requirements of the Directives. This responsibility obliges Member States to ensure that the notified bodies permanently have the technical qualifications required by the Directives and that the notified bodies keep the national authorities informed of the performance of their tasks.

5.18 With regard to conformity assessment, the subcontracting of work shall be subject to certain conditions which guarantee:

– the competence of the establishment operating as a subcontractor, on the basis of conformity with series EN45000 standards and the capability of the Member State notifying the subcontracting body to ensure effective monitoring of such compliance; and

– the ability of the body notified to exercise effective responsibility for work carried out under subcontract.

Notified bodies which can prove their conformity with harmonised standards under the EN45000 series, by submitting an accreditation certificate or other documentary evidence, are presumed to conform to the requirements of the new approach Directives. **5.19**

Finally, a list of notified bodies will be published by the Commission in the *Official Journal of the European Communities*, and updated regularly. **5.20**

The Decision also sets out the principal guidelines for affixing and use of the CE marking. These state that the CE marking symbolises *conformity to all the obligations incumbent on manufacturers for the product by virtue of the Community Directives providing for its affixing*. Thus, such conformity is not limited to the essential requirements relating to safety, public health, consumer protection etc. as certain directives may impose specific obligations not necessarily forming part of the essential requirements. These guidelines should be read in conjunction with Council Directive 93/68 which amended all of the new approach Directives as regards the CE marking. That Directive is published in the same Official Journal as Decision 93/465. Essentially, the CE marking when affixed to industrial products represents the fact that the person having affixed it or being responsible for affixing it has verified that the product conforms to all the Community total harmonisation provisions which apply to it and has been the subject of the appropriate conformity evaluation procedures. **5.21**

Where the industrial products are subject to other Directives concerning other aspects and which also provide for the CE marking, then the latter must indicate that the products are also presumed to conform to the provisions of those other Directives. **5.22**

However, where one or more of those Directives allow the manufacturer, for a transitional period, to choose which arrangements to apply, then the CE marking indicates conformity to the provisions *only* of those Directives applied by the manufacturer. In this case, the particulars of the Directives applied must be given in the documents or instructions accompanying the products, or, where appropriate, on the data plate itself. **5.23**

(ii) Modules for conformity assessment

There are nine Modules for conformity assessment (A, Aa, and B to H). Although the range of Modules set out in the Decision, and which may be specified in the various new approach Directives, seems confusing, the system is in fact fairly simple. Each Module is summarised below, but reference should be made to Annex 5 for the detailed description of each Module and the explanatory notes. An overview can be obtained from the diagram which is included in the Decision and which is set out on page 145. **5.24**

Module A – internal production control

Under this Module the manufacturer or his authorised representative **5.25** established within the EC ensures and declares that the products satisfy the requirements of the Directive that apply to them. The manufacturer affixes the CE marking to each product and draws up a written declaration of conformity. The manufacturer must also keep a file of technical documentation enabling conformity of the product with the requirements of the applicable Directive to be assessed. The dossier must be kept for a period of at least ten years after the last product has been manufactured.

5.26 Where neither the manufacturer nor his authorised representative is established in the EC, the obligation to keep the technical file available shall be the responsibility of the person placing the product on the EC's market.

5.27 It should be noted that specific Directives may use Module A with additional provisions containing supplementary requirements. These may include tests on specific aspects of the product and product checks at random intervals carried out by the appropriate notified body. Where this is the case, the Module is known as Module Aa.

Module B – (EC-type examination)

5.28 This Module applies during the design stage of production. Here a notified body ascertains and attests that a specimen, representative of the product envisaged, meets the provisions of the appropriate Directive. A request for EC-type examination is lodged by the manufacturer or his authorised representative in the EC with a notified body of his choice. A specimen ("type") may cover several versions of the same product, provided that the differences between the versions do not affect the level of safety and other essential requirements. The notified body examines the technical documentation and verifies that the type has been manufactured in conformity with the technical documentation and checks whether, where the manufacturer has chosen to apply the relevant standards under the Directive, these have actually been applied. Where the type meets the provisions of the applicable Directives, the notified body issues an EC-type examination certificate to the manufacturer. The notified bodies from each country must satisfy the EN45000 series of laboratory standards, in order that these test reports may be considered reliable.

5.29 Where a manufacturer applies Module B, he must also apply one of Module C, D, E or F during the production stage. The explanatory notes to the Modules state that Module C is designed to be used in combination with Module B, and Modules D, E and F will also normally be used in conjunction with Module B. However, in special cases (for example, when dealing with certain products of a very simple design) Modules D, E and F may be specified in the appropriate Directives for use on their own.

Module C – (conformity to type)

5.30 Under Module C (having satisfied Module B) the manufacturer issues a manufacturer's declaration of conformity to type. This specifies that the mass produced product does in fact comply with the essential requirements of the appropriate Directive and that it conforms with the type which has already been approved by the notified body under the Module B procedure. Module C also specifies certain possible supplementary requirements which may be provided for by particular Directives. These additional requirements are for tests on specific aspects of the product and also product checks at random intervals.

Finally, it should be noted that Module C does not require the manufacturer to implement any quality control systems (unlike Modules D and E). 5.31

Module D – (production quality assurance)

The manufacturer must have satisfied the requirements of Module B, except in the rare cases where Module D may be used on its own, in which case certain additional wording is incorporated from Module A. 5.32

Module D requires the manufacturer to operate a quality control system aimed at ensuring compliance of the products with the type as described in the EC type examination certificate and with the requirements of the Directives which apply to them. The quality control system is in turn assessed by the appropriate notified body. Quality control systems which implement the relevant harmonised standard are presumed to satisfy the requirements of Module D. The appropriate harmonised standard is EN29002, supplemented, if necessary, to take into account the specific nature of the products for which it is implemented. 5.33

Notified bodies shall periodically carry out audits to make sure that the manufacturer maintains and applies the quality system. As under Module C, the manufacturer issues a declaration of conformity to type. 5.34

Module E – (product quality assurance)

Where Module E may be used on its own (*i.e.* not in conjunction with Module B) certain modifications to the Module's requirements are provided for. 5.35

Where Module E is used with Module B, again the manufacturer must operate an approved quality control system for production and testing. The manufacturer declares conformity with the approved type and that they satisfy the requirements of the Directives which apply to them. He affixes the CE marking to each product and draws up a written declaration of conformity. If the manufacturer's quality control system incorporates harmonised standard EN29003, it will be presumed to satisfy the requirements of the Module. Again, the appropriate notified body keeps the manufacturer's quality control system under review. 5.36

Module F – (product verification)

Under this Module the manufacturer or his authorised representative established in the EC checks and tests that the products are inconformity with the type described in the EC type examination certificate and satisfy the requirements of the Directives that apply to them. The Module is modified in cases where it may be used on its own (*i.e.* without first using Module B). The notified body carries out examinations and tests in order to check the conformity of the product with the essential requirements of the Directives either by examination and testing of every product or by 5.37

examination and testing of products on a statistical basis, according to the manufacturer's choice (which may be limited in certain cases).

5.38 It should be noted that no quality assurance system need be implemented by the manufacturer in the case of Module F.

Module G – (unit verification)

5.39 Under this Module the manufacturer ensures and declares that the product concerned, which has been used in accordance with the certificate of conformity issued by the notified body, conforms to the requirements of the Directive that applies to it.

Module H – (full quality assurance)

5.40 Under Module H the manufacturer must operate an approved quality system for design, manufacture and final product inspection and testing described in the Module. The quality system is subject to surveillance by the notified bodies. The notified body shall presume that the quality system satisfies the relevant requirements if that system implements harmonised standard EN29001.

Contents of Chapter 6

The CE marking

Chapter 6
The CE marking

1. Introduction

An indispensible counterpart of the EC's new approach to harmonisation of technical standards and of its global approach to conformity assessment is that an easily identifiable mark of conformity should be affixed to the goods which satisfy the essential requirements of the new approach Directives. In this way, national safety inspectors can satisfy themselves that the goods need not be further subject to local conformity assessment. Thus, it is necessary to have an easily recognisable mark which acts as a kind of "passport" guaranteeing the goods to which it is affixed free access to all EC markets. The various new approach Directives contained references to a "CE mark" which was intended to perform this function of a "passport".[1] However, until 1993 there was no clear set of guidelines on the affixing of such a mark to the products covered. Thus, a number of inconsistencies arose in each of the new approach harmonising Directives. **6.1**

Significant difficulties and confusion were identified: **6.2**

– the meaning of the CE mark differed according to the various new approach Directives. For most of these Directives, the attaching of the CE mark indicated that the product satisfied the EC legislation itself and that the goods were found to be safe to use. However, in the toy safety Directive (Dir 88/378) and the simple pressure vessels (Dir 87/404) only those goods which were manufactured in accordance with the national standards transposing the European standards were entitled to carry the CE mark. In other words, the information which the CE mark intended to confer differed according to the EC Directive in question;
– whether or not the CE mark is to be reserved for goods which have undergone conformity assessment varied from Directive to Directive;
– there was no single format for the CE mark;
– in certain circumstances the manufacturers used a recognised national mark of conformity instead of the CE mark.

These types of differences or approach in the various EC Directives meant that very often national product inspectors might impose new conformity assessment procedures rather than take the risk of allowing "unsafe" products onto the market. This in its own way created an unncessary barrier to trade. (Furthermore, the CE mark in the various new **6.3**

[1] For the CE marking in the context of each new approach Directive, reference should be made to Annexes 6(i) *et seq*.

approach Directives set out in the Annex to this book were not reproduced in a consistent manner.)

6.4 In order to overcome some of the difficulties outlined above, the Commission put forward a proposal for a Council Regulation concerning the affixing and use of the CE marking of conformity on industrial products.[2]

2. Council Directive 93/68 amending the new approach Directives as regards the CE Marking

6.5 The proposal was adopted as Council Directive 93/68. The fourth recital to this Directive states that the provisions concerning the affixing and use of the CE marking requires that existing Directives undergo detailed amendment to bring them into line. The Directive therefore amends all of the new approach Directives (including the Low Voltage Directive 73/23, which is not strictly speaking a new approach Directive).

6.6 The amendments made to the new approach Directives have been incorporated into the texts of those Directives as set out in the Annexes to this book. The Member States were required to adopt the laws necessary to comply with the Directives by 1 July 1994 and apply those provisions with effect from 1 January 1995. However, until 1 January 1997 Member states should allow the placing on the market and the bringing into service of products which comply with the marking arrangements in force before 1 January 1995 (Article 14 of Directive Article 93/68).

6.7 The amendments made by Directive 93/68 to the new Approach Directives should be read in accordance with the Guidelines for the affixing and use of the CE marking set out in the Annex to Decision 93/465.

6.8 The CE Marking indicates that products may be placed freely on the Community Market, and conformity with all the provisions of the various directives. To that extent, the CE marking is *not* a "quality mark" and is intended for the appropriate authorities of the various Member States for the purposes of verifying conformity with the Directives, rather than for consumers. The CE marking is the mark which certifies that industrial products covered by the new Approach Directives conform to those Directives based upon the principles of the Global Approach. Additional marks may not be provided if they are liable to cause confusion with the CE marking.[3]

[2] COM (91) 145 final (OJ 1991 C160/14).
[3] See answer to Written Question E-2111/95, OJ (1995) C326/27.

Contents of Chapter 7

The New Approach Directives

Chapter 7 The New Approach Directives

1. Introduction

The text of the new approach Directives is set out in Annex 6, and this chapter summarises the scope, essential requirements, methods of proving conformity and other features of the Directives. Where harmonised standards have been written by the European standards organisations for the various Directives, they are set out in the annual catalogues of CEN and CENELEC specifying the document number, the year, the title of the standard, the EC Directive to which it applies and the relevant committee responsible for the standard. Apart from the CEN and CENELEC catalogues, European standards are published in the *Official Journal of the European Communities*. It should be noted that in the absence of European standards referable to the various new approach Directives, manufacturers may manufacture to any national standard which has been notified by a Member State to the EC Commission as being the standard satisfying the essential requirements of the relevant Directive. **7.1**

2. Directive 87/404 – simple pressure vessels

(i) Scope

The Directive[1] applies to simple pressure vessels manufactured in series and these are defined as any welded vessels subjected to an internal gauge pressure greater than 0.5 bar which are intended to contain air or nitrogen and which are not intended to be fired. Certain vessels are expressly excluded from the scope of the Directive, namely vessels specifically designed for nuclear use, vessels specifically intended for installation in or on the propulsion of ships and aircraft and fire extinguishers. **7.2**

(ii) Essential requirements

Member States must ensure that the vessels to which the Directive applies may be placed on the market and taken into service only if they do not compromise the safety of persons, domestic animals or property when **7.3**

[1] OJ 1987 L220/48, amended by Dir 90/488 (OJ 1990 L270/25 and Dir 93/68 (OJ L220/1)). See Annex 6(i) for the consolidated text.

properly installed and maintained and when used for the purposes for which they are intended. A list of essential safety requirements is set out in Annex I to the Directive. These requirements deal with the materials from which the vessels are manufactured, the design of the vessels and the manufacturing processes.

(iii) Certification procedures

7.4 The Directive provides for EC-type examination, EC verification by notified body and a manufacturer's declaration of conformity according to the type of vessel in question.

(iv) Entry into force

7.5 Member States were obliged to adopt the necessary laws to implement the Directive by January 1990, and to apply those laws from 1 July 1990. A transitional period expired in July 1992.

3. Directive 88/378 – toy safety

(i) Scope

7.6 The Directive[2] applies to toys which are defined to mean any product or material designed or clearly intended for use in play by children of less than 14 years of age. However, Annex I excludes some 21 products which are not to be regarded as toys including: sports equipment, fireworks, video toys, fashion jewellery and bicycles designed for sport or for travel on the public highway.

(ii) Essential requirements

7.7 Toys may be placed on the market only if they do not jeopardise the health and/or safety of users or third parties when they are used as intended or in a foreseeable way, bearing in mind the normal behaviour of children. The essential safety requirements for toys are set out in Annex II. The list is quite extensive and covers the physical and mechnical properties of types, flammability, chemical and electrical properties and hygiene.

(iii) Conformity

7.8 The Directive provides that manufacturers and importers who market toys which meet the relevant harmonised standards and apply the CE marking merely make available to the competent authorities a file of technical

[2] OJ 1988 L187/1. Amended by Dir 93/68 OJ 1993 L220/1. See Annex 6(ii) for the text.

documentation. Manufacturers or importers who have decided not to comply with the harmonised standards must follow the EC-type examination procedure and submit a model to a notified body which will then issue an EC-type certification, providing the toys meet the essential requirements of the Directive. As well as the CE marking, toys or their packaging must also bear the name and/or trademark and the address of the manufacturer or importer in order that those responsible for placing the product on the market are easily identifiable.

(iv) Entry into force

The Directive entered into force on 1 January 1990. **7.9**

4. Directive 89/392 – machinery (including mobile machinery and lifting equipment)

(i) Scope

The Directive[3] applies to machinery which is defined to mean an assembly of linked parts or components, at least one of which moves, with the appropriate actuators, control and power circuits etc, joined together for a specific application, in particular for the processing, treatment, moving or packaging of a material. It also covers interchangeable equipment modifying the function of a machine and which is marketed for the purpose of being assembled with a machine (except spare parts or tools). There is a long list of machines which are excluded from the scope of the Directive, e.g. pressure vessels (which are covered by Dir 87/404), vehicles and trailers (except those used in the mineral extraction industry), sea-going vessels and agricultural and forestry tractors as defined in Council Directive 74/150 (OJ 1974 L84/10). **7.10**

Where, in the case of machinery, the risks referred to in the Directive are wholly or partly covered by special EC Directives, then Directive 89/392 shall not apply, or shall cease to apply, in the case of that machinery and those risks on the entry into force of the specific Directives. **7.11**

Where, for machinery, the risks are of an electrical origin, that machinery will be covered exclusively by the low voltage Directive (OJ 1973 L77/29). **7.12**

(ii) Essential requirements

The essential health and safety requirements are set out in Annex I, which provides a detailed list along with certain preliminary observations. These **7.13**

[3] OJ 1989 L183/9, amended by Dir 91/368 (OJ 1991 L198/16). Dir 93/44 (OJ L175/12) and 93/68 (OJ 1993 L220/1).

include principles of safety integration, matters concerning materials and products, lighting, the design of the machine, the way in which it is controlled, mechnical hazards, guards and protective devices, electricity, noise, vibration, radiation and temperature hazards and matters concerning maintenance of the machinery. Annex I was substantially amended in 1993 and the amendments are incorporated in the text set out in Annex 6(iii).

(iii) Conformity

7.14 The Directive provides for a manufacturer's declaration of conformity, file examination and for the manufacturer to affix the CE mark as well as for submission of an example of the machine for EC-type examination set out in Annex VI to the Directive.

(iv) Entry into force

7.15 Member States were obliged to implement the Directive into national law by 1 January 1992 and to apply the measures in question with effect from 1 January 1993 except as regards certain roll-over protective structures (ROPS), falling object protective structures (FOPS) and self-propelled industrial trucks, for which the period shall apply from 1 July 1995.[4] There is a transitional period permitting Member States to allow, until 31 December 1994, the placing on the market of machinery in conformity with national regulations in force in their territory on 31 December 1992. In the case of self-propelled industrial trucks, ROPS and FOPS, the period is extended until 31 December 1995. The complex transitional arrangements are summarised in a Commission Notice published in September 1994 in Official Journal C253/3.

5. Directive 89/336 – electromagnetic compatibility (EMC)

(i) Scope

7.16 The Directive[5] applies to all apparatus likely to cause electromagnetic disturbance or the performance of which is liable to be affected by such disturbance. "Apparatus" means all electrical and electronic appliances together with equipment and installations containing electrical and/or

[4] The particular equipment referred to is that covered by Dir 86/295 (OJ 1986 L186/1), 86/296 (OJ 1986 L186/10) and 86/663 (OJ 1986 L384/12).
[5] OJ 1989 L139/19, as amended by Dir 92/31, and Dir 93/68 (OJ 1993 L220/1).

electronic components. Electromagnetic disturbance means any electromagnetic phenomenon which may degrade the performance of a device, unit of equipment or system.

It is true to say that there is some uncertainty in industries whose products are covered by the EMC Directive as to the meaning of terms such as "apparatus". Although "apparatus" is defined in Article 1, there is some confusion over the exact meaning of "installation". 7.17

Insofar as protection requirements specified in the EMC Directive are harmonised, in the case of certain apparatus covered by other specific EC Directives, the EMC Directive will not apply or at least it will cease to apply with regard to such apparatus or protection requirements upon the entry into force of those Directives. 7.18

(ii) Essential requirements

Apparatus covered by the Directive must be constructed so that the electromagnetic disturbance generated does not exceed a level allowing radio and telecommunications equipment and other apparatus to operate as intended. Furthermore, the apparatus must have an adequate level of intrinsic immunity of electromagnetic disturbance to enable it to operate as intended. The principal protection requirements of the Directive are set out in Annex III. This states that the maximum electromagnetic disturbance generated by the apparatus shall be such as not to hinder the use, in particular, of various apparatus including: domestic radio and television receivers, mobile radio equipment, industrial manufacturing equipment, information technology equipment, telecommunications networks and apparatus and lights and fluorescent lamps. 7.19

(iii) Conformity

The Directive provides for a manufacturer's self declaration of conformity and the provisions governing the declaration of conformity and the CE mark is set out in Annex I. An EC declaration of conformity may also be made by the manufacturer after he has obtained a certificate issued by a competent body following the submission of a technical file when the manufacturer has *not* applied the harmonised standards (Article 10, paragraph 2). Other provisions in the special case of apparatus designed for transmitting radio communications, after the manufacturer has obtained EC-type certification issued by a notified body is provided for in Article 10, paragraph 5. 7.20

(iv) Entry into force

Because the relevant harmonised standards were not available on 1 January 1992, (the date by which the Member States were obliged to implement the EMC Directive) a transitional period was inserted into the Directive (by Dir 7.21

92/31, OJ 1992 L126/11) allowing Member States up until 31 December 1995 to authorise the placing on the market and/or the putting into service of apparatus referred to in the Directive conforming to national regulations in force on their territory on 30 June 1992.

6. Directive 90/384 – non-automatic weighing instruments

(i) Scope

7.22 The Directive (OJ 1990 L189/1)[6] applies to non-automatic weighing instruments which are defined as weighing instruments requiring the intervention of an operator during weighing. The Directive distinguishes between two categories of instruments. In the first place, those which determine mass for commercial transactions or for the calculation of taxes, penalties, remuneration, price, determination of mass for weighing patients for the purposes of diagnosis and medical treatment, the making up of medicines on prescription and the determination of price on the basis of mass for the purposes of direct sales to the public. The other category of instrument use is all other applications other than falling within the first category.

(ii) Essential requirements

7.23 The essential requirements for the first category of instrument use is set out in Annex I to the Directive. These include metrological requirements dealing with units of mass, accuracy, classifications, design and construction of the instrument (including levelling, zeroing and other matters). The essential requirements set out in Annex 1 are designed to protect the public against incorrect results carried out by the machines covered by the Directive.

(iii) Conformity

7.24 The conformity of instruments to the essential requirements set out in Annex I may be certified by either of a number of procedures as selected by the applicant. These are EC-type examination (as mentioned in Annex II) followed either by the EC declaration of conformity type (incorporating guarantee of production quality – equivalent to Module D of Council Dec 93/465, OJ 1990 L380/13) or by EC verification (equivalent to Module F of Council Dec 93/465).

[6] As amended by Dir 93/68 1993 (OJ L220/1).

(iv) Entry into force

Member States are obliged to publish the laws implementing the Directive by 1 July 1992, and to apply those laws from 1 January 1993. However, there is a derogation stating that Member States may permit, during a period of ten years from the date on which their national legislation implementing the Directive applied, the placing on the market of instruments which conform to the rules in force before that date. **7.25**

7. Directive 90/385 – active implantable medical devices

(i) Scope

The Directive (OJ 1990 L189/17) applies to active implantable medical devices which are defined as any medical device relying for its functioning on a source of electrical energy or any source of power other than that directly generated by the human body or gravity, and which is intended to be totally or partially introduced, surgically or medically, into the human body or by medical intervention into a natural orifice and which is intended to remain after the procedure. **7.26**

The Directive constitutes a "specific Directive" within the meaning of Article 2 of the EMC Directive[7], and therefore the Directive takes precedence over the EMC Directive in that regard. **7.27**

(ii) Essential requirements

Annex I to the Directive lays down some detailed essential requirements regarding design and construction of the goods and listing a number of hazards against which the Directive is intended to protect users. **7.28**

(iii) Conformity

In order to affix the CE mark, the manufacturer must follow the procedure relating to the EC declaration of conformity set out in Annex II of the Directive, or follow the procedure relating to EC-type examination coupled with the procedure relating to EC verification set out in Annex IV or the procedure relating to the EC declaration of conformity to type. **7.29**

(iii) Entry into force

Member States were obliged to adopt the necessary laws to implement the **7.30**

[7] Dir 89/336. See para 7.16 *et seq.*

Directive by 1 July 1992 and those provisions were to apply from 1 January 1993.

7.31 However, there is a transitional period allowing Member States, up to the period to 31 December 1994, to permit the placing on the market of devices complying with national rules in force in their territory on 31 December 1992.

8. Directive 93/42 – medical devices

(i) Scope

7.32 The Recitals to this Directive refer to the fact that Council Directive 90/385 concerning active implantable devices was the first case of the application of the new approach to the field of medical devices. In the interests of uniform EC rules applicable to all medical devices, the Directive is largely based on the provisions of Dir 90/385, and contains amendments to that Directive.

7.33 The Directive applies to medical devices which are defined as any instruments, apparatus, appliance, material or other article, including software, for use alone or in combination, intended by the manufacturer to be used for human beings solely or principally for the purpose of diagnosis, prevention, monitoring, treatment or alleviation of disease, injury or handicap, investigation of the anatomy or of physiological process, the control of conception and which does not achieve its principal intended action in or on the human body by pharmacological, immunalogical or metabolic means, but which may be assisted in its functions by such means.

7.34 As with the Directive on implantable medical devices, the Directive contains a significant number of definitions.

(ii) Essential requirements

7.35 Annex I to the Directive lays down the essential requirements which the devices covered must meet. These concern the design and construction of the devices (chemical and physical properties, infection and microbiological contamination, construction and environmental properties, protection against radiation, etc).

7.36 Medical devices are divided into various classes (Classes I, II(a), II(b) and III). The rules for the classification of goods are set out in Annex IX to the Directive.

(iii) Conformity assessment procedures

7.37 The various conformity assessment procedures depend upon the class into which the various devices fall, and these are set out in Article 11 of the Directive.

(iv) Entry into force

Member States are obliged to adopt and publish the laws which implement the Directive not later than 1 July 1994, and those national laws must apply from 1 January 1995. 7.38

There is a transitional period for devices which must be subject to an EC declaration of conformity (full quality assurance system), EC-type examination, EC verification, EC declaration of conformity (production quality assurance) and EC declaration of conformity (product quality assurance) (corresponding to Modules C, D and E of Council Dec 93/465). In these cases Member states must accept the placing on the market and putting into service of devices which conform to the rules in force in their territory on 31 December 1994 during the period up to 13 June 1998. 7.39

9. Directive 90/396 – gas burning appliances

(i) Scope

The Directive (OJ 1990 L196/15) applies to appliances burning gaseous fuels used for cooking, heating, hot water production, refrigeration, lighting or washing and having, where applicable, a normal water temperature not exceeding 105°C. Forced draught burners and heating bodies to be equipped with such burners are also considered as covered by the Directive. It also applies to safety devices, controlling devices or regulating devices and sub-assemblies other than forced draught burners and heating bodies to be equipped with such burners separately marketed for trade use and design to be incorporated into an appliance burning gaseous fuel or assembled to constitute such an appliance (and these are known as "fittings"). Appliances specifically designed for use in industrial processes carried out on industrial premises are excluded from the scope of the Directive. 7.40

(ii) Essential requirements

The essential requirements which the appliances and fittings must satisfy are set in Annex I of the Directive. Apart from some general conditions (for example that all appliances must be accompanied by technical instructions intended for the installer, and also instructions concerning the use and servicing of the appliance, intended for the user) the essential requirements concern the materials used, design and contruction (for example gas release, ignition, combustion and temperatures). 7.41

(iii) Conformity

7.42 The means of certification of conformity of series manufactured appliances is to be an EC-type examination, corresponding to Module B of Council Decision 93/465 and (prior to the goods being placed on the market) at the choice of the manufacturer either:

- an EC declaration of conformity to type, corresponding to Module C of Council Decision 93/465; or
- an EC declaration of conformity to type (guarantee of production quality), corresponding to Module D of Council Decision 93/465; or
- the EC declaration of conformity to type (guarantee of product quality), corresponding to Module E of Council Decision 93/465; or
- EC verification, corresponding to Module F of Council Decision 93/465

(iv) Entry into force

7.43 Member States are obliged to adopt and publish the laws necessary to implement the Directive before 1 July 1991. Those laws should have applied from 1 January 1992.

There is a transitional period which states that during the period up to 31 December 1995 Member States *may* permit the placing on the market and/or the putting into service of appliances and fittings which comply with the regulations in force in the Member States before 1 January 1992.

10. Directive 89/686 – personal protective equipment

(i) Scope

7.44 The Directive (OJ 1989 L399/18) applies to personal protective equipment (PPE) which is defined as any device or appliance designed to be worn or held by an individual for protection against one or more health and safety hazards. The definition is quite wide and covers, for example, a unit constituted by several devices or appliances which have been integrally combined by the manufacturer for the protection of an individual against one or more potentially simultaneous risks.

7.45 The Directive does not apply to PPE covered by another Directive designed to achieve the same objectives as the Directive, or PPE expressly excluded from the scope of the Directive in Annex I. Excluded PPE types are: PPE designed and manufactured specifically for use by the Armed Forces or in the maintenance of law and order (*e.g.* helmets and shields), PPE for civil defence (aerosol cannisters, etc), PPE designed and manufactured for private use against adverse atmospheric conditions (headgear, seasonal clothing, footwear, umbrellas, etc), PPE intended for the protection or rescue of persons on vessels or aircraft, not worn all the time (*e.g.* lifebelts).

(ii) Essential requirements

The essential requirements which PPE must satisfy are set out in Annex II to the Directive. The first rather vague general requirement is that PPE must provide adequate protection against all risks encountered. The health and safety requirements concern design principles: ergonomics, levels and classes of protection, innocuousness of PPE, comfort and efficiency and information. In addition there are further requirements common to several classes or types of PPE (for example, PPE enclosing the parts of the body to be protected, PPE for the face, eyes and respiratory tracts, PPE for use in explosive atmospheres, PPE for use in very dangerous situations). Additional requirements refer to specific or particular risks. **7.46**

(iii) Conformity

Prior to the series production of PPE other than those referred to in paragraph 3 of Article 8, the manufacturer or his authorised representative in the EC will submit a model for EC-type examination (Module B). EC-type examination will not be required in the case of PPE models of simple design where the designer assumes the user can himself assess the level of protection provided against the minimal risk concerned the effects of which, when they are gradual, can be safely identified by the user in good time. This category covers PPE intended to protect the wearer against, for example, mechnical action whose effects are superficial (gardening gloves, thimbles, etc), sunlight (*e.g.* sunglasses) and minor impacts and vibrations which do not affect vital areas of the body (*e.g.* light footwear). **7.47**

(iv) Entry into force

Member States were obliged by 31 November 1991 to adopt and publish the laws necessary to implement the Directive, and those laws were to apply from 1 July 1992. A transitional period lasted until 30 June 1995. **7.48**

11. Directive 91/263 – telecommunications terminal equipment

(i) Scope

The Directive (OJ 1991 L128/1)[8] applies to terminal equipment which is **7.49**

[8] Amended by Dir 93/97 OJ L290/1 and Dir 93/68 1993 OJ L220/1.

defined as equipment intended to be connected to the public telecommunications network (*i.e.* to be connected directly to the termination of a public telecommunications network or to interwork with a public telecommunications network being connected directly or indirectly to the termination of a public telecommunications network) in order to send process or receive information.

7.50 The Directive also amends Directive 89/336 (the EMC Directive) by deleting Article 10(4) of that Directive. The effect of this is to delete the provisions of the EMC Directive insofar as they refer to the definition of telecommunications terminal equipment and to the conformity assessment procedures to be applied for such equipment. Those procedures are now governed exclusively by Directive 91/263. The Directive was amended by Directive 93/97 which concerns ground satellite communications station equipment.

(ii) Essential requirements

7.51 The essential requirements are listed in Article 4 which states that terminal equipment shall satisfy the following essential requirements:

- user safety insofar as this requirement is not covered by Directive 73/23 (the low voltage Directive);
- safety of employees of public telecommunications network operators insofar as this requirement is not covered by the low voltage Directive;
- electromagnetic compatibility requirements insofar as they are specific to terminal equipment;
- protection of the public telecommunications network;
- interworking of terminal equipment.

(iii) Conformity

7.52 According to the choice of the manufacturer, terminal equipment covered by the Directive will be subject either to an EC-type examination (Module B) or the EC declaration of conformity (full quality assurance, equivalent to Module H of Council Dec 93/465). An EC-type examination will be accompanied by a declaration issued according to the EC declaration of conformity to type procedure described in Annexes II and III of the Directive, equivalent to Modules C and D of Decision 93/465.

(iv) Entry into force

7.53 Member States are obliged to implement the Directive not later than 6 November 1992.

12. Directive 89/106 – construction products

(i) Scope

The Directive (OJ 1989 L40/12) applies to construction products insofar as the essential requirements in respect of construction works relate to them. A construction product means any product that is produced for incorporation in a permanent manner in construction works, including both buildings and civil engineering works. **7.54**

(ii) Essential requirements

The essential requirements applicable to works which may influence the technical characteristics of a product are set out in terms of objectives in Annex I to the Directive. The products must be suitable for construction works and the requirements in Annex I relate to mechnical resistance and stability, safety in case of fire, hygiene, health and the environment, safety in use, protection against noise and energy economy and heat retention. **7.55**

(iii) Conformity

Article 7 of the Directive refers to harmonised standards to be established by the European standards organisations, but Article 8 goes on to refer to European technical approval. European technical approval is a favourable technical assessment of the fitness for use of a product for an intended use, based on the fulfilment of the essential requirements for building works for which the product is used. Such approval may be granted to: **7.56**

– products for which there is neither a harmonised standard, nor a recognised national standard, nor a mandate for a harmonised standard, and for which the Commission considers that a standard could not, or not yet, be elaborated; and
– products which differ significantly from harmonised or recognised standards.

European technical approval is normally to be granted for a five year period, although this may be extended. The approval for a product is based upon examinations, tests and an assessment on a basis of certain interpretive documents referred to in the Directive. **7.57**

Attestation of conformity is dealt with in Articles 13, 14 and 15 of the Directive.

(iv) Entry into force

7.58 Member States were obliged to bring into force the national laws to implement the Directive within 30 months of its notification (it was notified to the Member States on 27 December 1988).

7.59 Although the Directive has notionally been in force since June 1991, its implementation has been deferred as the implementation depends upon the existence of harmonised technical specifications and standards from CEN/CENELEC and European technical approvals to be drawn up by the European Organisation for Technical Approvals, following the European Commission's adoption of various interpretative documents and the choice of conformity certification procedures for each product or group of products.

7.60 The interpretive documents are intended to give concrete form to the essential requirements set out in Annex I of the Directive. The principal objective of the interpretive documents is to exhibit a link between the essential requirements and the mandates which the Commission gives to the European standardisation bodies and to the European Organisation for Technical Approval (EOTA) to establish guidelines for European technical approvals. Examples of such an interpretive document can be seen in Official Journal OJ (1994) C62.

13. Directive 92/42 – hot water boilers

(i) Scope

7.61 The Directive (OJ 1992 L167/17) determines the efficiency requirements applicable to new hot water boilers fired by liquid or gaseous fuels with a rated output of no less than 4 kw and no more than 400 kw ("boilers"). Certain equipment is excluded from the scope of the Directive, including hot water boilers capable of being fired by different fuels including solid fuels, and cookers and appliances designed mainly to heat the premises in which they are installed and, as a subsidiary function, to supply hot water for central heating and sanitory hot water.

(ii) Essential requirements

7.62 The efficiency requirements are stated in Article 5 of the Directive, and the useful efficiency requirements are set out in a table in the Directive.

(iii) Conformity

7.63 The conformity of series produced boilers is to be certified by:

– examination of the efficiency of a boiler type in accordance with Module B of Council Decision 90/683, as set out in Annex III of the Directive;
– a declaration of conformity to the approved type in accordance with Module C, D or E of Council Decision 90/683, and as more particularly described in Annex IV to the Directive.

(iv) Entry into force

Member States are obliged to implement the Directive by 1 January 1993 and to apply their national laws from at least 1 January 1994. 7.64

There is a transitional provision which states that up until 31 December 1997 Member States must permit the placing on the market and putting into service of appliances complying with the national rules and schemes in force within their territories on 21 May 1992. 7.65

14. Directive 73/23 – low voltage

(i) Scope

Directive 73/23[9] is not strictly speaking a "new approach" Directive, as it pre-dates the European Community's new approach to standardisation. However, the Directive is now deemed to be a new approach Directive, and, for example, was amended by Directive 93/68 which harmonised the CE marking in all of the new approach directives. It also makes reference to harmonised standards, and therefore is of a similar format to the new approach Directives. 7.66

The Directive covers any electrical equipment which is intended to be used at a nominal voltage of between 50 and 1000 volts for alternating current and between 75 and 1500 volts for direct current, other than certain equipment and phenomena which are mentioned in Annex 2. Excluded items are : electrical equipment for use in explosive atmospheres, and for radiology and medical purposes or for goods and passenger lifts ; electricity meters ; plugs and sockets outlets for domestic use, electric fence controllers, radio electrical interference and specialised electrical equipment for use on ships, aircraft or railways. (Most of these exclusions are dealt with by other Directives.) 7.67

(ii) Essential requirements

The "essential requirements" are listed in Annex 1 which lays down the principal elements for the safety objectives for electrical equipment covered by the Directive. These conditions concern the protection of persons and domestic animals against danger of physical injury or other harm which 7.68

[9] OJ(1973) L77/29

might be caused by electrical contact direct or indirect ; a requirement that temperatures, marks or radiation which would cause a danger are not produced, and other similar hazards.

7.69 To the extent that few harmonised standards have been drawn up, the directive dictates that Member States must also allow the placing on the market of electrical equipment which complies with the safety provisions of the International Commission on the Rules for the Approval of Electrical Equipment or of the International Electro Technical Commission (IEC).

(iii) Conformity

7.70 The directive provides either for marks of conformity or certificates of conformity or a declaration of conformity issued by the manufacturer.

7.71 Article 8 also provides for a special certification procedure.

(iv) Entry into force

7.72 The directive entered force in 1974.

15. Proposal COM (95) 130 – *in vitro* diagnostic medical devices

(i) Scope

7.73 The recitals to this proposal, which were put forward in April 1995, point out that most medical devices are covered by Directive 90/385 (implantable medical devices) and Directive 93/42 relating to other medical devices with the exception of *in vitro* diagnostic medical devices, and that the proposal is intended to extend harmonisation to such diagnostic medical devices. The proposal is therefore based largely upon the provisions of Directives 90/385 and 93/42. As the electro-magnetic compatibility aspects form an integral part of the essential requirements of the proposal, Directive 89/336 (the electro-magnetic compatibility directive) does not apply to products covered by this Directive.

7.74 The Directive applies to *in vitro* diagnostic medical devices and their accessories, and both are referred to by the word "devices". An *in vitro* diagnostic medical device means any medical device (as defined) which is a reagent, reagent product, calibrator, control material, kit, instrument, apparatus, equipment or system whether used alone or in combination, intended by the manufacturer to be used *in vitro* for the examination of specimens including blood and tissue donations, derived from the human body, solely or principally for the purpose of providing information concerning a physiological state, state of health or disease or congenital abnormality or to determine the safety and compatibility with potential recipients. Products for general laboratory use are not devices unless such products, in view of their characteristics, are specifically intended by the manufacturer to be used for *in vitro* diagnostic examination.

For the purposes of the proposal, the "manufacturer" is deemed to be the natural or legal person with responsibility for the design, manufacture, packaging and labelling of the device before it is placed on the market under his own name, regardless of whether these operations are carried out by that person himself or on his behalf by a third party. The obligations to be met by manufacturers also apply to natural or legal persons who assemble, package, process, fully refurbish and/or label one or more ready-made products and assign to them intended purposes as devices with a view to their being placed on the market under their name. **7.75**

The intended purpose of a particular device is the use for which the device is intended according to the data supplied by the manufacturer on the labelling, in the instructions and/or in the promotional materials. **7.76**

(ii) Essential requirements

The essential requirements are listed in Annex 1, which sets out various general requirements as well as specific requirements regarding the design and manufacture of these products. The detailed requirements concerning the design and manufacture relate to chemical and physical properties, infections and microbial contamination, construction and environmental properties, matters concerning the measuring function of devices, protection against radiation, matters concerning the connection of devices to energy sources, requirements concerning self-testing, and information to be supplied by the manufacturer. **7.77**

(iii) Conformity

The conformity assessment procedures are set out in Article 9 of the proposal. For all devices other than those covered by Annex 2 (reagents and reagent products for blood grouping, and for the detection in human specimens of markers of HIV infection and hepatitis B and C) follow the procedure set out in Annex 3 (EC declaration of conformity). There are additional requirements for devices for self-testing. For all devices covered by Annex 2 other than those intended for performance evaluation, the manufacturer must, in order to affix the CE marking, either follow the full quality assurance procedure set out in Annex 4, or follow the EC-type examination procedure set out in Annex 5, coupled with an EC verification, or the procedure concerning production quality assurance as set out in Annex 7. **7.78**

As regards devices for performance evaluation, the manufacturer must follow the procedure referred to in Annex 8, which sets out a statement and procedures concerning such products. **7.79**

(iv) Entry into force

7.80 The current draft of the proposal requires the Members States to adopt the necessary laws to implement the Directive no later than 1 April 1998, and there is a four-year transitional period (calculated from the date of adoption of the Directive), during which Member States must accept the placing on the market and the putting into service of devices which conform to the rules in force on their territory on the date of adoption of the Directive.

7.81 The current proposal also contains various amendments to be made to Directives 93/42 and 89/392.

16. Directive 93/15 – explosives for civil use

(i) Scope

7.82 The Directive applies to explosives considered to be such in the United Nation's Recommendation on the Transport of Dangerous Goods and which fall within class 1 of those Recommendations. The Directive does not cover explosives including ammunition, intended for use in accordance with national law by the armed forces or the police, pyrotechnical articles and ammunition (except where specified in the Directive).

7.83 The Directive really has two aims. The first is to harmonise provisions relating to the placing on the market of explosives for civil use, and the second is to set up a system governing the supervision of transfers of explosives in the EC (Articles 9, 10 and 11).

(ii) Essential requirements

7.84 The essential requirements are set out in Annex 1 to the Directive. The general requirement is that each explosive must be designed, manufactured and supplied in such a way as to present a minimal risk to the safety of human life and health, and to prevent damage to property and the environment under normal, forseeable conditions, in particular as regards the safety rules and standard practices included until such time as it is used. Each explosive must attain the performance characteristics specified by the manufacturer in order to ensure maximum safety and reliability. In addition, certain special requirements are set out in Part II of Annex I.

(iii) Conformity

7.85 The procedures for the attestation of the conformity of explosive shall be either:

EC type examination (Module B of Decision 93/465) and, *either* :

– type conformity (Module C) or
– production quality assurance (Module D) or product quality assurance (Module E) or
– product verification (Module F) *or*
– unit verification (Module G)

The full details of conformity assessment are set out in Annex II.

(iv) Entry into force

Those parts of the Directive concerning the system for supervising **7.86** transfers of explosives applied from 1 October 1993, whereas the rules regarding placing such products on the market applied from 1 January 1995. There is a much longer transitional period than under most of the other new approach Directives, and this runs until 31 December 2002. During the transitional period, Member States must allow the placing on the markets of explosives which complied with the national regulations in force in their territory before 31 December 1994.

17. Directive 94/9 – equipment and systems for use in potentially explosive atmospheres

(i) Scope

The Directive covers equipment and protective systems which are defined **7.87** as machines, apparatus, fixed or mobile devices, control components and instrumentation and design units which are intended to halt incipient explosions immediately and/or to limit the effective range of explosion, insofar as that equipment is intended for use in potentially explosive atmospheres. Those are atmospheres involving the mixture with air, under atmospheric conditions, of flammable substances in the form of gases, vapours, mist or dust in which, after ignition has occurred, combustion spreads to the entire unburned mixture. Various items are excluded from the scope of the Directive (for example personal protective equipment which is covered by Directive 89/686). Furthermore, the Directive does not apply to "military equipment", nor means of transport.

Annex I to the Directive lays down certain criteria determining the **7.88** classification of "equipment groups" into categories.

(ii) Essential requirements

The essential health and safety requirements relating to the design and **7.89** construction of equipment and protective systems intended for use in potentially explosive atmospheres are set out in Annex II. The common requirements are fairly detailed and cover matters such as : principles of integrated explosion safety, special checking and maintenance conditions, surrounding area conditions, markings, instructions, the selection of materials, the design and construction of the equipment, matters concerning potential ignition sources, hazards arising from external effects and requirements in respect of safety related devices. There are also certain supplementary requirements in respect of certain equipment and also in respect of protective systems.

(iii) Conformity

7.90 The conformity assessment procedures are set out in Article 8 and Annexes III to IX and depend upon the equipment group under consideration. Reference should be made to the Annexes for the precise details.

(iv) Entry into force

7.91 Members states were required to implement the Directive by 1 September 1995, and apply the implementing laws with effect from 1 March 1996. However, there is a transitional period allowing the placing on the market and putting into service of equipment and protective systems conforming with the national regulations in force in Member states' territory on 23 March 1994. The transitional period lasts until 30 June 2003.

18. Directive 94/25 – recreational craft

(i) Scope

7.92 This Directive applies only to recreational craft of a minimum length of 2.5 metres and a maximum length of 24 metres, and includes recreational craft, partly completed boats and components as set out in Annex 2 when separate and when installed. (The components listed in Annex 2 comprise items such ignition protective equipment for inboard and stern drive engines, start-in gear protection devices for outboard engines, steering wheels and steering mechanisms, fuel tanks and fuel hoses). A recreational craft means any boat of any type within the size of hull length mentioned, regardless of the means of propulsion, insofar as the craft is intended for sports and leisure purposes. The fact that the same boat could be used for charter or for recreational boating training shall not prevent it being covered by the Directive when it is placed on the market for recreational purposes.

7.93 There are a number of exclusions from the scope of the Directive and these include: craft intended solely for racing (for example rowing boats), canoes, kayaks, gondolas, and pedalos, sailing surf boards, powered surf boats, original and individual replicas of historical craft designed before 1950 and craft built for "own use".

(ii) Essential requirements

The essential requirements are listed in Annex I. The Annex distinguishes between various boat design categories (ocean, offshore, inshore and those for sheltered waters, according to wind force and significant wave height. **7.94**

There are a number of general requirements (such as each craft must be marked with a hull identification number containing the manufacturer's code, the country of manufacture and other information) and depending on the design category, each craft shall be designed to minimise the risk of falling overboard, and for motorboats, must have good visibility from the main steering position. **7.95**

In addition, the essential requirements go on to lay down various matters concerning the integrity and structural requirements for recreational craft. These concern matters such as stability and free board, buoyancy and flotation, openings in hull, deck and superstructure and other matters. The Annex also contains matters relating to the handling characteristics and the requirements of various installations such as the ventilation of the engine compartment and other such matters. **7.96**

(iii) Conformity

The conformity assessment procedures are set out in Article 8, and depend upon the categories as set out in Annex I – "ocean", "offshore", "inshore", "sheltered waters". **7.97**

The appropriate Modules (as specified in Council Decision 93/465) are clearly set out in Article 8. **7.98**

(iv) Entry into force

Member states are obliged to implement the Directive by 16 December 1995, and apply the provisions of their national law from 16 June 1996. Furthermore, there is a transitional period during which Member States must allow the placing on the market and putting into service of recreational craft which comply with the rules in force in their territory on the date of adoption of the Directive (16 June 1994) for a period of four years from that date (*i.e.*, until 15 June 1998). **7.99**

Contents of Chapter 8
Implications for Non-EC Manufacturers

Chapter 8
Implications for Non-EC Manufacturers

1. Introduction

As stated earlier[1], the European standardisation bodies (in particular CEN and CENELEC) comprise national standards organisations of the European countries, and it is sometimes felt by non-European manufacturers that they may be disadvantaged in the sense that they have little influence on the standardisation process within Europe. For example, the United States is represented in the ISO and IEC, but not in CEN and CENELEC. Although the move at an international level does appear to be towards the greater exchange of information on standardisation work programmes, there is still a perceived fear that the question of standardisation and conformity assessment may lead to a "Fortress Europe". **8.1**

On the question of negotiations with third countries on mutual recognition in relation to conformity assessment, the EC Commission in June 1992 put forward a Communication to the Council of Ministers which is examined below. **8.2**

2. Negotiation of agreements between EC and certain third countries on mutual recognition in relation to conformity assessment

The Council Resolution of 21 December 1989 (OJ 1990 C10/1) provides that, in its relations with third countries, the EC will endeavour to promote international trade in regulated products, in particular by concluding mutual recognition agreements on the basis of Article 113 of the Treaty. In general, mutual recognition agreements will permit bodies notified by the Member States to certify that EC products comply with the requirements of the third country concerned. Similarly, the third country will be able to have the conformity of its products with the EC's requirements assessed by the certification bodies in its territory. Thus, the object of mutual recognition agreements is the mutual acceptance by the parties to the agreements of certificates, marks of conformity and test reports issued by the relevant **8.3**

---

[1] See generally Chap 4.

bodies with regard to conformity assessment in the fields covered by the agreement.[2]

8.4 In its Communication to the Council, the Commission states that mutual recognition agreements, whose objective is to facilitate trade between the parties at this stage are not aimed at the mutual acceptance by the parties of the other party's standards or technical regulations, nor at mutual recognition of the equivalence of these standards or technical regulations. Mutual recognition agreements concern the assessment of the conformity of products originating in the territories of the parties to the agreements, but they do constitute the first step towards the introduction of a harmonised system of standardisation and certification between the parties to the agreement. Moreover, they facilitate matters for the parties by permitting mutual acceptance of the results of conformity assessment. Each party is able to assess the conformity of its products to the other party's requirements in laboratories and certification bodies *in its own territory* and the assessment will be accepted by the other party.

8.5 The Commission's Communication essentially summarises the background against which the EC may negotiate and enter into mutual recognition agreements, and the sphere of their application.

8.6 Products coming from third countries currently already have access to EC certification bodies on a non-discriminatory basis under the principles of the GATT Agreement on Technical Barriers to Trade. The Commission's Communication states that there are two ways in which third country certification bodies and testing laboratories may take part in EC conformity assessment procedures required by the new approach Directives. In the first place, a notified body under a new approach Directive may subcontract part of its work to another body (which may or may not be situated in the EC) within clearly defined limits and on the basis of regular checks on the competence of the other body. The Commission has set down guiding principles for sub-contracting by notified bodies in a Working Paper.[3]

8.7 In the second place, the ability to delegate responsibility for issuing certificates of conformity to a certification organisation in a third country is subject to the conclusion of a mutual recognition agreement between the EC and the third country in question.

3. Conditions for conclusion of mutual recognition agreements

8.8 The Council Resolution of 21 December 1989 on a global approach to

---

[2] In such cases responsibility for conformity assessment work conducted in accordance with the EC Directives (including the issue of certificates of conformity) remains with the EC notified body.

[3] Reference Certif 90/5 – Rev 2.

conformity assessment laid down the conditions to be fulfilled for mutual recognition agreements to be concluded. The main condition is that the competence of the third country's economic operators is and remains on a par with that required of their Community counterparts. This is to be assessed on the basis of the following criteria:

– reference to the EN29000 and EN45000 European standards relating to quality assurance and the requirements to be met by testing laboratories and certification bodies. By referring to these standards it can be presumed that the conditions relating to technical competence are met. The same criteria applies to the notified bodies in the Member States of the EC;
– each party to the agreement must have an agency with the necessary authority and competence to notify the competent bodies;
– the notifying authority (*e.g.* a Ministry or Government department) also must have the power to withdraw notification where the notified bodies cease to meet the relevant criteria.

The Commission's Communication emphasises that a mutual recognition **8.9**
agreement between the EC and a third country must be based on mechanisms designed to inspire confidence. One essential feature of this is that the agreement must have a mechanism for ensuring the continued competence of the notified bodies in question.

Emphasis is placed in the Commission's Communication on the need to **8.10**
ensure equal ease of market access for the manufacturers in the territories of each party to the agreement. To make an accurate assessment of the legal and practical conditions governing market access in the field covered by the agreement, the parties should, before the agreement is signed, carry out a review of their respective practises, relating in particular to the nature of the third country's technical rules, conformity assessment procedures, geographical scope and other administrative conditions. The Communication warns that in general, negotiations can only be opened with third countries whose level of technical and industrial competence is on a par with the Community's.

Interest in negotiating mutual recognition agreements has been shown **8.11**
by a number of third countries and also by European trade and industry. In particular, European industry has indicated an interest in the EC's opening negotiations for mutual recognitions with the following countries: United States, Canada, Japan, Australia, New Zealand, Singapore, China, Taiwan, Korea, Thailand, Philippines, Malaysia, Czechoslovakia, Hungry, Yugoslavia, Poland, Romania, Brazil, Mexico and India.[4]

---

[4] The situation with regard to the EFTA countries is that the certification regime is covered by the agreement on the European Economic Area.

4. Procedures for negotiation of mutual recognition agreements

8.12 The procedures for negotiating mutual recognition agreements differ according to whether the subject matter falls within a "harmonised field" (*i.e.* an area in which there are existing EC Directives) or whether the field is non-harmonised. Under the procedures laid down in Article 113 of the EC Treaty, the Commission conducts negotiations in close consultation with the Member States. In so doing, the Commission may ask the Member States to provide competent national experts working on conformity assessment in the fields covered by the draft agreement in order to cover technical aspects of negotiations. The Commission has the exclusive right to negotiate these agreements once it has been authorised so to do by the Council of Ministers.

8.13 If a Member State which has non-harmonised national regulations wishes to regulate its relations with a third country in that field, it should inform the Commission. After consulting with the other Member States, the Commission shall ask the Council either to authorise the Member State concerned to negotiate it by lateral agreement, or to authorise the Commission to negotiate it if there is a general interest in the matter being dealt with at an EC level. The Council Decision authorising bilateral negotiations by Member States will also lay down the procedure for approving the agreement in question.

8.14 Agreements between Member States and third countries will have the effect of treating products from the third country fulfilling the conditions of the agreement in the same way as products from the Member State concerned for the purposes of conformity assessment.

8.15 Access to the Member States' markets for products from third countries which have been assessed for conformity with the Member State which is a party to the agreement will only be possible where the products have first been placed on the market of the Member State in question.

5. Future action

8.16 The Commission's Communication ends with a recommendation that the Council of Ministers adopts a Decision authorising the Commission to negotiate a mutual recognition agreement relating to conformity assessment with certain third countries.

8.17 The countries which are presently included in the list are: United States, Japan, Canada, Australia, New Zealand, Hong Kong, Israel, Singapore, Philippines and the Republic of Korea. In addition, Turkey, South Africa, Malaysia, Thailand, Indonesia and China are referred to on the assumption that these countries sign the GATT Code on Technical Barriers to Trade (which they did). The proposed Council Decision contains an Annex which lays down certain Directives for the conduct of the negotiations.

A Decision was adopted on 26 September 1992 by the Council of Ministers concerning the negotiation of mutual recognition agreements. 8.18

Bibliography

Centre for European Policy Studies (CEPS), *The European Commission without technical barriers* CEPS Standards Programme: Page No 1 (CEPS Working Party Report No 5 – March 1992).

CEPS, *Fixing European standards: moving beyond the Green Paper* (May 1992). Nicholas and Repussard "Common Standards for Enterprises" (published in the *Deadline 92* Series) 1988.

DTI Report "Keeping Your Product on the Market" (December 1991).

Gormley, *Prohibiting Restrictions on Trade within the EEC.*

Nicolas and Repussard "Common Standards for Enterprises" (European Commission publication 1994).

Oliver, *Free Movement of Goods in the EEC* (3rd ed) (1995).

COUNCIL DIRECTIVE 83/189/EEC

of 28 March 1983

laying down a procedure for the provision of information in the field of technical standards and regulations*†

THE COUNCIL OF THE EUROPEAN COMMUNITIES,

Having regard to the Treaty establishing the European Economic Community, and in particular Articles 100 and 213 thereof,

Having regard to the proposal from the Commission,

Having regard to the opinion of the European Parliament,

Having regard to the opinion of the Economic and Social Committee,

Whereas the prohibition of quantitative restrictions on the movement of goods and of measures having an equivalent effect is one of the basic principles of the Community;

Whereas barriers to trade resulting from technical regulations relating to products may be allowed only where they are necessary in order to meet essential requirements and have an objective in the public interest of which they constitute the main guarantee;

Whereas it is essential for the Commission to have the necessary information at its disposal before the adoption of technical provisions; whereas, consequently, the Member States which are required to facilitate the achievement of its task pursuant to Article 5 of the Treaty must notify it of their projects in the field of technical regulations;

Whereas all the Member States must also be informed of the technical regulations contemplated by any one Member State;

Whereas the Commission and the Member States must also be allowed sufficient time in which to propose amendments to a contemplated measure, in order to remove or reduce any barriers which it might create to the free movement of goods;

Whereas the Member States concerned must take account of these amendments when formulating the definitive text of the measure envisaged;

Whereas the Commission must also have the option of proposing or adopting a Community Directive governing the subject of the national measure contemplated;

Whereas, in the two instances set out above, the Member State in question must, pursuant to the general obligations laid down in Article 5 of the Treaty, defer implementation of the contemplated measure for a period sufficient to allow either a joint examination of the proposed amendments or the preparation of the proposal for a Council Directive or of the Commission Directive; whereas the time limits laid down in the Agreement of the representatives of the Governments of the Member States meeting within the Council of 28 May 1969 providing for standstill and notification to the Commission, as amended by the Agreement of 5 March 1973, have proved inadequate in the cases concerned and should accordingly be extended;

Whereas the procedure concerning the standstill arrangement and notification of the Commission contained in the abovementioned Agreement of 28 May 1969 remains applicable to products subject to that procedure which are not covered by this Directive;

---

*Consolidated text.
†OJ 1983 L109/8.
Amended by: Dir 88/182 and Dir 94/10 and Decision 92/400.

Whereas, in practice, national technical standards may have the same effects on the free movement of goods as technical regulations;

Whereas it would therefore appear necessary to inform the Commission of draft standards under similar conditions to those which apply to technical regulations; whereas, pursuant to Article 213 of the Treaty, the Commission may, within the limits and under the conditions laid down by the Council in accordance with the provisions of the Treaty, collect any information and carry out any checks required for the performance of the tasks entrusted to it;

Whereas it is also necessary for the Member States and the standards institutions to be informed of standards contemplated by standards institutions in the other Member States;

Whereas it is necessary to set up a Standing Committee, the members of which will be appointed by the Member States with the task of helping the Commission to examine draft national standards and cooperating in its efforts to lessen any adverse effects thereof on the free movement of goods,

HAS ADOPTED THIS DIRECTIVE:

Article 1

For the purposes of this Directive, the following meanings shall apply:

(1) "Product", any industrially manufactured product, and any agricultural product."

(2) "technical specification", a specification contained in a document which lays down the characteristics required of a product such as levels of quality, performance, safety or dimensions, including the requirements applicable to the product as regards the name under which the product is sold, terminology, symbols, testing and test methods, packaging, marking or labelling and conformity assessment procedures;

The term "technical specification" also covers production methods and processes used in respect of agricultural products as referred to in Article 38 (1) of the Treaty, products intended for human and animal consumption, and medicinal products as defined in Article 1 of Directive 65/65/EEC,[1] as well as production methods and processes relating to other products, where these have an effect on their characteristics.

(3) "other requirement", a requirement, other than a technical specification, imposed on a product for the purpose of protecting, in particular, consumers or the environment, and which affects its life cycle after it has been placed on the market, such as conditions of use, recycling, reuse or disposal, where such conditions can significantly influence the composition or nature of the product or its marketing;

(4) "standard", a technical specification approved by a recognized standardization body for repeated or continuous application, with which compliance is not compulsory and which is one of the following:

– international standard: a standard adopted by an international standardization organization and made available to the public,

– European standard: a standard adopted by a European standardization body and made available to the public,

– national standard: a standard adopted by a national standardization body and made available to the public;

(5) "standards programme", a work programme of a recognized standardisation body listing the subjects on which standardization work is being carried out;

(6) "draft standard", documents containing the text of the technical specifications concerning a given subject, which is being considered for adoption in accordance with the national

[1] OJ No 22, 9. 2. 1965, p. 369/65. Directive as last amended by Directive 93/39/EEC (OJ No L 214, 24, 8. 1993, p. 22).

standards procedure, as that document stands after the preparatory work and as circulated for public comment or scrutiny;
(7) "European Standardization Body", a body referred to in Annex I.
(8) "national standardization body", a body referred to in Annex II;
(9) "technical regulation", technical specifications and other requirements, including the relevant administrative provisions, the observance of which is compulsory, *de jure* or *de facto*, in the case of marketing or use in a Member State or a major part thereof, as well as laws, regulations or administrative provisions of Member States, except those provided for in Article 10, prohibiting the manufacture, importation, marketing or use of a product.

De facto technical regulations include:
– laws, regulations or administrative provisions of a Member State which refer either to technical specifications or other requirements or to professional codes or codes of practice which in turn refer to technical specifications or other requirements and compliance with which confers a presumption of conformity with the obligations imposed by the aforementioned laws, regulations or administrative provisions,
– voluntary agreements to which a public authority is a contracting party and which provide, in the public interest, for compliance with technical specifications or other requirements, excluding public procurement tender specifications,
– technical specifications or other requirements which are linked to fiscal or financial measures affecting the consumption of products by encouraging compliance with such technical specifications or other requirements; technical specifications or other requirements linked to national social-security systems are not included.

This comprises technical regulations imposed by the authorities designated by the Member States and appearing on a list to be drawn up by the Commission before the entry into force of this Directive in the framework of the Committee referred to in Article 5.

The same procedure shall be used for amending this list.
(10) "draft technical regulations", the text of a technical specification or other requirement, including administrative provisions formulated with the aim of enacting it or of ultimately having it enacted as a technical regulation, the text being at a stage of preparation at which substantial amendments can still be made.

This Directive shall not apply to those measures Member States consider necessary under the Treaty for the protection of persons, in particular workers, when products are used, provided that such measures do not affect the products.

Article 2

1. The Commission and the standardization bodies referred to in Annexes I and II shall be informed of the new subjects for which the national bodies referred to in Annex II have decided, by including them in their standards programme, to prepare or amend a standard, unless it is an identical or equivalent transposition of an international or European standard.
2. The information referred to in paragraph 1 shall indicate, in particular, whether the standard concerned:
– will transpose an international standard without being the equivalent,
– will be a new national standard, or
– will amend a national standard.

After consulting the Committee referred to in Article 5, the Commission may draw up rules for the consolidated presentation of this information and a plan and criteria governing the presentation of this information in order to facilitate its evaluation.
3. The Commission may ask for all or part of the standards programmes to be communicated to it.

It shall make this information available to the Member States in a form which allows the different programmes to be assessed and compared.

4. Where appropriate, the Commission shall amend Annex II on the basis of communications from the Member States.
5. The Council shall decide, on the basis of a proposal from the Commission, on any amendment to Annex I.

Article 3

The standardization bodies referred to in Annexes I and II, and the Commission, shall be sent all draft standards on request; they shall be kept informed by the body concerned of the action taken on any comments they have made relating to drafts.

Article 4

1. Member States shall take all necessary steps to ensure that their standardization bodies:
– communicate information in accordance with Articles 2 and 3,
– publish the draft standards in such a way that comments may also be obtained from parties established in other Member States.
– grant the other bodies referred to in Annex II the right to be involved passively or actively (by sending an observer) in the planned activities,
– do not object to a subject for standardization in their work programme being discussed at European level in accordance with the rules laid down by the European standardization bodies and undertake no action which may prejudice a decision in this regard.
2. Member States shall refrain in particular from any act of recognition, approval or use by reference of a national standard adopted in breach of Articles 2, 3 and 4.

Article 5

A Standing Committee shall be set up consisting of representatives appointed by the Member States who may call on the assistance of experts or advisers; its chairman shall be a representative of the Commission.

The Committee shall draw up its own rules of procedure.

Article 6

1. The Committee shall meet at least twice a year with the representatives of the standards institutions referred to in List 1.
2. The Commission shall submit to the Committee a report on the implementation and application of the abovementioned procedures and proposals aimed at eliminating existing or foreseeable barriers to trade.
3. The Committee shall express its opinion on the communications and proposals referred to in paragraph 2 and may in this connection propose, in particular, that the Commission:
– request the European standards institutions to draw up a European standard within a given time limit,
– ensure where necessary, in order to avoid the risk of barriers to trade, that initially the Member States concerned decide amongst themselves on appropriate measures,
– take all appropriate measures.
– identify the areas where harmonisation appears necessary and, should the case arise, undertake appropriate harmonisation in a given sector.
4. The Committee must be consulted by the Commission:

(a) before any amendment is made to the lists in the Annex (Article 2 (1));
(b) when drawing up the rules for the codified presentation of information and the plan and criteria for the presentation of standards programmes (Article 2 (2));
(c) when deciding on the actual system whereby the exchange of information provided for in this Directive is to be effected and on any change to it;
(d) when reviewing the operation of the system set up by this Directive (Article 11).
(e) on the requests to the standards institutions referred in the first indent of paragraph 3.

5. The Committee may be consulted by the Commission on any preliminary draft technical regulation received by the latter.

6. Any question regarding the implementation of this Directive may be submitted to the Committee at the request of its chairman or of a Member State.
7. The proceedings of the Committee and the information to be submitted to it shall be confidential.

However, the Committee and the national authorities may, provided that the necessary precautions are taken, consult, for an expert opinion, natural or legal persons, including persons in the private sector.

Article 7

1. Member States shall take all appropriate measures to ensure that, during the preparation of a European standard referred to in the first indent of Article 6 (3) or after its approval, their standardization bodies do not take any action which could prejudice the harmonization intended and, in particular, that they do not publish in the field in question a new or revised national standard which is not completely in line with an existing European standard.
2. Paragraph 1 shall not apply to the work of standards institutions undertaken at the request of the public authorities to draw up technical specifications or a standard for specific products for the purpose of enacting a technical regulation for such products.

Member States shall communicate all requests of the kind referred to in the preceding subparagraph to the Commission as draft technical regulations, in accordance with Article 8 (1), and shall state the grounds for their enactment.

Article 8

1. Subject to Article 10, Member States shall immediately communicate to the Commission any draft technical regulation, except where it merely transposes the full text of an international or European standard, in which case information regarding the relevant standard shall suffice; they shall also let the Commission have a statement of the grounds which make the enactment of such a technical regulation necessary, where these have not already been made clear in the draft.

Where appropriate, and unless it has already been sent with a prior communication, Member States shall simultaneously communicate the text of the basic legislative or regulatory provisions principally and directly concerned, should knowledge of such text be necessary to assess the implications of the draft technical regulations.

Member States shall communicate the draft again under the above conditions if they make changes to the draft that have the effect of significantly altering its scope, shortening the timetable originally envisaged for implementation, adding specifications or requirements, or making the latter more restrictive.

Where, in particular, the draft seeks to limit the marketing or use of a chemical substance, preparation or product on grounds of public health or of the protection of consumers or the environment Member States shall also forward either a summary or the references of all relevant data relating to the substance, preparation or products concerned and to known and available substitutes, where such information may be available, and communicate the anticipated effects of the measure on public health and the protection of the consumer and the environment, together with an analysis of the risk carried out as appropriate in accordance with the general principles for the risk evaluation of chemical substances as referred to in Article 10 (4) of Regulation (EEC) No 793/93[2] in the case of an existing substance or in Article 3 (2) of Directive 92/32/EEC[3] in the case of a new substance.

The Commission shall immediately notify the other Member States of the draft and all documents which have been forwarded to it; it may also refer this draft, for an opinion, to the Committee referred to in Article 5 and,

[2] OJ No L 84, 5. 4. 1993, p. 1.
[3] OJ No L 154, 5. 6. 1992, p. 1.

where appropriate, to the committee responsible for the field in question.

With respect to the technical specifications or other requirements referred to in Article 1 (9), third indent, the detailed comments or opinions of the Commission or the Member States may concern only the aspect which may hinder trade and not the fiscal or financial aspects of the measure.

2. The Commission and the Member States may make comments to the Member State which has forwarded a draft technical regulation; that Member State shall take such comments into account as far as possible in the subsequent preparation of the technical regulation.

3. Member States shall communicate the definitive text of a technical regulation to the Commission without delay.

4. Information supplied under this Article shall not be confidential except at the express request of the notifying Member State. Any such request shall be supported by reasons.

In cases of this kind, if necessary precautions are taken, the Commission and the national authorities may seek expert advice from physical or legal persons in the private sector.

5. When draft technical regulations form part of measures which are required to be communicated to the Commission at the draft stage under another Community act, Member States may make a communication within the meaning of Article 8 (1) of this Directive under that other act, provided that they formally indicate that the said communication also constitutes a communication for the purposes of this Directive.

The absence of a reaction from the Commission under this Directive to a draft technical regulation shall not prejudice any decision which might be taken under other Community acts.

Article 9

1. Member States shall postpone the adoption of a draft technical regulation for three months from the date of receipt by the Commission of the communication referred to in Article 8 (1).

2. Member States shall postpone:

– for four months the adoption of draft technical regulation in the form of a voluntary agreement within the meaning of Article 1 (9), second indent,

– without prejudice to paragraphs 3, 4 and 5, for six months the adoption of any other draft technical regulation,

from the date of receipt by the Commission of the communication referred to in Article 8 (1) if the Commission or another Member State delivers a detailed opinion, within three months of that date, to the effect that the measure envisaged may create obstacles to the free movement of goods within the internal market.

The Member State concerned shall report to the Commission on the action it proposes to take on such detailed opinions. The Commission shall comment on this reaction.

3. Member States shall postpone the adoption of a draft technical regulation for 12 months from the date of receipt by the Commission of the communication referred to in Article 8(1) if, within the three months following that date, the Commission announces its intention to propose or adopt a Directive, Regulation or Decision on the matter in accordance with Article 189 of the Treaty.

4. Member States shall postpone the adoption of a draft technical regulation for 12 months from the date of receipt by the Commission of the communication referred to in Article 8 (1) if, within the three months following that date, the Commission announces its finding that the draft technical regulation concerns a matter which is covered by a proposal for a Directive, Regulation or Decision presented to the Council in accordance with Article 189 of the Treaty.

5. If the Council adopts a common position during the standstill period referred to in paragraphs 3 and 4, that period shall, subject to Article 9 (6), be extended to 18 months.

6. The obligations referred to in paragraphs 3, 4 and 5 shall lapse:

– when the Commission informs the Member States that it no longer intends to propose or adopt a binding Community act,
– when the Commission informs the Member States of the withdrawal of its draft or proposal,
– when the Commission or the Council has adopted a binding Community act.
7. Paragraphs 1 to 5 shall not apply in those cases where, for urgent reasons, occasioned by serious and unforseeable circumstances, relating to the protection of public health or safety, the protection of animals or the preservation of plants, a Member State is obliged to prepare technical regulations in a very short space of time in order to enact and introduce them immediately without any consultations being possible. The Member State shall give, in the communication referred to in Article 8, the reasons which warrant the urgency of the measures taken. The Commission shall give its views on the communication as soon as possible. It shall take appropriate action in cases where improper use is made of this procedure. The European Parliament shall be kept informed by the Commission.

Article 10

1. Articles 8 and 9 shall not apply to those laws, regulations and administrative provisions of the Member States or voluntary agreements by means of which Member States:
– comply with binding Community acts which result in the adoption of technical specifications;
– fulfil the obligations arising out of international agreements which result in the adoption of common technical specifications in the Community;
– make use of safeguard clauses provided for in binding Community acts;
– apply Article 8 (1) of Council Directive 92/59/EEC of 29 June 1992 on general product safety[4];
– restrict themselves to implementing a judgment of the Court of Justice of the European Communities;
– restrict themselves to amending a technical regulation within the meaning of Article 1 (9) of this Directive, in accordance with a Commission request, with a view to removing an obstacle to trade.
2. Article 9 shall not apply to the laws, regulations and administrative provisions of the Member States prohibiting manufacture insofar as they do not impede the free movement of products.
3. Article 9 (3) to (6) shall not apply to the voluntary agreements referred to in Article 1 (9), second indent.
4. Article 9 shall not apply to the technical specification or other requirements referred to in Article 1 (9), third indent.

Article 11

The Commission shall report every two years to the European Parliament, the Council and the Economic and Social Committee on the results of the application of this Directive. Lists of standardization work entrusted to the European standardization organizations pursuant to this Directive, as well as statistics on the notifications received shall be published on an annual basis in the *Official Journal of the European Communities*.

Article 12

When Member States adopt a technical regulation, it shall contain a reference to this Directive or shall be accompanied by such reference on the occasion of its official publication. The methods of making such reference shall be laid down by Member States.

Article 13

This Directive is addressed to the Member States.

[4] OJ N0 L 228, 11. 8. 1992, p. 24.

Annex I
European standardization bodies

CEN
Comité Européen de Normalisation
Cenelec
Comité Europèen de Normalisation
Electrotechnique

ETSI
European Telecommunications
Standards Institute

Annex II
National standardization bodies

1. BELGIUM
IBN/BIN
Institut belge de normalisation
Belgisch Institut voor Normalisatie

CEB/BEC
Comité électrotechnique belge
Belgisch Elektrotechnisch Comité

2. DENMARK
DS
Dansk Standardiseringsråd

DEK
Dansk Elektrotecknisk Komite

3. GERMANY
DIN
Deutsche Institut für Normung e.V.

DKE
Deutsche Elektrotechnische
Kommission im DIN und VDE

4. GREECE
ELOT
Hellenic Organization for
Standardization

5. SPAIN
AENOR
Asociación Española de
Normalización y Certificación

6. FRANCE
AFNOR
Association française de
normalisation

UTE
Union technique de l'électricité –
Bureau de normalisation auprès de
l'AFNOR

7. IRELAND
NASI
National Standards Authority of
Ireland

ETCI
Electro-Technical Council of Ireland

8. ITALY
UNI
Ente Nazionale Italiano di
Unificazione

CEI
Comitato Elettrotecnico Italiano

9. LUXEMBOURG
ITM
Inspection du travail et des mines

SEE
Service de l'énergie de l'Etat

10. NETHERLANDS
NNI
Nederlands Normalisatie Institut

NEC
Nederlands Elektrotechnisch Comité

11. PORTUGAL
IPQ
Instituto Português de Qualidade

12. UNITED KINGDOM
BSI
British Standards Institution

BEC
British Electrotechnical Committee

LIST OF MEMBERS AND AFFILIATES OF THE COMITÉ EUROPÉEN DE NORMALISATION (CEN)

CEN National Members

AENOR
Associación Española de Normalización y Certificación
tel+34 1 410 48 51 fax+34 1 410 49 76

AFNOR
Association française de normalisation
tel+33 1 42 91 55 55 fax+33 1 42 91 56 56

BSI
British Standards Institution
tel+44 71 629 90 00 fax+44 71 629 05 06

DIN
Deutsches Institut für Normung e.V
tel+49 30 26 01–1 fax+49 30 26 01–2 31

DS
Dansk Standardiseringsråd
tel+45 39 77 01 01 fax+45 39 77 02 02

ELOT
Ellinikos Organismos Typopoiisis
tel+30 1 201 50 25 fax+30 1 202 59 17

IBN/BIN
Institut Belge de Normalisation/Belgisch Instituut voor Normalisatie
tel+32 2 734 92 05 fax+32 2 733 42 64

IPQ
Instituto Português da Qualidade
tel+35 1 52 37 35 fax+35 1 53 00 33

ITM
Inspection du Travail et des Mines (Luxembourg)
tel+352 49 921 21 06 fax+352 49 14 47

NNI
Nederlands Normalisatie-instituut
tel+31 15 69 03 90 fax+31 15 69 01 90

NSAI
National Standards Authority of Ireland
tel+353 1 37 01 01 fax+353 1 36 98 21

NSF
Norges Standardiseringsforbund
tel+47 2 46 60 94 fax+47 2 46 44 57

ON
Österreichisches Normungsinstitut
tel+43 222 26 75 35 fax+43 222 26 75 52

SFS
Suomen Standardisoimisliitto r.y.
tel+358 0 64 56 01 fax+358 0 64 31 47

SIS
Standardiseringskommissionen i Sverige
tel+46 8 631 52 00 fax+46 8 11 70 35

SNV
Schweizerische Normen-Vereinigung
tel+41 1 384 47 47 fax+41 1 384 47 74

STRI
Technological Institute of Iceland
tel+354 1 68 70 00 fax+354 1 68 74 09

UNI
Ente Nazionale Italiano di Unificazione
tel+39 2 70 02 41 fax+39 2 70 10 61 06

CEN Affiliates

BDS
Committee for Standardization, Certification and Metrology (Bulgaria)
tel+359 2 85 91 fax+359 2 80 14 02

CSN
Federal Office for Standards and Measurement (Czechoslovakia)
tel+42 2 235 43 33 fax+42 2 26 57 95

CYS
Cyprus Organization for Standards and Control of Quality
tel+357 2 30 34 41 fax+357 2 36 61 20

IRS
Institut roumain de normalisation

tel+40 36 10 400 11 40 43
fax+40 36 10 400 11 37 83

MSZH
Magyar Szabvanyugyi Hivatal
tel+36 1 118 30 11 fax+36 1 118 51 25

PKNMiJ
Polish Committee for Standardization, Measures and Quality Control
tel+48 22 20 54 34 fax+48 22 20 83 78

TSE
Turk Standardlari Enstitusu
tel+90 4 117 83 30 tax+90 4 125 43 99

Central Secretariat
rue de Stassart 36
B – 1050 Brussels
tel+32 2 550 0852
fax+32 2 550 0819
teletex 206–2210097=CENCEL
telex+172210097

LIST OF MEMBERS OF THE COMITÉ EUROPÉEN DE NORMALISATION ELECTROTECHNIQUE (CENELEC)

AUSTRIA
Österreichisches Elektrotechnisches Komitee
(ÖEK) beim Österreichischen Verband für Elektrotechnik (ÖVE)
Eschenbachgasse 9
A – 1010 WIEN
Tel: (Int+43) 222/587 63 73

BELGIUM
Comité Electrotechnique Belge (CEB)
Belgisch Elektrotechnisch Comité (BEC)
3 Galerie Ravenstein, bte 11
B – 1000 BRUXELLES
Tel: (Int+32) 2/512 00 28

DENMARK
Dansk Elektroteknisk Komite (DEK)
Strandgade 36, st.
DK – 1401 KØBENHAVN K
Tel: (Int+45) 1/57 50 50

FINLAND
Finnish Electrotechnical Standards Association
(SESKO)
P.O. Box 134
SF – 00211 HELSINKI 21
Tel: (Int+358) 0/69631

FRANCE
Union Technique de l'Electricité
(UTE)
Cédex 64
F – 92052 PARIS LA DEFENSE
Tel: (Int+33) 1/47 68 50 20

GERMANY
Deutsche Elektrotechnische Kommission im
DIN und VDE (DKE)
Stresemannallee 15
D – 6000 FRANKFURT AM MAIN 70
Tel: (Int+49) 69/6308–0

GREECE
Hellenic Organization for Standardization
(ELOT)
Didotou 15
GR – ATHENS 106 80
Tel: (Int+30) 1/360 9517 or 9947

ICELAND
The Icelandic Council for Standardization
Technological Institute of Iceland
Keldnaholt
IS – 110 REYKJAVIK
Tel: (Int+354) 1/68 7000

IRELAND
Electro-Technical Council of Ireland
(ETCI)
Ballymun Road
IRL – DUBLIN 9
Tel: (Int+353) 1/37 67 73

ITALY
Comitato Elettrotecnico Italiano
(CEI)
Viale Monza 259
I – 20126 MILANO
Tel: (Int+39) 2/25 77 31

LUXEMBOURG
Service de l'Energie de l'Etat
B.P. No 10
L – 2010 LUXEMBOURG
Tel: (Int+352) 44 20 30, ext 20

NETHERLANDS
Nederlands Electrotechnisch Comité
(NEC)
Kalfjeslaan 2,
Postbus 5059
NL – 2600 GB DELFT
Tel: (Int+31) 15/690 390

NORWAY
Norsk Elektroteknisk Komite (NEK)
Harbitzalléen 2A, Skøyen,

Postboks 280
N – 0212 OSLO 2
Tel: (Int+47) 2/52 69 50

PORTUGAL
Portuguese Institute for Quality
Rua José Estêvão, 83A
P – 1199 LISBOA Codex
Tel: (Int+351) 1/53 98 91

SPAIN
Comite Español del CENELEC
(Asociación Electrotécnica y
Electronica Española – AEE)
Avenida de Brasil 7
E – MADRID 20
Tel: (Int+34) 1/270 44 00

SWEDEN
Swedish Electrotechnical Commission
(SEK)
Kistagangen 19,
Box 1284
S – 164 28 KISTA STOCKHOLM
Tel: (Int+46) 8/750 78 20

SWITZERLAND
Comité Electrotechnique Suisse (CES)
Postfach
CH – 8034 ZÜRICH
Tel: (Int+41) 1/384 91 11

UNITED KINGDOM
British Electrotechnical Committee
British Standards Institution
2, Park Street
UK – LONDON W1A 2BS
Tel: (Int+44) 1/629 9000

DETAILS OF THE EUROPEAN TELECOMMUNICATIONS STANDARDS INSTITUTE (ETSI)

Postal Address: ETSI 06921 Sophia Antipolis Cedex France

Office Address: Route des Lucioles Sophia Antipolos, Valbonne, France

Telephone: + 33 92 94 42 00

Fax: + 33 93 65 47 16

A full list of current members can be obtained on request, as can the ETSI reference documents catalogue. The current catalogue references 419 available standards. The catalogue gives details of membership.

COUNCIL DECISION 93/465/EEC

of 22 July 1993

concerning the modules for the various phases of the conformity assessment procedures and the rules for the affixing and use of the CE conformity marking, which are intended to be used in the technical harmonization Directives

THE COUNCIL OF THE EUROPEAN COMMUNITIES,

Having regard to the Treaty establishing the European Economic Community, and in particular Article 100A thereof,

Having regard to the proposal from the Commission,[1]

In cooperation with the European Parliament,[2]

Having regard to the opinion of the Economic and Social Committee,[3]

Whereas the Council Decision 90/683/EEC of 13 December 1990 concerning the modules for the various phases of the conformity assessment procedures which are intended to be used in the technical harmonization Directives[4] needs to be substantially amended in various places; whereas it is necessary, in a spirit of clarity and rationality, to consolidate its provisions by means of this Decision;

Whereas the Council adopted a Resolution on 21 December 1989 concerning a global approach to conformity;[5]

Whereas the introduction of harmonised methods for the assessment of conformity and the adoption of a common doctrine for their implementation are likely to facilitate the adoption of future technical harmonization Directives concerning the placing on the market of industrial products and thus be conducive to the implementation of the internal market;

Whereas such methods should ensure that products are in full conformity with the essential requirements laid down in the technical harmonization Directives, in order to provide, in particular, for the health and safety of users and consumers;

Whereas such conformity should be assured without imposing unnecessarily onerous conditions on manufacturers, and by means of clear and comprehensible procedures;

Whereas limited flexibility should be introduced as regards use of additional modules, or variations in the modules, when the specific circumstances of a particular sector or directive so warrant, but not to such a degree as to undercut the purpose of the current Decision and only when explicitly justified;

Whereas in the abovementioned Resolution of 21 December 1989 the Council approved as a guiding principle and adoption of common rules on the use of the CE marking;

Whereas in its Decision 90/683/EEC the Council laid down that the industrial products covered by the technical harmonization Directives can be placed on the market only after the manufacturer has affixed the CE marking to them;

Whereas a single CE marking should be used in order to facilitate controls

[1] OJ No C 160, 20.6.1991, p. 14; and OJ No C 28, 2.2.1993, p. 16.
[2] OJ No C 125, 18.5.1992, p. 178. OJ No C 115, 26.4.1993, p. 117; and Decision of 14 July 1993 (not yet published in the Official Journal).
[3] OJ No C 14, 20.1.1992, p. 15; and OJ No C 129, 10.5.1993, p. 3.
[4] OJ No L 380, 31.12.1990, p. 13.
[5] OJ No C 10, 16.1.1990, p. 1.

on the Community market by inspectors and to clarify the obligations of economic operators in respect of marking under the various Community regulations;

Whereas the aim of the CE marking is to symbolise the conformity of a product with the levels of protection of collective interests imposed by the total harmonization Directives and to indicate that the economic operators has undergone all the evaluation procedures laid down by Community law in respect of his product,

HAS DECIDED AS FOLLOWS:

Article 1

1. The procedures for conformity assessment which are to be used in the technical harmonization Directives relating to the marketing of industrial products will be chosen from among the modules listed in the Annex and in accordance with the criteria set out in this Decision and in the general guidelines in the Annex.

These procedures may only depart from the modules when the specific circumstances of a particular sector or Directive so warrant. Such departures from the modules must be limited in extent and must be explicitly justified in the relevant Directive.

2. This Decision lays down rules for affixing the CE conformity marking provided for in Community legislation concerning the design, manufacture, placing on the market, entry into service or use of industrial products.

3. The Commission shall report periodically on the functioning of this Decision, and on whether conformity assessment and CE marking procedures are working satisfactorily or need to be modified.

No later than the end of the transitional period in 1997, or earlier if the matter is found to be urgent, the Commission shall also report back on any special problems raised by the incorporation of Council Directive 73/23/EEC of 19 February 1973 relating to electrical equipment designed for use within certain voltage limits[6] within the scope of CE marking procedures, and, in particular, whether safety is being compromised. It shall also review any problems raised by the issue of overlapping Council directives, and whether any further Community measures are required.

Article 2

1. Decision 90/683/EEC is hereby repealed.

2. References to the Decision repealed shall be construed as references to this Decision.

Done at Brussels, 22 July 1993.

For the Council
The President
M. OFFECIERS-VAN DE WIELE

[6] OJ No L 77, 26.3.1973, p. 29.

Annex
Conformity assessment procedures and CE marking in the technical harmonisation Directives

I. General guidelines

A. The principal guidelines for the use of conformity assessment procedures in technical harmonization Directives are the following:

(a) the essential objective of a conformity assessment procedure is to enable the public authorities to ensure that products placed on the market conform to the requirements as expressed in the provisions of the Directives, in particular with regard to the health and safety of users and consumers;

(b) conformity assessment can be subdivided into modules which relate to the design phase of products and to their production phase;

(c) as a general rule a product should be subject to both phases before being able to be placed on the market if the results are positive;[1]

(d) there are a variety of modules which cover the two phases in a variety of ways. The Directives must set the range of possible choices which can be considered by the Council to give the public authorities the high level of safety they seek, for a given product or product sector;

(e) in setting the range of possible choices open to the manufacturer, the Directives, will take into consideration, in particular, such issues as the appropriateness of the modules to the type of products, the nature of the risks involved, the economic infrastructures of the given sector (*e.g.* existence or non-existence of third parties), the types and importance of production, etc. The factors that have been taken into account must be explicitly spelled out by the Commission in these Directives;

(f) the Directives will, in setting the range of possible modules for a given product or product sector, attempt to leave as wide a choice to the manufacturer as is consistent with ensuring compliance with the requirements.

The Directives will set out the criteria governing the conditions in which the manufacturer chooses the most appropriate modules for his production from the modules laid down by the Directives;

(g) the Directives should avoid imposing unnecessarily modules which would be too onerous relative to the objectives of the Directive concerned;

(h) notified bodies should be encouraged to apply the modules without unnecessary burden for the economic operators. The Commission, in cooperation with the Member States, must ensure that close cooperation is organised between the notified bodies in order to ensure consistent technical application of the modules;

(i) in order to protect the manufacturers, the technical documentation provided to notified bodies has to be limited to that which is required solely for the purpose of assessment of conformity. Legal protection of confidential information is required;

(j) whenever Directives provide the manufacturer with the possibility of using modules based on quality

[1] The specific Directives may provide for different arrangements.

assurance techniques, the manufacturer must also be able to have recourse to a combination of modules not using quality assurance, and *vice versa*, except where compliance with the requirements laid down by the Directives requires the exclusive application of a certain procedure;

(k) for the purposes of operating the modules, Member States must notify on their own responsibility bodies under their jurisdiction which they have chosen from the technically competent bodies complying with the requirements of the Directives. This responsibility involves the obligation for the Member States to ensure that the notified bodies permanently have the technical qualifications required by the Directives and that the latter keep their competent national authorities informed of the performance of their tasks. Where a Member State withdraws its notification of a body, it must take appropriate steps to ensure that the dossiers are processed by another notified body to ensure continuity;

(l) in addition, with regard to conformity assessment, the sub-contracting of work shall be subject to certain conditions guaranteeing:
– the competence of the establishment operating as sub-contractor, on the basis of conformity with series EN 45,000 standards, and the capability of the Member State that has notified the sub-contracting body to ensure effective monitoring of such compliance,
– the ability of the body notified to exercise effective responsibility for the work carried out under sub-contract;

(m) notified bodies which can prove their conformity with harmonised standards (EN 45,000 series), by submitting an accreditation certificate or other documentary evidence, are presumed to conform to the requirements of the directives. Member States having notified bodies unable to prove their conformity with the harmonised standards (EN 45,000 series) may be requested to provide the Commission with the appropriate supporting documents on the basis of which notification was carried out;

(n) a list of notified bodies must be published by the Commission in the *Official Journal of the European Communities* and constantly updated.

B. The principal guidelines for the affixing and use of the CE marking are the following:

(a) The CE marking symbolises conformity to all the obligations incumbent on manufacturers for the product by virtue of the Community Directives providing for its affixing.

Thus, such conformity is not limited to the essential requirements relating to safety, public health, consumer protection etc., as certain Directives may impose specific obligations not necessarily forming part of the essential requirements.

(b) The CE marking affixed to industrial products symbolises the fact that the natural or legal person having affixed or been responsible for the affixing of the said marking has verified that the product conforms to all the Community total harmonization provisions which apply to it and has been the subject of the appropriate conformity evaluation procedures.

(c) Where the industrial products are subject to other Directives concerning other aspects and which also provide for the affixing of the CE marking, the latter must indicate that the products are also presumed to conform to the provisions of those other Directives.

However, where one or more of these Directives allow the manufacture, during a transitional period, to choose which arrangements to apply, the CE marking indicates conformity to the provisions only of those Directives applied by the manufacturer. In this case, particulars of the Directives applied, as published in the *Official Journal of the European Communities*, must be given in the documents, notices or instructions accompanying the products or, where appropriate, on the data plate.

(d) 1. The CE conformity marking must consist of the initials "CE" taking the following form:

If the CE marking is reduced or enlarged the proportions given in the above graduated drawing must be respected.

2. Where the Directive concerned does not impose specific dimensions, the CE marking must have a height of at least 5 mm.

3. The CE marking must be affixed to the product or to its data plate. However, where this is not possible or not warranted on account of the nature of the product, it must be affixed to the packaging, if any, and to the accompanying documents, where the Directive concerned provides for such documents.

4. The CE marking mujst be affixed visibly, legibly and indelibly.

(e) Any industrial product covered by the technical harmonization directives based on the principles of the global approach must bear the CE marking, save where the specific Directives provide otherwise; such exceptions constitute derogations not from the markinig requirement but from the administrative procedures for conformity evaluation, which may in certain cases be considered too cumbersome. Appropriate grounds must accordingly be given for any exception to or derogation from the marking requirement.

The CE marking is the only marking which certifies that the industrial products conform to the Directives based on the principles of the global approach.

Member States must refrain from introducing into their national regulations any reference to a conformity marking other than the CE marking in connection with conformity to all the provisions contained in the directives on CE marking.

(f) The CE marking must be affixed at the end of the production control phase.

(g) The CE conformity marking must be followed by the identification number of the notified body within the meaning of paragraph I.A where the said body is involved in the production control phase within the meaning of this Decision.

Such identification numbers must be assigned by the Commission as part of the body notification procedure. The Commission must publish lists of the notified bodies in the *Official Journal of European Communities*; such lists must be updated regularly.

A notified body must be assigned the same number when it is notified under several directives. The Commission must ensure that each notified body receives a single identification number, however many Directives it is notified under.

(h) It is necessary to lay down provisions concerning the use of certain products. In this case, the CE marking and the identification number of the notified body may be followed by a pictogram or any other mark indicating, for example, the category of use.

(i) The affixing for any other marking liable to deceive third parties as to the meaning and form of the CE marking must be prohibited.

(j) A product may bear different marks, for example marks indicating conformity to national or European standards or with traditional optional Directives, provided such marks are not liable to cause confusion with the CE marking.

Such marks may therefore only be affixed to the product, its packaging or the documentation accompanying the product on condition that the legibilty and visibility of the CE marking are not thereby reduced.

(k) The CE marking must be affixed by the manufacturer or his agent established within the Community. In exceptional, duly warranted cases, the specific Directives may provide that the CE marking can be affixed by the person responsible for placing the product on the Community market.

The identication number of the notified body must be affixed under its responsibility either by the body itself or by the manufacturer or his agent established within the Community.

(l) Member States must take all provisions of national law necessary to exclude any possibility of confusion and to prevent abuse of the CE marking.

Without prejudice to the provision in the Directive concerned relating to the application of the safeguard clause, where a Member State establishes that the CE marking has been affixed unduly, the manufacturer, his agent or, exceptionally, where the specific Directives so provide, the person responsible for placing the product in question on the Community market is obliged to make the product comply and to end the infringement under conditions imposed by the Member State. Where non-compliance continues, the Member State must take all appropriate measures to restrict or prohibit the placing on the market of the product in question or to ensure that it is withdrawn from the market in accordance with the procedures laid down in the safeguard clauses.

II. Modules for conformity assessment

Explanatory notes

Specific Directives may allow the CE marking to be affixed to the packaging or the accompanying documentation, instead of to the product itself.

The declaration of conformity or the certificate of conformity (whichever of the two applies in the Directive concerned) must cover either individual or several products and shall either accompany the product(s) covered or be kept by the manufacturer. The appropriate solution for the Directive concerned will be specified.

References to Articles refer to the standard paragraphs of Annex II.B to the Council Resolution of 7 May 1985 (OJ 1985 C136, 4.6.1985, p. 1), which have become standard articles in the "new approach" Directives.

The development of computerised communication of certificates and other documents issued by notified bodies is envisaged within INSIS.

Specific Directives may use modules A, C and H with additional provisions containing supplementary requirements which figure in the boxes in the modules.

Module C is designed to be used in combination with module B (EC type-examination). Modules D, E and F will also normally be used in combination with module B; however, in special cases (for example, when dealing with certain products of very simple design and construction) they may be used on their own.

Module A (internal production control)

1. This module describes the procedure whereby the manufacturer or his authorised representative established within the Community, who carries out the obligations laid down in point 2, ensures and declares that the products concerned satisfy the requirements of the Directive that apply to them. The manufacturer or his authorised manufacturer established within the Community must affix the CE marking to each product and draw up a written declaration of conformity.

2. The manufacturer must establish the technical documentation described in paragraph 3 and he or his authorised representative established with the Community must keep it for a period ending at least 10 years[2] after the last product has been manufactured at the disposal of the relevant national authorities for inspection purposes.

Where neither the manufacturer nor his authorised representative is established within the Community, the obligation to keep the technical documentation available is the responsibility of the person who places the product on the Community market.

3. Technical documentation must enable the conformity of the product with the requirements of the Directive to be assessed. It must, as far as relevant for such assessment, cover the design, manufacture and operation of the product.[3]

4. The manufacturer or his authorised representative must keep a copy of the declaration of conformity with the technical documentation.

5. The manufacturer must take all measures necessary in order that the manufacturing process ensures compliance of the manufactured products with the technical documentation referred to in point 2 and with the requirements of the Directive that apply to them.

---

[2] The specific directives may alter this period.

[3] The content of the technical documentation shall be laid down Directive by Directive in accordance with the products concerned.

For example, the documentation shall contain so far as relevant for assessment:

- a general description of the product,
- conceptual design and manufacturing drawings and schemes of components, sub-assemblies, circuits, etc,
- descriptions and explanations necessary for the understanding of said drawings and schemes and the operation of the product,
- a list of the standards referred to in Article 5, applied in full or in part, and descriptions of the solutions adopted to meet the essential requirements of the Directive where the standards referred to in Article 5 have not been applied,
- results of design calculations made, examinations carried out, etc,
- test reports.

Module Aa

This module consists of module A, plus the following supplementary requirements:

For each product manufactured one or more tests on one or more specific aspects of the product must be carried out by the manufacturer or on his behalf.[4] The tests are carried out on the responsibility of a notified body chosen by the manufacturer.

On the responsibility of the notified body, the manufacturer must affix the former's identification number during the manufacturing process.

---

[4] If this option is adopted in a specific Directive, the products concerned and the tests to be carried out must be specified.

or:

A notified body chosen by the manufacturer must carry out or have carried out product checks at random intervals. An adequate sample of the final products, taken on site by the notified body, must be examined and appropriate tests as set out in the relevant standard(s) referred to in Article 5, or equivalent tests, must be carried out to check the conformity of the product with the relevant requirements of the Directive.
In those cases where one or more of the products checked do not conform the notified body must take appropriate measures.
The product checking must include the following aspects:
(Relevant aspects must be specified here such as for example the statistical method to be applied, the sampling plan with its operational characteristics, etc.)
On the responsibility of the notified body, the manufacturer must affix the former's identification number during the manufacturing process.

Module B (EC type-examination)

1. This module describes that part of the procedure by which a notified body ascertains and attests that a specimen, representative of the production envisaged, meets the provisions of the Directive that apply to it.

2. The application for the EC type-examination must be lodged by the manufacturer or his authorised representative established within the Community with a notified body of his choice.
The application must include:
– the name and address of the manufacturer and, if the application is lodged by the authorised representative, his name and address in addition,
– a written declaration that the same application has not been lodged with any other notified body,
– the technical documentation, as described in point 3.
The applicant must place at the disposal of the notified body a specimen, representative of the production envisaged and hereinafter called "type".[5] The notified body may request further specimens if needed for carrying out the test programme.

3. The technical documentation must enable the conformity of the product with the requirements of the Directive to be assessed. It must, as far as relevant for such assessment, cover the design, manufacture and operation of the product.[6]

4. The notified body must:
4.1 examine the technical documentation, verify that the type has been manufactured in conformity with the

[5] A type may cover several versions of the product provided that the differences between the versions do not affect the level of safety and the other requirements concerning the performance of the product.

[6] The content of the technical documentation must be laid down Directive by Directive in accordance with the products concerned.
For example, the documentation shall contain as far as is relevant for assessment:
– a general type-description,
– conceptual design and manufacturing drawings and schemes of components, sub-assemblies, circuits, etc,
– descriptions and explanations necessary for the understanding of said drawings and schemes and the operation of the product,
– a list of the standards referred to in Article 5, applied in full or in part, and descriptions of the solutions adopted to meet the essential requirements of the Directive where the standards referred to in Article 5 have not been applied,
– results of design calculations made, examinations carried out, etc,
– test reports.

technical documentation and identify the elements which have been designed in accordance with the relevant provisions of the standards referred to in Article 5, as well as the components which have been designed without applying the relevant provisions of those standards;

4.2. perform or have performed the appropriate examinations and necessary tests to check whether, where the standards referred to in Article 5 have not been applied, the solutions adopted by the manufacturer meet the essential requirements of the Directive;

4.3. perform or have performed the appropriate examinations and necessary tests to check whether, where the manufacturer has chosen to apply the relevant standards, these have actually been applied;

4.4. agree with the applicant the location where the examinations and necessary tests will be carried out.

5. Where the type meets the provisions of the Directive, the notified body must issue an EC type-examination certificate to the applicant. The certificate must contain the name and address of the manufacturer, conclusions of the examination, conditions for its validity and the necessary data for identification of the approved type.[7]

A list of the relevant of the technical documentation must be annexed to the certificate and a copy kept by the notified body.

If the manufacturer is denied a type certification, the notified body must provide detailed reasons for such denial.

Provision must be made for an appeals procedure.

6. The applicant must inform the notified body that holds the technical documentation concerning the EC type-examination certificate of all modifications to the approved product which must receive additional approval where such changes may affect the conformity with the essential requirements or the prescribed conditions for use of the product. This additional approval is given in the form of an addition to the original EC type-examination certificate.

7. Each notified body must communicate to the other notified bodies the relevant information concerning the EC type-examination certificates and additions issued and withdrawn.[8]

8. The other notified bodies may receive copies of the EC type-examination certificates and/or their additions. The Annexes to the certificates must be kept at the disposal of the other notified bodies.

9. The manufacturer or his authorised representative must keep with the technical documentation copies of EC type-examination certificates and their additions for a period ending at least 10 years[9] after the last product has been manufactured.

Where neither the manufacturer nor his authorised representative is established within the Community, the obligation to keep the technical documentation available is the responsibility of the person who places the product on the Community market.

Module C (conformity to type)

1. This module describes that part of the procedure whereby the manufacturer or his authorised representative established within the Community ensures and declares that the products concerned are in conformity with the type as described in the EC type-examination certificate and satisfy the requirements of the Directive that applies to them. The manufacturer or his authorised representative established within the Community must affix the CE marking to each product and draw up a written declaration of conformity.

2. The manufacturer must take all

[7] The specific Directives may provide for the certificate to have a period of validity.

[8] The specific Directives may provide for different arrangements.

[9] The specific Directives may alter this period.

measures necessary to ensure that the manufacturing process assures compliance of the manufactured products with the type as described in the EC type-examination certificate and with the requirements of the Directive that apply to them.
3. The manufacturer or his authorised representative must keep a copy of the declaration of conformity for a period ending at least 10 years[10] after the last product has been manufactured.

Where neither the manufacturer nor his authorised representative is established within the Community, the obligation to keep the technical documentation available is the responsibility of the person who places the product on the Community market.

Possible supplementary requirements:

[10] The specific Directives may alter this period.

For each product manufactured one or more tests on one or more specific aspects of the product must be carried out by the manufacturer or on his behalf.* The tests must be carried out on the responsibility of a notified body, chosen by the manufacturer.
On the responsibility of the notified body, the manufacturer must affix the former's identification symbol during the manufacturing process.

* If this option is adopted in a specific Directive, the products concerned and the tests to be carried out must be specified.

or:

A notified body chosen by the manufacturer must carry out or have carried out product checks at random intervals. An adequate sample of the final products, taken on site by the notified body, must be examined and appropriate tests as set out in the relevant standard(s) referred to in Article 5, or equivalent tests, must be carried out to check the conformity of production with the relevant requirements of the Directive. In those cases where one or more of the products checked do not conform, the notified body must take appropriate measures.
The product checking must include the following aspects:

(Relevant aspects must be specified here such as for example the statistical method to be applied, the sampling plan with its operational characteristics, etc)

On the responsibility of the notified body, the manufacturer must affix the former's identification symbol during the manufacturing process.

Module D[11], (production quality assurance)

1. This module describes the procedure whereby the manufacturer who satisfies the obligations of point 2 ensures and declares that the products concerned [are in conformity with the type as described in the EC type-examination certificate and] satisfy the requirements of the Directive that apply to them. The manufacturer or his authorised representative established

[11] Where this module is used without module B:
– points 2 and 3 of module A must be added between points 1 and 2 in order to incorporate the need for technical documentation,
– the words in square brackets must be deleted.

within the Community must affix the CE marking to each product and draw up a written declaration of conformity. The CE marking must be accompanied by the identification symbol of the notified body responsible for EC monitoring as specified in point 4.

2. The manufacturer must operate an approved quality system for production, final product inspection and testing as specified in paragraph 3 and is subject to monitoring as specified in point 4.

3. Quality system

3.1. The manufacturer must lodge an application for assessment of his quality system with a notified body of his choice, for the products concerned.

The application must include:

– all relevant information for the product category envisaged,

– the documentation concerning the quality system,

– if applicable, the technical documentation of the approved type and a copy of the EC type-examination certificate.

3.2. The quality system must ensure compliance of the products [with the type as described in the EC type-examination certificate and] with the requirements of the Directive that apply to them.

All the elements, requirements and provisions adopted by the manufacturer shall be documented in a systematic and orderly manner in the form of written policies, procedures and instructions. The quality system documentation must permit a consistent interpretation of the quality programmes, plan, manuals and records.

It must contain in particular an adequate description of:

– the quality objectives and the organizational structure, responsibilities and powers of the management with regard to product quality,

– the manufacturing, quality control and quality assurance techniques, processes and systematic actions that will be used,

– the examinations and tests that will be carried out before, during and after manufacture, and the frequency with which they will be carried out,

– the quality records, such as inspection reports and test data, calibration data, qualification reports of the personnel concerned, etc.,

– the means to monitor the achievement of the required product quality and the effective operation of the quality system.

3.3. The notified body must assess the quality system to determine whether it satisfies the requirements referred to in point 3.2. It presumes conformity with these requirements in respect of quality systems that implement the relevant harmonised standard.[12]

The auditing team must have at least one member with experience of evaluation in the product technology concerned. The evaluation procedure must include an inspection visit to the manufacturer's premises.

The decision must be notified to the manufacturer. The notification must contain the conclusions of the examination and the reasoned assessment decision.

3.4. The manufacturer must undertake to fulfil the obligations arising out of the quality system as approved and to uphold it so that it remains adequate and efficient.

The manufacturer or his authorised representative shall keep the notified body that has approved the quality system informed of any intended updating of the quality system.

The notified body shall evaluate the modifications proposed and decide whether the amended quality system will still satisfy the requirements referred to in paragraph 3.2 or whether a re-assessment is required.

It must notify its decision to the manufacturer. The notification must contain the conclusions of the examination and the reasoned assessment decision.

[12] This harmonised standard will be EN 29,002, supplemented, if necessary, to take into account the specific nature of the products for which it is implemented.

4. Surveillance under the responsibility of the notified body

4.1. The purpose of surveillance is to make sure that the manufacturer duly fulfils the obligations arising out of the approved quality system.

4.2. The manufacturer must allow the notified body entrance for inspection purposes to the locations of manufacture, inspection and testing, and storage and must provide it with all necessary information, in particular:

– the quality system documentation,

– the quality records, such as inspection reports and test data, calibration data, qualification reports of the personnel concerned, etc.

4.3. The notified body must periodically[13] carry out audits to make sure that the manufacturer maintains and applies the quality system and must provide an audit report to the manufacturer.

4.4. Additionally the notified body may pay unexpected visits to the manufacturer. During such visits the notified body may carry out, or cause to be carried out, tests to verify that the quality system is functioning correctly, if necessary. The notified body must provide the manufacturer with a visit report and, if a test has taken place, with a test report.

5. The manufacturer must, for a period ending at least 10 years[14] after the last product has been manufactured, keep at the disposal of the national authorities:

– the documentation referred to in the second indent of point 3.1,

– the updating referred to in the second paragraph of point 3.4,

– the decisions and reports from the notified body which are referred to in the final paragraph of point 3.4, points 4.3 and 4.4.

6. Each notified body must give the other notified bodies the relevant information concerning the quality system approvals issued and withdrawn.[15]

Module E[16] (product quality assurance)

1. This module describes the procedure whereby the manufacturer who satisfies the obligations of point 2 ensures and declares that the products concerned [are in conformity with the type as described in the EC type-examination certificate and] satisfy the requirements of the Directive that apply to them. The manufacturer or his authorised representative established within the Community must affix the CE mark to each product and draw up a written declaration of conformity. The CE mark must be accompanied by the identification symbol of the notified body responsible for surveillance as specified in point 4.

2. The manufacturer must operate an approved quality system for final product inspection and testing as specified in paragraph 3 and must be subject to surveillance as specified in point 4.

3. Quality system

3.1. The manufacturer must lodge an application for assessment of his quality system for the products concerned, with a notified body of his choice.

The application must include:

– all relevant information for the product category envisaged,

– the quality system's documentation,

– if applicable, the technical documentation of the approved type and a copy of the EC type-examination certificate.

3.2. Under the quality system, each product must be examined and appro-

[13] In the specific Directives, the frequency may be specified.

[14] The specific Directives may alter this period.

[15] The specific Directives may provide for different arrangements.

[16] When this module is used without module B:

– points 2 and 3 of module A must be added between points 1 and 2 in order to incorporate the need for technical documentation.

– the words in square brackets must be deleted.

priate tests as set out in the relevant standard(s) referred to in Article 5 or equivalent tests shall be carried out in order to ensure its conformity with the relevant requirements of the Directive. All the elements, requirements and provisions adopted by the manufacturer must be documented in a systematic and orderly manner in the form of written policies, procedures and instructions. This quality system documentation must ensure a common understanding of the quality programmes, plans, manuals and records.

It must contain in particular an adequate description of:

– the quality objectives and the organizational structure, responsibilities and powers of the management with regard to product quality,
– the examinations and tests that will be carried out after manufacture,
– the means to monitor the effective operation of the quality system,
– quality records, such as inspection reports and test data, calibration data, qualification reports of the personnel concerned, etc.

3.3. The notified body must assess the quality system to determine whether it satisfies the requirements referred to in point 3.2. It presumes conformity with these requirements in respect of quality systems that implement the relevant harmonised standard.[17]

The auditing team must have at least one member experienced as an assessor in the product technology concerned. The assessment procedure must include an assessment visit to the manufacturer's premises.

The decision must be notified to the manufacturer. The notification must contain the conclusions of the examination and the reasoned assessment decision.

3.4. The manufacturer must undertake to discharge the obligations arising from the quality system as approved and to maintain it in an appropriate and efficient manner.

The manufacturer or his authorised representative must keep the notified body which has approved the quality system informed of any intended updating of the quality system.

The notified body must evaluate the modifications proposed and decide whether the modified quality system will still satisfy the requirements referred to in paragraph 3.2 or whether a re-assessment is required.

It must notify its decision to the manufacturer. The notification must contain the conclusions of the examination and the reasoned assessment decision.

4. Surveillance under the responsibility of the notified body

4.1. The purpose of surveillance is to make sure that the manufacturer duly fulfils the obligations arising out of the approved quality system.

4.2. The manufacturer must allow the notified body entrance for inspection purposes to the locations of inspection, testing and storage and shall provide it with all necessary information, in particular:

– the quality system documentation,
– the technical documentation,
– the quality records, such as inspection reports and test data, calibration data, qualification reports of the personnel concerned, etc.

4.3. The notified body must periodically[18] carry out audits to ensure that the manufacturer maintains and applies the quality system and must provide an audit report to the manufacturer.

4.4. Additionally, the notified body may pay unexpected visits to the manufacturer. At the time of such visits, the notified body may carry out tests or have them carried out in order to check the proper functioning of the quality system where necessary; it must provide the manufacturer with a visit

[17] This harmonised standard will be EN 29,003, supplemented if necessary to allow for the specific features of the products for which it is implemented.

[18] The intervals between audits may be specified in the specific Directives.

report and, if a test has been carried out, with a test report.

5. The manufacturer must, for a period ending at least 10 years[19] after the last product has been manufactured, keep at the disposal of the national authorities:

– the documentation referred to in the third indent of point 3.1,

– the updating referred to in the second paragraph of point 3.4,

– the decisions and reports from the notified body which are referred to in the final paragraph of point 3.4, points 4.3 and 4.4.

6. Each notified body must forward to the other notified bodies the relevant information concerning the quality system approvals issued and withdrawn.[20]

Module F[21] (product verification)

1. This module describes the procedure whereby a manufacturer or his authorised representative established within the Community checks and attests that the products subject to the provisions of point 3 [are in conformity with the type as described in the EC-type examination certificate and] satisfy the requirements of the directive that apply to them.

2. The manufacturer must take all measures necessary in order that the manufacturing process ensures conformity of the products [with the type as described in the EC type-examination certificate and] with the requirements of the Directive that apply to them. He shall affix the CE marking to each product and shall draw up a declaration of conformity.

3. The notified body must carry out the appropriate examinations and tests in order to check the conformity of the product with the requirements of the Directive either by examination and testing of every product as specified in point 4 or by examination and testing of products on a statistical basis, as specified in point 5, at the choice of the manufacturer.[22]

3a. The manufacturer or his authorised representative must keep a copy of the declaration of conformity for a period ending at least 10 years[23] after the last product has been manufactured.

4. Verification by examination and testing of every product

4.1. All products must be individually examined and appropriate tests as set out in the relevant standard(s) referred to in Article 5 or equivalent tests shall be carried out in order to verify their conformity with [the type as described in the EC-type examination certificate and] the requirements of the Directive that apply to them.

4.2. The notified body must affix or cause to be affixed, its identification symbol to each approved product and draw up a written certificate of conformity relating to the tests carried out.

4.3. The manufacturer or his authorised representative must ensure that he is able to supply the notified body's certificates of conformity on request.

5. Statistical verification

5.1. The manufacturer must present his products in the form of homogeneous lots and shall take all measures necessary in order that the manufacturing process ensures the homogeneity of each lot produced.

5.2. All products must be available for verification in the form of homogeneous lots. A random sample shall

---

[19] The specific Directives may alter this period.

[20] The specific Directives may provide for different arrangements.

[21] Where this module is used without module B:
– it must be supplemented by points 2 and 3 of module A (between points 1 and 2), so as to introduce the need for technical documentation,
– the next in square brackets must be deleted.

[22] The manufacturer's discretion may be limited in the specific Directives.

[23] The specific Directives may alter this period.

be drawn from each lot. Products in a sample shall be individually examined and appropriate tests as set out in the relevant standard(s) referred to in Article 5, or equivalent tests, shall be carried out to ensure their conformity with the requirements of the Directive which apply to them and to determine whether the lot is accepted or rejected.

5.3. The statistical procedure must use the following elements:

(Relevant elements must be specified here such as, for example, the statistical method to be applied, the sampling plan with its operational characteristics, etc.)

5.4. In the case of accepted lots, the notified body must affix, or cause to be affixed, its identification symbol to each product and shall draw up a written certificate of conformity relating to the tests carried out. All products in the lot may be put on the market except those products from the sample which were found not to be in conformity.

If a lot is rejected, the notified body or the competent authority must take appropriate measures to prevent the putting on the market of that lot. In the event of frequent rejection of lots the notified body may suspend the statistical verification.

The manufacturer may, under the responsibility of the notified body, affix the latter's identification symbol during the manufacturing process.

5.5. The manufacturer or his authorised representative must ensure that he is able to supply the notified body's certificates of conformity on request.

Module G (unit verification)

1. This module describes the procedure whereby the manufacturer ensures and declares that the product concerned, which has been issued with the certificate referred to in point 2, conforms to the requirements of the Directive that apply to it. The manufacturer or his authorised representative established within the Community must affix the CE marking to the product and draw up a declaration of conformity.

2. The notified body must examine the individual product and carry out the appropriate tests as set out in the relevant standard(s) referred to in Article 5, or equivalent tests, to ensure its conformity with the relevant requirements of the Directive.

The notified body must affix, or cause to be affixed, its identification symbol on the approved product and shall draw up a certificate of conformity concerning the tests carried out.

3. The aim of the technical documentation is to enable conformity with the requirements of the Directive to be assessed and the design, manufacture and operation of the product to be understood.[24]

Module H (full quality assurance)

1. This module describes the procedure whereby the manufacturer who satisfies the obligations of paragraph 2 ensures and declares that the products concerned satisfy the requirements of the Directive that apply to them. The manufacturer or his authorised representative established within the Community must affix the CE marking to each product and draw up a written declaration of conformity. The CE marking must be accompanied by the identification symbol of the

[24] The content of the technical documentation shall be laid down Directive by Directive in accordance with the products concerned. As an example, the documentation shall contain so far as relevant for assessment:
- a general description of the product,
- conceptual design and manufacturing drawings and schemes of components, sub-assemblies, circuits, etc,
- descriptions and explanations necessary for the understanding of said drawings and schemes and the operation of the product,
- a list of the standards referred to in Article 5, applied in full or in part, and descriptions of the solutions adopted to meet the essential requirements of the Directive where the standards referred to in Article 5 have not been applied,
- results of design calculations made, examinations carried out, etc,
- test reports.

notified body responsible for the surveillance as specified in point 4.
2. The manufacturer must operate an approved quality system for design, manufacture and final product inspection and testing as specified in point 3 and shall be subject to surveillance as specified in point 4.

3. Quality system

3.1. The manufacturer must lodge an application for assessment of his quality system with a notified body.

The application must include:
– all relevant information for the product category envisaged,
– the quality system's documentation.

3.2. The quality system must ensure compliance of the products with the requirements of the Directive that apply to them.

All the elements, requirements and provisions adopted by the manufacturer must be documented in a systematic and orderly manner in the form of written policies, procedures and instructions. This quality system documentation shall ensure a common understanding of the quality policies and procedures such as quality programmes, plans, manuals and records.

It must contain in particular an adequate description of:
– the quality objectives and the organizational structure, responsibilities and powers of the management with regard to design and product quality,
– the technical design specifications, including standards, that will be applied and, where the standards referred to in Article 5 will not be applied in full, the means that will be used to ensure that the essential requirements of the Directive that apply to the products will be met,
– the design control and design verification techniques, processes and systematic actions that will be used when designing the products pertaining to the product category covered,
– the corresponding manufacturing, quality control and quality assurance techniques, processes and systematic actions that will be used,
– the examinations and tests that will be carried out before, during and after manufacture, and the frequency with which they will be carried out,
– the quality records, such as inspection reports and test data, calibration data, qualification reports of the personnel concerned, etc.,
– the means to monitor the achievement of the required design and product quality and the effective operation of the quality system.

3.3. The notified body must assess the quality system to determine whether it satisfies the requirements referred to in point 3.2. It shall presume compliance with these requirements in respect of quality systems that implement the relevant harmonised standard.[25]

The auditing team must have at least one member experienced as an assessor in the product technology concerned. The evaluation procedure shall include an assessment visit to the manufacturer's premises.

The decision must be notified to the manufacturer. The notification must contain the conclusions of the examination and the reasoned assessment decision.

3.4. The manufacturer must undertake to fulfil the obligations arising out of the quality system as approved and to uphold it so that it remains adequate and efficient.

The manufacturer or his authorised representative must keep the notified body that has approved the quality system informed of any intended updating of the quality system.

The notified body must evaluate the modifications proposed and decide whether the amended quality system will still satisfy the requirements referred to in paragraph 3.2 or whether a re-assessment is required.

It must notify its decision to the manufacturer. The notification shall contain the conclusions of the examin-

[25] This harmonised standard shall be EN 29,001, completed if necessary to take into consideration the specificity of the products for which it is implemented.

ation and the reasoned assessment decision.

4. EC surveillance under the responsibility of the notified body

4.1. The purpose of surveillance is to make sure that the manufacturer duly fulfils the obligations arising out of the approved quality system.

4.2. The manufacturer must allow the notified body entrance for inspection purposes to the locations of design, manufacture, inspection and testing, and storage, and shall provide it with all necessary information, in particular:

– the quality system documentation,

– the quality records as foreseen by the design part of the quality system, such as results of analyses, calculations, tests, etc,

– the quality records as foreseen by the manufacturing part of the quality system, such as inspection reports and test data, calibration data, qualification reports of the personnel concerned, etc.

4.3. The notified body must periodically[26] carry out audits to make sure that the manufacturer maintains and applies the quality system and shall provide an audit report to the manufacturer.

[26] In the specific Directives, the frequency may be specified.

4.4. Additionally the notified body may pay unexpected visits to the manufacturer. At the time of such visits, the notified body may carry out tests or have them carried out in order to check the proper functioning of the quality system where necessary; it must provide the manufacturer with a visit report and, if a test has been carried out, with a test report.

5. The manufacturer must, for a period ending at least 10 years[27] after the last product has been manufactured, keep at the disposal of the national authorities:

– the documentation referred to in the second indent of the second subparagraph of point 3.1,

– the updating referred to in the second subparagraph of point 3.4,

– the decisions and reports from the notified body which are referred to in the final subparagraph of point 3.4, points 4.3 and 4.4.

6. Each notified body must forward to the other notified bodies the relevant information concerning the quality system approvals issued and withdrawn.[28]

Possible supplementary requirements:

[27] The specific Directives may alter this period.

[28] The specific Directives may provide for different arrangements.

Design examination

1. The manufacturer must lodge an application for examination of the design with a single notified body.

2. The application must enable the design, manufacture and operation of the product to be understood, and shall enable conformity with the requirements of the Directive to be assessed.

It must include:

- the technical design specifications, including standards, that have been applied,
- the necessary supporting evidence for their adequacy, in particular where the standards referred to in Article 5 have not been applied in full. This supporting evidence must include the results of tests carried out by the appropriate laboratory of the manufacturer or on his behalf.

3. The notified body must examine the application and where the design meets the provisions of the Directive that apply to it must issue an EC design examination certificate to the applicant. The certificate shall contain the conclusions of the examination, conditions for its validity, the necessary data for identification of the approved design and, if relevant, a description of the product's functioning.

4. The applicant must keep the notified body that has issued the EC design examination certificate informed of any modification to the approved design. Modifications to the approved design must receive additional approval from the notified body that issued the EC design examination certificate where such changes may affect the conformity with the essential requirements of the Directive or the prescribed conditions for use of the product. This additional approval is given in the form of an addition to the original EC design examination certificate.

5. The notified bodies must forward to the other notified bodies the relevant information concerning:

- the EC design examination certificates and additions issued,
- the EC design approvals and additional approvals withdrawn.*

---

*The specific Directives may provide for different arrangements.

Conformity assessment procedures in Community legislation

<table>
<tr><th></th><th>A. (Internal control of production)</th><th colspan="4">B. (type examination)</th><th>G. (unit verification)</th><th>H. (full quality assurance)</th></tr>
<tr><td>DESIGN</td><td>Manufacturer

Keeps technical documentation at the disposal of national authorities

Aa
Intervention of notified body</td><td colspan="4">Manufacturer submits to notified body
– Technical documentation
– Type

Notified body
– Ascertains conformity with essential requirements
– Carries out tests, if necessary
– Issues EC type-examination certificate</td><td>Manufacturer
– Submits technical documentation</td><td>EN 29001
Manufacturer
– Operates an approved quality system (QS) for design
Notified body
– Carries out surveillance of the QS
– Verifies conformity of the design (1)
– Issues EC design examination certificate (1)</td></tr>
<tr><th></th><th></th><th>C. (conformity to type)</th><th>D. (production quality assurance)</th><th>E. (product quality assurance)</th><th>F. (product verification)</th><th></th><th></th></tr>
<tr><td>PRODUCTION</td><td>A.
Manufacturer
– Declares conformity with essential requirements
– Affixes the CE marking

Aa
Notified body
– Test on specific aspects of the product (1)
– Product checks at random intervals (1)</td><td>Manufacturer
– Declares conformity with approved type
– Affixes the CE marking

Notified body
– Tests on specific aspects of the product (1)
– Product checks at random intervals (1)</td><td>EN 29002
Manufacturer
– Operates an approved quality system (QS) for production and testing
– Declares conformity with approved type
– Affixes the CE marking

Notified body
– Approves the QS
– Carries out surveillance of the QS</td><td>EN 29003
Manufacturer
– Operates an approved quality system (QS) for inspection and testing
– Declares conformity with approved type, or to essential requirements
– Affixes the CE marking

Notified body
– Approves the QS
– Carries out surveillance of the QS</td><td>Manufacturer
– Declares conformity with approved type, or with essential requirements
– Affixes the CE marking

Notified body
– Verifies conformity
– Issues certificate at conformity</td><td>Manufacturer
– Submits product
– Declares conformity
– Affixes the CE marking

Notified body
– Verifies conformity with essential requirements
– Issues certificate of conformity</td><td>Manufacturer
– Operates an approved QS for production and testing
– Declares conformity
– Affixes the CE marking

Notified body
– Carries out surveillance of the QS</td></tr>
</table>

(1) Supplementary requirements which may be used in specific Directives

COUNCIL DIRECTIVE 87/404/EEC

of 25 June 1987

on the harmonisation of the laws of the Member States relating to simple pressure vessels*†

THE COUNCIL OF THE EUROPEAN COMMUNITIES,

Having regard to the Treaty establishing the European Economic Community, and in particular Article 100 thereof,

Having regard to the proposal from the Commission,

Having regard to the opinion of the European Parliament,

Having regard to the opinion of the Economic and Social Committee,

Whereas Member States have the responsibility of ensuring the safety on their territory of persons, domestic animals and property with regard to the hazards resulting from the leakage or bursting of simple pressure vessels;

Whereas, in each Member State, mandatory provisions define in particular the safety level required of simple pressure vessels by specifying design and operating characteristics, conditions of installation and use and inspection procedures before and after placing on the market; whereas these mandatory provisions do not necessarily lead to different safety levels from one Member State to another but do, by their disparity, hinder trade within the Community;

Whereas the national provisions ensuring such safety must be harmonised in order to guarantee the free movement of simple pressure vessels without lowering existing and justified levels of protection in the Member States;

Whereas Community legislation as it stands at present provides that, notwithstanding one of the fundamental rules of the Community, namely the free movement of goods, barriers to intra-Community movement resulting from disparities in national laws on the marketing of products have to be accepted in so far as those provisions may be recognised as necessary to satisfy essential requirements; whereas the harmonisation of laws in the present case must therefore be confined to those provisions needed to satisfy the essential safety requirements for simple pressure vessels; whereas, because they are essential, these requirements must replace the corresponding national provisions;

Whereas this Directive therefore contains only mandatory and essential requirements; whereas, to facilitate proof of conformity with the essential requirements, it is necessary to have harmonised standards at European level in particular as to the design, operation and installation of simple pressure vessels so that products complying with them may be assumed to conform to the safety requirements; whereas these standards harmonised at European level are drawn up by private bodies and must remain non-mandatory texts; whereas for that purpose the European Committee for Standardization (CEN) and the European Committee for Electrotechnical Standardisation (CENELEC) are reçognised as the competent bodies for the adoption of harmonised standards in accordance with the general guidelines for cooperation between the Commission and those two bodies signed on 13 November 1984; whereas, for the purposes of this Directive,

---

*Amended by Dir 90/488 and Dir 93/68.
†OJ 1987 L220/48.

a harmonised standard is a technical specification (European standard or harmonisation document) adopted by one or both of those bodies upon a remit from the Commission in accordance with the provisions of Council Directive 83/189/EEC of 28 March 1983 (OJ 1983 L109/8) laying down a procedure for the provision of information in the field of technical standards and regulations and the abovementioned general guidelines;

Whereas a check on compliance with the relevant technical requirements is necessary to provide effective protection for users and third parties; whereas the existing inspection procedures differ from one Member State to another; whereas, in order to avoid multiple inspections, which are in effect barriers to the free movement of vessels, arrangements should be made for the mutual recognition of inspection procedures by the Member States; whereas, in order to facilitate the mutual recognition of inspection procedures, harmonised Community procedures should be set up and the criteria for appointing the bodies responsible for carrying out tests, surveillance and verification should be harmonised;

Whereas the presence on a simple pressure vessel of the EC mark indicates that it satisfies the provisions of this Directive and therefore makes it unnecessary, on import and placing into service of the vessel, to repeat the inspections already carried out; whereas nevertheless simple pressure vessels might represent a safety hazard; whereas provision should therefore be made for a procedure to reduce this hazard,

HAS ADOPTED THIS DIRECTIVE:

Chapter I
Scope, placing on the market and free movement

Article 1

1. This Directive applies to simple pressure vessels manufactured in series.

2. For the purposes of this Directive, "simple pressure vessel" means any welded vessel subjected to an internal gauge pressure greater than 0,5 bar which is intended to contain air or nitrogen and which is not intended to be fired.

Moreover,

– the parts and assemblies contributing to the strength of the vessel under pressure shall be made either of non-alloy quality steel or of non-alloy aluminium or non-age hardening aluminium alloys,

– the vessel shall be made of:
  – either a cylindrical part of circular cross-section closed by outwardly dished and/or flat ends which revolve around the same axis as the cylindrical part,
  – or two dished ends revolving around the same axis,

– the maximum working pressure of the vessel shall not exceed 30 bar and the product of that pressure and the capacity of the vessel (PS.V) shall not exceed 10,000 bar/litre,

– the minimum working temperature must be no lower than −50 °C and the maximum working temperature shall not be higher than 300 °C for steel and 100 °C for aluminium or aluminium alloy vessels.

3. The following vessels shall be excluded from the scope of the Directive:

– vessels specifically designed for nuclear use, failure of which may cause an emission of radioactivity,

– vessels specifically intended for installation in or the propulsion of ships and aircraft,

– fire extinguishers.

Article 2

1. Member States shall take all necessary steps to ensure that the vessels referred to in Article 1, hereinafter referred to as "vessels", may be placed on the market and taken into service only if they do not compromise the safety of persons, domestic animals or property when properly installed and

maintained and when used for the purposes for which they are intended.
2. The provisions of this Directive do not affect the right of Member States to specify – with due observance of the Treaty – the requirements they deem necessary to ensure that workers are protected when using vessels, provided it does not mean the vessels are modified in a way unspecified in this Directive.

Article 3

1. Vessels in respect of which the product of PS and V exceeds 50 bar/litre must satisfy the essential safety requirements set out in Annex I.
2. Vessels in respect of which the product of PS and V is 50 bar/litre or less must be manufactured in accordance with sound engineering practice in one of the Member States and bear markings as laid down in section 1 of Annex II, with the exception of the EC mark referred to in Article 16.

Article 4

Member States shall not impede the placing on the market and the taking into service in their territory of vessels which satisfy the requirements of this Directive.

Article 5

1. Member States shall presume that vessels bearing the CE marking comply with all the provisions of this Directive, including the conformity assessment procedures referred to in Chapter II.

Conformity of vessels with the national standards which transpose the harmonised standards the reference numbers of which have been published in the *Official Journal of the European Communities* shall result in a presumption of conformity to the essential safety requirements referred to in Article 3. Member States shall publish the reference numbers of such national standards.
2. Member States shall presume that vessels in respect of which the manufacturer has not applied, or has applied only in part, the standards referred to in paragraph 1, or for which no such standards exist, comply with the essential requirements referred to in Article 3 where, after receipt of an EEC type-examination certificate, their conformity with the approved model has been certified by the affixation of the CE marking.
3. (a) Where the vessels are subject to other Directives covering other aspects and which also provide for the affixing of the CE marking, the latter shall indicate that the vessels in question are also presumed to conform to the provisions of those other directives.
(b) However, where one or more of those Directives allow the manufacturer, during a transitional period, to choose which arrangements to apply, the CE marking shall indicate conformity only to the Directives applied by the manufacturer. In this case, particulars of the Directives applied, as published in the *Official Journal of the European Communities*, must be given in the documents, notices or instructions required by the Directives and accompanying such vessels.

Article 6

1. Where a Member State or the Commission considers that the harmonised standards referred to in Article 5 (1) do not entirely meet the essential requirements referred to in Article 3, the Commission or the Member State concerned shall bring the matter before the Standing Committee set up under Directive 83/189/EEC, hereinafter referred to as "the committee", giving the reasons therefore. The committee shall deliver an opinion without delay. In the light of the committee's opinion, the Commission shall inform the Member States whether or not it is necessary to withdraw those standards from the publications referred to in Article 5 (1).

Article 7

1. Where a Member State finds that

vessels bearing the CE marking and used in accordance with their intended purpose might compromise the safety of persons, domestic animals or property, it shall take all appropriate measures to withdraw those products from the market or to prohibit or restrict their being placed on the market.

The Member State concerned shall immediately inform the Commission of any such measure, indicating the reasons for its decision, and in particular whether non-conformity is due to:

(a) failure to meet the essential requirements referred to in Article 3, where the vessel does not meet the standards referred to in Article 5 (1);
(b) incorrect application of the standards referred to in Article 5 (1);
(c) shortcomings in the standards referred to in Article 5 (1) themselves.

2. The Commission shall enter into consultation with the parties concerned as soon as possible. Where, after such consultation, the Commission finds that any measure as referred to in paragraph 1 is justified, it shall immediately so inform the Member State that took the action and the other Member States. Where the decision referred to in paragraph 1 is attributed to shortcomings in the standards, the Commission, after consulting the parties concerned, shall bring the matter before the Committee within two months if the Member State which has taken the measures intends to maintain them and shall initiate the procedures referred to in Article 6.

3. Where a vessel which does not comply bears the CE marking, the competent Member State shall take appropriate action against whomsoever has affixed the mark and shall inform the Commission and the other Member States thereof.

4. The Commission shall ensure that the Member States are kept informed of the progress and outcome of this procedure.

Chapter II
Certification procedures

Article 8

1. Prior to production of pressure vessels of which the product of PS and V exceeds 50 bar/litre, manufactured:

(a) in accordance with the standards referred to in Article 5 (1), the manufacturer, or his authorised representative established within the Community, shall at his own choice:
  – either inform an approved inspection body as referred to in Article 9, which after examining the design and manufacturing schedule referred to in Annex II 3, shall draw up a certificate of adequacy attesting that the schedule is satisfactory, or
  – submit a prototype vessel for the EC type-examination referred to in Article 10;
(b) not, or only partly, in accordance with the standards referred to in Article 5 (1), the manufacturer, or his authorised representative established within the Community, must submit a prototype vessel for the EC type-examination referred to in Article 10.

2. Vessels manufactured in accordance with the standards referred to in Article 5 (1) or with the approved prototype shall, prior to their being placed on the market, be subject:

(a) to the EC verification referred to in Article 11 where the product of PS and V exceeds 3,000 bar/litre;
(b) at the choice of the manufacturer, where the product of PS and V does not exceed 3,000 bar/litre but exceeds 50 bar/litre:
  – either to the EC declaration of conformity referred to in Article 12, or
  – to the EC verification referred to in Article 11.

3. The records and correspondence relating to the certification procedures referred to in paragraphs 1 and 2 shall be drafted in an official language of the Member State in which the

approved body is established or in a language accepted by that body.

Article 9

1. Member States shall notify the Commission and the other Member States of the approved bodies which they have appointed to carry out the procedures referred to in Article 8 (1) and (2) together with the specific tasks which these bodies have been appointed to carry out and the identification numbers assigned to them beforehand by the Commission.

The Commission shall publish in the *Official Journal of the European Communities* a list of the notified bodies with their identification numbers and the tasks for which they have been notified. The Commission shall ensure that this list is kept up to date.
2. Annex III sets out the minimum criteria which Member States must meet as regards approval of these bodies.
3. A Member State which has approved a body must withdraw approval if it finds that the body no longer meets the criteria listed in Annex III. It shall immediately inform the Commission and the other Member States accordingly.

EC type-examination

Article 10

1. EC type-examination is the procedure by which an approved inspection body ascertains and certifies that a prototype vessel satisfies the provisions of this Directive which apply to it.
2. The application for EC type-examination shall be lodged by the manufacturer or by his authorised representative with a single approved inspection body in respect of a prototype vessel or of a prototype representing a family of vessels. That authorised representative must be established in the Community.

The application shall include:
– the name and address of the manufacturer or of his authorised representative and the place of manufacture of the vessels,
– the design and manufacturing schedule referred to in Annex II 3.

It shall be accompanied by a vessel which is representative of the production envisaged.
3. The approved body shall carry out the EC type-examination in the manner described below.

It shall examine not only the design and manufacturing schedule in order to check its conformity, but also the vessel submitted.

When examining the vessel, the body shall:

(a) verify that the vessel has been manufactured in conformity with the design and manufacturing schedule and may safely be used under its intended working conditions;
(b) perform appropriate examinations and tests to check that the vessel complies with the essential requirements applicable to it.

4. If the prototype complies with the provisions applicable to it the body shall draw up an EC type-examination certificate which shall be forwarded to the applicant. That certificate shall state the conclusions of the examination, indicate any conditions to which its issue may be subject and be accompanied by the descriptions and drawings necessary for identification of the approved prototype.

The Commission, the other approved bodies and the other Member States may obtain a copy of the certificate and, on a reasoned request, a copy of the design and manufacturing schedule and the reports on the examinations and tests carried out.
5. A body which refuses to issue an EC type-examination certificate shall so inform the other approved bodies. A body which withdraws an EC type-examination certificate shall so inform the Member State which approved it. The latter shall inform the other Member States and the Commission thereof, giving the reasons for the decision.

EC verification

Article 11

1. EC verification is the procedure whereby a manufacturer or his authorised representative established within the Community ensures and declares that the vessels which have been checked in accordance with paragraph 3 are in conformity to the type described in the EC type-examination certificate or with the design and manufacturing schedule referred to in Annex II section 3 having received a certificate of adequacy.

2. The manufacturer shall take all the necessary measures for the manufacturing process to ensure that the vessels conform to the type described in the EC type-examination certificate or to the design and manufacturing schedule referred to in Annex II section 3. The manufacturer or his authorised representative established within the Community shall affix the CE marking to each vessel and draw up a declaration of conformity.

3. The approved body shall carry out the appropriate examination and tests in order to check the conformity of the vessels with the requirements of this Directive by examination and testing of vessels in accordance with the following paragraphs:

3.1. The manufacturer shall present his vessels in the form of uniform batches and shall take all necessary measures in order that the manufacturing process ensures the uniformity of each batch produced.

3.2. These batches shall be accompanied by the EC type-examination certificate referred to in Article 10, or, where the vessels are not manufactured in accordance with an approved prototype, by the design and manufacturing schedule referred to in Annex II section 3. In this case the approved body shall, prior to EC verification, examine the schedule in order to certify its conformity.

3.3. When a batch is examined, the inspection body shall ensure that the vessels have been manufactured and checked in accordance with the design and manufacturing schedule and perform a hydrostatic test or a pneumatic test of equivalent effect on each vessel in the batch at the pressure Ph equal to 1,5 times the vessel's design pressure in order to check its soundness. The pneumatic test shall be subject to acceptance of the test safety procedures by the Member State in which the test is performed.

Moreover, the inspection body shall carry out tests on test-pieces taken from a representative production test-piece or from a vessel, as the manufacturer chooses, in order to examine the weld quality. The tests shall be carried out on longitudinal welds. However, where differing weld techniques are used for longitudinal and circular welds, the tests shall be repeated on the circular welds.

For those vessels referred to in Annex I, section 2.1.2 these tests on test-pieces shall be replaced by a hydrostatic test on five vessels taken at random from each batch in order to check that they conform to the requirement of Annex I, section 2.1.2.

3.4. In the case of accepted batches, the approved body shall affix, or cause to be affixed, its identification number to each recipient and draw up a written certificate of conformity relating to the tests carried out. All recipients in the batch may be placed on the market except for those which have not successfully undergone a hydrostatic test or a pneumatic test.

If a batch is rejected, the notified body or the competent authority shall take appropriate measures to prevent the putting on the market of that batch. In the event of frequent rejection of batches, the notified body may suspend the statistical verification.

The manufacturer may, under the responsibility of the notified body, affix the latter's identification number during the manufacturing process.

3.5. The manufacturer or his authorised representative must be able to supply on request the approved body's certificates of conformity referred to in subparagraph 3.4.

EC declaration of conformity

Article 12

1. A manufacturer fulfilling the obligations arising out of Article 13 shall affix the CE marking provided for in Article 16 to vessels which he declares to be in conformity:
– to the design and manufacturing schedule referred to in Annex II section 3, and on which a certificate of adequacy has been drawn up, or
– an approved prototype.
By this EC declaration of conformity procedure the manufacturer becomes subject to EC surveillance, in cases where the product of PS and V exceeds 200 bar/litre.
2. The purpose of EC surveillance is to ensure, as required by Article 14 (2), that the manufacturer duly fulfils the obligations arising out of Article 13 (2). Surveillance shall be performed by the approved body which issued the EC type-examination certificate referred to in Article 10 where the vessels have been manufactured in accordance with an approved prototype or, if this is not the case, by the approved body to which the design and manufacturing schedule was sent in accordance with Article 8 (1) (a), first indent.

Article 13

1. Where a manufacturer makes use of the procedure referred to in Article 12, he must, before commencing manufacture, send the approved body which issued the EC type-examination certificate or the certificate of adequacy a document describing the manufacturing processes and all of the pre-determined systematic measures taken to ensure conformity of the pressure vessels to the standards referred to in Article 5 (1) or the approved prototype.

This document shall include:

(a) a description of the means of manufacture and checking appropriate to the construction of the vessels:

(b) an inspection document describing the appropriate examinations and tests to be carried out during manufacture, together with the procedures therefore and the frequency with which they are to be performed;

(c) an undertaking to carry out the examinations and tests in accordance with the inspection document referred to above and to have a hydrostatic test or, subject to the agreement of the Member State, a pneumatic test carried out on each vessel manufactured at a test pressure equal to 1,5 times the design pressure.

These examinations and tests shall be carried out under the responsibility of qualified staff who are sufficiently independent from production personnel, and shall be covered by a report;

(d) the addresses of the places of manufacture and storage and the date on which manufacture is to commence.

2. In addition, when the product of PS and V exceeds 200 bar/litre, manufacturers shall authorise access to the said place of manufacture and storage by the body responsible for EC surveillance, for inspection purposes, and shall allow that body to select sample vessels and shall provide it with all necessary information, and in particular:
– the design and manufacturing schedule,
– the inspection report,
– the EC type-examination certificate or certificate of adequacy, where appropriate,
– a report on the examinations and tests carried out.

Article 14

1. The approved body which issued the EC type-examination certificate or certificate of adequacy must, before the date on which any manufacture begins, examine both the document referred to in Article 13 (1) and the design and manufacturing schedule referred to in Annex II, section 3, in order to certify their conformity where vessels are not manufactured in

accordance with an approved prototype.
2. In addition, where the product of PS and V exceeds 200 bar/litre, that body must during manufacture:
– ensure that the manufacturer actually checks series-produced vessels in accordance with Article 13 (1) (c),
– take random samples at the places of manufacture or at the place of storage of vessels for inspection purposes.

The body shall supply the Member State which approved it, and, on request, the other approved bodies, the other Member States and the Commission, with a copy of the inspection report.

Chapter III
EC mark

Article 15

Without prejudice to Article 7.

(a) where a Member State establishes that the CE marking has been affixed unduly, the manufacturer or his authorised representative established within the Community shall be obliged to make the product conform as regards the provisions concerning the CE marking and to end the infringement under the conditions imposed by the Member State;

(b) where non-conformity continues, the Member State must take all appropriate measures to restrict or prohibit the placing on the market of the product in question or to ensure that it is withdrawn from the market in accordance with the procedures laid down in Article 7.

Article 16

1. The CE marking and the inscriptions provided for in Annex II, section 1, shall be affixed in a visible, easily legible and indelible form to the vessel or to a data plate attached to the vessel in such a way that it cannot be removed.

The CE conformity marking shall consist of the initials "CE" in the form shown in the specimen in Annex II. The CE marking shall be followed by the distinguishing number referred to in Article 9 (1) of the approved inspection body responsible for EC verifications or EC surveillance.
2. The affixing to markings on the vessels which are likely to deceive third parties as to the meaning and form of the CE marking shall be prohibited. Any other marking may be affixed to the vessels or the data plate provided that the visibility and legibility of the CE marking is not thereby reduced.

Chapter IV
Final provisions

Article 17

Any decision taken pursuant to this Directive and resulting in restrictions on the placing on the market and/or taking into service of a vessel shall state the exact grounds on which it is based. Such a decision shall be notified without delay to the party concerned, who shall at the same time be informed of the judicial remedies available to him under the laws in force in the Member State in question and of the time limits to which such remedies are subject.

Article 18

1. Before 1 January 1990, Member States shall adopt and publish the laws, regulations and administrative provisions necessary in order to comply with this Directive. They shall forthwith inform the Commission thereof.

They shall apply such provisions from 1 July 1990.

Member States shall, for the period up to 1 July 1992, permit the placing on the market and/or in service of vessels conforming to the rules in force in their territories before the date of application of this Directive.

Annex I

The essential safety requirements for the vessels are set out below:

1. Materials

Materials must be selected according to the intended use of the vessels and in accordance with 1.1 to 1.4.

1.1 Pressurised parts

The materials referred to in Article 1 used for manufacturing the pressurised parts must be:
– capable of being welded,
– ductile and tough, so that a rupture at minimum working temperature does not give rise to either fragmentation or brittle-type fracture,
– not adversely affected by ageing.

For steel vessels, the materials must in addition meet the requirements set out in section 1.1.1 and, for aluminium or aluminium alloy vessels, those set out in section 1.1.2.

They must be accompanied by an inspection slip drawn up by the producer of the materials as described in Annex II.

1.1.1. *Steel vessels*

Non-alloy quality steels meet the following requirements:

(a) they must be non-effervescent and be supplied after normalisation treatment, or in an equivalent state;

(b) the content per product of carbon must be less than 0.25 % and that of sulphur and phosphorus must each be less than 0.05 %.

(c) they must have the following mechanical properties per product:
– the maximum tensile strength $R_{m,max}$ must be less than 580 N/mm^2,
– the elongation after rupture must be:
– if test pieces are taken parallel to the direction of rolling:
thickness ≥3 mm: A ≥22%,
thickness < 3 mm: $A_{80\ mm}$ ≥17%,
– if test pieces are taken perpendicular to the direction of rolling:
thickness ≥3 mm: A ≥20%,
thickness < 3 mm: $A_{80\ mm}$ ≥15 %,
– the average failure energy KCV for three longitudinal test pieces at minimum working temperature must not be less than 35 J/cm^2. Not more than one of the three figures may be less than 35 J/cm^2, with a minimum of 25 J/cm^2.
In the case of steels used in the manufacture of vessels whose minimum working temperature is lower than −10 °C and whose wall thickness exceeds five mm, this property must be checked.

1.1.2. *Aluminium vessels*

Non-alloy aluminium must have an aluminium content of at least 99.5% and those alloys described in Article 1 (2) must display adequate resistance to intercrystalline corrosion at maximum working temperature.

Moreover these materials must satisfy the following requirements:

(a) they must be supplied in an annealed state; and

(b) must have the following mechanical characteristics per product:
– the maximum tensile strength $R_{m,max}$ must be no more than 350 N/mm^2,
– the elongation after rupture must be:
– A ≥16% if the test piece is taken parallel to the direction of rolling,
– A ≥14% if the test piece is taken perpendicular to the direction of rolling,

1.2. Welding materials

The welding materials used to manufacture the welds on or of the vessel must be appropriate to and compatible with the materials to be welded.

1.3. Accessories contributing towards the strength of the vessel

These accessories (*e.g.* bolts and nuts) must be made of a material specified in 1.1 or of other kinds of steel, aluminium or an appropriate aluminium alloy compatible with materials used for the manufacture of pressurised parts.

The latter materials must at minimum working temperature have an appropriate elongation after rupture and toughness.

1.4. Non-pressurised parts

All unpressurised parts of welded vessels must be of materials which are compatible with that of the components to which they are welded.

2. *Vessel design*

The manufacturer must, when designing the vessel, define the use to which it will be put, and select:

– the minimum working temperature T_{min},

– the maximum working temperature T_{max},

– the maximum working pressure PS.

However, should a minimum working temperature exceeding −10 °C be selected, the qualities required of the materials must be satisfied at −10 °C.

The manufacturer must also take account of the following provisions:

– it must be possible to inspect the inside of vessels,

– it must be possible to drain the vessels,

– the mechanical qualities shall be maintained throughout the period of use of the vessel for the intended purpose,

– the vessels shall, bearing in mind their prescribed use, be adequately protected against corrosion,

and the fact that under the conditions of use envisaged

– the vessels will not be subjected to stresses likely to impair their safety in use,

– internal pressure will not permanently exceed the maximum working pressure PS; however, it may momentarily do so by up to 10%.

Circular and longitudinal seams must be made using full penetration welds or welds of equivalent effectiveness. Convex ends other than hemispherical ones shall have a cylindrical edge.

2.1. Wall thickness

If the product PS.V is not more than 3,000 bar/litre, the manufacturer must select one of the methods described in 2.1.1 and 2.1.2 for determining vessel wall thickness; if the product of PS and V is more than 3,000 bar/litre, or if the maximum working temperature exceeds 100 °C, such thickness must be determined by the method described in 2.1.1.

The actual wall thickness of the cylindrical section and ends shall, however, be not less than 2 mm in the case of steel vessels and not less than 3 mm in the case of aluminium or aluminium alloy vessels.

2.1.1. *Calculation method*

The minimum thickness of pressurised parts must be calculated having regard to the intensity of the stresses and to the following provisions:

– the calculation pressure to be taken into account must not be less than the maximum working pressure selected,

– the permissible general membrane stress must not exceed the lower of the values 0.6 R_{ET} or 0.3 R_m. The manufacturer must use the R_{ET} and R_m minimum values guaranteed by the material manufacturer in order to determine the permissible stress.

However, where the cylindrical portion of the vessel has one or more longitudinal welds made using a non-automatic welding process, the thickness calculated as above must be multiplied by the coefficient 1.15.

2.1.2. *Experimental method*

Wall thickness must be so determined as to enable the vessels to resist at ambient temperature a pressure equal to at least five times the maximum working pressure, with a permanent circumferential deformation factor of no more than 1%.

3. *Manufacturing processes*

Vessels shall be constructed and subjected to production checks in accordance with the design and manufacturing record referred to in Annex II, section 3.

3.1. Preparation of the component parts

Preparation of the component parts (*e.g.* forming and chamfering) must not give rise to surface defects or cracks or changes in the mechanical characteristics likely to be detrimental to the safety of the vessels.

3.2. Welds on pressurised parts

The characteristics of welds and adjacent zones must be similar to those of the welded materials and shall be free of any surface or internal defects detrimental to the safety of the vessels.

Welds must be performed by qualified welders or operators possessing the appropriate level of competence, in accordance with approved welding processes. Such approval and qualification tests must be carried out by approved inspection bodies.

The manufacturer must also, during manufacture, ensure consistent weld quality by conducting appropriate tests using adequate procedures. These tests must be the subject of a report.

4. *Placing in services of the vessels*

Vessels must be accompanied by the instructions drawn up by the manufacturer, as referred to in Annex II, section 2.

Annex II

1. *CE marking and inscriptions*

1. (a) CE conformity marking

– The CE conformity marking shall consist of the initials "CE" in the following form:

– If the CE marking is reduced or enlarged the proportions given in the above graduated drawing must be respected.

– The various components of the CE marking must have substantially the same vertical dimension, which may not be less than 5 mm.

1. (b) Inscriptions

The vessel or data plate must bear at least the following information:

– the maximum working pressure (PS in bar),

– the minimum working temperature T_{min} in °C

– the capacity of the vessel (V in l),

– the name or mark of the manufacturer

– the type and serial or batch identification of the vessel,

– the last two digits of the year in which the CE marking was affixed.

Where a data plate is used, it must be so designed that it cannot be re-used and must include a vacant space to enable other information to be provided.

2. *Instructions*

The instructions must provide the following information:

– the particulars given in 1 above

except for the vessel's serial identification,
– the intended use of the vessel,
– the maintenance and installation requirements for vessel safety.

They must be in the official language or languages of the country of destination.

3. Design and manufacturing schedules

The design and manufacturing schedules must contain a description of the techniques and operations employed in order to meet the essential requirements referred to in Article 3 or the standards referred to in article 5 (1) and in particular:

(a) a detailed manufacturing drawing of the vessel type;
(b) the instructions;
(c) a document describing:
– the materials selected,
– the welding processes selected,
– the checks selected,
– any pertinent details as to the vessel design.

When the procedures laid down in Articles 11 to 14 are used, the schedule must also include:

(i) the certificates relating to the suitable qualification of the welding operations and of the welders or operators;
(ii) the inspection slip for the materials used in the manufacture of parts and assemblies contributing to the strength of the pressure vessel;
(iii) a report on the examinations and tests performed or a description of the proposed checks.

4. Definitions and symbols

4.1 Definitions

(a) The design pressure "P" is the gauge pressure chosen by the manufacturer and used to determine the thickness of the pressurized parts.
(b) The maximum working pressure "PS" is the maximum gauge pressure which may be exerted under normal conditions of use.
(c) The minimum working temperature T_{min} is the lowest stabilised temperature in the wall of the vessel under normal conditions of use.
(d) The maximum working temperature T_{max} is the highest stabilised temperature which the wall of the vessel may attain under normal conditions of use.
(e) The yield strength "R_{ET}" is the value at the maximum working temperature T_{max}
– of the upper yield point ReH, for a material with both a lower and an upper yield point,
– of the proof stress $R_p0,2$,
– or of the proof stress $R_p1,0$ in the case of non-alloy aluminium.
(f) Families of vessels:
Vessels form part of the same family if they differ from the prototype only in diameter, provided that the permissible requirements referred to in sections 2.1.1 or 2.1.2 of Annex 1 are complied with and/or in the length of their cylindrical portion within the following limits:
– where a prototype has one or more shell rings in addition to the ends, variants in the family must have at least one shell ring,
– where a prototype has just two dished ends, variants in the family must have no shell rings.
Variations in length causing the apertures and/or penetrations to be modified must be shown in the drawing for each variant.
(g) A batch of vessels consists at the most of 3,000 vessels of the model of the same type.
(h) There is series manufacture within the meaning of this Directive if more than one vessel of the same type is manufactured during a given period by a continuous manufacturing process, in accordance with a common design and using the same manufacturing processes.
(i) Inspection slip: document by

which the producer certifies that the products delivered meet the requirements of the order and in which he sets out the results of the routine in-plant inspection test, in particular chemical composition and mechanical characteristics performed on products made by the same production process as the supply, but not necessarily on the products delivered.

4.2. Symbols

A	elongation after rupture ($L_o=5.65\sqrt{S_o}$)	%
A 80 mm	elongation after rupture ($L_o=80$ mm)	%
KCV	rupture energy	J/cm²
P	design pressure	bar
PS	working pressure	bar
P_h	hydrostatic or pneumatic test pressure	bar
$R_p0.2$	proof stress at 0.2%	N/mm²
R_{ET}	yield strength at maximum working temperature	N/mm²
R_{eH}	upper yield point	N/mm²
R_m	tensile strength	N/mm²
T_{max}	maximum working temperature	°C
T_{min}	minimum working temperature	°C
V	capacity of the vessel	L
$R_{m,\,max}$	maximum tensile strength	N/mm²
$R_p1.0$	proof stress at 1.0%	N/mm²

Annex III

Minimum criteria to be taken into account by Member States when appointing inspection bodies

1. The inspection body, its director and the staff responsible for carrying out the verification tests shall not be the designer, manufacturer, supplier or installer of vessels which they inspect, nor the authorised representative of any of those parties. They shall not become directly involved in the design, construction, marketing or maintenance of the vessels, nor represent the parties engaged in these activities. This does not preclude the possibility of exchanges of technical information between the manufacturer and the inspection body.
2. The inspection body and its staff must carry out the verification tests with the highest degree of professional integrity and technical competence and must be free from all pressures and inducements, particularly financial, which might influence their judgment or the results of the inspection, especially from persons or groups of persons with an interest in the result of verifications.
3. The inspection body must have at its disposal the necessary staff and possess the necessary facilities to enable it to perform properly the administrative and technical tasks connected with verification; it must also have access to the equipment required for special verification.
4. The staff responsible for inspection must have:
– sound technical and professional training,
– satisfactory knowledge of the requirements of the tests they carry out and adequate experience of such tests.
– the ability to draw up the certificates, records and reports required to authenticate the performance of the tests.
5. The impartiality of inspection staff must be guaranteed. Their remune-

ration must not depend on the number of tests carried out nor on the results of such tests.

6. The inspection body must take out liability insurance unless its liability is assumed by the State in accordance with national law, or the Member State itself is directly responsible for the tests.

7. The staff of the inspection body is bound to observe professional secrecy with regard to all information gained in carrying out its tasks (except *vis-à-vis* the competent administrative authorities of the State in which its activities are carried out) under this Directive or any provision of national law giving effect to it.

COUNCIL DIRECTIVE 88/378/EEC

of 3 May 1988

on the approximation of the laws of the Member States concerning the safety of toys*†

THE COUNCIL OF THE EUROPEAN COMMUNITIES,

Having regard to the Treaty establishing the European Economic Community, and in particular Article 100A thereof,

Having regard to the proposal from the Commission,

In cooperation with the European Parliament,

Having regard to the opinion of the Economic and Social Committee,

Whereas the laws, regulations and administrative provisions in force in the various Member States relating to the safety characteristics of toys differ in scope and content; whereas such disparities are liable to create barriers to trade and unequal conditions of competition within the internal market without necessarily affording consumers in the common market, especially children, effective protection against the hazards arising from the products in question;

Whereas these obstacles to the attainment of an internal market in which only sufficiently safe products would be sold should be removed; whereas, for this purpose, the marketing and free movement of toys should be made subject to uniform rules based on the objectives regarding protection of consumer health and safety as set out in the Council resolution of 23 June 1986 concerning the future orientation of the policy of the European Economic Community for the protection and promotion of consumer interests;

Whereas, to facilitate proof of conformity with the essential requirements, it is necessary to have harmonised standards at European level which concern, in particular, the design and composition of toys so that products complying with them may be assumed to conform to the essential requirements; whereas these standards harmonised at European level are drawn up by private bodies and must remain non-mandatory texts; whereas for that purpose the European Committee for Standardisation (CEN) and the European Committee for Electrotechnical Standardisation (CENELEC) are recognised as the competent bodies for the adoption of harmonised standards in accordance with the general guidelines for cooperation between the Commission and those two bodies signed on 13 November 1984; whereas, for the purposes of this Directive, a harmonised standard is a technical specification (European standard or harmonisation document) adopted by one or both of those bodies upon a remit from the Commission in accordance with the provisions of Council Directive 83/189/EEC of 28 March 1983 (OJ 1983 L109/8) laying down a procedure for the provision of information in the field of technical standards and regulations, as last amended by the Act of Accession of Spain and Portugal, and on the basis of the general guidelines;

Whereas, in accordance with the Council Resolution of 7 May 1985 on a new approach to technical harmonisation and standards, the harmonisation to be achieved should consist of establishing the essential safety requirements to be satisfied by all toys if they are to be placed on the market;

---

*OJ 1988 L187/1.
†Amended by Dir 93/68 (OJ 1993 L220).

Whereas, in view of the size and mobility of the toy market and the diversity of the products concerned, the scope of this Directive should be determined on the basis of a sufficiently broad definition of "toys"; whereas, nevertheless, it should be made clear that some products are not to be regarded as toys for the purposes of this Directive either because they are not in fact intended for children or because they call for supervision or special conditions of use;

Whereas toys placed on the market should not jeopardise the safety and/or health either of users or of third parties; whereas the standard of safety of toys should be determined in relation to the criterion of the use of the product as intended, but allowance should also be made for any foreseeable use, bearing in mind the normal behaviour of children who do not generally show the same degree of care as the average adult user;

Whereas the standard of safety of the toy must be considered when it is marketed, bearing in mind the need to ensure that this standard is maintained throughout the foreseeable and normal period of use of the toy;

Whereas compliance with the essential requirements is likely to guarantee consumer health and safety; whereas all toys placed on the market must comply with these requirements and, if they do, no obstacle must be put in the way of their free movement;

Whereas toys may be presumed to comply with these essential requirements where they are in conformity with the harmonised standards, reference numbers of which have been published in the *Official Journal of the European Communities*;

Whereas toys that conform to a model approved by an approved body may also be regarded as complying with the essential requirements; whereas such conformity must be certified by the affixing of a European mark;

Whereas certification procedures must be established to define the way in which national approved bodies have to approve models of toys not in conformity with standards and issue type-examination certificates for them and for toys in conformity with standards, a model of which is submitted to them for approval;

Whereas adequate information for the Member States, the Commission and all the approved bodies must be provided for at the various stages of the certification and inspection procedures;

Whereas Member States must appoint bodies, called "approved bodies", for the purposes of applying the system introduced for toys; whereas adequate information on these bodies must be provided and they must all comply with minimum criteria for their approval;

Whereas cases might arise where a toy does not satisfy the essential safety requirements; whereas, in such cases, the Member State which ascertains this fact must take all appropriate measures to withdraw the products from the market or to prohibit their being placed on the market; whereas a reason must be given for this decision and, where the reason is a shortcoming in the harmonised standards, these, or a part thereof, must be withdrawn from the list published by the Commission;

Whereas the Commission is to ensure that the harmonised European standards in all the areas covered by the essential requirements listed in Annex II are drawn up in sufficient time to enable Member States to adopt and publish the necessary provisions by 1 July 1989; whereas the national provisions adopted on the basis of this Directive should consequently become effective on 1 January 1990;

Whereas provision must be made for suitable action to be taken against anyone wrongfully affixing a mark of conformity;

Whereas checks on the safety of toys

already on the market must be carried out by the competent authorities of the Member States;

Whereas, for some categories of toys that are particularly dangerous or intended for very young children, warnings or details of precautions to be taken must also be given;

Whereas the Commission must receive regular information on activities carried out under this Directive by the approved bodies;

Whereas those to whom any decision taken under this Directive is addressed must know the reason for that decision and the remedies open to them;

Whereas the opinion of the Scientific Advisory Committee for the evaluation of the toxicity and ecotoxicity of chemical compounds has been taken into account with respect to the health-based limits of bioavailability of metallic compounds in toys to children,

HAS ADOPTED THIS DIRECTIVE:

Article 1

1. This Directive shall apply to toys. A "toy" shall mean any product or material designed or clearly intended for use in play by children of less than 14 years of age.
2. The products listed in Annex 1 shall not be regarded as toys for the purposes of this Directive.

Article 2

1. Toys may be placed on the market only if they do not jeopardise the safety and/or health of users or third parties when they are used as intended or in a foreseeable way, bearing in mind the normal behaviour of chilren.
2. In the condition in which it is placed on the market, taking account of the period of foreseeable and normal use, a toy must meet the safety and health conditions laid down in this Directive.
3. For the purposes of this Directive, the expression "placed on the market" shall cover both sale and distribution free of charge.

Article 3

Member States shall take all steps necessary to ensure that toys cannot be placed on the market unless they meet the essential safety requirements set out in Annex II.

Article 4

Member States shall not impede the placing on the market on their territory of toys which satisfy the provisions of this Directive.

Article 5

1. Member States shall presume that toys bearing the CE marking laid down in Article 11 comply with all the provisions of this Directive, including the conformity assessment procedures referred to in Articles 8, 9 and 10.

Conformity of toys with the national standards which transpose the harmonised standards the reference numbers of which have been published in the *Official Journal of the European Communities* shall result in a presumption of conformity to the essential safety requirements referred to in Article 3. Member States shall publish the reference numbers of such national standards.
2. Member States shall presume that toys in respect of which the manufacturer has not applied the standards referred to in paragraph 1, or has applied them only in part, or for which no such standards exist, satisfy the essential requirements referred to in Article 3 where, after receipt of an EEC type-examination certificate, their conformity with the approved model has been certified by the affixation of the EC mark.
3. (a) Where the toys are subject to other Directives covering other aspects and which also provide for the affixing of the CE marking, the latter shall indicate that the toys in question are also presumed to conform to the provisions of those other Directives.
3. (b) However, where one or more of these Directives allow the manufacturer, during a transitional period, to choose which arrangements to apply,

the CE marking shall indicate conformity only to the Directives applied by the manufacturer. In this case, particulars of the Directives applied, as published in the *Official Journal of the European Community*, must be given in the documents, notices or instructions required by the Directives and accompanying such toys or, failing that, on their packaging.

Article 6

1. Where a Member State or the Commission considers that the harmonised standards referred to in Article 5 (1) do not entirely satisfy the essential requirements referred to in Article 3, the Commission or the Member State shall refer the matter to the Standing Committee set up under Directive 83/189/EEC, hereinafter referred to as "the committee", setting out its reasons. The committee shall issue an opinion as a matter of urgency.
After receiving the committee's opinion, the Commission shall notify the Member States whether or not the standards concerned or a part thereof have to be withdrawn from the publications referred to in Article 5 (1).
2. The Commission shall inform the European standardisation body concerned and, if necessary, issue a new standardisation brief.

Article 7

1. Where a Member State ascertains that toys bearing the CE marking which are used as intended or in accordance with Article 2 are likely to jeopardise the safety and/or health of consumers and/or third parties, it shall take all appropriate measures to withdraw the products from the market, or to prohibit or restrict their placing on the market. The Member State shall inform the Commission immediately of this measure and indicate the reasons for its decision, stating in particular whether the non-compliance results from:
(a) failure to meet the essential requirements referred to in Article 3, if the toy does not meet the standards referred to in Article 5 (1);
(b) incorrect application of the standards referred to in Article 5 (1);
(c) shortcomings in the standards referred to in Article 5 (1).
2. The Commission shall enter into consultation with the parties concerned as soon as possible. Where, after such consultation, the Commission finds that any measure as referred to in paragraph 1 is justified, it shall forthwith so inform the Member State that took the action and the other Member States. Where the decision referred to in paragraph 1 is attributed to shortcomings in the standards, the Commission, after consulting the parties concerned, shall bring the matter before the Committee within two months if the Member State which has taken the measures intends to maintain them and shall initiate the procedures referred to in Article 6.
3. Where the toy which does not comply with the requirements bears the CE marking, the competent Member State shall take appropriate measures and inform the Commission, which shall inform the other Member States.
4. The Commission shall ensure that the Member States are kept informed of the progress and outcome of this procedure.

Article 8

1. (a) Before being placed on the market, toys manufactured in accordance with the harmonised standards referred to in Article 5 (1) must have affixed to them the CE marking by which the manufacturer or his authorised representative established within the Community confirms that the toys comply with those standards;
(b) The manufacturer or his authorised representative established within the Community shall keep the following information available for inspection:
– a description of the means (such as the use of a test report or technical file)

whereby the manufacturer ensures conformity of production with the standards referred to in Article 5 (1) and, as appropriate: an EC type-certificate drawn up by an approved body; copies of the documents the manufacturer has submitted to the approved body; a description of the means whereby the manufacturer ensures conformity with the approved model,

– the addresses of the places of manufacture and storage,

– detailed information concerning the design and manufacture.

Where neither the manufacturer nor his authorised representative are established within the Community, the above obligation to keep a dossier available shall be the responsibility of the person who places the toy on the Community market.

2. (a) Toys which do not conform in whole or in part to the standards referred to in Article 5 (1) must have affixed to them, before being placed on the market, the CE marking by which the manufacturer or his authorised representative established within the Community confirms that the toy concerned conforms to the model examined in accordance with the procedures laid down in Article 10 which an approved body has stated complies with the essential requirements referred to in Article 3;

(b) the manufacturer or his authorised representative established within the Community shall keep the following information available for inspection:

– a detailed description of manufacture,

– a description of the means (such as the use of a test report or technical file) whereby the manufacturer ensures conformity with the approved model,

– the addresses of the places of manufacture and storage,

– copies of the documents the manufacturer has submitted to an approved body in accordance with Article 10 (2),

– the test certificate for the sample or a certified copy thereof.

Where neither the manufacturer nor his authorised representative is established within the Community, the above obligation to keep a dossier available shall be the responsibility of the person who places the toy on the market in the Community.

3. In the event of non-observance of the obligations laid down in paragraphs 1 (b) and 2 (b), the competent Member State shall take appropriate measures to ensure that those obligations are observed.

Where non-observance of the obligations is obvious, it may in particular require the manufacturer or his authorised representative established within the Community to have a test performed at his own expense within a specified period by an approved body in order to verify compliance with the harmonised standards and essential safety requirements.

Article 9

1. The minimum criteria which Member States must meet in order to appoint the approved bodies referred to in this Directive are contained in Annex III.

2. Member States shall notify the Commission and the other Member States of the bodies which they have appointed to carry out the EC type-examination referred to in Article 8 (2) and Article 10 together with the specific tasks which these bodies have been appointed to carry out and the identification numbers assigned to them beforehand by the Commission.

The Commission shall publish in the *Official Journal of the European Communities* a list of the notified bodies with their identification number and the tasks for which they have been notified. The Commission shall ensure that this list is kept up to date.

3. A Member State which has approved a body shall withdraw approval if it finds that the body no longer meets the criteria listed in Annex III. It shall forthwith inform the Commission thereof.

Article 10

1. EC type-examination is the procedure by which an approved body

ascertains and certifies that a model of a toy satisfies the essential requirements referred to in Article 3.

2. The application for EC type-examination shall be lodged with an approved body by the manufacturer or by his authorised representative established within the Community.

The application shall include:

– a description of the toy,

– the name and address of the manufacturer or of his authorised representative or representatives, and the place of manufacture of the toy,

– comprehensive manufacturing and design data; and shall be accompanied by a model of the toy to be manufactured.

3. The approved body shall carry out the EC type-examination in the manner described below:

– it shall examine the documents supplied by the applicant and establish whether they are in order,

– it shall check that the toy would not jeopardise safety and/or health, as provided for in Article 2,

– it shall carry out the appropriate examinations and tests – using as far as possible the harmonised standards referred to in Article 5 (1) – in order to check whether the model meets the essential requirements referred to in Article 3,

– it may ask for further examples of the model.

4. If the model complies with the essential requirements referred to in Article 3, the approved body shall draw up an EC type-examination certificate which shall be notified to the applicant. This certificate shall state the conclusions of the examination, indicate any conditions attaching to it and be accompanied by the descriptions and drawings of the approved toy.

The Commission, the other approved bodies and the other Member States may obtain on request a copy of the certificate and, on reasoned request, a copy of the design and manufacturing schedule and the reports on the examinations and tests carried out.

5. An approved body which refuses to issue an EC type-examination certificate shall so inform the Member State which approved it and the Commission, giving the reasons for refusal.

Article 11

1. The CE marking referred to in Articles 5, 7 and 8 and the name and/or trade name and/or mark and address of the manufacturer or his authorised representative or the importer into the Community shall as a rule be affixed either to the toy or on the packaging in a visible, easily legible and indelible form. In the case of small toys and toys consisting of small parts these particulars may be affixed in the same way to the packaging, to a label or to a leaflet. Where the said particulars are not affixed to the toy, the consumer's attention must be drawn to the advisability of keeping them.

2. The CE conformity marking shall consist of the initials "CE" taking the form of the specimen given in Annex V.

3. The affixing of markings on the toys which are likely to deceive third parties as to the meaning and form of the CE marking shall be prohibited. Any other marking may be affixed to the toys, their packaging or a label provided that the visibility and legibility of the EC marking is not thereby reduced.

4. The particulars referred to in paragraph 1 may be abbreviated provided that the abbreviation enables the manufacturer, his authorised representative or the importer into the Community to be identified.

5. Annex IV sets out the warnings and indications of precautions to be taken during use that have to be given for certain toys. Member States may require that these warnings and precautions, or some of them, together with the information specified in paragraph 4, be given in their own national language or languages when the toys are placed on the market.

Article 12

1. Member States shall take the necess-

ary measures to ensure that sample checks are carried out on toys which are on their market, so as to verify their conformity with this Directive.

The authority responsible for inspection:

– shall obtain access, on request, to the place of manufacture or storage and to the information referred to in Article 8 (1) (b) and (2) (b),

– may ask the manufacturer, his authorised representative or the person responsible for marketing the toy established within the Community to supply the information as provided for in Article 8 (1) (b) and (2) (b) within a period specified by the Member State,

– may select a sample and take it away for examination and testing.

1a. Without prejudice to Article 7:

(a) where a Member State establishes that the CE marking has been affixed unduly, the manufacturer or his authorised representative established within the Community shall be obliged to make the product conform as regards the provisions concerning the CE marking and to end the infringement under the conditions imposed by the Member State;

(b) where non-conformity continues, the Member State must take all appropriate measures to restrict or prohibit the placing on the market of the product in question or to ensure that it is withdrawn from the market in accordance with the procedures laid down in Article 7.

2. Every three years, Member States shall send the Commission a report on the application of this Directive.

3. The Member States and the Commission shall take the necessary measures to guarantee confidentiality with regard to the forwarding of the copies relating to the EC type-examination referred to in Article 10 (4).

Article 13

Member States shall regularly inform the Commission of the activities carried out in pursuance of this Directive by the bodies they have approved so that the Commission may ensure that the inspection procedures are implemented correctly and without discrimination.

Article 14

Any decision taken pursuant to this Directive and involving restrictions on the placing of the toy on the market shall state the exact grounds on which it is based. It shall be notified at the earliest opportunity to the party concerned, who shall at the same time be informed of the remedies available to him under the laws in force in the Member State in question and of the time limits applying to such remedies.

Article 15

1. Member States shall adopt and publish by 30 June 1989 the provisions necessary to comply with this Directive. They shall forthwith inform the Commission thereof.

They shall apply these provisions from 1 January 1990.

2. Member States shall communicate to the Commission the texts of the provisions of national law which they adopt in the field covered by this Directive.

Article 16

This Directive is addressed to the Member States.

Annex I
Products not regarded as toys for the purpose of this Directive

(Article 1 (1))

1. Christmas decorations

2. Detailed scale models for adult collectors

3. Equipment intended to be used collectively in playgrounds

4. Sports equipment

5. Aquatic equipment intended to be used in deep water

6. Folk dolls and decorative dolls and other similar articles for adult collectors

7. "Professional" toys installed in public places (shopping centres, stations, etc)

8. Puzzles with more than 500 pieces or without picture, intended for specialists

9. Air guns and air pistols

10. Fireworks, including percussion caps[1]

11. Slings and catapults

12. Sets of darts with metallic points

13. Electric ovens, irons or other functional products operated at a nominal voltage exceeding 24 volts

14. Products containing heating elements intended for use under the supervision of an adult in a teaching context

15. Vehicles with combustion engines

16. Toy steam engines

17. Bicycles designed for sport or for travel on the public highway

18. Video toys that can be connected to a video screen, operated at a nominal voltage exceeding 24 volts

19. Babies' dummies

20. Faithful reproductions of real fire arms

21. Fashion jewellery for children

---

[1] With the exception of percussion caps specifically designed for use in toys without prejudice to more stringent provisions already existing in certain Member States.

Annex II
Essential safety requirements for toys

1. General principles

1. In compliance with the requirements of Article 2 of the Directive, the users of toys as well as third parties must be protected against health hazards and risk of physical injury when toys are used as intended or in a forseeable way, bearing in mind the normal behaviour of children. Such risks are those:

(a) which are connected with the design, construction or composition of the toy;

(b) which are inherent in the use of the toy and cannot be completely eliminated by modifying the toy's construction and composition without altering its function or depriving it of its essential properties.

2. (a) The degree of risk present in the use of a toy must be commensurate with the ability of the users, and where appropriate their supervisors, to cope with it. This applies in particular to toys which, by virtue of their functions, dimensions and characteristics, are intended for use by children of under 36 months.

(b) To observe this principle, a minimum age for users of toys and/or the need to ensure that they are used only under adult supervision must be specified where appropriate.

3. Labels on toys and/or their packaging and the instructions for use which accompany them must draw the attention of users or their supervisors fully and effectively to the risks involved in using them and to the ways of avoiding such risks.

II. Particular risks

1. Physical and mechanical properties

(a) Toys and their parts and, in the case of fixed toys, their anchorages, must have the requisite mechanical strength and, where appropriate, stability to withstand the stresses to which they are subjected during use without breaking or becoming liable to distortion at the risk of causing physical injury.

(b) Accessible edges, protrusions, cords, cables and fastening on toys must be so designed and constructed that the risks of physical injury from contact with them are reduced as far as possible.

(c) Toys must be so designed and constructed as to minimise the risk of physical injury which could be caused by the movement of their parts.

(d) Toys, and their component parts, and any detachable parts of toys which are clearly intended for use by children under 36 months must be of such dimensions as to prevent their being swallowed and/or inhaled.

(e) Toys, and their parts and the packaging in which they are contained for retail sale must not present risk of strangulation or suffocation.

(f) Toys intended for use in shallow water which are capable of carrying or supporting a child on the water must be designed and constructed so as to reduce as far as possible, taking into account the recommended use of the toy, any risk of loss of buoyancy of the toy and loss of support afforded to the child.

(g) Toys which it is possible to get inside and which thereby constitute an enclosed space for occupants must have a means of exit which the latter can open easily from the inside.

(h) Toys conferring mobility on their users must, as far as possible, incorporate a braking system which is suited to the type of toy and is commensurate with the kinetic energy developed by it. Such a system must be easy for the user to operate without risk of ejection or physical injury for the user or for third parties.

(i) The form and composition of projectiles and the kinetic energy they may develop when fired from a toy designed for that purpose must be such that, taking into account the nature of the toy, there is no unreasonable risk of physical injury to the user or to third parties.

(j) Toys containing heating elements must be so constructed as to ensure that:
– the maximum temperature of any accessible surfaces does not cause burns when touched,
– liquids and gases contained within toys do not reach temperatures or pressures which are such that their escape from a toy, other than for reasons essential to the proper functioning of the toy,

might cause burns, scalds or other physical injury.

2. Flammability

(a) Toys must not constitute a dangerous flammable element in the child's environment. They must therefore be composed of materials which:
1. do not burn if directly exposed to a flame or spark or other potential seat of fire; or
2. are not readily flammable (the flame goes out as soon as the fire cause disappears); or
3. if they do ignite, burn slowly and present a low rate of spread of the flame; or
4. irrespective of the toy's chemical composition, are treated so as to delay the combustion process.
Such combustible materials must not constitute a risk of ignition for other materials used in the toy.

(b) Toys which, for reasons essential to their functioning, contain dangerous substances or preparations as defined in Council Directive 67/548/EEC (JO 1967 196/1), in particular materials and equipment for chemistry experiments, model assembly, plastic or ceramic moulding, enamelling, photography or similar activities, must not contain, as such, substances or preparations which may become flammable due to the loss of non-flammable volatile components.

(c) Toys must not be explosive or contain elements or substances likely to explode when used as specified in Article 2 (1) of the Directive. This provision does not apply to toy percussion caps, for which reference should be made to point 10 of Annex I and the related footnote.

(d) Toys and, in particular, chemical games and toys, must not contain as such substances or preparations:
– which, when mixed, may explode:
– through chemical reaction, or through heating,
– when mixed with oxidizing substances,
– which contain volatile components which are flammable in air and liable to form flammable or explosive vapour/air mixture.

3. Chemical properties

1. Toys must be so designed and constructed that, when used as specified in Article 2 (1) of the Directive, they do not present health hazards or risks of physical injury by ingestion, inhalation or contact with the skin, mucous tissues or eyes.

   They must in all cases comply with the relevant Community legislation relating to certain categories of products or to the prohibition, restriction of use or labelling of certain dangerous substances and preparations.

2. In particular, for the protection of children's health, bioavailability resulting from the use of toys must not, as an objective, exceed the following levels per day:
   0.2 μg for antimony,
   0.1 μg for arsenic,
   25.0 μg for barium,
   0.6 μg for cadmium,
   0.3 μg for chromium,
   0.7 μg for lead,
   0.5 μg for mercury,
   5.0 μg for selenium,
   or such other values as may be laid down for these or other substances in Community legislation based on scientific evidence.

   The bioavailability of these substances means the soluble extract having toxicological significance.

3. Toys must not contain dangerous substances or preparations within the meaning of Directives 67/548/EEC and 88/379/EEC (OJ 1988 L187/14) in amounts which may harm the health of children using them. At all events it is strictly forbidden to include, in a toy, dangerous substances or preparations if they are intended to be used as such while the toy is being used.

However, where a limited number of substances or preparations are essential to the functioning of certain toys, in particular materials and equipment for chemistry experiments, model assembly, plastic or ceramic moulding, enamelling, photography or similar activities, they are permitted up to a maximum concentration level to be defined for each substance or preparation by mandate to the European Committee for Standardisation (CEN) according to the procedure of the committee set up by Directive 83/189/EEC, provided the permitted substances and preparations comply with the Community classification rules in respect of labelling, without prejudice to point 4 of Annex IV.

4. Electrical properties

(a) Electric toys must not be powered by electricity of a nominal voltage exceeding 24 volts and no part of the toy may exceed 24 volts.

(b) Parts of toys which are connected to, or liable to come into contact with a source of electricity capable of causing electric shock, together with the cables or other conductors through which electricity is conveyed to such parts, must be properly insulated and mechanically protected so as to prevent the risk of such shock.

(c) Electric toys must be so designed and constructed as to ensure that the maximum temperatures reached by all directly accessible surfaces are not such as to cause burns when touched.

5. Hygiene

Toys must be so designed and manufactured as to meet the requirements of hygiene and cleanliness in order to avoid any risk of infection, sickness and contamination.

6. Radioactivity

Toys must not contain radioactive elements or substances in forms or proportions likely to be detrimental to a child's health. Council Directive 80/836/Euratom shall apply (OJ 1980 L246/1).

Annex III

Conditions to be fulfilled by the approved bodies

(Article 9 (1))

The bodies designated by the Member States must fulfil the following minimum conditions:

1. availability of personnel and of the necessary means and equipment;
2. technical competence and professional integrity of personnel;
3. independence, in carrying out the tests, preparing the reports, issuing the certificates and performing the surveillance provided for in this Directive, of staff and technical personnel in relation to all circles, groups or persons directly or indirectly concerned with toys;
4. maintenance of professional secrecy by personnel;
5. subscription of a civil liability insurance unless that liability is covered by the state under national law.

Fulfilment of the conditions under 1 and 2 shall be verified at intervals by the competent authorities of the Member States.

Annex IV
Warnings and indications of precautions to be taken when using toys

(Article 11 (5))

Toys must be accompanied by appropriate clearly legible warnings in order to reduce inherent risks in their use as described in the essential requirements, and specifically:

1. *Toys not intended for children under 36 months*

Toys which might be dangerous for children under 36 months of age shall bear a warning, for example: "Not suitable for children under 36 months" or "Not suitable for children under three years" together with a brief indication, which may also appear in the instructions for use, of the specific risks calling for this restriction.

This provision does not apply to toys which, on account of their function, dimensions, characteristics, properties or other cogent grounds, are manifestly unsuitable for children under 36 months.

2. *Slides, suspended swings and rings, trapezes, ropes and similar toys attached to a crossbeam*

Such toys shall be accompanied by instructions drawing attention to the need to carry out checks and maintenance of the main parts (suspensions, fixings, anchorages, etc) at intervals, and pointing out that, if these checks are not carried out, the toy may cause a fall or overturn.

Instructions must also be given as to correct assembly of the toy, indicating those parts which can present dangers if it is incorrectly assembled.

3. *Functional toys*

Functional toys or their packaging shall bear the marking "Warning: to be used under the direct supervision of an adult".

In addition, these toys shall be accompanied by directions giving working instructions as well as the precautions to be taken by the user, with the warning that failure to take these precautions would expose the user to the hazards – to be specified – normally associated with the appliance or product of which the toy is a scale model or an imitation. It will also be indicated that the toy must be kept out of the reach of very young children.

"Functional toys" means toys which are used in the same way as, and are often scale models of, appliances or installations intended for adults.

4. *Toys containing inherently dangerous substances or preparations. Chemical toys*

(a) Without prejudice to the application of the provisions laid down in Community Directives on the classification, packaging and labelling of dangerous substances or preparations, the instructions for use of toys containing inherently dangerous substances or preparations shall bear a warning of the dangerous nature of these substances or preparations and an indication of the precautions to be taken by the user in order to avoid hazards associated with them, which shall be specified concisely according to the type of toy. The first aid to be given in the event of serious accidents resulting from the use of this type of toy shall also be mentioned. It shall also be stated that the toys must be kept out of reach of very young children.

(b) In addition to the instructions provided for in (a), chemical toys

shall bear the following marking on their packaging:
"Warning: for children over[1] years of age only. For use under adult supervision".

In particular, the following are regarded as chemical toys: chemistry sets, plastic embedding sets, miniature workshops for ceramics, enamelling or photography and similar toys.

5. *Skates and skateboards for children*

If these products are offered for sale as toys they shall bear the marking:
"Warning: protective equipment should be worn."

[1] Age to be decided by the manufacturer.

Moreover, the instructions for use shall contain a reminder that the toy must be used with caution, since it requires great skill, so as to avoid falls or collisions causing injury to the user and third parties. Some indication shall also be given as to recommended protective equipment (helmets, gloves, knee-pads, elbow-pads, etc).

6. *Toys intended for use in water*

The toys intended for use in water defined in Section II.1 (f) of the Annex II shall contain the warning in accordance with CEN's brief to adapt standard EN/71, parts 1 and 2:
"Warning! Only to be used in water in which the child is within its depth and under supervision".

Annex V

CE conformity markings

– The CE conformity marking shall consist of the initials "CE" taking the following form:

– If the CE marking is reduced or enlarged the proportions given in the above graduated drawing must be respected.

– The various components of the CE marking must have substantially the same vertical dimension, which may not be less than 5 mm.

COUNCIL DIRECTIVE 89/392/EEC

of 14 June 1989

on the approximation of the laws of the Member States relating to machinery*†

THE COUNCIL OF THE EUROPEAN COMMUNITIES,

Having regard to the Treaty establishing the European Economic Community, and in particular Article 100A thereof,

Having regard to the proposal from the Commission,[1]

In cooperation with the European Parliament,[2]

Having regard to the opinion of the Economic and Social Committee,[3]

Whereas Member States are responsible for ensuring the health and safety on their territory of their people and, where appropriate, of domestic animals and goods and, in particular, of workers notably in relation to the risks arising out of the use of machinery;

Whereas, in the Member States, the legislative systems regarding accident prevention are very different; whereas the relevant compulsory provisions, frequently supplemented by *de facto* mandatory technical specifications and/or voluntary standards, do not necessarily lead to different levels of health and safety, but nevertheless, owing to their disparities, constitute barriers to trade within the Community; whereas, furthermore, conformity certification and national certification systems for machinery differ considerably;

Whereas the maintenance or improvement of the level of safety attained by the Member States constitutes one of the essential aims of this Directive and of the principle of safety as defined by the essential requirements;

Whereas existing national health and safety provisions providing protection against the risks caused by machinery must be approximated to ensure free movement of machinery without lowering existing justified levels of protection in the Member States; whereas the provisions of this Directive concerning the design and construction of machinery, essential for a safer working environment shall be accompanied by specific provisions concerning the prevention of certain risks to which workers can be exposed at work, as well as by provisions based on the organisation of safety of workers in the working environment;

Whereas the machinery sector is an important part of the engineering industry and is one of the industrial mainstays of the Community economy;

Whereas paragraphs 65 and 68 of the White Paper on the completion of the internal market, approved by the European Council in June 1985, provide for a new approach to legislative harmonisation;

Whereas the social cost of the large number of accidents caused directly by the use of machinery can be reduced by inherently safe design and construction of machinery and by proper installations and maintenance;

Whereas the field of application of this Directive must be based on a general

---

*As amended to cover mobile machinery and light equipment. See Directive 91/368/EEC (OJ 1991 L198/16).

†OJ 1989 L183/9.

[1] OJ No C 29, 3. 2. 1988, p. 1, and OJ No C 214, 16. 8. 1988, p. 23.

[2] OJ No C 326, 19. 12. 1988, p. 143, and OJ No C 158, 26. 6. 1989.

[3] OJ No C337, 31. 12. 1988, p. 30.

definition of the term "machinery" so as to allow the technical development of products; whereas the development of "complex installations" and the risks they involve are of an equivalent nature and their express inclusion in the Directive is therefore justified;

Whereas specific Directives containing design and construction provisions for certain categories of machinery are now envisaged; whereas the very broad scope of this Directive must be limited in relation to these Directives and also existing Directives where they contain design and construction provisions;

Whereas Community law, in its present form, provides – by way of derogation from one of the fundamental rules of the Community, namely the free movement of goods – that obstacles to movement within the Community resulting from disparities in national legislation relating to the marketing of products must be accepted in so far as the provisions concerned can be recognised as being necessary to satisfy imperative requirements; whereas, therefore, the harmonisation of laws in this case must be limited only to those requirements necessary to satisfy the imperative and essential health and safety requirements relating to machinery; whereas these requirements must replace the relevant national provisions because they are essential;

Whereas the essential health and safety requirements must be observed in order to ensure that machinery is safe; whereas these requirements must be applied with discernment to take account of the state of the art at the time of construction and of technical and economic requirements;

Whereas the putting into service of machinery within the meaning of this Directive can relate only to the use of the machinery itself as intended by the manufacturer; whereas this does not preclude the laying-down of conditions of use external to the machinery, provided that it is not thereby modified in a way not specified in this Directive;

Whereas, for trade fairs, exhibitions, etc., it must be possible to exhibit machinery which does not conform to this Directive; whereas, however, interested parties should be properly informed that the machinery does not conform and cannot be purchased in that condition;

Whereas, therefore, this Directive defines only the essential health and safety requirements of general application, supplemented by a number of more specific requirements for certain categories of machinery; whereas, in order to help manufacturers to prove conformity to these essential requirements and in order to allow inspection for conformity to the essential requirements, it is desirable to have standards harmonised at European level for the prevention of risks arising out of the design and construction of machinery; whereas these standards harmonised at European level are drawn up by private-law bodies and must retain their non-binding status; whereas for this purpose the European Committee for Standardisation (CEN) and the European Committee for Electrotechnical Standardisation (CENELEC) are the bodies recognised as competent to adopt harmonised standards in accordance with the general guidelines for cooperation between the Commission and these two bodies signed on 13 November 1984; whereas, within the meaning of this Directive, a harmonised standard is a technical specification (European standard or harmonisation document) adopted by either or both of these bodies, on the basis of a remit from the Commission in accordance with the provisions of Council Directive 83/189/EEC of 28 March 1983 (OJ 1983 L109/8) laying down a procedure for the provision of information in the field of technical standards and regulations, as last amended by Directive 88/182/EEC (OJ 1988 L81/75), and on the basis of general guidelines referred to above;

Whereas the legislative framework needs to be improved in order to ensure an effective and appropriate contribution by employers and employees to the standardisation process; whereas such improvement should be completed at the latest by the time this Directive is implemented;

Whereas, as is currently the practice in Member States, manufacturers should retain the responsibility for certifying the conformity of their machinery to the relevant essential requirements; whereas conformity to harmonised standards creates a presumption of conformity to the relevant essential requirements; whereas it is left to the sole discretion of the manufacturer, where he feels the need, to have his products examined and certified by a third party;

Whereas, for certain types of machinery having a higher risk factor, a stricter certification procedure is desirable; whereas the EC type-examination procedure adopted may result in an EC declaration being given by the manufacturer without any stricter requirement such as a guarantee of quality, EC verification or EC supervision;

Whereas it is essential that, before issuing an EC declaration of conformity, the manufacturer or his authorised representative established in the Community should provide a technical construction file; whereas it is not, however, essential that all documentation be permanently available in a material manner but it must be made available on demand; whereas it need not include detailed plans of the subassemblies used in manufacturing the machines, unless knowledge of these is indispensable in order to ascertain conformity with essential safety requirements;

Whereas it is necessary not only to ensure the free movement and putting into service of machinery bearing the CE marking and having an EC conformity certificate but also to ensure free movement of machinery not bearing the CE marking where it is to be incorporated into other machinery or assembled with other machinery to form a complex installation;

Whereas the Member States' responsibility for safety, health and the other aspects covered by the essential requirements on their territory must be recognised in a safeguard clause providing for adequate Community protection procedures;

Whereas the addresses of any decision taken under this Directive must be informed on the reasons for such a decision and the legal remedies open to them;

Whereas the measures aimed at the gradual establishment of the internal market must be adopted by 31 December 1992; whereas the internal market consists of an area without internal frontiers within which the free movement of goods, persons, services and capital is guaranteed,

HAS ADOPTED THIS DIRECTIVE:

Chapter 1
Scope, placing on the market and freedom of movement

Article 1

1. This Directive applies to machinery and lays down the essential health and safety requirements therefore, as defined in Annex 1.

It shall also apply to safety components placed on the market.

2. For the purposes of this Directive, "machinery" means an assembly of linked parts or components, at least one of which moves, with the appropriate actuators, control and power circuits, etc, joined together for a specific application, in particular for the processing, treatment, moving or packaging of a material.

The term "machinery" also covers an assembly of machines which, in order to achieve the same end, are arranged and controlled so that they function as an integral whole.

"Machinery" also means interchangeable equipment modifying the function of a machine, which is placed on the market for the purpose of being assembled with a machine or a series of different machines or with a tractor by the operator himself in so far as this equipment is not a spare part or a tool;

For the purposes of this Directive, *safety components* means a component, provided that it is not interchangeable equipment, which the manufacturer of his authorised representative established in the Community places on the market to fulfil a safety function when in use and the failure or malfunctioning of which endagers the safety or health of exposed persons.

3. The following are excluded from the scope of this Directive:
– machinery whose only power source is directly applied manual effort, unless it is a machine used for lifting or lowering loads,
– machinery for medical use used in direct contact with patients,
– special equipment for use in fairgrounds and/or amusement parks,
– steam boilers, tanks and pressure vessels,
– machinery specially designed or put into service for nuclear purposes which, in the event of failure, may result in an emission of radioactivity,
– radioactive sources forming part of a machine,
– firearms,
– storage tanks and pipelines for petrol, diesel fuel, inflammable liquids and dangerous substances.
– means of transport, *i.e.* vehicles and their trailers intended solely for transporting passengers by air or on road, rail or water networks, as well as means of transport in so far as such means are designed for transporting goods by air, on public road or rail networks or on water. Vehicles used in the mineral extraction industry shall not be excluded,
– seagoing vessels and mobile offshore units together with equipment on board such vessels or units,
– cableways, including funicular railways, for the public or private transportation of persons,
– agricultural and forestry tractors, as defined in Article 1 (1) of Council Directive 74/150/EEC of 4 March 1974 on the approximation of the laws of the Member States relating to the type-approval of wheeled agricultural or forestry tractors,[4] as latest amended by Directive 88/297/EEC,[5]
– machines specially designed and constructed for military or police purposes.
– lifts which permanently serve specific levels of buildings and constructions, having a car moving between guides which are rigid and inclined at an angle of more than 15 degrees to the horizontal and designed for the transport of:
  – persons,
  – persons or goods,
  – goods alone if the car is accessible, that is to say, a person may enter it without difficulty, and fitted with controls situated inside the car or within reach of a person inside,
– means of transport of persons using rack and pinion rail mounted vehicles,
– mine winding gear,
– theatre elevators,
– construction site hoists intended for lifting persons and goods.

4. Where, for machinery or safety components, the risks referred to in this Directive are wholly or partly covered by special Community Directives, this Directive shall not apply, or shall cease to apply, in the case of such machinery or safety components and of such risks on the implementation of these specific Directives.

5. Where, for machinery, the risks are mainly of electrical origin, such machinery shall be covered exclusively by Council Directive 73/23/EEC of 19 February 1973 on the harmonization of the laws of the Member States relating to electrical equipment designed for use within certain voltage limits.

Article 2

1. Member States shall take all appropriate measures to ensure that machi-

---

[4] OJ No L 109, 26. 4. 1983, p. 8.
[5] OJ No L 81, 26. 3. 1988, p. 75.

nery or safety components covered by this Directive may be placed on the market and put into service only if it does not endanger the health or safety of persons and, where appropriate, domestic animals or property, when properly installed and maintained and used for its intended purpose.
2. This Directive shall not affect Member States' entitlement to lay down, in due observance of the Treaty, such requirements as they may deem necessary to ensure that persons and in particular workers are protected when using the machines in question, provided that this does not mean that the machinery is modified in a way not specified in the Directive.
3. At trade fairs, exhibitions, demonstrations, etc., Member States shall not prevent the showing of machinery or safety components which do not conform to the provisions of this Directive, provided that a visible sign clearly indicates that such machinery or safety components do not conform and that they are not for sale until they have been brought into conformity by the manufacturer or his authorised representative established in the Community. During demonstrations, adequate safety measures shall be taken to ensure the protection of persons.

Article 3

Machinery and safety components covered by this Directive shall satisfy the essential health and safety requirements set out in Annex I.

Article 4

1. Member States shall not prohibit, restrict or impede the placing on the market and putting into service in their territory of machinery and safety components which comply with the provisions of this Directive.
2. Member States shall not prohibit, restrict or impede the placing on the market of machinery where the manufacturer or his authorised representative established in the Community declares in accordance with Annex II.B that it is intended to be incorporated into machinery or assembled with other machinery to constitute machinery covered by this Directive except where it can function independently. "Interchangeable equipment", within the meaning of the third subparagraph of Article 1 (2), shall be regarded as machinery and accordingly must in all cases bear the CE marking and be accompanied by the EC declaration of conformity referred to in Annex II (A).
3. Member States may not prohibit, restrict or impede the placing on the market of safety components as defined in Article 1 (2) where they are accompanied by the EC declaration of conformity by the manufacturer or his authorised representative established in the Community as referred to in Annex II, point C.

Article 5

1. Member States shall regard the following as conforming to all the provisions of this Directive, including the procedures for checking the conformity provided for in Chapter II:
– machinery bearing the CE marking and accompanied by the EC declaration of conformity referred to in Annex II.A.,
– safety components accompanied by the EC declaration of conformity referred to in Annex II.C.

In the absence of harmonised standards, Member States shall take any steps they deem necessary to bring to the attention of the parties concerned the existing national technical standards and specifications which are regarded as important or relevant to the proper implementation of the essential safety and health requirements in Annex I.
2. Where a national standard transposing a harmonised standard, the reference for which has been published in the *Official Journal of the European Communities*, covers one or more of the essential safety requirements, machinery or safety components constructed in accordance with this standard shall be presumed to comply with the relevant essential requirements.

Member States shall publish the references of national standards transposing harmonised standards.

3. Member States shall ensure that appropriate measures are taken to enable the social partners to have an influence at national level on the process of preparing and monitoring the harmonised standards.

Article 6

1. Where a Member State or the Commission considers that the harmonised standards referred to in Article 5 (2) do not entirely satisfy the essential requirements referred to in Article 3, the Commission or the Member State concerned shall bring the matter before the Committee set up under Directive 83/189/EEC, giving the reasons therefore. The Committee shall deliver an opinion without delay.

Upon receipt of the Committee's opinion, the Commission shall inform the Member States whether or not it is necessary to withdraw those standards from the published information referred to in Article 5 (2).

2. A standing committee shall be set up, consisting of representatives appointed by the Member States and chaired by a representative of the Commission.

The standing committee shall draw up its own rules of procedure.

Any matter relating to the implementation and practical application of this Directive may be brought before the standing committee, in accordance with the following procedure:

The representative of the Commission shall submit to the committee a draft of the measures to be taken. The committee shall deliver its opinion on the draft, within a time limit which the chairman may lay down according to the urgency of the matter, if necessary by taking a vote.

The opinion shall be recorded in the minutes; in addition, each Member State shall have the right to ask to have its position recorded in the minutes.

The Commission shall take the utmost account of the opinion delivered by the committee. It shall inform the committee of the manner in which its opinion has been taken into account.

Article 7

1. Where a Member State ascertains that

– machinery bearing the EC marking, or

– safety components accompanied by the EC declaration of conformity,

used in accordance with their intended purpose is liable to endanger the safety of persons, and, where appropriate, domestic animals or property, it shall take all appropriate measures to withdraw such machinery or safety components from the market, to prohibit the placing on the market, putting into service or use thereof, or to restrict free movement thereof.

The Member State shall immediately inform the Commission of any such measure, indicating the reasons for its decision and, in particular, whether non-conformity is due to:

(a) failure to satisfy the essential requirements referred to in Article 3;
(b) incorrect application of the standards referred to in Article 5 (2);
(c) shortcomings in the standards referred to in Article 5 (2) themselves.

2. The Commission shall enter into consultation with the parties concerned without delay. Where the Commission considers, after this consultation, that the measure is justified, it shall immediately so inform the Member State which took the initiative and the other Member States. Where the Commission considers, after this consultation, that the action is unjustified, it shall immediately so inform the Member State which took the initiative and the manufacturer or his authorised representative established within the Community. Where the decision referred to in paragraph 1 is based on a shortcoming in the standards, and where the Member State at the origin of the decision maintains its position, the Commission shall immediately

inform the Committee in order to initiate the procedures referred to in Article 6 (1).

3. Where

– machinery which does not comply bears the CE marking,

– a safety component which does not comply is accompanied by an EC declaration of conformity,

the competent Member States shall take appropriate action against whomsoever has affixed the marking or drawn up the declaration and shall so inform the Commission and other Member States.

4. The Commission shall ensure that the Member States are kept informed of the progress and outcome of this procedure.

Chapter II
Certification procedure

Article 8

1. The manufacturer or his authorised representative established in the Community must, in order to certify that machinery and safety components are in conformity with this Directive, draw up for all machinery or safety component manufactured an EC declaration of conformity based on the model given in Annex II, A or C as appropriate.

In addition, for machinery alone, the manufacturer or his authorised representatives established in the Community must affix to the machine the CE marking referred to in Article 10.

2. Before placing on the market, the manufaturer, or his authorised representative established in the Community, shall:

(a) if the machinery is not referred to in Annex IV, draw up the file provided for in Annex V;

(b) if the machinery is referred to in Annex IV and its manufacturer does not comply, or only partly complies, with the standards referred to in Article 5 (2) or if there are no such standards, submit an example of the machinery for the EC type-examination referred to in Annex VI;

(c) if the machinery is referred to in Annex IV and is manufactured in accordance with the standards referred to in Article 5 (2):
– either draw up the file referred to in Annex VI and forward it to a notified body, which will acknowledge receipt of the file as soon as possible and keep it,
– submit the file referred to in Annex VI to the notified body, which will simply verify that the standards referred to in Article 5 (2) have been correctly applied and will draw up a certificate of adequacy for the file,
– or submit the example of the machinery for the EC type-examination referred to in Annex VI.

3. Where the first indent of paragraph 2 (c) applies, the provisions of the first sentence of paragraph 5 and paragraph 7 of Annex VI shall also apply.

Where the second indent of 2 (c) applies, the provisions of paragraphs 5, 6 and 7 of Annex VI shall also apply.

4. Where paragraph 2 (a) and the first and second indents of paragraph 2 (c) apply, the EC declaration of conformity shall solely state conformity with the essential requirements of the Directive.

Where paragraph 2 (b) and the third indent of para 2 (c) apply, the EC declaration of conformity shall state conformity with the example that underwent EC type-examination.

4. Safety components shall be subject to the certification procedures applicable to machinery pursuant to paragraphs 2, 3 and 4. Furthermore, during EC type-examination, the notified body shall verify the suitability of the safety component for fulfilling the safety functions declared by the manufacturer.

5. (a) Where the machinery is subject to other Directives concerning other aspects and which also provide for the affixing of the CE marking, the latter shall indicate that the machinery is also presumed to conform to the provisions of those other Directives.

(b) However, where one or more of these Directives allow the manufacturer, during a transitional period, to choose which arrangements to apply, the CE marking shall indicate conformity only to the Directives applied by the manufacturer. In this case, particulars of the Directives applied, as published in the *Official Journal of the European Communities*, must be given in the documents, notices or instructions required by the Directives and accompanying such machinery.

6. Where neither the manufacturer nor his authorised representative established in the Community fulfils the obligations of the preceding paragraphs, these obligations shall fall to any person placing the machinery on the market in the Community. The same obligations shall apply to any person assembling machinery or parts thereof or safety components of various origins or constructing machinery for his own use.

7. The obligations laid down in paragraph 6 shall not apply to persons who assemble with a machine or tractor interchangeable equipment as provided for in Article 1, provided that the parts are compatible and each of the constituent parts of the assembled machine bears the CE marking and is accompanied by the EC declaration of conformity.

Article 9

1. Member States shall notify the Commission and the other Member States of the approved bodies which they have appointed to carry out the procedures referred to in Article 8 together with the specific tasks which these bodies have been appointed to carry out and the identification numbers assigned to them beforehand by the Commission.

The Commission shall publish in the *Official Journal of the European Communities* a list of the notified bodies and their identification numbers and the tasks for which they have been notified. The Commission shall ensure that this list is kept up to date.

2. Member States shall apply the criteria laid down in Annex VII in assessing the bodies to be indicated in such notification. Bodies meeting the assessment criteria laid down in the relevant harmonised standards shall be presumed to fulfil those criteria.

3. A Member State which has approved a body must withdraw its notification if it finds that the body no longer meets the criteria referred to in Annex VII. It shall immediately inform the Commission and the other Member States accordingly.

Chapter III
CE marking

Article 10

1. The CE conformity marking shall consist of the initials "CE". The form of the marking to be used is shown in Annex III.

Annex III shows the model to be used.

2. The CE marking shall be affixed to machinery distinctly and visibly in accordance with point 1.7.3 of Annex I.

3. The affixing of markings or inscriptions on the machinery which are likely to deceive third parties as to the meaning and form of the CE marking shall be prohibited. Any other marking may be affixed to the machinery provided that the visibility and legibility of the CE marking is not thereby reduced.

4. Without prejudice to Article 7:

(a) where a Member State establishes that the CE marking has been affixed unduly, the manufacturer or his authorised representative established within the Community shall be obliged to make the product conform as regards the provisions concerning the CE marking and to end the infringement under the conditions imposed by the Member State;

(b) where non-conformity continues, the Member State must take all appropriate measures to restrict or prohibit the placing on the market of the product in question or to ensure that it is withdrawn from the market in accordance with the procedures laid down in Article 7.

Chapter IV
Final provisions

Article 11

Any decision taken pursuant to this Directive which restricts the placing on the market and putting into service of machinery or safety component shall state the exact grounds on which it is based. Such a decision shall be notified as soon as possible to the party concerned, who shall at the same time be informed of the legal remedies available to him under the laws in force in the Member State concerned and of the time limits to which such remedies are subject.

Article 12

The Commission will take the necessary steps to have information on all the relevant decisions relating to the management of this Directive made available.

Article 13

1. Before 1 January 1992 Member States shall adopt and publish the laws, regulations and administrative provisions necessary in order to comply with this Directive. They shall forthwith inform the Commission thereof.

When Member States adopt these measures, they shall contain a reference to this Directive or shall be accompanied by such reference on the occasion of their official publication. The methods of making such a reference shall be laid down by the Member States.

The Member States shall apply the measures in question with effect from 1 January 1993, except as regards the equipment referred to in Directives 86/295/EEC, 86/296/EEC and 86/663/EEC, for which these measures shall apply from 1 July 1995.

2. Furthermore, Member States shall allow, for the period until 31 December 1994, except as regards the equipment referred to in Directives 86/295/EEC, 86/296/EEC and 86/663/EEC, for which this period shall end on 31 December 1995, the placing on the market and putting into service of machinery in conformity with the national regulations in force in their territory on 31 December 1992.

Directives 86/295/EEC, 86/296/EEC and 86/663/EEC shall not impede implementation of paragraph 1 as from 1 July 1995.

3. Member States shall communicate to the Commission the texts of the provisions of national law which they adopt in the field governed by this Directive.

4. The Commission shall, before 1 January 1994, examine the progress made in the standardisation work relating to this Directive and propose any appropriate measures.

Article 14

This Directive is addressed to the Member States.

Done at Luxembourg, 14 June 1989.

For the Council
The President
P. SOLBES

Extract from Directive 91/368

Article 2

The following are hereby repealed as from 31 December 1994:

– Articles 2 and 3 of Council Directive 73/631/EEC of 19 November 1973 (OJ 1973 L335/51) on the approximation of the laws, regulations and administrative provisions of the Member States relating to the certification and marking of wire ropes, chains and hooks, as lasted amended by Directive 76/434/EEC (OJ 1976 L122/20),

– Commission Directive 76/434/EEC of 13 April 1976 adapting to technical progress the Council Directive of 19 November 1973 on the approximation of the laws of the Member States relating to the certification and marking of wire ropes, chains and hooks.

The following are hereby repealed as from 31 December 1995:
– Council Directive 86/295/EEC of 26 May 1986 on the approximation of the laws of the Member States relating to roll-over protective structures (ROPS) for certain construction plant,
– Council Directive 86/296/EEC of 26 May 1986 on the approximation of the laws of the Member States relating to falling-object protective structures (FOPS) for certain construction plant,
– Council Directive 86/663/EEC of 22 December 1986 on the approximation of the laws of the Member States relating to self-propelled industrial trucks, as last amended by Directive 89/240/EEC.

Annex I

Essential health and safety requirements relating to the design and construction of machinery and safety components

For the purposes of this Annex, "machinery" means either "machinery" or "safety component" as defined in Article 1 (2).

Preliminary observations

1. The obligations laid down by the essential health and safety requirements apply only when the corresponding hazard exists for the machinery in question when it is used under the conditions foreseen by the manufacturer. In any event, requirements 1.1.2, 1.7.3 and 1.7.4 apply to all machinery covered by this Directive.
2. The essential health and safety requirements laid down in this Directive are mandatory. However, taking into account the state of the art, it may not be possible to meet the objectives set by them. In this case, the machinery must as far as possible be designed and constructed with the purpose of approaching those objectives.
3. The essential health and safety requirements have been grouped according to the hazards which they cover.

Machinery presents a series of hazards which may be indicated under more than one heading in this Annex.

The manufacturer is under an obligation to assess the hazards in order to identify all those which apply to his machine; he must then design and construct it taking account of his assessment.

1. Essential health and safety requirements

1.1. General remarks

1.1.1. *Definitions*

For the purpose of this Directive
1. "danger zone" means any zone within and/or around machinery in which an exposed person is subject to a risk to his health or safety;
2. "exposed person" means any person wholly or partially in a danger zone;
3. "operator" means the person or persons given the task of installing, operating, adjusting, maintaining, cleaning, repairing or transporting machinery.

1.1.2. *Principles of safety integration*

(a) Machinery must be so constructed that it is fitted for its function, and can be adjusted and maintained without putting persons at risk when these operations are carried out under the conditions foreseen by the manufacturer.

The aim of measures taken must be to eliminate any risk of accident throughout the foreseeable lifetime of the machinery, including the phases of assembly and dismantling, even where risks of accident arise from foreseeable abnormal situations.

(b) In selecting the most appropriate methods, the manufacturer must apply the following principles, in the order given:
– eliminate or reduce risks as far as possible (inherently safe machinery design and construction),
– take the necessary protection measures in relation to risks that cannot be eliminated,
– inform users of the residual risks due to any shortcomings of the protection measures adopted, indicate whether any particular training is required and specify any need to provide personal protection equipment.

(c) When designing and constructing machinery, and when drafting the instructions, the manufacturer must envisage not only the normal use of the machinery but also uses which could reasonably be expected.
The machinery must be designed to prevent abnormal use if such use would engender a risk. In other cases the instructions must draw the user's attention to ways – which experience has shown might occur – in which the machinery should not be used.

(d) Under the intended conditions of use, the discomfort, fatigue and psychological stress faced by the operator must be reduced to the minimum possible taking ergonomic principles into account.

(e) When designing and constructing machinery, the manufacturer must take account of the constraints to which the operator is subject as a result of the necessary or foreseeable use of personal protection equipment (such as footwear, gloves, etc).

(f) Machinery must be supplied with all the essential special equipment and accessories to enable it to be adjusted, maintained and used without risk.

1.1.3. *Materials and products*

The materials used to construct machinery or products used and created during its use must not endanger exposed persons' safety or health.

In particular, where fluids are used, machinery must be designed and constructed for use without risks due to filling, use, recovery or draining.

1.1.4. *Lighting*

The manufacturer must supply integral lighting suitable for the operations concerned where its lack is likely to cause a risk despite ambient lighting of normal intensity.

The manufacturer must ensure that there is no area of shadow likely to cause nuisance, that there is no irritating dazzle and that there are no dangerous stroboscopic effects due to the lighting provided by the manufacturer.

Internal parts requiring frequent inspection, and adjustment and maintenance areas, must be provided with appropriate lighting.

1.1.5. *Design of machinery to facilitate its handling*

Machinery or each component part thereof must:
– be capable of being handled safely,
– be packaged or designed so that it can be stored safely and without damage (*e.g.* adequate stability, special supports, etc).

Where the weight, size or shape of machinery or its various component parts prevents them from being moved by hand, the machinery or each component part must:
– either be fitted with attachments for lifting gear, or
– be designed so that it can be fitted with such attachments (*e.g.* threaded holes), or
– be shaped in such a way that standard lifting gear can easily be attached.

Where machinery or one of its component parts is to be moved by hand, it must:
– either be easily movable, or

– be equipped for picking up (*e.g.* hand-grips, etc.) and moving in complete safety.
Special arrangements must be made for the handling of tools and/or machinery parts, even if lightweight, which could be dangerous (shape, material, etc).

1.2. Controls

1.2.1. *Safety and reliability of control systems*

Control systems must be designed and constructed so that they are safe and reliable, in a way that will prevent a dangerous situation arising. Above all they must be designed and constructed in such a way that:
– they can withstand the rigours of normal use and external factors,
– errors in logic do not lead to dangerous situations.

1.2.2. *Control devices*

Control devices must be:
– clearly visible and identifiable and appropriately marked where necessary,
– positioned for safe operation without hesitation or loss of time, and without ambiguity,
– designed so that the movement of the control is consistent with its effect,
– located outside the danger zones, except for certain controls where necessary, such as emergency stop, console for training of robots,
– positioned so that their operation cannot cause additional risk,
– designed or protected so that the desired effect, where a risk is involved, cannot occur without an intentional operation,
– made so as to withstand forseeable strain; particular attention must be paid to emergency stop devices liable to be subjected to considerable strain.
Where a control is designed and constructed to perform several different actions, namely where there is no one-to-one correspondence (*e.g.* keyboards, etc), the action to be performed must be clearly displayed and subject to confirmation where necessary.

Controls must be so arranged that their layout, travel and resistance to operation are compatible with the action to be performed, taking account of ergonomic principles. Constraints due to the necessary or foreseeable use of personal protection equipment (such as footwear, gloves, etc) must be taken into account.

Machinery must be fitted with indicators (dials, signals, etc) as required for safe operation. The operator must be able to read them from the control position.

From the main control position the operator must be able to ensure that there are no exposed persons in the danger zones.

If this is impossible, the control system must be designed and constructed so that an acoustic and/or visual warning signal is given whenever the machinery is about to start. The exposed person must have the time and the means to take rapid action to prevent the machinery starting up.

1.2.3. *Starting*

It must be possible to start machinery only by voluntary actuation of a control provided for the purpose.

The same requirement applies:
– when restarting the machinery after a stoppage, whatever the cause,
– when effecting a significant change in the operating conditions (*e.g.* speed, pressure, etc),
unless such restarting or change in operating conditions is without risk to exposed persons.

This essential requirement does not apply to the restarting of the machinery or to the change in operating conditions resulting from the normal sequence of an automatic cycle.

Where machinery has several starting controls and the operators can therefore put each other in danger, additional devices (*e.g.* enabling devices or selectors allowing only one part of the starting mechanism to be actuated at any one time) must be fitted to rule out such risks.

It must be possible for automated plant functioning in automatic mode to be restarted easily after a stoppage once the safety conditions have been fulfilled.

1.2.4. *Stopping device*

Normal stopping. Each machine must be fitted with a control whereby the machine can be brought safely to a complete stop.

Each workstation must be fitted with a control to stop some or all of the moving parts of the machinery, depending on the type of hazard, so that the machinery is rendered safe. The machinery's stop control must have priority over the start controls.

Once the machinery or its dangerous parts have stopped, the energy supply to the actuators concerned must be cut off.

Emergency stop. Each machine must be fitted with one or more emergency stop devices to enable actual or impending danger to be averted. The following exceptions apply:

- machines in which an emergency stop device would not lessen the risk, either because it would not reduce the stopping time or because it would not enable the special measures required to deal with the risk to be taken,
- hand-held portable machines and hand-guided machines.

This device must:

- have clearly identifiable, clearly visible and quickly accessible controls,
- stop the dangerous process as quickly as possible, without creating additional hazards,
- where necessary, trigger or permit the triggering of certain safeguard movements.

Once active operation of the emergency stop control has ceased following a stop command, that command must be sustained by engagement of the emergency stop device until that engagement is specially overridden; it must not be possible to engage the device without triggering a stop command; it must be possible to disengage the device only by an appropriate operation, and disengaging the device must not restart the machinery but only permit restarting.

Complex installations. In the case of machinery or parts of machinery designed to work together, the manufacturer must so design and construct the machinery that the stop controls, including the emergency stop, can stop not only the machinery itself but also all equipment upstream and/or downstream if its continued operation can be dangerous.

1.2.5. *Mode selection*

The control mode selected must override all other control systems with the exception of the emergency stop.

If machinery has been designed and built to allow for its use in several control or operating modes presenting different safety levels (*e.g.* to allow for adjustment, maintenance, inspection, etc), it must be fitted with a mode selector which can be locked in each position. Each position of the selector must correspond to a single operating or control mode.

The selector may be replaced by another selection method which restricts the use of certain functions of the machinery to certain categories of operator (*e.g.* access codes for certain numerically controlled functions, etc).

If, for certain operations, the machinery must be able to operate with its protection devices neutralised, the mode selector must simultaneously:

- disable the automatic control mode,
- permit movements only by controls requiring sustained action,
- permit the operation of dangerous moving parts only in enhanced safety conditions (*e.g.* reduced speed, reduced power, step-by-step, or other adequate provision) while preventing hazards from linked sequences,
- prevent any movement liable to pose a danger by acting voluntarily or involuntarily on the machine's internal sensors.

In addition, the operator must be

able to control operation of the parts he is working on at the adjustment point.

1.2.6. *Failure of the power supply*

The interruption, re-establishment after an interruption or fluctuation in whatever manner of the power supply to the machinery must not lead to a dangerous situation.

In particular:
- the machinery must not start unexpectedly,
- the machinery must not be prevented from stopping if the command has already been given,
- no moving part of the machinery or piece held by the machinery must fall or be ejected,
- automatic or manual stopping of the moving parts whatever they may be must be unimpeded,
- the protection devices must remain fully effective.

1.2.7. *Failure of the control circuit*

A fault in the control circuit logic, or failure of or damage to the control circuit must not lead to dangerous situations.

In particular:
- the machinery must not start unexpectedly,
- the machinery must not be prevented from stopping if the command has already been given,
- no moving part of the machinery or piece held by the machinery must fall or be ejected,
- automatic or manual stopping of the moving parts whatever they may be must be unimpeded,
- the protection devices must remain fully effective.

1.2.8. *Software*

Interactive software between the operator and the command or control system of a machine must be user-friendly.

1.3. Protection against mechanical hazards

1.3.1. *Stability*

Machinery, components and fittings thereof must be so designed and constructed that they are stable enough, under the foreseen operating conditions (if necessary taking climatic conditions into account) for use without risk of overturning, falling or unexpected movement.

If the shape of the machinery itself or its intended installation does not offer sufficient stability, appropriate means of anchorage must be incorporated and indicated in the instructions.

1.3.2. *Risk of break-up during operation*

The various parts of machinery and their linkages must be able to withstand the stresses to which they are subject when used as foreseen by the manufacturer.

The durability of the materials used must be adequate for the nature of the work place foreseen by the manufacturer, in particular as regards the phenomena of fatigue, ageing, corrosion and abrasion.

The manufacturer must indicate in the instructions the type and frequency of inspection and maintenance required for safety reasons. He must, where appropriate, indicate the parts subject to wear and the criteria for replacement.

Where a risk of rupture or disintegration remains despite the measures taken (*e.g.* as with grinding wheels) the moving parts must be mounted and positioned in such a way that in case of rupture their fragments will be contained.

Both rigid and flexible pipes carrying fluids, particularly those under high pressure, must be able to withstand the foreseen internal and external stresses and must be firmly attached and/or protected against all manner of external stresses and strains; precautions must be taken to ensure that no risk is posed by a rupture (sudden movement, high-pressure jets, etc).

Where the material to be processed is fed to the tool automatically, the following conditions must be fulfilled to avoid risks to the persons exposed (*e.g.* tool breakage):

– when the workpiece comes into contact with the tool the latter must have attained its normal working conditions,

– when the tool starts and/or stops (intentionally or accidentally) the feed movement and the tool movement must be coordinated.

1.3.3. *Risks due to falling or ejected objects*

Precautions must be taken to prevent risks from falling or ejected objects (*e.g.* workpieces, tools, cuttings, fragments, waste, etc).

1.3.4. *Risks due to surfaces, edges or angles*

In so far as their purpose allows, accessible parts of the machinery must have no sharp edges, no sharp angles, and no rough surfaces likely to cause injury.

1.3.5. *Risks related to combined machinery*

Where the machinery is intended to carry out several different operations with the manual removal of the piece between each operation (combined machinery), it must be designed and constructed in such a way as to enable each element to be used separately without the other elements constituting a danger or risk for the exposed person.

For this purpose, it must be possible to start and stop separately any elements that are not protected.

1.3.6. *Risks relating to variations in the rotational speed of tools*

When the machine is designed to perform operations under different conditions of use (*e.g.* different speeds or energy supply), it must be designed and constructed in such a way that selection and adjustment of these conditions can be carried out safely and reliably.

1.3.7. *Prevention of risks related to moving parts*

The moving parts of machinery must be designed, built and laid out to avoid hazards or, where hazards persist, fixed with guards or protective devices in such a way as to prevent all risk of contact which could lead to accidents.

All necessary steps must be taken to prevent accidental blockage of moving parts involved in the work. In cases where, despite the precautions taken, a blockage is likely to occur, specific protection devices or tools, the instruction handbook and possibly a sign on the machinery should be provided by the manufacture to enable the equipment to be safely unblocked.

1.3.8. *Choice of protection against risks related to moving parts*

Guards or protection devices used to protect against the risks related to moving parts must be selected on the basis of the type of risk. The following guidelines must be used to help make the choice.

A. Moving transmission parts. Guards designed to protect exposed persons against the risks associated with moving transmission parts (such as pulleys, belts, gears, rack and pinions, shafts, etc.) must be:

– either fixed, complying with requirements 1.4.1 and 1.4.2.1, or

– movable, complying with requirements 1.4.1 and 1.4.2.2.A.

Movable guards should be used where frequent access is foreseen.

B. Moving parts diretly involved in the process. Guards or protection devices designed to protect exposed persons against the risks associated with moving parts contributing to the work (such as cutting tools, moving parts of presses, cylinders, parts in the process of being machined, etc.) must be:

– wherever possible fixed guards com-

plying with requirements 1.4.1 and 1.4.2.1,
– otherwise, movable guards complying with requirements 1.4.1 and 1.4.2.2.B or protection devices such as sensing devices (*e.g.* non-material barriers, sensor mats), remote-hold protection devices (*e.g.* two-hand controls), or protection devices intended automatically to prevent all or part of the operator's body from encroaching on the danger zone in accordance with requirements 1.4.1 and 1.4.3.

However, when certain moving parts directly involved in the process cannot be made completely or partially inaccessible during operation owing to operations requiring nearby operator intervention, where technically possible such parts must be fitted with:
– fixed guards, complying with requirements 1.4.1 and 1.4.2.1. preventing access to those sections of the parts that are not used in the work,
– adjustable guards, complying with requirements 1.4.1 and 1.4.2.3 restricting access to those sections of the moving parts that are strictly for the work.

1.4. Required characteristics of guards and protection devices

1.4.1. *General requirement*

Guards and protection devices must:
– be of robust construction,
– not give rise to any additional risk,
– not be easy to by-pass or render non-operational,
– be located at an adequate distance from the danger zone,
– cause minimum obstruction to the view of the production process,
– enable essential work to be carried out on installation and/or replacement of tools and also for maintenance by restricting access only to the area where the work has to be done, if possible without the guard or protection device having to be dismantled.

1.4.2. *Special requirements for guards*

1.4.2.1. *Fixed guards*. Fixed guards must be securely held in place.

They must be fixed by systems that can be opened only with tools.

Where possible, guards must be unable to remain in place without their fixings.

1.4.2.2. *Movable guards*

A. Type A movable guards must:
– as far as possible remain fixed to the machinery when open,
– be associated with a locking device to prevent moving parts starting up as long as these parts can be accessed and to give a stop command whenever they are no longer closed.

B. Type B movable guards must be designed and incorporated into the control system so that:
– moving parts cannot start up while they are within the operator's reach,
– the exposed person cannot reach moving parts once they have started up,
– they can be adjusted only by means of an intentional action, such as the use of a tool, key, etc,
– the absence or failure of one of their components prevents starting or stops the moving parts,
– protection against any risk of ejection is proved by means of an appropriate barrier.

1.4.2.3. *Adjustable guards restricting access*

Adjustable guards restricting access to those areas of the moving parts strictly necessary for the work must:
– be adjustable manually or automatically according to the type of work involved,
– be readily adjustable without the use of tools,
– reduce as far as possible the risk of ejection.

1.4.3. *Special requirements of protection devices*

Protection devices must be designed and incorporated into the control system so that:
– moving parts cannot start up while they are within the operators' reach,
– the exposed person cannot reach moving parts once they have started up,

– they can be adjusted only by means of an intentional action, such as the use of a tool, key, etc,
– the absence or failure of one of their components prevents starting or stops the moving parts.

1.5. Protection against other hazards

1.5.1. *Electricity supply*

Where machinery has an electricity supply it must be designed, constructed and equipped so that all hazards of an electrical nature are or can be prevented.

The specific rules in force relating to electrical equipment designed for use within certain voltage limits must apply to machinery which is subject to those limits.

1.5.2. *Static electricity*

Machinery must be so designed and constructed as to prevent or limit the build-up of potentially dangerous electrostatic charges and/or be fitted with a discharging system.

1.5.3. *Energy supply other than electricity*

Where machinery is powered by an energy other than electricity (*e.g.* hydraulic, pneumatic or thermal energy, etc), it must be so designed, constructed and equipped as to avoid all potential hazards associated with these types of energy.

1.5.4. *Errors of fitting*

Errors likely to be made when fitting or refitting certain parts which could be a source of risk must be made impossible by the design of such parts or, failing this, by information given on the parts themselves and/or the housings. The same information must be given on moving parts and/or their housings where the direction of movement must be known to avoid a risk. Any further information that may be necessary must be given in the instructions.

Where a faulty connection can be the source of risk, incorrect fluid connections, including electrical conductors, must be made impossible by the design or, failing this, by information given on the pipes, cables, etc and/or connector blocks.

1.5.5. *Extreme temperatures*

Steps must be taken to eliminate any risk of injury caused by contact with or proximity to machinery parts or materials at high or very low temperatures.

The risk of hot or very cold material being ejected should be assessed. Where this risk exists, the necessary steps must be taken to prevent it or, if this is not technically possible, to render it non-dangerous.

1.5.6 *Fire*

Machinery must be designed and constructed to avoid all risk of fire or overheating posed by the machinery itself or by gases, liquids, dusts, vapours or other substances produced or used by the machinery.

1.5.7 *Explosion*

Machinery must be designed and constructed to avoid any risk of explosion posed by the machinery itself or by gases, liquids, dusts, vapours or other substances produced or used by the machinery.

To that end the manufacturer must take steps to:
– avoid a dangerous concentration of products,
– prevent combustion of the potentially explosive atmosphere,
– minimise any explosion which may occur so that it does not endanger the surroundings.

The same precautions must be taken if the manufacturer foresees the use of the machinery in a potentially explosive atmosphere.

Electrical equipment forming part of the machinery must conform, as far as the risk from explosion is concerned, to the provision of the specific Directives in force.

1.5.8. *Noise*

Machinery must be so designed and constructed that risks resulting from the emission of airborne noise are reduced to the lowest level taking account of technical progress and the availability of means of reducing noise, in particular at source.

1.5.9. *Vibration*

Machinery must be so designed and constructed that risks resulting from vibrations produced by the machinery are reduced to the lowest level, taking account of technical progress and the availability of means of reducing vibration, in particular at source.

1.5.10. *Radiation*

Machinery must be so designed and constructed that any emission of radiation is limited to the extent necessary for its operation and that the effects on exposed persons are non-existent or reduced to non-dangerous proportions.

1.5.11. *External radiation*

Machinery must be so designed and constructed that external radiation does not interfere with its operation.

1.5.12. *Laser equipment*

Where laser equipment is used, the following provisions should be taken into account:
– laser equipment on machinery must be designed and constructed so as to prevent any accidental radiation,
– laser equipment on machinery must be protected so that effective radiation, radiation produced by reflection or diffusion and secondary radiation do not damage health,
– optical equipment for the observation or adjustment of laser equipment on machinery must be such that no health risk is created by the laser rays.

1.5.13. *Emissions of dust, gases, etc*

Machinery must be so designed, constructed and/or equipped that risks due to gases, liquids, dust, vapours and other waste materials which it produces can be avoided.

Where a hazard exists, the machinery must be so equipped that the said substances can be contained and/or evacuated.

Where machinery is not enclosed during normal operation, the devices for containment and/or evacuation must be situated as close as possible to the source emission.

1.5.14. *Risk of being trapped in a machine*

Machinery must be designed, constructed or fitted with a means of preventing an exposed person from being enclosed within it or, if that is impossible, with a means of summoning help.

1.5.15. *Risk of slipping, tripping or falling*

Parts of the machinery where persons are liable to move about or stand must be designed and constructed to prevent persons slipping, tripping or falling in or off these parts.

1.6. Maintenance

1.6.1 *Machinery maintenance*

Adjustment, lubrication and maintenance points must be located outside danger zones. It must be possible to carry out adjustment, maintenance, repair, cleaning and servicing operations while machinery is at a standstill.

If one or more of the above conditions cannot be satisfied for technical reasons, these operations must be possible without risk (see 1.2.5).

In the case of automated machinery and, where necessary, other machinery, the manufacturer must make provision for a connecting device for mounting diagnostic fault–finding equipment.

Automated machine components which have to be changed frequently, in particular for a change in manufacture or where they are liable to wear or likely to deteriorate following an accident, must be capable of being removed and replaced easily and in safety. Access to the components must enable these tasks to be carried out with the necessary technical means (tools, measuring instruments, etc) in accordance with an operating method specified by the manufacturer.

1.6.2. *Access to operating position and servicing points*

The manufacturer must provide means of access (stairs, ladders, catwalks, etc) to allow access in safety to all areas used for production, adjustment and maintenance operations.

1.6.3. *Isolation of energy sources*

All machinery must be fitted with means to isolate it from all energy sources. Such isolators must be clearly identified. They must be capable of being locked if reconnection could endanger exposed persons. In the case of machinery supplied with electricity through a plug capable of being plugged into a circuit, separation of the plug is sufficient.

The isolator must be capable of being locked also where an operator is unable, from any of the points to which he has access, to check that the energy is still cut off.

After the energy is cut off, it must be possible to dissipate normally any energy remaining or stored in the circuits of the machinery without risk to exposed persons.

As an exception to the above requirements, certain circuits may remain connected to their energy sources in order, for example, to hold parts, protect information, light interiors, etc. In this case, special steps must be taken to ensure operator safety.

1.6.4. *Operator intervention*

Machinery must be so designed, constructed and equipped that the need for operator intervention is limited.

If operator intervention cannot be avoided, it must be possible to carry it out easily and in safety.

1.6.5. *Cleaning of internal parts*

The machinery must be designed and constructed in such a way that it is possible to clean internal parts which have contained dangerous substances or preparations without entering them; any necessary unblocking must also be possible from the outside. If it is absolutely impossible to avoid entering the machinery, the manufacturer must take steps during its construction to allow cleaning to take place with the minimum of danger.

1.7. Indicators

1.7.0. *Information devices*

The information needed to control machinery must be unambiguous and easily understood.

It must not be excessive to the extent of overloading the operator.

Where the health and safety of exposed persons may be endangered by a fault in the operation of unsupervised machinery, the machinery must be equipped to give an appropriate acoustic or light signal as a warning.

1.7.1. *Warning devices*

Where machinery is equipped with warning devices (such as signals, etc), these must be unambiguous and easily perceived.

The operator must have facilities to check the operation of such warning devices at all times.

The requirements of the specific Directives concerning colours and safety signals must be complied with.

1.7.2. *Warning of residual risks*

Where risks remain despite all the measures adopted or in the case of potential risks which are not evident (*e.g.* electrical cabinets, radioactive sources, bleeding of a hydraulic circuit, hazard in an unseen area, etc), the manufacturer must provide warnings.

Such warnings should preferably use readily understandable pictograms and/or be drawn up in one of the languages of the country in which the machinery is to be used, accompanied, on request, by the languages understood by the operators.

1.7.3. Marking

All machinery must be marked legibly and indelibly with the following minimum particulars:
– name and address of the manufacturer,
– the CE marking (see Annex III),
– designation of series or type,
– serial number, if any,
– the year of construction.

Furthermore, where the manufacturer constructs machinery intended for use in a potentially explosive atmosphere, this must be indicated on the machinery.

Machinery must also bear full information relevant to its type and essential to its safe use (*e.g.* maximum speed of certain rotating parts, maximum diameter of tools to be fitted, mass, etc).

Where a machine part must be handled during use with lifting equipment, its mass must be indicated legibly, indelibly and unambiguously.

The interchangeable equipment referred to in Article 1 (2), third subparagraph must bear the same information.

1.7.4. *Instructions*

(a) All machinery must be accompanied by instructions including at least the following:
– a repeat of the information with which the machinery is marked (see 1.7.3), together with any appropriate additional information to facilitate maintenance (*e.g.* addresses of the importer, repairers, etc),
– foreseen use of the machinery within the meaning of 1.1.2 (c),
– workstation(s) likely to be occupied by operators,
– instructions for safe:
  – putting into service,
  – use,
  – handling, giving the mass of the machinery and its various parts where they are regularly to be transported separately,
  – assembly, dismantling,
  – adjustment,
  – maintenance (servicing and repair),
– where necessary, training instructions,
– where necessary the essential characteristics of tools which may be fitted to the machinery,
– where necessary, the essential characteristics of tools which may be fitted to the machinery.

Where necessary, the instructions should draw attention to ways in which the machinery should not be used.

(b) The instructions must be drawn up in one of the Community languages by the manufacturer or his authorised representative established in the Community. On being put into service, all machinery must be accompanied by a translation of the instructions in the language or languages of the country in which the machinery is to be used and by the instructions in the original language. This translation must be done either by the manufacturer or his authorised representative established in the Community or by the person introducing the machinery into the language area in question. By way of derogation from this requirement, the maintenance instructions for use by specialised personnel employed by the manufacturer or his authorised representative established in the Community may be drawn up in only one of the Community languages understood by that personnel.

(c) The instructions must contain the drawings and diagrams necessary for putting into service, maintenance, inspection, checking of correct operation and, where appropriate, repair of the machinery, and all useful instructions in particular with regard to safety.

(d) Any literature describing the machinery must not contradict the instructions as regards safety aspects. The technical documentation describing the machinery must give information regarding the airborne noise emissions referred to in (f) and, in the case of hand-held and/or hand-guided machinery, information regarding vibration as referred to in 2.2.

(e) Where necessary, the instructions must give the requirements relating to installation and assembly for reducing noise or vibration (*e.g.* use of dampers, type and mass of foundation block, etc).

(f) The instructions must give the following information concerning airborne noise emissions by the machinery, either the actual value or a value established on the basis of measurements made on identical machinery:
– equivalent continuous A-weighted sound pressure level at workstations, where this exceeds 70 dB(A); where this level does not exceed 70 dB(A), this fact must be indicated,
– peak C-weighted instantaneous sound pressure value at workstations, where this exceeds 63 Pa (130 dB in relation to 20 μPa),
– sound power level emitted by the machinery where the equivalent continuous A-weighted sound pressure level at workstations exceeds 85 dB(A).

In the case of very large machinery, instead of the sound power level, the equivalent continuous sound pressure levels at specified positions around the machinery may be indicated.

Where the harmonised standards are not applied, sound levels must be measured using the most appropriate method for the machinery.

The manufacturer must indicate the operating conditions of the machinery during measurement and what methods have been used for the measurement.

Where the workstation(s) are undefined or cannot be defined, sound pressure levels must be measured at a distance of 1 metre from the surface of the machinery and at height of 1.60 metres from the floor or access platform. The position and value of the maximum sound pressure must be indicated.

(g) If the manufacturer foresees that the machinery will be used in a potentially explosive atmosphere, the instructions must give all the necessary information.

(h) In the case of machinery which may also be intended for use by non-professional operators, the wording and layout of the instructions for use, whilst respecting the other essential requirements mentioned above, must take into account the level of general education and acumen that can reasonably be expected from such operators.

2. Essential health and safety requirements for certain categories of machinery

2.1. Agri-foodstuffs machinery

Where machinery is intended to prepare and process foodstuffs (*e.g.* cooking, refrigeration, thawing, washing, handling, packaging, storage, transport or distribution), it must be so designed and constructed as to avoid any risk of infection, sickness or contagion and the following hygiene rules must be observed:

(a) materials in contact, or intended to come into contact, with the foodstuffs must satisfy the conditions set down in the relevant Directives. The machinery must be so designed and constructed that these materials can be clean before each use;

(b) all surfaces including their joinings must be smooth, and must have neither ridges nor crevices which could harbour organic materials;

(c) assemblies must be designed in such a way as to reduce projections, edges and recesses to a minimum. They should preferably be made by welding or continuous bonding. Screws, screwheads and rivets may not be used except where technically unavoidable;

(d) all surfaces in contact with the foodstuffs must be easily cleaned and disinfected, where possible after removing easily dismantled parts. The inside surfaces must have curves of a radius sufficient to allow thorough cleaning;

(e) liquid deriving from foodstuffs as well as cleaning, disinfecting and rinsing fluids should be able to be discharged from the machine without impediment (possibly in a "clean" position);

(f) machinery must be so designed and constructed as to prevent any liquids or living creatures, in particular insects, entering, or any organic matter accumulating in areas that cannot be cleaned (*e.g.* for machinery not mounted on feet or casters, by placing a seal between the machinery and its base, by the use of sealed units, etc);

(g) machinery must be so designed and constructed that no ancillary substances (*e.g.* lubricants, etc) can come into contact with foodstuffs. Where necessary, machinery must be designed and constructed so that continuing compliance with this requirement can be checked.

Instructions.

In addition to the information required in section 1, the instructions must indicate recommended products and methods for cleaning, disinfecting and rinsing (not only for easily accessible areas but also where areas to which access is impossible or unadvisable, such as piping, have to be cleaned *in situ*).

2.2. Portable hand-held and/or hand-guided machinery

Portable hand-held and/or hand-guided machinery must conform to the following essential health and safety requirements:

– according to the type of machinery, it must have a supporting surface of sufficient size and have a sufficient number of handles and supports of an appropriate size and arranged to ensure the stability of the machinery under the operating conditions foreseen by the manufacturer,

– except where technically impossible or where there is an independent control, in the case of handles which cannot be released on complete safety, it must be fitted with start and stop controls arranged in such a way that the operator can operate them without releasing the handles,

– it must be designed, constructed or equipped to eliminate the risks of accidental starting and/or continued operation after the operator has released the handles. Equivalent steps must be taken if this requirement is not technically feasible,

– portable hand-held machinery must be designed and constructed to allow, where necessary, a visual check of the contact of the tool with the material being processed.

Instructions.

The instructions must give the following information concerning vibrations transmitted by hand-held and hand-guided machinery:

– the weighted root mean square acceleration value to which the arms are subjected, if it exceeds 2.5 m/s^2 as determined by the appropriate test code. Where the acceleration does not exceed 2.5 m/s^2, this must be mentioned.

If there is no applicable test code, the manufacturer must indicate the measurement methods and conditions under which measurements were made.

2.3. Machinery for working wood and analogous materials

Machinery for working wood and machinery for working materials with physical and technological characteristics similar to those of wood, such as cork, bone, hardened rubber, hardened plastic material and other similar stiff material must conform to the following essential health and safety requirements:

(a) the machinery must be designed, constructed or equipped so that the piece being machined can be placed and guided in safety; where the piece is hand-held on a workbench the latter must be sufficiently stable during the work and must not impede the movement of the piece;

(b) where the machinery is likely to be used in conditions involving the risk of ejection of pieces of wood, it must be designed, constructed or equipped to eliminate this ejection, or, if this is not the case, so that the ejection does not engender risks for the operator and/or exposed persons;

(c) the machinery must be equipped with an automatic brake that stops the tool in a sufficiently short time if there is a risk of contact with the tool whilst it runs down;

(d) where the tool is incorporated into a non-fully automated machine, the latter must be so designed and constructed as to eliminate or reduce the risk of serious accidental injury, for example by using cylindrical cutter blocks, restricting depth of cut, etc.

3. Essential health and safety requirements to offset the particular hazards due to the mobility of machinery

Machinery presenting hazards due to mobility must be designed and constructed to meet the requirements set out below.

Risks due to mobility always exist in the case of machinery which is self-propelled, towed or pushed or carried by other machinery or tractors, is operated in working areas and whose operation requires either mobility while working, be it continuous or semi-continuous movement, between a succession of fixed working positions.

Risks due to mobility may also exist in the case of machinery operated without being moved, but equipped in such a way as to enable it to be moved more easily from one place to another (machinery fitted with wheels, rollers, runners, etc. or placed on gantries, trolleys, etc.).

In order to verify that rotary cultivators and power harrows do not present unacceptable risks to the exposed persons, the manufacturer or his authorised representative established within the Community must, for each type of machinery concerned, perform the appropriate tests or have such tests performed.

3.1. General

3.1.1. *Definition*

"Driver" means an operator responsible for the movement of machinery. The driver may be transported by the machinery or may be on foot, accompanying the machinery, or may be guiding the machinery by remote control (cables, radio, etc.).

3.1.2. *Lighting*

If intended by the manufacturer to be used in dark places, self-propelled machinery must be fitted with a lighting device appropriate to the work to be carried out, without prejudice to any other regulations applicable (road traffic regulations, navigation rules, etc.).

3.1.3. *Design of machinery to facilitate its handling*

During the handling of the machine and/or its parts, there must be no possibility of sudden movements or of hazards due to instability as long as the machine and/or its parts are handled in accordance with the manufacturer's instructions.

3.2. Work stations

3.2.1. *Driving position*

The driving must be designed with due regard to ergonomic principles. There may be two or more driving positions and, in such cases, each driving position must be provided with all the requisite controls. Where there is more than one driving position, the machinery must be designed so that the use of one of them precludes the use of the others, except in emergency stops. Visibility from the driving position must be such that the driver can in complete safety for himself and the exposed persons, operate the machinery and its tools in their intended conditions of use. Where necessary, appropriate devices must be provided to remedy hazards due to inadequate direct vision.

Machinery must be so designed and constructed that, from the driving position, there can be no risk to the driver and operators on board from inadvertent contact with the wheels or tracks.

The driving position must be designed and constructed so as to avoid any health risk due to exhaust gases and/or lack of oxygen.

The driving position of ride-on drivers must be so designed and constructed that a driver's cab may be fitted as long as there is room. In that case, the cab must incorporate a place for the instructions needed for the driver and/or operators. The driving position must be fitted with an adequate cab where there is a hazard due to a dangerous environment.

Where the machinery is fitted with a cab, this must be designed, constructed and/or equipped to ensure that the driver has good operating conditions and is protected against any hazards that might exist (for instance: inadequate heating and ventilation, inadequate visibility, excessive noise and vibration, falling objects, penetration by objects, rolling over, etc.). The exit must allow rapid evacuation. Moreover, an emergency exit must be provided in a direction which is different from the usual exit.

The materials used for the cab and its fittings must be fire-resistant.

3.2.2. *Seating*

The driving seat of any machinery must enable the driver to maintain a stable position and be designed with due regard to ergonomic principles.

The seat must be designed to reduce vibrations transmitted to the driver to the lowest level that can be reasonably achieved. The seat mountings must withstand all stresses to which they can be subjected, notably in the event of rollover. Where there is no floor beneath the driver's feet, the driver must have footrests covered with a slip-resistant material.

Where machinery is fitted with provisions for a rollover protection structure, the seat must be equipped with a safety belt or equivalent device which keeps the driver in his seat without restricting any movements necessary for driving or any movements caused by the suspension.

3.2.3. *Other places*

If the conditions of use provide that operators other than the driver are occasionally or regularly transported by the machinery, or work on it, appropriate places must be provided which enable them to be transported or to work on it without risk, particularly the risk of falling.

Where the working conditions so permit, these work places must be equipped with seats.

Should the driving position have to be fitted with a cab, the other places must also be protected against the hazards which justified the protection of the driving position.

3.3. Controls

3.3.1. *Control devices*

The driver must be able to actuate all control devices required to operate the machinery from the driving position, except for functions which can be safely activated only by using control devices located away from the driving position. This refers in particular to working positions other than the driving position, for which operators other than the driver are responsible or for which the driver has to leave his driving position in order to carry out the manoeuvre in safety.

Where there are pedals they must be so designed, constructed and fitted to allow operation by the driver in safety with the minimum risk of confusion;

they must have a slip-resistant surface and be easy to clean.

Where their operation can lead to hazards, notably dangerous movements, the machinery's controls, except for those with preset positions, must return to the neutral position as soon as they are released by the operator.

In the case of wheeled machinery, the steering system must be designed and constructed to reduce the force of sudden movements of the steering wheel or steering lever caused by shocks to the guide wheels.

Any control that locks the differential must be so designed and arranged that it allows the differential to be unlocked when the machinery is moving.

The last sentence of section 1.2.2 does not apply to the mobility function.

3.2.3. *Starting/moving*

Self-propelled machinery with a ride-on driver must be so equipped as to deter unauthorised persons from starting the engine.

Travel movements of self-propelled machinery with a ride-on driver must be possible only if the driver is at the controls.

Where, for operating purposes, machinery must be fitted with devices which exceed its normal clearance zone (*e.g.* stabilisers, jib, etc), the driver must be provided with the means of checking easily, before moving the machinery, that such devices are in a particular position which allows safe movement.

This also applies to all other parts which, to allow safe movement, have to be in particular positions, locked if necessary.

Where it is technically and economically feasible, movement of the machinery must depend on safe positioning of the aforementioned parts.

It must not be possible for movement of the machinery to occur while the engine is being started.

3.3.3. *Travelling function*

Without prejudice to the provisions of road traffic regulations, self-propelled machinery and its trailers must meet the requirements for slowing down, stopping, braking and immobilisation so as to ensure safety under all the operating, loading, speed, ground and gradient conditions allowed for by the manufacturer and corresponding to conditions encountered in normal use.

The driver must be able to slow down and stop self-propelled machinery by means of a main device. Where safety so requires in the event of a failure of the main device, or in the absence of the energy supply to actuate the main device, an emergency device with fully independent and easily accessible controls must be provided for slowing down and stopping.

Where safety so requires, a parking device must be provided to render stationary machinery immobile. This device may be combined with one of the devices referred to in the second paragraph, provided that it is purely mechanical.

Remote-controlled machinery must be designed and constructed to stop automatically if the driver loses control.

Section 1.2.4 does not apply to the travelling function.

3.3.4. *Movement of pedestrian-controlled machinery*

Movement of pedestrian-controlled self-propelled machinery must be possible only through sustained action on the relevant control by the driver. In particular, it must not be possible for movement to occur while the engine is being started.

The control systems for pedestrian-controlled machinery must be designed to minimise the hazards arising from inadvertent movement of the machine towards the driver. In particular:

(a) crushing;
(b) injury from rotating tools.

Also, the speed of normal travel of

the machine must be compatible with the pace of a driver on foot.

In the case of machinery on which a rotary tool may be fitted, it must not be possible to actuate that tool when the reversing control is engaged, except where movement of the machinery results from movement of the tool. In the latter case, the reversing speed must be such that it does not endanger the driver.

3.3.5. *Control circuit failure*

A failure in the power supply to the power-assisted steering, where fitted, must not prevent machinery from being steered during the time required to stop it.

3.4. Protection against mechanical hazards

3.4.1. *Uncontrolled movements*

When a part of a machine has been stopped, any drift away from the stopping position, for whatever reason other than action at the controls, must be such that it is not a hazard to exposed persons.

Machinery must be so designed, constructed and where appropriate placed on its mobile support as to ensure that when moved the uncontrolled oscillations of its centre of gravity do not affect its stability or exert excessive strain on its structure.

3.4.2. *Risk of break-up during operation*

Parts of machinery rotating at high speed which, despite the measures taken, may break up or disintegrate, must be mounted and guarded in such a way that, in case of breakage, their fragments will be contained or, if that is not possible, cannot be projected towards the driving and/or operation positions.

3.4.3. *Rollover*

Where, in the case of self-propelled machinery with a ride-on driver and possibly ride-on operators, there is a risk of rolling over, the machinery must be designed for and be fitted with anchorage points allowing it to be equipped with a rollover protective structure (ROPS).

This structure must be such that in case of rolling over if affords the ride-on driver and where appropriate the ride-on operators an adequate deflection-limiting volume (DLV).

In order to verify that the structure complies with the requirement laid down in the second paragraph, the manufacturer or his authorised representative established within the Community must, for each type of structure concerned, perform appropriate tests or have such tests performed.

In addition, the earth-moving machinery listed below with a capacity exceeding 15 kW must be fitted with a rollover protective structure:
- crawler loaders or wheel loaders,
- backhoe loaders,
- crawler tractors or wheel tractors,
- scrapers, self-loading or not,
- graders,
- articulated steer dumpers.

3.4.4. *Falling objects*

Where, in the case of machinery with a ride-on driver and possibly ride-on operators, there is a risk due to falling objects or material, the machinery should be designed for, and fitted with, if its size allows, anchorage points allowing it to be equipped with a falling-object protective structure (FOPS).

This structure must be such that in the case of falling objects or material, it guarantees the ride-on operators an adequate deflection-limiting volume (DLV).

In order to verify that the structure complies with the requirement laid down in the second paragraph, the manufacturer or his authorised representative established within the Community must, for each type of structure concerned, perform appropriate tests or have such tests performed.

3.4.5. *Means of access*

Handholds and steps must be

designed, constructed and arranged in such a way that the operators use them instinctively and do not use the controls for that purpose.

3.4.6. *Towing devices*

All machinery used to tow or to be towed must be fitted with towing or coupling devices designed, constructed and arranged to ensure easy and safe connection and disconnection, and to prevent accidental disconnection during use.

In so far as the towbar load requires, such machinery must be equipped with a support with a bearing surface suited to the load and the ground.

3.4.7. *Transmission of power between self-propelled machinery (or tractor) and recipient machinery*

Transmission shafts with universal joints linking self-propelled machinery (or tractor) to the first fixed bearing of recipient machinery must be guarded on the self-propelled machinery side and the recipient machinery side over the whole length of the shaft and associated universal joints.

On the side of the self-propelled machinery (or tractor), the power take-off to which the transmission shaft is attached must be guarded either by a screen fixed to the self-propelled machinery (or tractor) or by any other device offering equivalent protection.

On the towed machinery side, the input shaft must be enclosed in a protective casing fixed to the machinery.

Torque limiters or freewheels may be fitted to universal joint transmissions only on the side adjoining the driven machine. The universal-joint transmission shaft must be marked accordingly.

All towed machinery whose operation requires a transmission shaft to connect it to self-propelled machinery or a tractor must have a system for attaching the transmission shaft so that when the machinery is uncoupled the transmission shaft and its guard are not damaged by contact with the ground or part of the machinery.

The outside parts of the guard must be so designed, constructed and arranged that they cannot turn with the transmission shaft. The guard must cover the transmission shaft to the ends of the inner jaws in the case of simple universal joints and at least to the centre of the outer joint or joints in the case of "wide-angle" universal joints.

Manufacturers providing means of access to working positions near to the universal joint transmission shaft must ensure that shaft guards as described in the sixth paragraph cannot be used as steps unless designed and constructed for that purpose.

3.4.8. *Moving transmission parts*

By way of derogation from section 1.3.8.A, in the case of internal combustion engines, removable guards preventing access to the moving parts in the engine compartment need not have locking devices if they have to be opened either by the use of a tool or key or by a control located in the driving position if the latter is in a fully enclosed cab with a lock to prevent unauthorised access.

3.5. Protection against other hazards

3.5.1. *Batteries*

The battery housing must be constructed and located and the battery installed so as to avoid as far as possible the chance of electrolyte being ejected on to the operator in the event of rollover and/or to avoid the accumulation of vapours in places occupied by operators.

Machinery must be so designed and constructed that the battery can be disconnected with the aid of an easily accessible device provided for that purpose.

3.5.2. *Fire*

Depending on the hazards anticipated by the manufacturer when in use, machinery must, where its size permits:

– either allow easily accessible fire extinguishers to be fitted,

– or be provided with built-in extinguisher systems.

3.5.3. *Emissions of dust, gases, etc*

Where such hazards exist, the containment equipment provided for in section 1.5.13 may be replaced by other means, for example precipitation by water spraying.

The second and third paragraphs of section 1.5.13 do not apply where the main function of the machinery is the spraying of products.

3.6. Indications

3.6.1. *Signs and warning*

Machinery must have means of signalling and/or instruction plates concerning use, adjustment and maintenance, wherever necessary, to ensure the health and safety of exposed persons. They must be chosen, designed and constructed in such a way as to be clearly visible and indelible.

Without prejudice to the requirements to be observed for travelling on the public highway, machinery with a ride-on driver must have the following equipment:

– an acoustic warning device to alert exposed persons,

– a system of light signals relevant to the intended conditions of use such as stop lamps, reversing lamps and rotating beacons. The latter requirement does not apply to machinery intended solely for underground working and having no electrical power.

Remote-controlled machinery which under normal conditions of use exposes persons to the hazards of impact or crushing must be fitted with appropriate means to signal its movements or with means to protect exposed persons against such hazards. The same applies to machinery which involves, when in use, the constant repetition of a forward and backward movement on a single axis where the back of the machine is not directly visible to the driver.

Machinery must be so constructed that the warning and signalling devices cannot all be disabled unintentionally. Where this is essential for safety, such devices must be provided with the means to check that they are in good working order and their failure must be made apparent to the operator.

Where the movement of machinery or its tools is particularly hazardous, signs on the machinery must be provided to warn against approaching the machinery while it is working; the signs must be legible at a sufficient distance to ensure the safety of persons who have to be in the vicinity.

3.6.2. *Marking*

The minimum requirements set out in 1.7.3 must be supplemented by the following:

– nominal power expressed in kW,

– mass in kg of the most usual configuration and, where appropriate:

– maximum drawbar pull provided for by the manufacturer at the coupling hook, in N,

– maximum vertical load provided for by the manufacturer on the coupling hook, in N.

3.6.3. *Instruction handbook*

Apart from the minimum requirements set out in 1.7.4, the instruction handbook must contain the following information:

(a) regarding the vibrations emitted by the machinery, either the actual value or a figure calculated from measurements performed on identical machinery:

– the weighted root mean square acceleration value to which the arms are subjected, if it exceeds 2.5 m/s^2; should it not exceed 2.5 m/s^2, this must be mentioned,

– the weighted root mean square acceleration value to which the body (feet or posterior) is subjected, if it exceeds 0.5 m/s^2; should it not exceed 0.5 m/s^2, this must be mentioned.

Where the harmonised standards are not applied, the vibration must be measured using the most appropriate method for the machinery concerned.

The manufacturer must indicate

the operating conditions of the machinery during measurement and which methods were used for taking the measurements;

(b) in the case of machinery allowing several uses depending on the equipment used, manufacturers of basic machinery to which interchangeable equipment may be attached and manufacturers of the interchangeable equipment must provide the necessary information to enable the equipment to be fitted and used safely.

4. Essential health and safety requirements to offset the particular hazards due to a lifting operation

Machinery presenting hazards due to lifting operations – mainly hazards of load falls and collisions or hazards of tipping caused by a lifting operation – must be designed and constructed to meet the requirements below.

Risks due to a lifting operation exist particularly in the case of machinery designed to move a unit load involving a change in level during the movement. The load may consist of objects, materials or goods.

4.1. General remarks

4.1.1. *Definitions*

(a) lifting accessories:
"lifting accessories" means components or equipment not attached to the machine and placed between the machinery and the load or on the load in order to attach it;

(b) separate lifting accessories:
"separate lifting accessories" means accessories which help to make up or use a slinging device, such as eyehooks, shackles, rings, eyebolts, etc;

(c) guided load:
"guided load" means the load where the total movement is made along rigid or flexible guides, whose position is determined by fixed points;

(d) working coefficient:
"working coefficient" means the arithmetic ratio between the load guaranteed by the manufacturer up to which a piece of equipment, an accessory or machinery is able to hold it and the maximum working load marked on the equipment, accessory or machinery respectively;

(e) test coefficient:
"test coefficient" means the arithmetic ratio between the load used to carry out the static or dynamic tests on a piece of equipment, an accessory or machinery and the maximum working load marked on the piece of equipment, accessory or machinery;

(f) static test:
"static test" means the test during which the machinery or the lifting accessory is first inspected and subjected to a force corresponding to the maximum working load multiplied by the appropriate static test coefficient and then reinspected once the said load has been released to ensure no damage has occurred;

(g) dynamic test:
"dynamic test" means the test during which the machinery is operated in all its possible configurations at maximum working load with account being taken of the dynamic behaviour of the machinery in order to check that the machinery and safety features are functioning properly.

4.1.2. *Protection against mechanical hazards*

4.1.2.1. Risks due to lack stability
Machinery must be so designed and constructed that the stability required in 1.3.1 is maintained both in service and out of service, including all stages of transportation, assembly and dismantling, during foreseeable component failures and also during the tests carried out in accordance with the instruction handbook.

To that end, the manufacturer or his

authorised representative established within the Community must use the appropriate verification methods; in particular, for self-propelled industrial trucks with lift exceeding 1.80 m, the manufacturer or his authorised representative established within the Community must, for each type of industrial truck concerned, perform a platform stability test or similar test, or have such tests performed.

4.1.2.2. Guide rails and rail tracks
Machinery must be provided with devices which act on the guide rails or tracks to prevent derailment.

However, if derailment occurs despite such devices, or if there is a failure of a rail or of a running component, devices must be provided which prevent the equipment, component or load from falling or the machine overturning.

4.1.2.3. Mechanical strength
Machinery, lifting accessories and removable components must be capable of withstanding the stresses to which they are subjected, both in and, where applicable, out of use, under the installation and operating conditions provided for by the manufacturer, and in all relevant configurations, with due regard, where appropriate, to the effects of atmospheric factors and forces exerted by persons. This requirement must also be satisfied during transport, assembly and dismantling.

Machinery and lifting accessories must be designed and constructed so as to prevent failure from fatigue or wear, taking due account of their intended use.

The materials used must be chosen on the basis of the working environments provided for by the manufacturer, with special reference to corrosion, abrasion, impacts, cold brittleness and ageing.

The machinery and the lifting accessories must be designed and constructed to withstand the overload in the static tests without permanent deformation or patent defect. The calculation must take account of the values of the static test coefficient chosen to guarantee an adequate level of safety: that coefficient has, as a general rule, the following values:

(a) manually-operated machinery and lifting accessories: 1.5;
(b) other machinery: 1.25.

Machinery must be designed and constructed to undergo, without failure, the dynamic tests carried out using the maximum working load multiplied by the dynamic test coefficient. This dynamic test coefficient is chosen so as to guarantee an adequate level of safety: the coefficient is, as a general rule, equal to 1.1.

The dynamic tests must be performed on machinery ready to be put into service under normal conditions of use. As a general rule, the tests will be performed at the nominal speeds laid down by the manufacturer. Should the control circuit of the machinery allow for a number of simultaneous movements (for example, rotation and displacement of the load), the tests must be carried out under the least favourable conditions, *i.e.* as a general rule, by combining the movements concerned.

4.1.2.4. Pulleys, drums, chains or ropes
Pulleys, drums and wheels must have a diameter commensurate with the size of rope or chains with which they can be fitted.

Drums and wheels must be so designed, constructed and installed that the ropes or chains with which they are equipped can wind round without falling off.

Ropes used directly for lifting or supporting the load must not include any splicing other than at their ends (splicings are tolerated in installations which are intended from their design to be modified regularly according to needs for use). Complete ropes and their endings have a working coefficient chosen so as to guarantee an adequate level of safety; as a general rule, this coefficient is equal to five.

Lifting chains have a working coefficient chosen so as to guarantee an adequate level of safety; as a general rule, this coefficient is equal to four.

In order to verify that an adequate working coefficient has been attained, the manufacturer or his authorised representative established within the Community must, for each type of chain and rope used directly for lifting the load, and for the rope ends, perform the appropriate tests or have such tests performed.

4.1.2.5. Separate lifting accessories
Lifting accessories must be sized with due regard to fatigue and ageing processes for a number of operating cycles consistent with their expected life-span as specified in the operating conditions for a given application.
Moreover:

(a) the working coefficient of the metallic rope/rope-end combination is chosen so as to guarantee an adequate level of safety; this coefficient is, as a general rule, equal to five. Ropes must not comprise any splices or loops other than at their ends;
(b) where chains with welded links are used, they must be of the short-link type. The working coefficient of chains of any type is chosen so as to guarantee an adequate level of safety; this coefficient is, as a general rule, equal to four;
(c) the working coefficient for textile ropes or slings is dependent on the material, method of manufacture, dimensions and use. This coefficient is chosen so as to guarantee an adequate level of safety; it is, as a general rule, equal to seven, provided the materials used are shown to be of very good quality and the method of manufacture is appropriate to the intended use. Should this not be the case, the coefficient is, as a general rule, set at a higher level in order to secure an equivalent level of safety:
    Textile ropes and slings must not include any knots, connections or splicing other than at the ends of the sling, except in the case of an endless sling;
(d) all metallic components making up, or used with, a sling must have a working coefficient chosen so as to guarantee an adequate level of safety; this coefficient is, as a general rule, equal to four;
(e) the maximum working capacity of a multi-legged sling is determined on the basis of the safety coefficient of the weakest leg, the number of legs and a reduction factor which depends on the slinging configuration;
(f) in order to verify that an adequate working coefficient has been attained, the manufacturer or his authorised representative established within the Community must, for each type of component referred to in (a), (b), (c) and (d) perform the appropriate tests or have such tests performed.

4.1.2.6. Control of movements
Devices for controlling movements must act in such a way that the machinery on which they are installed is kept safe:

(a) machinery must be so designed or fitted with devices that the amplitude of movement of its components is kept within the specified limits. The operation of such devices must, where appropriate, be preceded by a warning;
(b) where several fixed or rail-mounted machines can be manoeuvred simultaneously in the same place, with risks of collision, such machines must be so designed and constructed as to make it possible to fit systems enabling these risks to be avoided,
(c) the mechanisms of machinery must be so designed and constructed that the loads cannot creep dangerously or fall freely and unexpectedly, even in the event of partial or total failure of the power supply or when the operator stops operating the machine;
(d) it must not be possible, under normal operating conditions, to lower the load solely by friction brake, except in the case of machinery, whose function requires it to operate in that way;
(e) holding devices must be so

designed and constructed that inadvertent dropping of the loads is avoided.

4.1.2.7. Handling of loads
The driving position of machinery must be located in such a way as to ensure that widest possible view of trajectories of the moving parts, in order to avoid possible collisions with persons or equipment or other machinery which might be manoeuvring at the same time and liable to constitute a hazard.

Machinery with guided loads fixed in one place must be designed and constructed so as to prevent exposed persons from being hit by the load or the counter-weights.

4.1.2.8. Lightning
Machinery in need of protection against the effects of lightning while being used must be fitted with a system for conducting the resultant electrical charges to earth.

4.2. Special requirements for machinery whose power source is other than manual effort

4.2.1. *Controls*

4.2.1.1. Driving position
The requirements laid down in section 3.2.1 also apply to non-mobile machinery.

4.2.1.2. Seating
The requirements laid down in section 3.2.2, first and second paragraphs, and those laid down in section 3.2.3 also apply to non-mobile machinery.

4.1.2.3. Control devices
The devices controlling movements of the machinery or its equipment must return to their neutral position as soon as they are released by the operator. However, for partial or complete movements in which there is no risk of the load or the machinery colliding, the said devices may be replaced by controls authorising automatic stops at preselected levels without holding a hold-to-run control device.

4.1.2.4. Loading control
Machinery with a maximum working load of not less than 1,000 kilograms or an overturning moment of not less than 40,000 Nm must be fitted with devices to warn the driver and prevent dangerous movements of the load in the event of:
- overloading the machinery
  - either as a result of maximum working loads being exceeded, or
  - as a result of the moments due to the loads being exceeded,
- the moments conducive to overturning being exceeded as a result of the load being lifted.

4.2.2. *Installation guided by cables*

Cable carriers, tractors or tractor carriers must be held by counter-weights or by a device allowing permanent control of the tension.

4.2.3. *Risks to exposed persons. Means of access to driving position and intervention points*

Machinery with guided loads and machinery whose load supports follow a clearly defined path must be equipped with devices to prevent any risks to exposed persons.

Machinery serving specific levels at which operators can gain access to the load platform in order to stack or secure the load must be designed and constructed to prevent uncontrolled movement of the load platform, in particular while being loaded or unloaded.

4.2.4. *Fitness for purpose*

When machinery is placed on the market or is first put into service, the manufacturer or his authorised representative established within the Community must ensure, by taking appropriate measures or having them taken, that lifting accessories and machinery which are ready for use – whether manually or power-operated – can fulfil their specified functions safely. The said measures must take into account the static and dynamic aspects of the machinery.

Where the machinery cannot be assembled in the manufacturer's premises, or in the premises of his authorised representative established within the Community, appropriate measures must be taken at the place of use. Otherwise, the measures may be taken either in the manufacturer's premises or at the place of use.

4.3. Marking

4.3.1. *Chains and ropes*

Each length of lifting chain, rope or webbing not forming part of an assembly must bear a mark or, where this is not possible, a plate or irremovable ring bearing the name and address of the manufacturer or his authorised representative established in the Community and the identifying reference of the relevant certificate.

The certificate should show the information required by the harmonised standards or, should those not exist, at least the following information:
– the name of the manufacturer or his authorised representative established within the Community,
– the address within the Community of the manufacturer or his authorised representative, as appropriate,
– a description of the chain or rope which includes:
  – its nominal size,
  – its construction,
  – the material from which it is made, and
– any special metallurgical treatment applied to the material,
– if tested, the standard used,
– a maximum load to which the chain or rope should be subjected in service. A range of values may be given for specified applications.

4.3.2. *Lifting accessories*

All lifting accessories must show the following particulars:
– identification of the manufacturer,
– identification of the material (*e.g.* international classification) where this information is needed for dimensional compatibility,
– identification of the maximum working load,
– CE marking.

In the case of accessories including components such as cables or ropes, on which marking is physicaly impossible, the particulars referred to in the first paragraph must be displayed on a plate or by some other means and securely affixed to the accessory.

The particulars must be legible and located in a place where they are not liable to disappear as a result of machining, wear, etc, or jeopardise the strength of the accessory.

4.3.3. *Machinery*

In addition to the minimum information provided for in 1.7.3, each machine must bear, legibly and indelibly, information concerning the nominal load:
(i) displayed in uncoded form and prominently on the equipment in the case of machinery which has only one possible value;
(ii) where the nominal load depends on the configuration of the machine, each driving position must be provided with a load plate indicating, preferably in diagrammatic form or by means of tables, the nominal loads for each configuration.

Machinery equipped with a load support which allows access to persons and involves a risk of falling must bear a clear and indelible warning prohibiting the lifting of persons. This warning must be visible at each place where access is possible.

4.4. Instruction handbook

4.4.1. *Lifting accessories*

Each lifting accessory or each commercially indivisible batch of lifting accessories must be accompanied with an instruction handbook setting out at least the following particulars:
– normal conditions of use,
– instructions of use, assembly and maintenance,
– the limits of use (particularly for the accessories which cannot comply with 4.1.2.6 (e)).

4.4.2. *Machinery*

In addition to section 1.7.4, the instruction handbook must include the following information:

(a) the technical characteristics of the machinery, and in particular:
– where appropriate, a copy of the load table described in section 4.3.3 (ii),
– the reactions at the supports or anchors and characteristics of the tracks,
– where appropriate, the definition and the means of installation of the ballast;

(b) the contents of the logbook, if the latter is not supplied with the machinery;

(c) advice for use, particularly to offset the lack of direct sight of the load by the operator;

(d) the necessary instructions for performing the tests before first putting into service machinery which is not assembled on the manufacturer's premises in the form in which it is to be used.

5. Essential safety and health requirements for machinery intended for underground work

Machinery intended for underground work must be designed and constructed to meet the requirements below.

5.1. Risks due to lack of stability

Powered roof supports must be so designed and constructed as to maintain a given direction when moving and not slip before and while they come under load and after the load has been removed. They must be equipped with anchorages for the top plates of the individual hydraulic props.

5.2. Movement

Powered roof supports must allow for unhindered movement of exposed persons.

5.3. Lighting

The requirements laid down in the third paragraph of section 1.1.4 do not apply.

5.4. Control devices

The accelerator and brake controls for the movement of machinery running on rails must be manual. The deadman's control may be foot-operated, however.

The control devices of powered roof supports must be designed and laid out so that, during displacement operations, operators are sheltered by a support in place. The control devices must be protected against any accidental release.

5.5. Stopping

Self-propelled machinery running on rails for use in underground work must be equipped with a deadman's control acting on the circuit controlling the movement of the machinery.

5.6. Fire

The second indent of 3.5.2 is mandatory in respect of machinery which comprise highly flammable parts.

The braking system of machinery meant for use in underground working must be designed and constructed so as not to produce sparks or cause fires.

Machinery with heat engines for use in underground working must be fitted only with internal combustion engines using fuel with a low vaporising pressure and which exclude any spark of electrical origin.

5.7. Emissions of dust, gases etc

Exhaust gases from internal combustion engines must not be discharged upwards.

6. Essential health and safety requirements to offset the particular hazards due to the lifting or moving or persons

Machinery presenting hazards due to the lifting or moving of persons must be designed and constructed to meet the requirements set out below.

6.1. General

6.1.1. *Definition*

For the purpose of this Chapter, "carrier" means the device by which persons are supported in order to be lifted, lowered or moved.

6.1.2. *Mechanical strength*

The working coefficients defined in heading 4 are inadequate fcor machinery intended for the lifting or moving of persons and must, as a general rule, be doubled. The floor of the carrier must be designed and constructed to offer the space and strength corresponding to the maximum number of persons and the maximum working load set by the manufacturer.

6.1.3.. *Loading control for types of device moved by power other than human strength*

The requirement of 4.2.1.4 apply regardless of the maximum working load figure. This requirement does not apply to machinery in respect of which the manufacturer can demonstrate that there is no risk of overloading and/or overturning.

6.2. Controls

6.2.1. Where safety requirements do not impose other solutions:

The carrier must, as a general rule, be designed and constructed so that persons inside have means of controlling movements upwards and downwards and, if appropriate, of moving the carrier horizontally in relation to the machinery.

In operation, those controls must override the other devices controlling the same movement, with the exception of the emergency stop devices.

The controls for these movement must be of the maintained command type, except in the case of machinery serving specific levels.

6.2.2. If machinery for the lifting or moving or persons can be moved with the carrier in a position other than the rest position, it must be designed and constructed so that the person or persons in the carrier have the means of preventing hazards produced by the movement of the machinery.

6.2.3. Machinery for the lifting or moving of persons must be designed, constructed or equipped so that excess speeds of the carrier do not cause hazards.

6.3. Risks of persons falling from the carrier

6.3.1. If the measures referred to in 1.1.15 are not adequate, carriers must be fitted with a sufficient number of anchorage points for the number of persons possibly using the carrier, strong enough for the attachment of personal protective equipment against the danger of falling.

6.3.2. Any trapdoors in floors or ceilings or side doors must open in a direction which obviates any risk of falling should they open unexpectedly.

6.3.3. Machinery for lifting or moving must be designed and constructed to ensure that the floor of the carrier does not tilt to an extent which creates a risk of the occupants falling, including when moving.

The flor of the carrier must be slip-resistant.

6.4. Risks of the carrier falling or overturning

6.4.1. Machinery for the lifting or moving of persons must be designed and constructed to prevent the carrier falling or overturning.

6.4.2. Acceleration and braking of the carrier or carrying vehicle, under the control of the operator or triggered by a safety device and under the maximum load and speed conditions laid down by the manufacturer, must not cause any danger to exposed persons.

6.5. Markings

Where necessary to ensure safety, the carrier must bear the relevant essential information.

Annex II

A. Contents of the EC declaration of conformity for machinery[1]

The EC declaration of conformity must contain the following particulars:
– name and address of the manufacturer or his authorised representative established in the Community,[2]
– description of the machinery,[3]
– where appropriate, the name and address of the notified body and the number of the EC type-examination certificate,
– where appropriate, the name and address of the notified body to which the file has been forwarded in accordance with the first indent of Article 8 (2) (c),
– where appropriate, the name and address of the notified body which has carried out the verification referred to in the second indent of Article 8 (2) (c),
– where appropriate, a reference to the harmonised standards.
– all relevant provisions complied with by the machinery,
– where appropriate, name and address of the notified body and number of the EC type-examination certificate,
– where appropriate, the name and address of the notified body to which the file has been forwarded in accordance with the first indent of Article 8 (2) (c),
– where appropriate, the name and address of the notified body which has carried out the verification referred to in the second indent of Article 8 (2) (c),
– where appropriate, a reference to the harmonised standards,
– where appropriate, the national technical standards and specifications used,
– identification of the person empowered to sign on behalf of the manufacturer or his authorised representatives.

B. Contents of the declaration by the manufacturer or his authorised representatives established in the Community (Article 4 (2))

The manufacturer's declaration referred to in Article 4 (2) must contain the following particulars:
– name and address of the manufacturer or the authorised representative,
– description of the machinery or machinery parts,
– where appropriate, the name and address of the notified body which has carried out the verification referred to in the second indent of Article 8 (2) (c),
– where appropriate, a reference to the harmonised standards,
– a statement that the machinery must not be put into service until the machinery into which it is to be incorporated has been declared in conformity with the provisions of the Directive,
– identification of the person signing.

C. Contents of the EC declaration of conformity for safety components placed on the market separately.[1]

The EC declaration of conformity must contain the following particulars:
– name and address of the manufacturer or his authorised representative established in the Community,[2]

[1] This declaration must be drawn up in the same language as the original instructions (see Annex I, Section 1.7.4 (b)) and must be either typewritten or handwritten in block capitals. It must be accompanied by a translation in one of the official languages of the country in which the machinery is to be used. This translation must be done in accordance with the same conditions as for the translation of the instructions.

[2] Business name and full address, authorised representatives must also give the business name and address of the manufacturer.

[3] Description of the machinery (make, type, serial number, etc.).

– description of the safety component,[4]
– safety function fulfilled by the safety component, if not obviouis from the description,
– where appropriate, the name and address of the notified body and the number of the EC type-examination certificate,
– where appropriate, the name and address of the notified body to which the file was forwarded in accordance with the first indent of Article 8 (2) (c),
– where appropriate, the name and address of the notified body which carried out the verification referred to in the second indent of Article 8 (2) (c),
– where appropriate, a reference to the harmonised standards.
– where appropriate, the national technical standards and specifications used,
– identification of the person empowered to sign on behalf of the manufacturer or his authorised representative established in the Community.

[4] Description of the safety component (make, type, serial number, if any, etc.).

Annex III

CE conformity marking

– The CE conformity marking shall consist of the initials "CE" taking the following form:

– If the CE marking is reduced or enlarged the proportions given in the above drawing must be respected.
– The various components of the CE marking must have substantially the same vertical dimension, which may not be less than 5 mm. This minimum dimension may be waived for small-scale machinery.

Annex IV

Types of machinery and components for which the procedure referred to in article 8 (2) (b) and (c) must be applied

A. Machinery

1. Circular saws (single- or multi-blade) for working with wood and analogous materials or for working with meat and analogous materials.
1.1 Sawing machines with fixed tool during operation, having a fixed bed with manual feed of the workpiece or with a demountable power feed.
1.2 Sawing machines with fixed tool during operation, having a manually operated reciprocating saw-bench or carriage.
1.3 Sawing machines with fixed tool during operation, having a built-in mechanical feed device for the work-pieces, with manual loading and/or unloading.

1.4 Sawing machines with movable tool during operation, with a mechanical feed device and manual loading and/or unloading.
2. Hand-fed surface planing machines for woodworking.
3. Thicknessers for one-side dressing with manual loading and/or unloading for woodworking.
4. Band-saws with a fixed or mobile bed and band-saws with a mobile carriage with manual loading and/or unloading for working with wood analogous materials or for working with meat and analogous materials.
5. Combined machines of the types referred to in 1 to 4 and 7 for working with wood and analogous materials.
6. Hand-fed tenoning machines with several tool holders for woodworking.
7. Hand-fed vertical spindle moulding machines for working with wood and analogous materials.
8. Portable chain saws for woodworking.
9. Presses, including press-brakes, for the cold working of metals, with manual loading and/or unloading, whose movable working parts may have a travel exceeding 6 mm and a speed exceeding 30 mm/s.
10. Injection or compression plastics-moulding machines with manual loading or unloading.
11. Injection or compression rubber-moulding machines with manual loading or unloading.
12. Machinery for underground working of the following types:
– machinery on rails: locomotives and brake-vans,
– hydraulic-powered roof supports,
– internal combustion engines to be fitted to machinery for underground working.
13. Manually-loaded trucks for the collection of household refuse incorporating a compression mechanism.
14. Guards and detachable transmission shafts with universal joints as described in section 3.4.7.
15. Vehicles servicing lifts.
16. Devices for the lifting of persons involving a risk of falling from a vertical height of more than three metres.
17. Machines for the manufacture of pyrotechnics.

B. Safety components

1. Electro-sensitive devices designed specifically to detect persons in order to ensure their safety (non-material barriers, sensor mats, electromagnetic detectors, etc.).
2. Logic units, which ensure the safety functions of bi-manual controls.
3. Automatic movable screens to protect the presses referred to in 9, 10 and 11.
4. Roll-over protection structures (ROPS).
5. Falling-objects protective structures (FOPS).

Annex V
EC declaration of conformity

For the purposes of this Annex, "machinery" means either "machinery" or "safety component" as defined in Article 1 (2).

1. The EC declaration of conformity is the procedure by which the manufacturer, or his authorised representative established in the Community declares that the machinery being placed on the market complies with all the essential health and safety requirements applying to it.
2. Signature of the EC declaration of conformity authorises the manufacturer, or his authorised representative in the Community, to affix the EC mark to the machinery.
3. Before drawing up the EC declaration of conformity, the manufacturer, or his authorised representative in the Community, shall have ensured and be

able to guarantee that the documentation listed below is and will remain available on his premises for any inspection purposes:

(a) a technical construction file comprising:
   - an overall drawing of the machinery together with drawings of the control circuits,
   - full detailed drawings, accompanied by any calculation notes, test results, etc, required to check the conformity of the machinery with the essential health and safety requirements,
   - a list of:
     - the essential requirements of this Directive,
     - standards, and
     - other technical specifications, which were used when the machinery was designed,
   - a description of methods adopted to eliminate hazards presented by the machinery,
   - if he so desires, any technical report or certificate obtained from a competent body or laboratory,[1]
   - if he declares conformity with a harmonised standard which provides therefore, any technical report giving the results of tests carried out at his choice either by himself or by a competent body or laboratory,[1]
   - a copy of the instructions for the machinery;

(b) for series manufacture, the internal measures that will be implemented to ensure that the machinery remains in conformity with the provisions of the Directive.

The manufacturer must carry out necessary research or tests on components, fittings or the completed machine to determine whether by its design or construction, the machine is capable of being erected and put into service safely.

Failure to present the documentation in response to a duly substantiated request by the competent national authorities may constitute sufficient grounds for doubting the presumption of conformity with the requirements of the Directive.

4. (a) The documentation referred to in 3 above need not permanently exist in a material manner but it must be possible to assemble it and make it available within a period of time commensurate with its importance. It does not have to include detailed plans or any other specific information as regards the sub-assemblies used for the manufacture of the machinery unless a knowledge of them is essential for verification of conformity with the basic safety requirements.

(b) The documentation referred to in 3 above shall be retained and kept available for the competent national authorities for at least 10 years following the date of manufacture of the machinery or of the last unit produced, in the case of series manufacture.

(c) The documentation referred to in 3 above shall be drawn up in one of the official languages of the Communities, with the exception of the instructions for the machinery.

[1] A body or laboratory is presumed competent if it meets the assessment criteria laid down in the relevant harmonised standards.

Annex VI
EC type-examination

For the purposes of this Annex, "machinery" means either "machinery" or "safety component" as defined in Article 1 (2).

1. EC type-examination is the procedure by which a notified body ascertains and certifies that an example of machinery satisfies the provisions of this Directive which apply to it.
2. The application for EC type-examination shall be lodged by the manufacturer or by his authorised representative established in the Community, with a single notified body in respect of an example of the machinery.

The application shall include:
- the name and address of the manufacturer or his authorised representative established in the Community and the place of manufacture of the machinery,
- a technical file comprising at least:
  - an overall drawing of the machinery together with drawings of the control circuits,
  - full detailed drawings, accompanied by any calculation notes, test results, etc, required to check the conformity of the machinery with the essential health and safety requirements,
  - a description of methods adopted to eliminate hazards presented by the machinery and a list of standards used,
  - a copy of the instructions for the machinery,
  - for series manufacture, the internal measures that will be implemented to ensure that the machinery remains in conformity with the provisions of the Directive.

It shall be accompanied by a machine representative of the production planned or, where appropriate, a statement of where the machine may be examined.

The documentation referred to above does not have to include detailed plans or any other specific information as regards the sub-assemblies used for the manufacture of the machinery unless a knowledge of them is essential for verification of conformity with the basic safety requirements.

3. The notified body shall carry out the EC type-examination in the manner described below:
- it shall examine the technical construction file to verify its appropriateness and the machine supplied or made available to it.
- during the examination of the machine, the body shall
  (a) ensure that it has been manufactured in conformity which the technical construction file and may safely be used under its intended working conditions;
  (b) check that standards, if used, have been properly applied;
  (c) perform appropriate examinations and tests to check that the machine complies with the essential health and safety requirements applicable to it.

4. If the example complies with the provisions applicable to it the body shall draw up an EC type-examination certificate which shall be forwarded to the applicant. That certificate shall state the conclusions of the examination, indicate any conditions to which its issue may be subject and be accompanied by the descriptions and drawings necessary for identification of the approved example.

The Commission, the Member States and the other approved bodies may obtain a copy of the certificate and, on a reasoned request, a copy of the technical construction file and of the reports on the examinations and tests carried out.

5. The manufacturer or his authorised representative established in the

Community shall inform the notified body of any modifications, even of a minor nature, which he has made or plans to make to the machine to which the example relates. The notified body shall examine those modifications and inform the manufacturer or his authorised representative established in the Community whether the EC type-examination certificate remains valid.
6. A body which refuses to issue an EC type-examination certificate shall so inform the other notified bodies. A body which withdraws an EC type-examination certificate shall so inform the Member State which notified it. The latter shall inform the other Member States and the Commission thereof, giving the reasons for the decision.
7. The files and correspondence referring to the EC type-examination procedures shall be drawn up in an official language of the Member State where the notified body is established or in a language acceptable to it.

Annex VII

Minimum criteria to be taken into account by Member States for the notification of bodies

For the purposes of this Annex, "machinery" means either "machinery" or "safety component" as defined in Article 1 (2).

1. The body, its director and the staff responsible for carrying out the verification tests shall not be the designer, manufacturer, supplier or installer of machinery which they inspect, nor the authorised representative of any of these parties. They shall not become either involved directly or as authorised representatives in the design, construction, marketing or maintenance of the machinery. This does not preclude the possibility of exchanges of technical information between the manufacturer and the body.
2. The body and its staff shall carry out the verification tests with the highest degree of professional integrity and technical competence and shall be free from all pressures and inducements, particularly financial, which might influence their judgement or the results of the inspection, especially from persons or groups of persons with an interest in the result of verifications.
3. The body shall have at its disposal the necessary staff and possess the necessary facilities to enable it to perform properly the administrative and technical tasks connected with verification; it shall also have access to the equipment required for special verification.
4. The staff responsible for inspection shall have:
– sound technical and professional training,
– satisfactory knowledge of the requirements of the tests they carry out and adequate experience of such tests,
– the ability to draw up the certificates, records and reports required to authenticate the performance of the tests.
5. The impartiality of inspection staff shall be guaranteed. Their remuneration shall not depend on the number of tests carried out or on the results of such tests.
6. The body shall take out liability insurance unless its liability is assumed by the State in accordance with national law, or the Member State itself is directly responsible for the tests.
7. The staff of the body shall be bound to observe professional secrecy with regard to all information gained in carrying out its tasks (except *vis-à-vis* the competent administrative authorities of the State in which its activities are carried out) under this Directive or any provision of national law giving effect to it.

COUNCIL DIRECTIVE 89/336/EEC

of 3 May 1989

on the approximation of the laws of the Member States relating to electromagnetic compatibility*†

THE COUNCIL OF THE EUROPEAN COMMUNITIES

Having regard to the Treaty establishing the European Economic Community, and in particular Article 100A thereof,

Having regard to the proposal from the Commission,

In cooperation with the European Parliament,

Having regard to the opinion of the Economic and Social Committee,

Whereas it is necessary to adopt measures with the aim of progressively establishing the internal market over a period expiring on 31 December 1992; whereas the internal market comprises an area without internal frontiers in which the free movement of goods, persons, services and capital is ensured;

Whereas Member States are responsible for providing adequate protection for radiocommunications and the devices, apparatus or systems whose performance may be degraded by electromagnetic disturbance produced by electrical and electronic apparatus against the degradation caused by such disturbances;

Whereas Member States are also responsible for ensuring that electric energy distribution networks are protected from electromagnetic disturbance with can affect them and, consequently, equipment fed by them;

Whereas Council Directive 86/361/EEC of 24 July 1986 (OJ 1986 L217/21) on the initial stage of the recognition of type-approval for telecommunications terminal equipment covers in particular the signals emitted by such equipment when it is operating normally and the protection of public telecommunications networks from harm; whereas it is therefore still necessary to provide adequate protection for these networks, including the equipment connected to them, against temporary disturbances caused by signals of an accidental nature that may be emitted by this equipment;

Whereas in some Member States, mandatory provisions define in particular the permissible electromagnetic disturbance levels that this equipment is liable to cause and its degree of immunity to such signals; whereas these mandatory provisions do not necessarily lead to different protection levels from one Member State to another but do, by their disparity, hinder trade within the Community;

Whereas the national provisions ensuring such protection must be harmonised in order to guarantee the free movement of electrical and electronic apparatus without lowering existing and justified levels of protection in the Member States;

Whereas Community legislation as it stands at present provides that, notwithstanding one of the fundamental rules of the Community, namely the free movement of goods, barriers to intra-Community trade resulting from disparities in national laws on the marketing of products have to be accepted in so far as those provisions may be recognised as necessary to

---

* As amended by: Dir 91/263
Dir 92/31
Dir 93/68

† OJ 1989 L139/19.

satisfy essential requirements; whereas the harmonisation of laws in the case in point must therefore be confined to those provisions needed to comply with the protection requirements relating to electromagnetic compatibility; whereas these requirements must replace the corresponding national provisions;

Whereas this Directive therefore defines only protection requirements relating to electromagnetic compatibility; whereas, to facilitate proof of conformity with these requirements, it is important to have harmonised standards at European level concerning electromagnetic compatibility, so that products complying with them may be assumed to comply with the protection requirements; whereas these standards harmonised at European level are drawn up by private bodies and must remain non-binding texts; whereas for that purpose the European Committee for Electrotechnical Standardisation (CENELEC) is recognised as the competent body in the field of this Directive for the adoption of harmonised standards in accordance with the general guidelines for cooperation between the Commission and the European Committee for Standardisation (CEN) and CENELEC signed on 13 November 1984; whereas, for the purposes of this Directive, a harmonised standard is a technical specification (European standard or harmonisation document) adopted by CENELEC upon a remit from the Commission in accordance with the provisions of Council Directive 83/189/EEC of 28 March 1983 (OJ 1983 L109/8) laying down a procedure for the provision of information in the field of technical standards and regulations, as last amended by Directive 88/182/EEC (OJ 1988 L81/75), and pursuant to the abovementioned general guidelines;

Whereas, pending the adoption of harmonised standards for the purposes of this Directive, the free movement of goods should be facilitated by accepting, as a transitional measure, on a Community level, apparatus complying with the national standards adopted, in accordance with the Community inspection procedure ensuring that such national standards meet the protection objectives of this Directive;

Whereas the EC declaration of conformity concerning the apparatus constitutes a presumption of its conformity with this Directive; whereas this declaration must take the simplest possible form;

Whereas, for apparatus covered by Directive 86/361/EEC, in order to obtain efficient protection as regards electromagnetic compatibility, compliance with the provisions of this Directive should nevertheless be certified by marks or certificates of conformity issued by bodies notified by the Member States; whereas, to facilitate the mutual recognition of marks and certificates issued by these bodies, the criteria to be taken into consideration for appointing them should be harmonised.

Whereas it is nevertheless possible that equipment might disturb radiocommunications and telecommunications networks; whereas provision should therefore be made for a procedure to reduce this hazard;

Whereas this Directive applies to the appliances and equipment covered by Directives 76/889/EEC (OJ 1976 L336/1) and 76/890/EEC (OJ 1976 L336/22) which relate to the approximation of the laws of the Member States relating to radio interference caused by electrical household appliances, portable tools and similar equipment and to the suppression of radio interference with regard to fluorescent lighting luminaries fitted with starters; whereas those Directive should therefore be repealed,

HAS ADOPTED THIS DIRECTIVE:

Article 1

For the purposes of this Directive:

1. "apparatus" means all electrical and electronic appliances together with

equipment and installations containing electrical and/or electronic components.
2. "electromagnetic disturbance" means any electromagnetic phenomenon which may degrade the performance of a device, unit of equipment or system. An electromagnetic disturbance may be electromagnetic noise, an unwanted signal or a change in the propagation medium itself.
3. "immunity" means the ability of a device, unit of equipment or system to perform without degradation of quality in the presence of an electromagnetic disturbance.
4. "electromagnetic compatibility" means the ability of a device, unit of equipment or system to function satisfactorily in its electromagnetic environment without introducing intolerable electromagnetic disturbances to anything in that environment.
5. "competent body" means any body which meets the criteria listed in Annex II and is recognised as such.
6. "EC type-examination certificate" is a document in which a notified body referred to in Article 10 (6) certifies that the type of equipment examined complies with the provisions of this Directive which concern it.

Article 2

1. This Directive applies to apparatus liable to cause electromagnetic disturbance or the performance of which is liable to be affected by such disturbance.

It defines the protection requirements and inspection procedures relating thereto.

2. In so far as protection requirements specified in this Directive are harmonised, in the case of certain apparatus, by specific Directives, this Directive shall not apply or shall cease to apply with regard to such apparatus or protection requirements upon the entry into force of those specific Directives.
3. Radio equipment used by radio amateurs within the meaning of Article 1, definition 53, of the radio regulations in the International Telecommunications Convention, shall be excluded from the scope of this Directive, unless the apparatus is available commercially.

Article 3

Member States shall take all appropriate measures to ensure that the apparatus as referred to in Article 2 may be placed on the market or taken into service only if it bears the CE marking provided for in Article 10 indicating its conformity to all the provisions of this Directive, including the conformity assessment procedures laid down in Article 10, when it is properly installed and maintained and when it is used for the purposes for which it is intended.

Article 4

The apparatus referred to in Article 2 shall be so constructed that:

(a) the electromagnetic disturbance it generates does not exceed a level allowing radio and telecommunications equipment and other apparatus to operate as intended;
(b) the apparatus has an adequate level of intrinsic immunity of electromagnetic disturbance to enable it to operate as intended.

The principal protection requirements are set out in Annex III.

Article 5

Member States shall not impede for reasons relating to electromagnetic compatibility the placing on the market and the taking into service on their territory of apparatus covered by this Directive which satisfies the requirements thereof.

Article 6

1. The requirements of this Directive shall not prevent the application in any Member State of the following special measures:

(a) measures with regard to the taking into service and use of the apparatus taken for a specific site in order to overcome an existing or predicted electromagnetic compatibility problem;

(b) measures with regard to the installation of the apparatus taken in order to protect the public telecommunications networks or receiving or transmitting stations used for safety purposes.

2. Without prejudice to Directive 83/189/EEC, Member States shall inform the Commission and the other Member States of the special measures taken pursuant to paragraph 1.

3. Special measures that have been recognised as justified shall be contained in an appropriate notice made by the Commission in the *Official Journal of the European Communities.*

Article 7

1. Member States shall presume compliance with the protection requirements referred to in Article 4 in the case of apparatus which is in conformity;

(a) with the relevant national standards transposing the harmonised standards, the reference numbers of which have been published in the *Official Journal of the European Communities.* Member States shall publish the reference numbers of such national standards;

(b) or with the relevant national standards referred to in paragraph 2 in so far as, in the areas covered by such standards, no harmonised standards exist.

2. Member States shall communicate to the Commission the texts of their national standards, as referred to in paragraph 1 (b), which they regard as complying with the protection requirements referred to in Article 4. The Commission shall forward such texts forthwith to the other Member States. In accordance with the procedure provided for in Article 8 (2), it shall notify the Member States of those national standards in respect of which there is a presumption of conformity with the protection requirements referred to in Article 4.

Member States shall publish the reference numbers of those standards. The Commission shall also publish them in the *Official Journal of the European Communities.*

3. Member States shall accept that where the manufacturer has not applied, or has applied only in part, the standards referred to in paragraph 1, or where no such standards exist, apparatus shall be regarded as satisfying the protection requirements has been certified by the means of attestation provided for in Article 10 (2).

Article 8

1. Where a Member State or the Commission considers that the harmonised standards referred to in Article 7 (1) (a) do not entirely satisfy the requirements referred to in Article 4, the Member State concerned or the Commission shall bring the matter before the Standing Committee set up by Directive 83/189/EEC, hereinafter referred to as 'the Committee', giving the reasons therefore. The Committee shall deliver an opinion without delay.

Upon receipt of the Committee's opinion, the Commission shall inform the Member States as soon as possible whether or not it is necessary to withdraw in whole or in part those standards from the publications referred to in Article 7 (1) (a).

2. After receipt of the communication referred to in Article 7 (2), the Commission shall consult the Committee. Upon receipt of the latter's opinion, the Commission shall inform the Member States as soon as possible whether or not the national standard in question shall enjoy the presumption of conformity and, if so, that the references thereof shall be published nationally.

If the Commission or a Member State considers that a national standard no longer satisfies the necessary conditions for presumption of compliance with the protection requirements referred to in Article 4, the Commission shall consult the Committee, which shall give its opinion without delay. Upon receipt of the latter's opinion, the Commission shall inform the Member States as soon as possible whether or not the standard in question shall continue to enjoy a presumption of conformity and, if not, that it must be withdrawn in whole or in part from the publications referred to in Article 7 (2).

Article 9

1. Where a Member State ascertains that apparatus accompanied by one of the means of attestation provided for in Article 10 does not comply with the protection requirements referred to in Article 4, it shall take all appropriate measures to withdraw the apparatus from the market, prohibit its placing on the market or restrict its free movement.

The Member State concerned shall immediately inform the Commission of any such measure, indicating the reasons for its decision and, in particular, whether non-compliance is due to:

(a) failure to satisfy the protection requirements referred to in Article 4, where the apparatus does not meet the standards referred to in Article 7 (1);
(b) incorrect application of the standards referred to in Article 7 (1);
(c) shortcomings in the standards referred to in Article 7 (1) themselves.

2. The Commission shall consult the parties concerned as soon as possible. If the Commission finds, after such consultations, that the action is justified, it shall forthwith so inform the Member State that took the action and the other Member States.

Where the decision referred to in paragraph 1 is attributed to shortcomings in the standards, the Commission, after consulting the parties, shall bring the matter before the Committee within two months if the Member State which has taken the measures intends to uphold them, and shall initiate the procedures referred to in Article 8.

3. Where apparatus which does not comply is accompanied by one of the means of attestation referred to in Article 10, the competent Member State shall take appropriate action against the author of the attestation and shall inform the Commission and the other Member States thereof.

4. The Commission shall ensure that the Member States are kept informed of the progress and outcome of this procedure.

Article 10

1. In the case of apparatus for which the manufacturer has applied the standards referred to in Article 7 (1), the conformity of apparatus with this Directive shall be certified by an EC declaration of conformity issued by the manufacturer or his authorised representative established within the Community. The declaration shall be held at the disposal of the competent authority for ten years following the placing of the apparatus on the market.

The manufacturer or his authorised representative established within the Community shall also affix the EC conformity mark to the apparatus or else to the packaging, instructions for use or guarantee certificate.

Where neither the manufacturer nor his authorised representative is established within the Community, the above obligation to keep the EC declaration of conformity available shall be the responsibility of the person who places the apparatus on the Community market.

The provisions governing the EC declaration and the CE marking are set out in Annex I.

Member States shall take the necessary measures to prohibit the affixing to apparatus, its packaging, the instructions for use and the guarantee certificate of markings which are likely to deceive third parties as to the meaning and form of the CE marking. Any other marking may be affixed to the apparatus, its packaging, the instructions for use or the guarantee certificate provided that the visibility and legibility of the CE marking is not thereby reduced.

2. In the case of apparatus for which the manufacturer has not applied, or has applied only in part, the standards referred to in Article 7 (1) or failing

such standards, the manufacturer or his authorised representative established within the community shall hold at the disposal of the relevant competent authorities, as soon as the apparatus is placed on the market, a technical construction file. This file shall describe the apparatus, set out the procedures used to ensure conformity of the apparatus with the protection requirements referred to in Article 4 and include a technical report or certificate, one or other obtained from a competent body.

The file shall be held at the disposal of the competent authorities for ten years following the placing of the apparatus on the market.

Where neither the manufacturer nor his authorised representative is established within the Community, this obligation to keep a technical file available shall be the responsibility of the person who places the apparatus on the Community market.

The conformity of apparatus with that described in the technical file shall be certified in accordance with the procedure laid down in paragraph 1.

Member States shall presume, subject to the provisions of this paragraph, that such apparatus meets the protection requirements referred to in Article 4.

. . .

5. The conformity of apparatus designed for the transmission of radiocommunications, as defined in the International Telecommunication Union Convention, with the provisions of this Directive shall be certified in accordance with the procedure laid down in paragraph 1 once the manufacturer or his authorised representative established within the Community has obtained an EC type-examination certificate concerning this apparatus issued by one of the notified bodies referred to in paragraph 6 below.

This provision shall not apply to the above apparatus where it is designed and intended exclusively for radio amateurs within the meaning of Article 2 (3).

6. Member States shall notify the Commission and the other Member States of the competent authorities referred to in this Article and the bodies responsible for issuing the EC type-examination certificates referred to in paragraph 5, together with the specific tasks which these bodies have been appointed to carry out and the identification numbers assigned to them beforehand by the Commission.

The Commission shall publish in the *Official Journal of the European Communities* a list of the notified authorities and bodies with their identification numbers and the tasks for which they have been notified. The Commission shall ensure that this list is kept up to date.

Such notification shall state whether those bodies are competent for all apparatus covered by this Directive or whether their responsibility is limited to certain specific areas.

Member States shall apply the criteria listed in Annex II for the assessment of the bodies to be notified.

Bodies which comply with the assessment criteria fixed by the relevant harmonised standards shall be presumed to comply with the aforementioned criteria.

A Member State which has notified a body must withdraw approval if it finds that the body no longer meets the criteria listed in Annex II. It shall forthwith inform the Commission and the other Member States thereof.

7. Without prejudice to Article 9:

(a) where a Member State or a competent authority establishes that the CE marking has been affixed unduly, the manufacturer or his authorised representative established within the Community shall be obliged to make the product comply, as regards the provisions concerning the CE marking and to end the infringement under conditions imposed by the Member State;

(b) where non-compliance continues, the Member State must take all appropriate measures to restrict or prohibit the placing on the market of the product in question or to ensure that it is withdrawn from the market in accordance with the procedures laid down in Article 9.

Article 11

Directive 76/889/EEC and Directive 76/890/EEC shall be repealed as from 1 January 1992.

Article 12

1. By 1 July 1991, Member States shall adopt and publish the laws, regulations and administrative provisions necessary to comply with this Directive. They shall inform the Commission thereof.

They shall apply these provisions as from 1 January 1992.

However, Member States shall, for the period up to 31 December 1995, authorise the placing on the market and/or the putting into service of apparatus referred to in this Directive conforming to the national regulations in force in their territory on 30 June 1992.

2. Member States shall communicate to the Commission the texts of the provisions of national law which they adopt in the field covered by this Directive.

Article 13

This Directive is addressed to the Member States.

Annex I

1. EC declaration of conformity

The EC declaration of conformity must contain the following:

– description of the apparatus to which it refers,

– reference to the specification under which conformity is declared, and, where appropriate, to the national measures implemented to ensure the conformity of the apparatus with the provisions of the Directive,

– identification of the signatory empowered to bind the manufacturer or his authorised representative,

– where appropriate, reference to the EC type-examination certificate issued by a notified body.

2. CE conformity mark

– The CE conformity mark shall consist of the initials "CE" taking the following form:

– If the CE marking is reduced or enlarged the proportions given in the above graduated drawing must be respected.

– Where apparatus is the subject of other Directives covering other aspects and which also provide for the CE conformity marking, the latter shall indicate that the appliances are also presumed to conform to those other directives.

– However, where one or more of these Directives allow the manufacturer, during a transitional period, to choose which arrangements to apply, the CE marking shall indicate conformity only to the Directives applied by the manufacturer. In this case, particulars of the Directives applied, as published in the *Official Journal of the European Communities*, must be given in the documents, notices or instructions required by the Directives and accompanying such apparatus.

– The various components of the CE marking must have substantially the same vertical dimensions, which may not be less than 5 mm.

Annex II

Criteria for the assessment of the bodies to be notified

The bodies designated by the Member States must fulfil the following minimum conditions:

1. availability of personnel and of the necessary means and equipment;
2. technical competence and professional integrity of personnel;
3. independence, in carrying out the tests, preparing the reports, issuing the certificates and performing the verification function provided for in this Directive, of staff and technical personnel in relation to all circles, groups or persons directly or indirectly concerned with the product in question;
4. maintenance of professional secrecy by personnel;
5. possession of civil liability insurance unless such liability is covered by the State under national law.

Fulfilment of the conditions under points 1 and 2 shall be verified at intervals by the competent authorities of the Member States.

Annex III

Illustrative list of the principal protection requirements

The maximum electromagnetic disturbance generated by the apparatus shall be such as not to hinder the use of in particular the following apparatus:

(a) domestic radio and television receivers
(b) industrial manufacturing equipment
(c) mobile radio equipment
(d) mobile radio and commercial radiotelephone equipment
(e) medical and scientific apparatus
(f) information technology equipment
(g) domestic appliances and household electronic equipment
(h) aeronautical and marine radio apparatus
(i) educational electronic equipment
(j) telecommunications networks and apparatus
(k) radio and television broadcast transmitters
(l) lights and fluorescent lamps.

Apparatus, and especially the apparatus referred to in (a) to (l), should be constructed in such a way that it has an adequate level of electromagnetic immunity in the usual electromagnetic compatibility environment where the apparatus is intended to work so as to allow its unhindered operation taking into account the levels of disturbance generated by apparatus complying with the standards laid down in Article 7.

The information required to enable use in accordance with the intended purpose of the apparatus must be contained in the instructions accompanying the apparatus.

COUNCIL DIRECTIVE 90/384/EEC

of 20 June 1990

on the harmonisation of the laws of the Member States relating to non-automatic weighing instruments*†

THE COUNCIL OF THE EUROPEAN COMMUNITIES,

Having regard to the Treaty establishing the European Economic Community, and in particular Article 100A thereof,

Having regard to the proposal from the Commission,

In cooperation with the European Parliament,

Having regard to the opinion of the Economic and Social Committee,

Whereas Member States have the responsibility of protecting the public against incorrect results of weighing operations by means of non-automatic weighing instruments when used for certain categories of applications;

Whereas, in each Member State, mandatory provisions fix in particular the necessary performance requirements of non-automatic weighing instruments by specifying metrological and technical requirements, together with inspection procedures before and after going into service; whereas these mandatory provisions do not necessarily lead to different levels of protection from one Member State to another but do, by their disparity, impede trade within the Community;

Whereas the national provisions ensuring such protection must be harmonised in order to guarantee the free movement of non-automatic weighing instruments while ensuring a justified level of protection in the Community;

Whereas Community legislation as it stands at present provides that, notwithstanding one of the fundamental rules of the Community, namely the free movement of goods, barriers to intra-Community movement resulting from disparities in national laws on the use of products have to be accepted in so far as the provisions of those national laws are recognised as necessary to ensure that the products concerned meet essential requirements; whereas the harmonisation of laws in the present case must therefore be confined to those provisions needed to ensure that non-automatic weighing instruments satisfy the essential metrological and performance requirements must replace the corresponding national provisions;

Whereas this Directive therefore contains only mandatory and essential requirements; whereas, to facilitate proof of conformity with the essential requirements, it is necessary to have harmonised standards at European level, in particular as to the metrological, design and construction characteristics, so that instruments complying with those harmonised standards may be assumed to conform to the essential requirements; whereas these standards, harmonised at European level, are drawn up by private bodies and must remain non-mandatory texts; whereas for that purpose the European Committee for Standardisation (CEN) and the European Committee for Electrotechnical Standardisation (CENELEC) are recognised as the competent bodies for the adoption of harmonised standards in accordance with the general guidelines for cooperation between the Commission and those two bodies signed on 13 November 1984; whereas, within the meaning of this Directive, a harmonised standard is a technical specifica-

---

*OJ 1990 L189/1.

†Amended by Dir 93/68.

tion (European standard or harmonised document) adopted by one or both of those bodies upon a remit from the Commission in accordance with Council Directive 83/189/EEC of 28 March 1983 (OJ 1983 L109/8) laying down a procedure for the provision of information in the field of technical standards and regulations, as amended by Directive 88/182/EEC (OJ 1988 L81/75), and the abovementioned general guidelines;

Whereas assessment of conformity with the relevant metrological and technical provisions is necessary to provide effective protection for users and third parties; whereas, the existing conformity assessment procedures differ from one Member State to another; whereas, to avoid multiple assessments of conformity, which are in effect barriers to the free movement of the instruments, arrangements should be made for the mutual recognition of conformity assessment procedures by the Member States; whereas, to facilitate the mutual recognition of conformity assessment procedures, harmonised Community procedures should be set up, together with harmonised criteria for the designation of the bodies responsible for carrying out tasks pertaining to the conformity assessment procedures;

Whereas it is therefore essential to ensure that such designated bodies ensure a high level of quality throughout the Community;

Whereas the presence on a non-automatic weighing instrument of the CE marking of conformity or of the sticker bearing the letter "M" indicates that there is a presumption that it satisfies the provisions of this Directive and therefore makes it unnecessary to repeat the assessments of conformity already carried out;

Whereas the measures aimed at the gradual establishment of the internal market must be adopted by 31 December 1992; whereas the internal market consists of an area without internal frontiers within which the free movement of goods, persons, services and capital is guaranteed,

HAS ADOPTED THIS DIRECTIVE:

Chapter 1
Scope, placing on the market, free movement

Article 1

1. A weighing instrument is defined as a measuring instrument serving to determine the mass of a body by using the action of gravity on that body. A weighing instrument may also serve to determine other mass-related magnitudes, quantities, parameters or characteristics.

A non-automatic weighing instrument is defined as a weighing instrument requiring the intervention of an operator during weighting.

This Directive applies to all non-automatic weighing instruments, hereinafter referred to as "instruments".

2. A distinction is made in this Directive between two categories of instrument use:

(a) (1) determination of mass for commercial transactions;
 (2) determination of mass for the calculation of a toll, tariff, tax, bonus, penalty, remuneration, indemnity or similar type of payment;
 (3) determination of mass for the application of laws or regulations; expert opinion given in court proceedings;
 (4) determination of mass in the practice of medicine for weighing patients for the purposes of monitoring, diagnosis and medical treatment;
 (5) determination of mass for making up medicines on prescription in a pharmacy and determination of mass in analyses carried out in medical and pharmaceutical laboratories;
 (6) determination of price on the basis of mass for the purposes of direct sales to the public and the making-up of prepackages;

(b) all applications other than those listed in point 2 (a) of this article.

Article 2

1. Member States shall take all steps to ensure that instruments may not be placed on the market unless they meet the requirements of this Directive which apply to them.
2. Member States shall take all steps to ensure that instruments may not be brought into service for the uses set out in Article 1 (2) (a) unless they meet the requirements of this Directive which apply to them, including the conformity assessment procedures referred to in Chapter II, and accordingly bear the CE marking provided for in Article 10.

Article 3

Instruments used for the applications listed in Article 1 (2) (a) must satisfy the essential requirements set out in Annex I.

In cases where the instrument includes or is connected to devices which are not used for the applications listed in Article 1 (2) (a), such devices shall not be subject to the essential requirements.

Article 4

1. Member States shall not impede the placing on the market of instruments which meet the requirements of this Directive which apply to them.
2. Member States shall not impede the putting into service for the uses referred to in Article 1 (2) (a) of instruments which meet the requirements of this Directive which apply to them.

Article 5

1. Member States shall presume conformity with the essential requirements referred to in Article 3 in respect of instruments which comply with the relevant national standards implementing the harmonised standards that meet the essential requirements referred to in Article 3.
2. The Commission shall publish the references of the harmonised standards referred to in paragraph 1 in the *Official Journal of the European Communities*.

Member States shall publish the references of the national standards referred to in paragraph 1.

Article 6

Where a Member State or the Commission considers that the harmonised standards referred to in Article 5 (1) do not fully meet the essential requirements referred to in Article 3, the Commission or the Member State concerned shall bring the matter before the Standing Committee set up under Directive 83/189/EEC, hereinafter referred to as "the Committee", giving reasons for doing so. The Committee shall deliver an opinion without delay.

In the light of the Committee's opinion, the Commission shall inform the Member States whether or not it is necessary to withdraw those standards from the publications referred to in Article 5 (2).

Article 7

1. Where a Member State considers that instruments bearing the CE marking of conformity referred to in Annex 2, sections 2, 3 and 4, do not meet the requirements of this Directive when properly installed and used for the purposes for which they are intended, it shall take all appropriate measures to withdraw those instruments from the market or to prohibit or restrict their being put into service and/or placed on the market.

The Member State concerned shall immediately inform the Commission of any such measure, indicating the reasons for its decision, and in particular whether non-compliance is due to:

(a) failure to meet the essential requirements referred to in Article 3, where instruments do not meet the standards referred to in Article 5 (1);
(b) incorrect application of the standards referred to in Article 5 (1);
(c) shortcomings in the standards

referred to in Article 5 (1) themselves.

2. The Commission shall enter into consultation with the parties concerned as soon as possible.

After such consultation the Commission shall immediately inform the Member State, which took the action, of the result. Should it find that the measure is justified it shall immediately inform the other Member States.

If the decision is attributed to shortcomings in the standards, the Commission, after consulting the parties concerned, shall bring the matter before the Committee within two months if the Member State which has taken the measures intends to maintain them, and shall subsequently initiate the procedures referred to in Article 6.

3. Where an instrument which does not comply bears the CE marking of conformity, the competent Member State shall take appropriate action against whomsoever has affixed the mark and shall inform the Commission and the other Member States thereof.

4. The Commission shall ensure that the Member States are kept informed of the progress and outcome of this procedure.

Chapter II
Conformity assessment

Article 8

1. The conformity of instruments to the essential requirements set out in Annex I may be certified by either of the following procedures as selected by the applicant:

(a) EC type examination as referred to in Annex II.1, followed either by the EC declaration of type conformity (guarantee of production quality) as referred to in Annex II.2, or by the EC verification as referred to in Annex II.3.

However, EC type examination shall not be compulsory for instruments which do not use electronic devices and whose load-measuring device does not use a spring to balance the load;

(b) EC unit verification as referred to in Annex II.4.

2. The documents and correspondence relating to the procedures referred to in paragraph 1 shall be drafted in an official language of the Member State where the said procedures are to be carried out, or in a language accepted by the competent body.

3. (a) Where the instruments are subject to other Directives covering other aspects and which also provide for the affixing of the CE marking, the latter shall indicate that the instruments in question are also presumed to conform to the provisions of those other Directives.

3. (b) However, where one or more of the Directives which apply to the instruments allow the manufacturer, during a transitional period, to choose which arrangements to apply, the CE marking shall indicate conformity only to the Directives applied by the manufacturer. In this case, particulars of the Directives applied, as published in the *Official Journal of the European Communities*, must be given in the documents, notices or instructions required by the Directives and accompanying such instruments.

Article 9

1. Member States shall notify the Commission and the other Member States of the bodies which they have appointed to carry out the procedures referred to in Article 8 together with the specific tasks which these bodies have been appointed to carry out and the identification numbers assigned to them beforehand by the Commission.

The Commission shall publish in the *Official Journal of the European Communities* a list of the notified bodies and their identification numbers and the tasks for which they have been notified. The Commission shall ensure that this list is kept up to date.

2. Member States shall apply the minimum criteria set out in Annex V for the designation of bodies. Bodies

which satisfy the criteria fixed by the relevant harmonised standards shall be presumed to satisfy the criteria set out in Annex V.

3. A Member State which has designated a body shall cancel the designation if the body no longer meets the criteria for designation referred to in paragraph 2. It shall immediately inform the other Member States and the Commission thereof and withdraw the notification.

Chapter III
CE marking of conformity and inscriptions

Article 10

1. The CE marking of conformity and the required supplementary data as described in Annex IV.1 shall be affixed in a clearly visible, easily legible and indelible form to instruments for which EC conformity has been established.

2. The inscriptions referred to in Annex IV.2 shall be affixed in a clearly visible, easily legible and indelible form to all other instruments.

3. The affixing of markings on the instruments which are likely to deceive third parties as to the meaning and form of the CE marking shall be prohibited. Any other marking may be affixed to the instruments provided that the visibility and legibility of the CE markings is not thereby reduced.

Article 11

Without prejudice to Article 7.

(a) where a Member State establishes that the CE marking has been affixed unduly, the manufacturer or his authorised representative established within the Community shall be obliged to make the instrument conform as regards the provisions concerning the CE marking and to end the infringement under the conditions imposed by the Member State;

(b) where non-conformity continues, the Member State must take all appropriate measures to restrict or prohibit the placing on the market of the instrument in question or to ensure that it is withdrawn from the market in accordance with the procedures laid down in Article 7.

Article 12

Where an instrument which is used for any of the applications referred to in Article 1 (2) (a) includes or is connected to devices that have not been subject to conformity assessment as referred to in Article 8, each of these devices shall bear the symbol restricting its use as defined by Annex IV.3. This symbol shall be affixed to the devices in a clearly visible and indelible form.

Chapter IV
Final provisions

Article 13

Member States take all steps to ensure that instruments bearing the CE marking attesting conformity with the requirements of this Directive continue to conform to those requirements.

Article 14

Any decision taken pursuant to this Directive and resulting in restrictions on the putting into service of an instrument shall state the exact grounds on which it is based. Such a decision shall be notified without delay to the party concerned, who shall at the same time be informed of the judicial remedies available to him under the laws in force in the Member State in question and of the time limits to which such remedies are subject.

Article 15

1. Member States shall, before 1 July 1992, adopt and publish the laws, regulations and administrative provisions necessary in order to comply with this Directive. They shall forthwith inform the Commission thereof.

2. Member States shall apply such provisions from 1 January 1993.

3. However, by way of derogation from paragraph 2, Member States shall permit during a period of 10 years from the date on which they apply the provisions referred to in paragraph 1 the placing on the market and/or putting into service of instruments which conform to the rules in force before that date.
4. Member States shall communicate to the Commission the texts of the provisions of national law which they adopt in the field covered by this Directive.
5. Directive 73/360/EEC shall be repealed as from 1 January 1993, except as regards the application of paragraph 3.

Article 16

This Directive is addressed to the Member States.

Annex I

The essential requirements that must be met by the instruments referred to in Article 1 (2) (a) are set out below. The terminology used is that of the Organisation internationale de Métrologie Légale.

Preliminary observation

Where an instrument includes or is connected to more than one indicating or printing device used for the applications listed in Article 1 (2) (a), those devices which repeat the results of the weighing operation and which cannot influence the correct functioning of the instrument shall not be subject to the essential requirements if the weighing results are printed or recorded correctly and indelibly by a part of the instrument which meets the essential requirements and the results are accessible to both parties concerned by the measurement. However, in the case of instruments used for direct sales to the public, display and printing devices for the vendor and the customer must fulfil the essential requirements.

Metrological requirements

1. Units of mass

The units of mass used shall be the legal units within the meaning of Directive 80/181/EEC (OJ 1981 L39/39), as last amended by Directive 85/1/EEC (OJ 1985 L2/11).

Subject to compliance with this condition, the following units are permitted:

- SI units: kilogram, microgram, milligram, gram, tonne,
- Imperial units: pound, ounce (avoirdupois), Troy ounce,
- other non-SI units: metric carat, if weighing precious stones.

For instruments that make use of the Imperial units of mass referred to above, the relevant essential requirements specified below shall be converted to the said Imperial units, using simple interpolation.

2. Accuracy classes

2.1. The following accuracy classes have been defined:

I special
II high
III medium
IIII ordinary

The specifications of these classes are given in Table 1.

Table 1 *Accuracy classes*

Class	*Verification scale interval(e)*	*Minimum capacity (Min)*	*Number of verification scale intervals* $n = \frac{Max}{e}$	
		minimum value	*minimum value*	*maximum value*
I	0.001 g $\leqslant$ e	100 e	50,000	–
II	0.001 g $\leqslant$ e $\leqslant$ 0.005 g	20 e	100	100,000
	0.1 g $\leqslant$ e	50 e	5,000	100,000
III	0.1 g $\leqslant$ e $\leqslant$ 2 g	20 e	100	10,000
	5 g $\leqslant$ e	20 e	500	10,000
IIII	5 g $\leqslant$ e	10 e	100	1,000

The minimum capacity is reduced to 5e for instruments in classes II and III for determining a conveying tariff.

2.2. *Scale intervals*

2.2.1. The actual scale interval (d) and the verification scale interval (e) shall be in the form:
1×10^k, 2×10^k, or 5×10^k mass units, k being any integer or zero.
2.2.2. For all instruments other than those with auxiliary indicating devices:
d=e
2.2.3. For instruments with auxiliary indicating devices the following conditions apply:
$e = 1 \times 10^k$ g
$d < e \leqslant 10$ d
except for instruments of class 1 with d $< 10^{-4}$ g, for which e $= 10^{-3}$ g.

3. Classification

3.1. *Instruments with one weighing range*

Instruments equipped with an auxiliary indicating device shall belong to class I or class II. For these instruments the minimum capacity lower limits for these two classes are obtained from Table 1 by replacement in column 3 of the verification scale interval (e) by the actual scale interval (d).
If d $< 10^{-4}$ g, the maximum capacity of class 1 may be less than 50,000 e.

3.2. *Instruments with multiple weighing ranges*

Multiple weighing ranges are permitted, provided they are clearly indicated on the instrument. Each individual weighing range is classified according to 3.1. If the weighing ranges fall into different accuracy classes the instrument shall comply with the severest of the requirements that apply for the accuracy classes in which the weighing ranges fall.

3.3. *Multi-interval instruments*

3.3.1. Instruments with one weighing range may have several partial weighing ranges (multi-interval instruments).

Multi-interval instruments shall not be equipped with an auxiliary indicating device.
3.3.2. Each partial weighing range 1 of multi-interval instruments is defined by:
– its verification scale interval e_i
with $e_{(i+1)} > e_i$
– its maximum capacity Max_i
with $Max_r = Max$
– its minimum capacity Min_i
with $Min_i = Max_{(i-1)}$
and $Min_1 = Min$
where:
i = 1, 2, . . . r,
i = partial weighing range number,
r = the total number of partial weighing ranges.
All capacities are capacities of net load,

irrespective of the value of any tare used.

3.3.3. The partial weighing ranges are classified according to Table 2. All partial weighing ranges shall fall into the same accuracy class, this class being the instrument's accuracy class.

4. Accuracy

4.1. On implementation of the procedures laid down in Article 8, the error of indication shall not exceed the maximum permissible error of indication as shown in Table 3. In case of digital indication the error of indication shall be corrected for the rounding error.

The maximum permissible errors apply to the net and tare value for all possible loads, excluded preset tare values.

4.2. The maximum permissible errors in service are twice the maximum permissible errors fixed in section 4.1.

5. Weighing results of an instrument shall be repeatable, and shall be reproducible by the other indicating devices used and with other methods of balancing used.

The weighing results shall be sufficiently insensitive to changes in the position of the load on the load receptor.

6. The instrument shall react to small variations in the load.

7. Influence quantities and time

7.1. Instruments of classes II, III and IIII, liable to be used in a tilted position, shall be sufficiently insensitive to the degree of tilting that can exist in a normal installed condition.

7.2. The instruments shall meet the metrological requirements within the temperature range specified by the manufacturer. The value of this range shall be at least equal to:

5 °C for an instrument in class I,
15 °C for an instrument in class II,
30 °C for an instrument in class III or IIII.

Table 2 *Multi-interval instruments*

i = 1, 2, . . . r
i = partial weighing range number
r = total number of partial weighing ranges

Class	*Verification scale interval (e)*	*Minimum capacity (Min)*	*Number of verification scale intervals*	
		Minimum value	*Minimum value*([1]) $n = \frac{Max_i}{e_{(i+1)}}$	*Maximum value* $n = \frac{Max_i}{e_i}$
I	0.001 g $\leq e_i$	100 e_i	50,000	—
II	0.001 g $\leq e_i \leq$ 0.05 g	20 e_i	5,000	100,000
	0.1 g $\leq e_i$	50 e_i	5,000	100,000
III	0.1 g $\leq e_i$	20 e_i	500	10,000
IIII	5 g $\leq e_i$	10 e_i	50	1,000

Table 3 *Maximum permissible errors*

Load				*Maximum permissible error*
Class I	*Class II*	*Class III*	*Class IIII*	
0 ≤ m ≤ 50,000 e	0 ≤ m ≤ 5,000 e	0 ≤ m ≤ 500 e	0 ≤ m ≤ 50 e	± 0.5 e
50,000 e < m ≤ 200,000 e	5,000 e < m ≤ 20,000 e	500 e < m ≤ 2,000 e	50 e < m ≤ 200 e	± 1.0 e
200,000 e < m	20,000 e < m ≤ 100,000 e	2,000 e < m ≤ 10,000 e	200 e < m ≤ 1,000 e	±1.5 e

In the absence of a manufacturer's specification, the temperature range of −10 °C to +40 °C applies.

7.3. Instruments operated from a mains power supply shall meet the metrological requirements under conditions of power supply within the limits of normal fluctuation.

Instruments operated from battery power shall indicate whenever the voltage drops below the minimum required value and shall under those circumstances either continue to function correctly or be automatically put out of service.

7.4. Electronic instruments, except those in class I and in class II if e is less than 1 g, shall meet the metrological requirements under conditions of high relative humidity at the upper limit of their temperature range.

7.5. Loading an instrument in class II, III or IIII for a prolonged period of time shall have a negligible influence on the indication at load or on the zero indication immediately after removal of the load.

7.6. Under other conditions the instruments shall either continue to function correctly or be automatically put out of service.

Design and construction

8. General requirements

8.1. Design and construction of the instruments shall be such that the instruments will preserve their metrological qualities when properly used and installed, and when used in an environment for which they are intended. The value of the mass must be indicated.

8.2. When exposed to disturbances, electronic instruments shall not display the effects of significant faults, or shall automatically detect and indicate them.

Upon automatic detection of a significant fault, electronic instruments shall provide a visual or audible alarm that shall continue until the user takes corrective action or the fault disappears.

8.3. The requirements of 8.1 and 8.2 shall be met on a lasting basis during a period of time that is normal in view of the intended use of such instruments.

Digital electronic devices shall always exercise adequate control of the correct operation of the measuring process, of the indicating facility, and of all data storage and data transfer.

Upon automatic detection of a significant durability error, electronic instruments shall provide a visual or audible alarm that shall continue until the user takes corrective action or the error disappears.

8.4. When external equipment is connected to an electronic instrument through an appropriate interface the metrological qualities of the instrument shall not be adversely influenced.

8.5. The instruments shall have no characteristics likely to facilitate fraudulent use, whereas possibilities for unintentional misuse shall be minimal. Components that may not be dismantled or adjusted by the user shall be secured against such actions.

8.6. Instruments shall be designed to permit ready execution of the statutory controls laid down by this Directive.

9. Indication of weighing results and other weight values

The indication of the weighing results and other weight values shall be accurate, unambiguous and non-misleading and the indicating device shall permit easy reading of the indication under normal conditions of use.

The names and symbols of the units referred to in paragraph 1 of this Annex shall comply with the provisions of Directive 80/181/EEC (OJ 1980 L33/39) with the addition of the symbol for the metric carat which shall be the symbol "ct".

Indication shall be impossible above the maximum capacity (Max), increased by 9e.

An auxiliary indicating device is permitted only to the right of the decimal mark. An extended indicating device may be used only temporarily, and printing shall be inhibited during its functioning.

Secondary indications may be shown, provided that they cannot be mistaken for primary indications.

10. Printing of weighing results and other weight values

Printed results shall be correct, suitably identified and unambiguous. The printing shall be clear, legible, non-erasable and durable.

11. Levelling

When appropriate, instruments shall be fitted with a levelling device and a level indicator, sufficiently sensitive to allow proper installation.

12. Zeroing

Instruments may be equipped with zeroing devices. The operation of these devices shall result in accurate zeroing and shall not cause incorrect measuring results.

13. Tare devices and preset tare devices

The instruments may have one or more tare devices and a preset tare device. The operation of the tare devices shall result in accurate zeroing and shall ensure correct net weighing. The operation of the preset tare device shall ensure correct determination of the calculated net value.

14. Instruments for direct sales to the public with a maximum capacity not greater than 100 kg: additional requirements

Instruments for direct sale to the public shall show all essential information about the weighing operation and, in the use of price-indicating instruments, shall clearly show the customer the price calculation of the product to be purchased.

The price to pay, if indicated, shall be accurate.

Price-computing instruments shall display the essential indications long enough for the customer to read them properly.

Price-computing instruments may perform functions other than per-article weighing and price computation only if all indications related to all transactions are printed clearly, unambiguously and conveniently arranged on a ticket or label for the customer.

Instruments shall bear no characteristics that can cause, directly or indirectly, indications whose interpretation is not easy or straightforward.

Instruments shall safeguard customers against incorrect sales transactions due to their malfunctioning.

Auxiliary indicating devices and extended indicating devices are not permitted.

Supplementary devices are permitted only if they cannot lead to fraudulent use.

Instruments similar to those normally used for direct sales to the public which do not satisfy the requirements of this section must carry near to the display the indelible marking "Not to be used for direct sale to the public".

15. Price labelling instruments

Price labelling instruments shall meet the requirements of price indicating instruments for direct sale to the public, as far as applicable to the instrument in question. The printing of a price label shall be impossible below a minimum capacity.

Annex II

1. EC type-examination

1.1. EC type-examination is the procedure whereby a notified body verifies and certifies that an instrument, representative of the production envisaged, meets the requirements of this Directive that apply to it.

1.2. The application for type-examination shall be lodged by the manufacturer or his authorised representative established within the Community with a single notified body.

The application shall include:

- the name and address of the manufacturer and, if the application is lodged by the authorised representative, his name and address in addition,
- a written declaration that the application has not been lodged with any other notified body,
- the design documentation, as described in Annex III.

The applicant shall place at the disposal of the notified body an instrument, representative of the production envisaged, hereinafter called the "type".

1.3. The notified body shall:

1.3.1. examine the design documentation and verify that the type has been manufactured in accordance with that documentation;

1.3.2. agree with the applicant on the location where the examinations and/or tests shall be carried out;

1.3.3. perform or have performed the appropriate examinations and/or tests to check whether the solutions adopted by the manufacturer meet the essential requirements where the standards referred to in Article 5 have not been applied;

1.3.4. perform or have performed the appropriate examinations and/or tests to check whether, where the manufacturer has chosen to apply the relevant standards, these have been applied effectively, thereby assuring conformity with the essential requirements.

1.4. Where the type meets the provisions of this Directive, the notified body shall issue an EC type-approval certificate to the applicant. The certificate shall contain the conclusions of the examination, conditions (if any) for its validity, the necessary data for identification of the approved instrument and, if relevant, a description of its functioning. All the relevant technical elements such as drawings and layouts shall be annexed to the EC type-approval certificate.

The certificate shall have a validity period of 10 years from the date of its issue, and may be renewed for subsequent periods of 10 years each.

In the event of fundamental changes to the design of the instruments, *e.g.* as a result of the application of new techniques, the validity of the certificate may be limited to two years and extended by three years.

1.5. Each notified body shall periodically make available to all Member States the list of:

- applications received for EC type-examination,
- EC type-approval certificates issued,
- applications for type-certificates refused,
- additions and amendments relating to documents already issued.

Each notified body shall moreover inform all the Member States forthwith of withdrawals of EC type-approval certificates.

Each Member State shall make this information available to the bodies which it has notified.

1.6. The other notified bodies may receive a copy of the certificates together with the Annexes to them.

1.7. The applicant shall keep the notified body that has issued the EC type-approval certificate informed of any modification to the approved type.

Modifications to the approved type must receive additional approval from the notified body that issued the EC type-approval certificate where such changes influence conformity with the essential requirements of this Directive or the prescribed conditions for use of the instrument. This additional approval is given in the form of an addition to the original EC type-approval certificate.

2. EC declaration of type conformity (guarantee of production quality)

2.1. The EC declaration of type conformity (guarantee of production quality) is the procedure whereby the manufacturer who satisfies the obligations of paragraph 2.2. declares that the instruments concerned are, where applicable, in conformity with the type as described in the EC type-approval certificate and satisfy the requirements of this Directive that apply to them.

The manufacturer or his authorised representative established within the Community shall affix the CE marking to each instrument and the inscriptions provided for in Annex IV and shall draw up a written declaration of conformity.

The CE marking shall be accompanied by the identification number of the notified body responsible for the EC surveillance referred to in paragraph 2.4.

2.2. The manufacturer shall have adequately implemented a quality system as specified in paragraph 2.3 and shall be subject to EC surveillance as specified in paragraph 2.4.

2.3. *Quality system*

2.3.1. The manufacturer shall lodge an application for approval of this quality system with a notified body.

The application shall include:

- an undertaking to carry out the obligations arising from the approved quality system,
- an undertaking to maintain the approved quality system to ensure its continuing suitability and effectiveness.

The manufacturer shall make available to the notified body all relevant information, in particular the quality system's documentation and the design documentation of the instrument.

2.3.2. The quality system shall ensure conformity of the instruments with the type as described in the EC type-approval certificate and with the requirements of this Directive that apply to them.

All the elements, requirements and provisions adopted by the manufacturer shall be documented in a systematic and orderly manner in the form of written rules, procedures and instructions. This quality system documentation shall ensure a proper understanding of the quality programmes, plans, manuals and records.

It shall contain in particular an adequate description of:

- the quality objectives and the organisational structure, responsibilities and powers of the management with regard to product quality,
- the manufacturing process, the quality control and assurance techniques and the systematic measures that will be used,
- the examinations and tests that will be carried out before, during, and after manufacture, and the frequency with which they will be carried out,
- the means to monitor the achievements of the required product quality and the effective operation of the quality system.

2.3.3. The notified body shall examine and evaluate the quality system to determine whether it satisfies the requirements referred to in paragraph 2.3.2. It shall presume conformity with these requirements in respect of quality systems that implement the corresponding harmonised standard.

It shall notify its decision to the manufacturer and inform the other notified bodies thereof. The notification to the manufacturer shall contain the conclusions of the examination and, in the event of refusal, the justification for the decision.

2.3.4. The manufacturer or his authorised representative shall keep the notified body that has approved the quality system informed of any updating of the quality assurance system in relation to changes brought about by, *e.g.* new technologies and new quality concepts.

2.3.5. Any notified body that withdraws approval of a quality system shall so inform the other notified bodies.

2.4. *EC surveillance*

2.4.1. The purpose of EEC surveillance is to make sure that the manufacturer duly fulfils the obligations arising out of the approved quality system.

2.4.2. The manufacturer shall grant the notified body access for inspection purposes to the manufacture, inspection, testing and storage premises and shall provide it with all necessary information, in particular:

– the quality system documentation,
– the design documentation,
– the quality records, *e.g.* the inspection reports and tests and calibration data, reports on the qualifications of the personnel concerned, etc.

The notified body shall periodically carry out audits in order to ensure that the manufacturer is maintaining and applying the quality system; it shall provide the manufacturer with an audit report.

In addition, the notified body may carry out unscheduled visits to the manufacturer. During such visits, the notified body may carry out full or partial audits. It shall provide the manufacturer with a report on the visit, and, where appropriate, an audit report.

2.4.3. The notified body shall ensure that the manufacturer maintains and applies the approved quality system.

3. EC verification

3.1. EC verification is the procedure whereby the manufacturer or his authorised representative established within the Community ensures and declares that the instruments which have been checked in accordance with paragraph 3.3 are, where applicable, in conformity to the type described in the EC type-examination certificate and satisfy the requirements of the Directive which apply to them.

3.2. The manufacturer shall take all necessary measures in order that the manufacturing process ensures conformity of the instruments, where applicable, with the type as described in the EC type-examination certificate and with the requirements of this Directive which apply to them. The manufacturer or his authorised representative established within the Community shall affix the CE marking to each instrument and draw up a written declaration of conformity.

3.3. The notified body shall carry out the appropriate examinations and tests in order to check the conformity of the product to the requirements of this Directive by examination and testing of every instrument, as specified in paragraph 3.5.

3.4. For instruments not subject to EC type-approval, the documents relating to the design of the instrument, as set out in Annex III, must be accessible to the notified body should the latter so request.

3.5. Verification by checking and testing of each instrument.

3.5.1. All instruments shall be individually examined and appropriate tests, as set out in the relevant standard(s) referred to in Article 5, or equivalent tests, shall be carried out in order to verify their conformity, where applicable, to the type as described in the EC type-examination certificate and the requirements of this Directive that apply to them.

3.5.2. The notified body shall affix, or cause to be affixed, its identification number on each instrument the conformity of which to requirements has been established and draw up a written certificate of conformity relating to the tests carried out.

3.5.3. The manufacturer or his authorised representative shall ensure that he is able to supply the notified body's certificates of conformity on request.

4. EC unit verification

4.1. EC unit verification is the procedure whereby the manufacturer or his authorised representative established within the Community ensures and declares that the instrument, generally designed for a specific application, which has been issued with the certificate referred to in pargraph 4.2 conforms to the requirements of this Directive that apply to it. The manufacturer or his authorised representative shall affix the CE marking to the instrument and shall draw up a written declaration of conformity.

4.2. The notified body shall examine the instrument and carry out the appropriate tests, as set out in the respective standard(s) referred to in Article 5, or equivalent tests, in order to ensure its conformity to the relevant requirements of this Directive.

The notified body shall affix, or cause to be affixed, its identification number to the instrument the conformity of which to requirements has been established and shall draw up a written certificate of conformity concerning the tests carried out.

4.3. The aim of the technical documentation relating to the design of the instrument, as referred to in Annex III, is to enable conformity to the requirements of this Directive to be assessed and the design, manufacture and operation of the instrument to be understood. It must be accessible to the notified body.

4.4. The manufacturer or his authorised representative shall ensure that he is able to supply the notified body's certificates of conformity on request.

5. Common provisions

5.1. The EC declaration of type conformity (guarantee of production quality), the EC verification, and the EC unit verification, may be carried out at the manufacturer's works or any other location if transport to the place of use does not require dismantling of the instrument, if the taking into service at the place of use does not require assembly of the instrument or other technical installation work likely to affect the instrument's performance, and if the gravity value at the place of putting into service is taken into consideration or if the instrument's performance is insensitive to gravity variations. In all other cases, they shall be carried out at the place of use of the instrument.

5.2. If the instrument's performance is sensitive to gravity variations the procedures referred to in 5.1 may be carried out in two stages, where the second stage shall comprise all examinations and tests of which the outcome is gravity-dependent, and the first stage all other examinations and tests. The second stage shall be carried out at the place of use of the instrument. If a Member State has established gravity zones on its territory the expression "at the place of use of the instrument" may be read as "in the gravity zone of use of the instrument".

5.3.1. Where a manufacturer has opted for execution in two stages of one of the procedures mentioned in 5.1, and where these two stages will be carried out by different parties, an instrument which has undergone the first stage of the procedure shall bear the identification number of the notified body involved in that stage.

5.3.2. The party which has carried out the first stage of the procedure shall issue for each of the instruments a certificate containing the necessary data for identification of the instrument and specifying the examination and tests that have been carried out.

The party which carries out the second stage of the procedure shall carry out those examinations and tests that have not yet been carried out.

The manufacturer or his authorised representative shall ensure that he is able to supply the notified body's certificates of conformity on request.

5.3.3. The manufacturer who has opted for the EC declaration of type conformity (guarantee of production quality) in stage one may either use this same procedure in stage two or decide to continue in stage two with EC verification.

5.3.4. The CE marking shall be affixed to the instrument on completion of the second stage along with the identification number of the notified body which took part in the second stage.

Annex III
Design documentation

The technical documentation must render the design, manufacture and operation of the product intelligible and enable an assessment to be made of its conformity with the requirements of the Directive.

The documentation shall include in so far as relevant for assessment:

- a general description of the type,
- conceptual designs and manufacturing drawings and plans of components, sub-assemblies, circuits, etc,
- descriptions and explanations necessary for the understanding of the above, including the operation of the instrument,
- a list of the standards referred to in Article 5, applied in full or in part, and descriptions of the solutions adopted to meet the essential requirements where the standards referred to in Article 5 have not been applied,
- results of design calculations made and of examinations, etc,
- test reports,
- the EC type-approval certificates and the results of relevant tests on instruments containing parts identical to those in the design.

Annex IV

1. Instruments subject to EC conformity assessment procedure

1.1. These instruments must bear:

(a) – the CE conformity marking comprising the CE symbol as described in Annex VI, followed by the last two digits of year in which it was affixed.
 – the identification number(s) of the notified body/bodies that has/have carried out the EC surveillance or the EC verification.

The abovementioned mark and inscriptions shall be affixed to the instrument, distinctly grouped together;

(b) a green sticker at least 12.5 mm×12.5 mm square bearing a capital letter "M" printed in black;

(c) the following inscriptions:
- the number of the EC type-approval certificate, where appropriate,
- the manufacturer's mark or name,
- the accuracy class, enclosed in an oval or in two horizontal lines joined by two half circles,
- maximum capacity in the form Max . . .,
- minimum capacity in the form Min . . .,
- verification scale interval in the form e=,
- the last two digits of the year in which its CE marking was affixed,

plus, where applicable:
- serial number,
- for instruments consisting of separate but associated units: identification mark on each unit,
- scale interval if it is different from e, in the form d= . . .,
- maximum additive tare effect, in the form T=+ . . .,
- maximum subtractive tare effect if it is different from Max, in the form T=− . . .,
- tare interval if it is different from d, in the form d_T= . . .,
- maximum safe load if it is different from Max, in the form Lim . . .,
- the special temperature limits, in the form . . . °C/ . . . °C,

– ration between load recepter and load.

1.2. The instruments shall have adequate facilities for the affixing of the CE marking of conformity and/or inscriptions. These shall be such that it shall be impossible to remove the mark and inscriptions without damaging them, and that the mark and inscriptions shall be visible when the instrument is in its regular operating position.

1.3. Where a data plate is used it shall be possible to seal the plate unless it cannot be removed without being destroyed. If the data plate is sealable it shall be possible to apply a control mark to it.

1.4. The inscriptions Max, Min, *e*, *d*, shall also be shown near the display of the result if they are not already located there.

1.5. Each load measuring device which is connected or can be connected to one or more load receptors shall bear the relevant inscriptions relating to the said load receptors.

2. Other instruments

The other instruments must bear:

– the manufacturer's mark or name,
– the maximum capacity in the form Max . . .

These instruments may not bear the stickers provided for in 1.1. (b).

3. Restrictive use symbol specified in Article 12

This symbol shall be constituted by a capital letter "M" printed in black on a red background at least 25 mm×25 mm square with two intersecting diagonals forming a cross.

Annex V

Set out below are the minimum criteria to be applied by Member States when designating bodies for the carrying-out of tasks pertaining to the procedures referred to in Article 8.

1. The bodies shall dispose of the necessary personnel, means and equipment.
2. The personnel shall have technical competence and professional integrity.
3. The bodies shall work independently of all circles, groups or persons having direct or indirect interest in non-automatic weighing instruments regarding the carrying-out of the tests, the preparation of the reports, the issuing of the certificates and the surveillance requested by this Directive.
4. The personnel shall respect the professional secret.
5. The bodies shall have taken out a civil liability insurance if their civil liability is not covered by the State under national law.

The fulfilment of the conditions under points 1 and 2 shall be periodically verified by the Member States.

Annex VI

CE conformity marking

– The CE conformity marking shall consist of the initials "CE" taking the following form:

- If the CE marking is reduced or enlarged the proportions given in the above graduated drawing must be respected.
- The various components of the CE marking must have substantially the same vertical dimension, which may not be less than 5 mm.

COUNCIL DIRECTIVE 90/385/EEC

of 20 June 1990

on the approximation of the laws of the Member States relating to active implantable medical devices*†

THE COUNCIL OF THE EUROPEAN COMMUNITIES,

Having regard to the Treaty establishing the European Economic Community, and in particular Article 100A thereof,

Having regard to the proposal from the Commission,

In cooperation with the European Parliament,

Having regard to the opinion of the Economic and Social Committee,

Whereas in each Member State active implantable medical devices must give patients, users and other persons a high level of protection and achieve the intended level of performance when implanted in human beings;

Whereas several Member States have sought to ensure that level of safety by mandatory specification relating both to the technical safety features and the inspection procedures for such devices; whereas those specifications differ from one Member State to another;

Whereas national provisions ensuring that safety level should be harmonised in order to guarantee the free movement of active implantable medical devices without lowering existing and justified levels of safety in the Member States;

Whereas harmonised measures must be distinguished from measures taken by Member States to manage the financing of public health and sickness insurance schemes directly or indirectly concerning such devices; whereas, therefore, such provisions do not affect the right of Member States to implement the abovementioned measures in compliance with Community law;

Wheras maintaining or improving the level of protection achieved in Member States constitutes one of this Directive's essential objectives as defined by the essential requirements;

Whereas rules governing active implantable medical devices can be confined to those provisions needed to satisfy the essential requirements; whereas, because they are essential, these requirements must replace corresponding national provisions;

Whereas, in order to facilitate proof of conformity with these essential requirements and to permit monitoring of that conformity, it is desirable to have Europe-wide harmonised standards in respect of the prevention of risks in connection with the design, manufacture and packaging of active implantable medical devices; whereas such standards harmonised at European level are drawn up by private-law bodies and must retain their status as non-mandatory texts; whereas, to that end, the European Committee for Standardisation (CEN) and the European Committee for Electrotechnical Standardisation (CENELEC) are recognised as being the competent bodies to adopt harmonised standards in accordance with the general guidelines for cooperation between the Commission and these two bodies, signed on 13 November 1984; whereas, for the purposes of this Directive, a harmonised standard is a technical specification (European standard or harmonisation document) adopted by either or both of these bodies, as

*OJ 1990 L189/17.
† Amended by Dir 93/42 and Dir 93/68.

instructed by the Commission pursuant to the provisions of Council Directive 83/189/EEC of 28 March 1983 (OJ 1983 L109/8) laying down a procedure for the provision of information in the field of technical standards and regulations, as last amended by Directive 88/182/EEC (OJ 1988 L81/75), and under the abovementioned general guidelines;

Whereas evaluation procedures have to be established and accepted by common accord between the Member States in accordance with Community criteria;

Whereas the specific nature of the medical sector makes it advisable to make provision for the notified body and the manufacturer or his agent established in the Community to fix, by common accord, the time limits for completion of the evaluation and verification operations for the conformity of devices,

HAS ADOPTED THIS DIRECTIVE:

Article 1

1. This Directive shall apply to active implantable medical devices.

2. For the purposes of this Directive, the following definitions shall apply:

(a) "medical device" means any instrument, apparatus, appliance, material or other article, whether used alone or in combination, together with any accessories or software for its proper functioning, intended by the manufacturer to be used for human beings in the:
- diagnosis, prevention, monitoring, treatment or alleviation of disease or injury,
- investigation, replacement or modification of the anatomy or of a physiological process,
- control of conception,

and which does not achieve its principal intended action by pharmacological, chemical, immunological or metabolic means, but which may be assisted in its function by such means;

(b) "active medical device" means any medical device relying for its functioning on a source of electrical energy or any source of power other than that directly generated by the human body or gravity;

(c) "active implantable medical device" means any active medical device which is intended to be totally or partially introduced, surgically or medically, into the human body or by medical intervention into a natural orifice, and which is intended to remain after the procedure;

(d) "custom-made device" means any active implantable medical device specifically made in accordance with a medical specialist's written prescription which gives, under his responsibility, specific design characteristics and is intended to be used only for an individual named patient;

(e) "device intended for clinical investigation" means any active implantable medical device intended for use by a specialist doctor when conducting investigations in an adequate human clinical environment;

(f) "intended purpose" means the use for which the medical device is intended and for which it is suited according to the data supplied by the manufacturer in the instructions;

(g) "putting into service" means making available to the medical profession for implantation.

(h) "placing on the market" means the first making available in return for payment or free of charge of a device other than a device intended for clinical investigation, with a view to distribution and/or use on the Community market, regardless of whether it is new or fully refurbished;

(i) "manufacturer" means the natural or legal person with responsibility for the design, manufacture, packaging and labelling of a device before it is placed on the market under his own name, regardless of whether these operations are carried out by that person himself or on his behalf by a third party.

The obligations of this Directive to be met by manufacturers also apply to the natural or legal person who assembles, packages, processes, fully refurbishes and/or labels one or more ready-made products and/or assigns to them their intended purpose as a device with a view to their being placed on the market under his own name. This subparagraph does not apply to the person who, while not a manufacturer within the meaning of the first subparagraph, assembles or adapts devices already on the market to their intended purpose for an individual patient.

3. Where an active implantable medical device is intended to administer a substance defined as a medicinal product within the meaning of Council Directive 65/65/EEC of 26 January 1965 (OJ 1965, 369) on the approximation of provisions laid down by law, regulation or administrative action relating to proprietary medicinal products, as last amended by Directive 87/21/EEC (OJ 1987 L15/36), that substance shall be subject to the system of marketing authorisation provided for in that Directive.

4. Where an active implantable medical device incorporates, as an integral part, a substance which, if used separately, may be considered to be a medicinal product within the meaning of Article 1 of Directive 65/65/EEC, that device must be evaluated and authorised in accordance with the provisions of this Directive.

5. This Directive constitutes a specific Directive within the meaning of Article 2 (2) of Council Directive 89/336/EEC of 3 May 1989 (OJ 1989 L139/19) on the approximation of the laws of the Member States relating to electromagnetic compatibility.

Article 2

Member States shall take all necessary steps to ensure that the devices referred to in Article 1 (2) (c) and (d) may be placed on the market and put into service only if they do not compromise the safety and health of patients, users and, where applicable, other persons when properly implanted, maintained and used in accordance with their intended purposes.

Article 3

The active implantable medical devices referred to in Article 1 (2) (c), (d) and (e), hereinafter referred to as "devices", must satisfy the essential requirements set out in Annex 1, which shall apply to them account being taken of the intended purpose of the devices concerned.

Article 4

1. Member States shall not prevent the placing on the market or the putting into service within their territory of devices complying with the provisions of this Directive and bearing the CE marking provided for in Article 12 which indicate that they have been the subject of an evaluation of their conformity in accordance with Article 9.

2. Member States shall not create any obstacles to:

- devices intended for clinical investigations being made available to specialist doctors for that purpose if they satisfy the conditions laid down in Article 10 and in Annex 6,
- custom-made devices being placed on the market and put into service if they satisfy the conditions laid down in Annex 6 and are accompanied by the statement referred to in that Annex.

These devices shall not bear the CE marking.

3. At trade fairs, exhibitions, demonstrations, etc, Member States shall not prevent the showing of devices which do not conform to this Directive, provided that a visible sign clearly indicates that such devices do not conform and cannot be put into service until they have been made to comply by the manufacturer or his authorised rep-

resentative established within the Community.

4. When a device is put into service, Member States may require the information described in sections 13, 14 and 15 of Annex 1 to be in their national language(s).

5. (a) Where the devices are subject to other Directives concerning other aspects and which also provide for the affixing of the CE marking, the latter shall indicate that the devices are also presumed to conform to the provisions of the other Directives.

5. (b) However, where one or more of these Directives allow the manufacturer, during a transitional period, to choose which arrangements to apply, the CE marking shall indicate conformity to the provisions only of those Directives applied by the manufacturer. In this case, particulars of the Directive applied, as published in the *Official Journal of the European Communities*, must be given in the documents, notices or instructions required by the Directives and accompanying such devices; these documents, notices or instructions shall be accessible without it being necessary to destroy the packaging which keeps the device sterile.

Article 5

Member States shall presume compliance with the essential requirements referred to in Article 3 in respect of devices which are in conformity with the relevant national standards adopted pursuant to the harmonised standards the references of which have been published in the *Official Journal of the European Communities*; Member States shall publish the references of such national standards.

Article 6

1. Where a Member State or the Commission considers that the harmonised standards referred to in Article 5 do not entirely meet the essential requirements referred to in Article 3, the Commission or the Member State concerned shall bring the matter before the Standing Committee set up under Directive 83/189/EEC, giving the reasons therefore. The Committee shall deliver an opinion without delay.

In the light of the opinion of the Committee, the Commission shall inform Member States of the measures to be taken with regard to the standards and the publication referred to in Article 5.

2. A Standing Committee, hereinafter referred to as the "Committee", shall be set up, composed of the representatives of the Member States and chaired by the representative of the Commission.

The Committee shall draw up its rules of procedure.

Any matter relating to the implementation and practical application of this Directive may be brought before the Committee, in accordance with the procedure set out below.

The representative of the Commission shall submit to the Committee a draft of the measures to be taken. The Committee shall deliver its opinion according to the urgency of the matter, if necessary by taking a vote.

The opinion shall be recorded in the minutes; in addition, each Member State shall have the right to ask to have its position recorded in the minutes.

The Commission shall take the utmost account of the opinion delivered by the Committee. It shall inform the Committee of the manner in which its opinion has been taken into account.

Article 7

1. Where a Member State finds that the devices referred to in Article 1 (2) (c) and (d), correctly put into service and used in accordance with their intended purpose, may compromise the health and/or safety of patients, users or, where applicable, other persons, it shall take all appropriate measures to withdraw such devices from the market or prohibit or restrict their being placed on the market or their being put into service.

The Member State shall immediately inform the Commission of any such measure, indicating the reasons for its decision and, in particular, whether non-compliance with this Directive is due to:

(a) failure to meet the essential requirements referred to in Article 3, where the device does not meet in full or in part the standards referred to in Article 5;
(b) incorrect application of those standards;
(c) shortcomings in the standards themselves.

2. The Commission shall enter into consultation with the parties concerned as soon as possible. Where, after such consultation, the Commission finds that:

– the measures are justified, it shall immediately so inform the Member State which took the initiative and the other Member States; where the decision referred to in paragraph 1 is attributed to shortcomings in the standards, the Commission shall, after consulting the parties concerned, bring the matter before the Committee referred to in Article 6 (1) within two months if the Member State which has taken the decision intends to maintain it and shall initiate the procedures referred to in Article 6 (1),
– the measures are unjustified, it shall immediately so inform the Member State which took the initiative and the manufacturer or his authorised representative established within the Community.

3. Where a device which does not comply bears the CE marking, the competent Member State shall take appropriate action against whomsoever has affixed the mark and shall inform the Commission and the other Member States thereof.

4. The Commission shall ensure that the Member States are kept informed of the progress and outcome of this procedure.

Article 8

1. Member States shall take the necessary steps to ensure that information brought to their knowledge regarding the incidents mentioned below involving a device is recorded and evaluated in a centralised manner:

(a) any deterioration in the characteristics and performances of a device, as well as any inaccuracies in the instruction leaflet which might lead to or might have led to the death of a patient or to a deterioration in his state of health;
(b) any technical or medical reason resulting in withdrawal of a device from the market by the manufacturer.

2. Member States shall, without prejudice to Article 7, forthwith inform the Commission and the other Member States of the incidents referred to in paragraph 1 and of the relevant measures taken or contemplated.

Article 9

1. In the case of devices other than those which are custom-made or intended for clinical investigations, the manufacturer must, in order to affix the CE marking, at his own choice:

(a) follow the procedure relating to the EC declaration of conformity set out in Annex 2; or
(b) follow the procedure relating to EC type-examination set out in Annex 3, coupled with:
(i) the procedure relating to EC verification set out in Annex 4, or
(ii) the procedure relating to the EC declaration of conformity to type set out in Annex 5.

2. In the case of custom-made devices, the manufacturer must draw up the declaration provided for in Annex 6 before placing each device on the market.

3. Where appropriate, the procedures provided for in Annexes 3, 4 and 6 may be discharged by the manufacturer's authorised representative established in the Community.

4. The records and correspondence relating to the procedures referred to in paragraphs 1, 2 and 3 shall be in an official language of the Member State in which the said procedures will be

carried out and/or in an language acceptable to the notified body defined in Article 11.

5. During the conformity assessment procedure for a device, the manufacturer and/or the notified body shall take account of the results of any assessment and verification operations which, where appropriate, have been carried out in accordance with this Directive at an intermediate stage of manufacture.

6. Where the conformity assessment procedure involves the intervention of a notified body, the manufacturer, or his authorised representative established in the Community, may apply to a body of his choice within the framework of the tasks for which the body has been notified.

7. The notified body may require, where duly justified, any information or data which is necessary for establishing and maintaining the attestation of conformity in view of the chosen procedure.

8. Decisions taken by the notified bodies in accordance with Annexes II and III shall be valid for a maximum of five years and may be extended on application, made at a time agreed in the contract signed by both parties, for further periods of five years.

9. By derogation from paragraphs 1 and 2 the competent authorities may authorise, on duly justified request, the placing on the market and putting into service, within the territory of the Member State concerned, of individual devices for which the procedures referred to in paragraphs 1 and 2 have not been carried out and the use of which is in the interest of protection of health.

Article 9a

1. Where a Member State considers that the conformity of a device or family of devices should be established, by way of derogation from the provisions of Article 9, by applying solely one of the given procedures chosen from among those referred to in Article 9, it shall submit a duly substantiated request to the Commission and ask it to take the necessary measures. These measures shall be adopted in accordance with the procedure referred to in Article 7 (2) of Directive 93/42/EEC ([1]).

2. The Commission shall inform the Member States of the measures taken and, where appropriate, publish the relevant parts of these measures in the *Official Journal of the European Communities*.

Article 10

1. In the case of devices intended for clinical investigations, the manufacturer or his authorised representative established in the Community shall, at least 60 days before the commencement of the investigations, submit the statement referred to in Annex 6 to the competent authorities of the Member State in which the investigations are to be conducted.

2. The manufacturer may commence the relevant clinical investigations at the end of a period of 60 days after notification, unless the competent authorities have notified him within that period of a decision to the contrary, based on considerations of public health or public order.

Member States may however authorise manufacturers to start the clinical investigations in question before the expiry of the 60-day period, provided that the Ethical Committee concerned has delivered a favourable opinion with respect to the investigation programme in question.

2. (a) The authorisation referred to in the second subparagraph of paragraph 2 may be subject to approval by the competent authority.

3. The Member States shall, if necessary, take the appropriate steps to ensure public health and order.

Article 11

1. Member States shall notify the Commission and the other Member State of the bodies which they have appointed

[1] OJ No L169.

to carry out the procedures referred to in Article 9 together with the specific tasks which these bodies have been appointed to carry out and the identification numbers assigned to them beforehand by the Commission.

The Commission shall publish in the *Official Journal of the European Communities* a list of the notified bodies and their identification numbers and the tasks for which they have been notified. The Commission shall ensure that this list is kept up to date.

2. Member States shall apply the minimum criteria, set out in Annex 8, for the designation of bodies. Bodies that satisfy the criteria fixed by the relevant harmonised standards shall be presumed to satisfy the relevant minimum criteria.

3. A Member State that has notified a body shall withdraw that notification if it finds that the body no longer meets the criteria referred to in paragraph 2. It shall immediately inform the other Member States and the Commission thereof.

4. The notified body and the manufacturer or his agent established in the Community shall fix, by common accord, the time limits for completion of the evaluation and verification operations referred to in Annexes 2 to 5.

Article 12

1. Devices other than those which are custom made or intended for clinical investigations considered to meet the essential requirements referred to in Article 3 must bear the CE marking of conformity.

2. The CE marking of conformity, as shown in Annex 9, must appear in a visible, legible and indelible form on the sterile pack and, where appropriate, on the sales packaging, if any, and on the instruction leaflet.

It must be accompanied by the identification number of the notified body responsible for implementation of the procedures set out in Annexes 2, 4 and 5.

3. The affixing of markings on the devices which are likely to deceive third parties as to the meaning and form of the CE marking shall be prohibited. Any other marking may be affixed to the packaging or to the instruction leaflet accompanying the device provided that the visibility and legibility of the CE marking is not hereby reduced.

Article 13

Without prejudice to Article 7

(a) where a Member State establishes that the CE marking has been affixed unduly, the manufacturer or his authorised representative established within the Community shall be obliged to end the infringement under conditions imposed by the Member State;

(b) where non-compliance continues, the Member State must take all appropriate measures to restrict or prohibit the placing on the market of the device in question or to ensure that it is withdrawn from the market in accordance with the procedures laid down in Article 7.

Article 14

Any decision taken pursuant to this Directive and resulting in the refusal of or restrictions on the placing on the market and/or putting into service of a device shall state the exact grounds on which it is based. Such decision shall be notified without delay to the party concerned, who shall at the same time be informed of the remedies available to him under the laws in force in the Member State in question and of the time limits to which such remedies are subject.

In the event of a decision as referred to in the previous paragraph the manufacturer, or his authorised representative established in the Community, shall have an opportunity to put forward his viewpoint in advance, unless such consultation is not possible because of the urgency of the measures to be taken.

Article 15

Member States shall ensure that all the

parties involved in the application of this Directive are bound to observe confidentiality with regard to all information obtained in carrying out their tasks. This does not affect the obligations of Member States and notified bodies with regard to mutual information and the dissemination of warnings.

Article 16

1. Before 1 July 1992, Member States shall adopt and publish the laws, regulations and administrative provisions necessary in order to comply with this Directive. They shall forthwith inform the Commission thereof.

They shall apply such provisions from 1 January 1993.

2. Member States shall communicate to the Commission the texts of the provisions of national law which they adopt in the field covered by this Directive.

3. Member States shall, for the period up to 31 December 1994, permit the placing on the market and putting into service of devices complying with national rules in force in their territory on 31 December 1992.

Article 17

This Directive is addressed to the Member States.

Annex I

Essential requirements

I. General requirements

1. The devices must be designed and manufactured in such a way that, when implanted under the conditions and for the purposes laid down, their use does not compromise the clinical condition or the safety of patients. They must not present any risk to the persons implanting them or, where applicable, to other persons.

2. The devices must achieve the performances intended by the manufacturer, viz be designed and manufactured in such a way that they are suitable for one or more of the functions referred to in Article 1 (2) (a) as specified by him.

3. The characteristics and performances referred to in sections 1 and 2 must not be adversely affected to such a degree that the clinical condition and safety of the patients or, as appropriate, of other persons are compromised during the lifetime of the device anticipated by the manufacturer, where the device is subjected to stresses which may occur during normal conditions of use.

4. The devices must be designed, manufactured and packed in such a way that their characteristics and performances are not adversely affected in the storage and transport conditions laid down by the manufacturer (temperature, humidity, etc).

5. Any side effects or undesirable conditions must constitute acceptable risks when weighed against the performances intended.

II. Requirements regarding design and construction

6. The solutions adopted by the manufacturer for the design and construction of the devices must comply with safety principles taking account of the generally acknowledged state of the art.

7. Implantable devices must be designed, manufactured and packed in a non-reusable pack according to appropriate procedures to ensure they are sterile when placed on the market and, in the storage and transport conditions stipulated by the manufacturer, remain so until the packaging is removed and they are implanted.

8. Devices must be designed and manufactured in such a way as to

remove or minimise as far as possible:
- the risk of physical injury in connection with their physical, including dimensional, features,
- risks connected with the use of energy sources with particular reference, where electricity is used, to insulation, leakage currents and overheating of the devices,
- risks connected with reasonably foreseeable environmental conditions such as magnetic fields, external electrical influences, electrostatic discharge, pressure or variations in pressure and acceleration,
- risks connected with medical treatment, in particular those resulting from the use of defibrillators or high-frequency surgical equipment,
- risks connected with ionizing radiation from radioactive substances included in the device, in compliance with the protection requirements laid down in Directive 80/836/Euratom (OJ 1980 L246/1), as amended by Directives 84/467/Euratom (OJ 1984 L265/4) and 84/466/Euratom (OJ 1984 L265/1),
  - risks which may arise where maintenance and calibration are impossible, including:
  - excessive increase of leakage currents,
  - ageing of the materials used,
  - excess heat generated by the device,
  - decreased accuracy of any measuring or control mechanism.

9. The devices must be designed and manufactured in such a way as to guarantee the characteristics and performances referred to in I. "General requirements", with particular attention being paid to:
- the choice of materials used, particularly as regards toxicity aspects,
- mutual compatibility between the materials used and biological tissues, cells and body fluids, account being taken of the anticipated use of the device,
- compatibility of the devices with the substances they are intended to administer,
- the quality of the connections, particularly in respect of safety,
- the reliability of the source of energy,
- if appropriate, that they are leakproof,
- proper functioning of the programming and control systems, including software.

10. Where a device incorporates, as an integral part, a substance which, when used separately, is likely to be considered to be a medicinal product as defined in Article 1 of Directive 65/65/EEC, and whose action in combination with the device may result in its bioavailability, the safety, quality and usefulness of the substance, account being taken of the purpose of the device, must be verified by analogy with the appropriate methods specified in Directive 75/318/EEC (OJ 1975 L147/1), as last amended by Directive 89/341/EEC (OJ 1989 L142/11).

11. The devices and, if appropriate, their component parts must be identified to allow any necessary measure to be taken following the discovery of a potential risk in connection with the devices and their component parts.

12. Devices must bear a code by which they and their manufacturer can be unequivocably identified (particularly with regard to the type of device and year of manufacture); it must be possible to read this code, if necessary, without the need for a surgical operation.

13. When a device or its accessories bear instructions required for the operation of the device or indicate operating or adjustment parameters, by means of a visual system, such information must be understandable to the user and, as appropriate, the patient.

14. Every device must bear, legibly and indelibly, the following particulars, where appropriate in the form of generally recognised symbols:

14.1. On the sterile pack:
- the method of sterilisation,
- an indication permitting this packaging to be recognised as such,
- the name and address of the manufacturer,

- a description of the device,
- if the device is intended for clinical investigations, the words: "exclusively for clinical investigations",
- if the device is custom-made, the words 'custom-made device',
- a declaration that the implantable device is in a sterile condition,
- the month and year of manufacture,
- an indication of the time limit for implanting a device safely.

14.2. On the sales packaging:

- the name and address of the manufacturer,
- a description of the device,
- the purpose of the device,
- the relevant characteristics for its use,
- if the device is intended for clinical investigations, the words "exclusively for clinical investigations",
- if the device is custom-made, the words: "custom-made device",
- a declaration that the implantable device is in a sterile condition,
- the month and year of manufacture,
- an indication of the time limit for implanting a device safely,
- the conditions for transporting and storing the device.

15. When placed on the market, each device must be accompanied by instructions for use giving the following particulars:

- the year of authorisation to affix the CE marking,
- the details referred to in 14.1 and 14.2, with the exception of those referred to in the eighth and ninth indents,
- the performances referred to in section 2 and any undesirable side effects,
- information allowing the physician to select a suitable device and the corresponding software and accessories,
- information constituting the instructions for use allowing the physician and, where appropriate, the patient to use the device, its accessories and software correctly, as well as information on the nature, scope and times for operating controls and trials and, where appropriate, maintenance measures,
- information allowing, if appropriate, certain risks in connection with implantation of the device to be avoided,
- information regarding the risks of reciprocal interference[1] in connection with the presence of the device during specific investigations or treatment,
- the necessary instructions in the event of the sterile pack being damaged and, where appropriate, details of appropriate methods of resterilisation,
- an indication, if appropriate, that a device can be reused only if it is reconditioned under the responsibility of the manufacturer to comply with the essential requirements.

The instruction leaflet must also include details allowing the physician to brief the patient on the contra-indications and the precautions to be taken. These details should cover in particular:

- information allowing the lifetime of the energy source to be established,
- precautions to be taken as regards exposure, in reasonably foreseeable environmental conditions, to magnetic fields, external electrical influences, electrostatic discharge, pressure or variations in pressure, acceleration, etc,
- adequate information regarding the medicinal products which the device in question is designed to administer.

16. Confirmation that the device satisfies the requirements in respect of characteristics and performances, as referred to in I. "General requirements", in normal conditions of use, and the evaluation of the side effects or undesirable effects must be based on clinical data established in accordance with Annex 7.

---

[1] Risks of reciprocal interference means adverse effects on the device caused by instruments present at the time of investigations or treatment and vice versa.

Annex II
EC declaration of conformity

(Complete quality assurance system)

1. The manufacturer shall apply the quality system approved for the design, manufacture and final inspection of the products concerned as specified in sections 3 and 4 and shall be subject to EC surveillance as specified in section 5.

2. The declaration of conformity is the procedure by means of which the manufacturer who satisfies the obligations of section 1 ensures and declares that the products concerned meet the provisions of this Directive which apply to them.

The manufacturer or his authorised representative established within the Community shall affix the CE marking in accordance with Article 12 and shall draw up a written declaration of conformity.

This declaration shall cover one or more identified examples of the product and shall be kept by the manufacturer or his authorised representative established within the Community.

The CE marking shall be accompanied by the identification number of the notified body responsible.

3. Quality system

3.1. The manufacturer shall make an application for evaluation of his quality system to a notified body.

The application shall include:

- all the appropriate items of information for the category of products manufacture of which is envisaged,
- the quality-system documentation,
- an undertaking to fulfil the obligations arising from the quality system as approved,
- an undertaking to maintain the approved quality system in such a way that it remains adequate and efficacious,
- an undertaking by the manufacturer to institute and keep up-dated a post-marketing surveillance system. The undertaking shall include an obligation for the manufacturer to notify the competent authorities of the following incidents immediately on learning of them:
  (i) any deterioration in the characteristics or performances, and any inaccuracies in the instruction leaflet for a device which might lead to or have led to the death of a patient or a deterioration in his state of health;
  (ii) any technical or medical reason resulting in withdrawal of a device from the market by the manufacturer.

3.2. The application of the quality system must ensure that the products conform to the provisions of this Directive which apply to them at every stage, from design to final controls.

All the elements, requirements and provisions adopted by the manufacturer for his quality system shall be documented in a systematic and orderly manner in the form of written policies and procedures. This quality-system documentation must make possible a uniform interpretation of the quality policies and procedures such as quality programmes, quality plans, quality manuals and quality records.

It shall include in particular an adequate description of:

(a) the manufacturer's quality objectives;

(b) the organisation of the business and in particular:
   - the organisational structures, the responsibilities of the managerial staff and their organisational authority where quality of design and manufacture of the products is concerned,
   - the methods of monitoring the

efficient operation of the quality system and in particular its ability to achieve the desired quality of the design and of the products, including control of products which do not conform;

(c) the procedures for monitoring and verifying the design of the products and in particular:
   - the design specifications, including the standards which will be applied and a description of the solutions adopted to fulfil the essential requirements which apply to the products when the standards referred to in Article 5 are not applied in full,
   - the techniques of control and verification of the design, the processes and systematic actions which will be used when the products are being designed;

(d) the techniques of control and of quality assurance at the manufacturing stage and in particular:
   - the processes and procedures which will be used, particularly as regards sterilisation, purchasing and the relevant documents,
   - product-identification procedures drawn up and kept up to date from drawings, specifications or other relevant documents at every stage of manufacture;

(e) the appropriate tests and trials which will be effected before, during and after production, the frequency with which they will take place, and the test equipment used.

3.3. Without prejudice to Article 13 of this Directive, the notified body shall effect an audit of the quality system to determine whether it meets the requirements referred to in 3.2. It shall presume conformity with these requirements for the quality systems which use the corresponding harmonised standards.

The team entrusted with the evaluation shall include at least one member who has already had experience of evaluations of the technology concerned. The evaluation procedure shall include an inspection on the manufacturer's premises.

The decision shall be notified to the manufacturer after the final inspection. It shall contain the conclusions of the control and a reasoned evaluation.

3.4. The manufacturer shall inform the notified body which has approved the quality system of any plan to alter the quality system.

The notified body shall evaluate the proposed modifications and shall verify whether the quality system so modified would meet the requirements referred to in 3.2; it shall notify the manufacturer of its decision. This decision shall contain the conclusions of the control and a reasoned evaluation.

4. Examination of the design of the product

4.1. In addition to the obligations incumbent on him under section 3, the manufacturer shall make an application for examination of the design dossier relating to the product which he plans to manufacture and which falls into the category referred to in 3.1.

4.2. The application shall describe the design, manufacture, and performances of the product in question and shall include the necessary particulars which make it possible to evaluate whether it complies with the requirements of this Directive.

It shall include *inter alia*:

- the design specifications, including the standards which have been applied,
- the necessary proof of their appropriations, in particular where the standards referred to in Article 5 have not been applied in full. This proof must include the results of the appropriate tests carried out by the manufacturer or carried out under his responsibility,
- a statement as to whether or not the device incorporates, as an integral part, a substance as referred to in section 10 of Annex 1, whose action in combination with the device may

result in its bioavailability, together with data on the relevant trials conducted,
- the clinical data referred to in Annex 7,
- the draft instruction leaflet.

4.3. The notified body shall examine the application and, where the product complies with the relevant provisions of this Directive, shall issue the applicant with an EC design examination certificate. The notified body may require the application to be supplemented by further tests or proof so that compliance with the requirements of the Directive may be evaluated. The certificate shall contain the conclusions of the examination, the conditions of its validity, the data needed for indentification of the approved design and, where appropriate, a description of the intended use of the product.

4.4. The applicant shall inform the notified body which issued the EC design examination certificate of any modification made to the approved design. Modifications made to the approved design must obtain supplementary approval from the notified body which issued the EC design examination certificate where such modifications may affect conformity with the essential requirements of this Directive or the conditions prescribed for the use of the product. This supplementary approval shall be given in the form of an addendum to the EC design examination certificate.

5. Surveillance

5.1. The aim of surveillance is to ensure that the manufacturer duly fulfils the obligations arising from the approved quality system.

5.2. The manufacturer shall authorise the notified body to carry out all necessary inspections and shall supply it with all appropriate information, in particular:
- the quality-system documentation,
- the data stipulated in the part of the quality system relating to design, such as the results of analyses, calculations, tests, etc,
- the data stipulated in the part of the quality system relating to manufacture, such as reports concerning inspections, tests, standardisations/ calibrations and the qualifications of the staff concerned, etc.

5.3. The notified body shall periodically carry out as appropriate inspections and evaluations in order to ascertain that the manufacturer is applying the approved quality system, and shall supply the manufacturer with an evaluation report.

5.4. In addition, the notified body may make unannounced visits to the manufacturer, and shall supply him with an inspection report.

6. The notified body shall communicate to the other notified bodies all relevant information concerning approvals of quality systems issued, refused and withdrawn.

6. Administrative provisions

6.1 For at least five years from last date of manufacture of the product, the manufacturer shall keep available for the national authorities:
- the declaration of conformity,
- the documentation referred to in the second indent of section 3.1,
- the amendments referred to in section 3.4,
- the documentation referred to in section 4.2,
- the decisions and reports of the notified body referred to in sections 3.4, 4.3, 5.3 and 5.4.

6.2. On request, the notified body shall make available to the other notified bodies and the competent authority all relevant information on approvals of quality systems issued, refused or withdrawn.

6.3. Where neither the manufacturer nor his authorised representative are established in the Community, the task of keeping available for the authorities the technical documentation referred to in Article 4 (2) shall fall to the person responsible for placing the appliance on the Community market.

Annex III
EC type-examination

1. EC type-examination is the procedure whereby a notified body observes and certifies that a representative sample of the production envisaged satisfies the relevant provisions of this Directive.

2. The application for EC type-examination shall be made by the manufacturer, or by his authorised representative established in the Community, to a notified body.

The application shall include:

- the name and address of the manufacturer and the name and address of the authorised representative if the application is made by the latter,
- a written declaration specifying that an application has not been made to any other notified body,
- the documentation described in section 3 needed to allow an evaluation to be made of the conformity of a representative sample of the production in question, hereinafter referred to as "type", with the requirements of this Directive.

The applicant shall make a "type" available to the notified body. The notified body may request other samples as necessary.

3. The documentation must make it possible to understand the design, the manufacture and the performances of the product. The documentation shall contain the following items in particular:

- a general description of the type,
- design drawings, methods of manufacture envisaged, in particular as regards sterilisation, and diagrams of parts, sub-assemblies, circuits, etc,
- the descriptions and explanations necessary for the understanding of the abovementioned drawings and diagrams and of the operation of the product,
- a list of the standards referred to in Article 5, applied in full or in part, and a description of the solutions adopted to satisfy the essential requirements where the standards referred to in Article 5 have not been applied,
- the results of design calculations, investigations and technical tests carried out, etc,
- a statement as to whether or not the device incorporates, as an integral part, a substance as referred to in section 10 of Annex 1, whose action in combination with the device may result in its bioavailability, together with data on the relevant trials conducted,
- the clinical data referred to in Annex 7,
- the draft instruction leaflet.

4. The notified body shall:

4.1. examine and evaluate the documentation, verify that the type has been manufactured in accordance with that documentation; it shall also record the items which have been designed in accordance with the applicable provisions of the standards referred to in Article 5, as well as the items for which the design is not based on the relevant provisions of the said standards;

4.2. carry out or have carried out the appropriate inspections and the tests necessary to verify whether the solutions adopted by the manufacturer satisfy the essential requirements of this Directive where the standards referred to in Article 5 have not been applied;

4.3. carry out or have carried out the appropriate inspections and the tests necessary to verify whether, where the manufacturer has chosen to apply the relevant standards, these have actually been applied;

4.4. agree with the applicant on the place where the necessary inspections and tests will be carried out.

5. Where the type meets the provisions of this Directive, the notified body shall issue an EC type-examination certifi-

cate to the applicant. The certificate shall contain the name and address of the manufacturer, the conclusions of the control, the conditions under which the certificate is valid and the information necessary for identification of the type approved.

The significant parts of the documentation shall be attached to the certificate and a copy shall be kept by the notified body.

6. The applicant shall inform the notified body which issued the EC type-examination certificate of any modification made to the approved product.

Modifications to the approved product must receive further approval from the notified body which issued the EC type-examination certificate where such modifications may affect conformity with the essential requirements or with the conditions of use specified for the product. This new approval shall be issued, where appropriate, in the form of a supplement to the initial EC type-examination certificate.

7. Administrative provisions

7.1 On request, each notified body shall make available to the other notified bodies and the competent authority, all relevant information on EC type-examination certificates and addenda issued, refused or withdrawn.

7.2 Other notified bodies may obtain a copy of the EC type-examination certificates and/or the addenda to them. The annexes to the certificates shall be made available to the other notified bodies when a reasoned application is made and after the manufacturer has been informed.

7.3 The manufacturer or his authorised representative shall keep with the technical documentation a copy of the EC type-examination certificates and the supplements to them for a period of at least five years from the manufacture of the last appliance.

7.4 Where neither the manufacturer nor his authorised representative are established in the Community, the task of keeping the technical documentation available for the authorities shall fall to the person responsible for placing the appliance concerned on the Community market.

Annex IV
EC Verification

1. EC verification is the procedure whereby the manufacturer or his authorised representative established within the Community ensures and declares that the products subject to the provisions of section 3 are in conformity with the type as described in the EC type-examination certification and satisfy the requirements of this Directive that apply to them.

2. The manufacturer or his authorised representative established within the Community shall take all measures necessary in order that the manufacturing process ensures conformity of the products to the type as described in the EC type-examination certification and to the requirements of this Directive that apply to them. The manufacturer or his authorised representative established within the Community shall affix the CE marking to each product and draw up a written declaration of conformity.

3. The manufacturer shall, before the start of manufacture, prepare documents defining the manufacturing processes, in particular as regards sterilization, together with all the routine, pre-established provisions to be implemented to ensure uniformity of production and conformity of the

products with the type as described in the EC type-examination certificate as well as with the relevant requirements of this Directive.

4. The manufacturer shall undertake to institute and keep updated a post-marketing surveillance system. This undertaking shall include the obligation on the part of the manufacturer to notify the competent authorities of the following events immediately on learning of them:

(i) any change in the characteristics or performances and any inaccuracies in the instruction leaflet for a device which might lead to or have led to the death of a patient or deterioration in his state of health;

(ii) any technical or medical reason resulting in the withdrawal of a device from the market by a manufacturer.

5. The notified body shall carry out the appropriate examinations and tests in order to check the conformity of the product to the requirements of this Directive by examination and testing of products on a statistical basis, as specified in section 6. The manufacturer must authorise the notified body to evaluate the efficiency of the measures taken pursuant to section 3, by audit where appropriate.

6. Statistical verification

6.1. Manufacturers shall present the products manufactured in the form of uniform batches and shall take all necessary measures in order that the manufacturing process ensures the uniformity of each batch produced.

6.2. A random sample shall be taken from each batch. Products in a sample shall be individually examined and appropriate tests, as set out in the standard(s) referred to in Article 5, or equivalent tests shall be carried out to verify their conformity to the type as described in the EC type-examination certificate and thereby determine whether a batch is to be accepted or rejected.

6.3. Statistical control of products shall be based on attributes, entailing a sampling system with the following characteristics:

- a level of quality corresponding to a probability of acceptance of 95%, with a non-conformity percentage of between 0.29 ands 1%.
- a limit quality corresponding to a probability of acceptance of 5%, with a percentage of non-conformity of between 3 and 7%.

6.4. Where batches are accepted, the notified body shall affix, or cause to be affixed, its identification number to each product and draw up a written certificate of conformity relating to the tests carried out. All products in the batch may be placed on the market except for those products from the sample which were found not to be in conformity.

Where a batch is rejected, the notified body shall take appropriate measures to prevent the placing on the market of that batch. In the event of frequent rejection of batches the notified body may suspend the statistical verification.

The manufacturer may, under the responsibility of the notified body, affix the latter's identification number during the manufacturing process.

6.5. The manufacturer or his authorised representative shall ensure that he is able to supply the notified body's certificates of conformity on request.

Annex V
EC declaration of conformity to type

(Assurance of production quality)

1. The manufacturer shall apply the quality system approved for the design, manufacture and final inspection of the products concerned as specified in sections 3: he shall be subject to the surveillance referred to in section 4.

2. The declaration of conformity is the procedural element whereby the manufacturer who satisfies the obligations of section 1 guarantees and declares that the products concerned conform to the type described in the EC type-examination certificate and meet the provisions of the Directive which apply to them.

The manufacturer or his authorised representative established within the Community shall affix the CE marking in accordance with Article 12 and draw up a written declaration of conformity. This declaration shall cover one or more identified specimens of the product and shall be kept by the manufacturer. The CE marking shall be accompanied by the identification number of the notified body responsible.

3. Quality system

3.1. The manufacturer shall make an application for evaluation of his quality system to a notified body.

The application shall include:

- all the appropriate information concerning the products which it is intended to manufacture.
- the quality-system documentation,
- an undertaking to fulfil the obligations arising from the quality system as approved,
- an undertaking to maintain the approved quality system in such a way that it remains adequate and efficacious,
- where appropriate the technical documentation relating to the approved type and a copy of the EC type-examination certificate.
- an undertaking by the manufacturer to institute and keep up-dated a post-marketing surveillance system. The undertaking shall include an obligation for the manufacturer to notify the competent authorities of the following incidents immediately on learning of them:
  - (i) any deterioration in the characteristics or performances, and any inaccuracies in the instruction leaflet for a device which might lead to or have led to the death of a patient or a deterioration in his state of health;
  - (ii) any technical or medical reason resulting in withdrawal of a device from the market by the manufacturer.

3.2. The application of the quality system must ensure that the products conform to the type described in the EC type-examination certificate.

All the elements, requirements and provisions adopted by the manufacturer for his quality system shall be documented in a systematic and orderly manner in the form of written policies and procedures. This quality-system documentation must make possible a uniform interpretation of the quality policies and procedures such as quality programmes, quality plans, quality manuals and quality records.

It shall include in particular an adequate description of:

(a) the manufacturer's quality objectives;

(b) the organisation of the business and in particular:
  - the organisational structures, the responsibilities of the managerial staff and their organisational authority where manufacture of the products is concerned.

- the methods of monitoring the efficient operation of the quality system and in particular its ability to achieve the desired quality of the products, including control of products which do not conform;

(c) the techniques of control and of quality assurance at the manufacturing stage and in particular:
- the process and procedures which will be used, particularly as regards sterilisation, purchasing and the relevant documents,
- product identification procedures drawn up and kept up-to-date from drawings, specifications or other relevant documents at every stage of manufacture;

(d) the appropriate tests and trials which will be effected before, during and after production, the frequency with which they will take place, and the test equipment used.

3.3. Without prejudice to Article 13, the notified body shall effect an audit of the quality system to determine whether it meets the requirements referred to in 3.2. It shall presume conformity with these requirements for the quality systems which use the corresponding harmonised standards.

The team entrusted with the evaluation shall include at least one member who has already had experience of evaluations of the technology concerned. The evaluation procedure shall include an inspection on the manufacturer's premises.

The decision shall be notified to the manufacturer after the final inspection. It shall contain the conclusions of the control and a reasoned evaluation.

3.4. The manufacturer shall inform the notified body which has approved the quality system of any plan to alter that system.

The notified body shall evaluate the proposed modifications and shall verify whether the quality system so modified would meet the requirements referred to in 3.2; it shall notify the manufacturer of its decision. This decision shall contain the conclusions of the control and a reasoned evaluation.

4. Surveillance

4.1. The aim of surveillance is to ensure that the manufacturer duly fulfils the obligations which arise from the approved quality system.

4.2. The manufacturer shall authorise the notified body to carry out all necessary inspections and shall supply it with all appropriate information, in particular:
- the quality-system documentation,
- the data stipulated in the part of the quality system relating to manufacture, such as reports concerning inspections, test, standardisations, calibrations and the qualifications of the staff concerned, etc.

4.3. The notified body shall periodically carry out appropriate inspections and evaluations in order to ascertain that the manufacturer is applying the approved quality system, and shall supply the manufacturer with an evaluation report.

4.4. In addition, the notified body may make unnounced visits to the manufacturer, and shall supply him with an inspection report.

5. The notified body shall communicate to the other notified bodies all relevant information concerning approvals of quality systems issued, refused or withdrawn.

Annex VI
Statement concerning devices intended for special purposes

1. The manufacturer or his authorised representative established within the Community shall draw up for custom-made devices or for devices intended for clinical investigations the statement comprising the elements stipulated in section 2.

2. The statement shall comprise the following information:

2.1. For custom-made devices:

- data allowing the device in question to be identified,
- a statement affirming that the device is intended for exclusive use by a particular patient, together with his name,
- the name of the doctor who drew up the prescription and, if applicable, the name of the clinic concerned,
- the particular features of the device as described by the medical prescription concerned.
- a statement affirming that the device complies with the essential requirements given with the essential requirements given in Annex I and where applicable, indicating which essential requirements have not been wholly met, together with the grounds.

2.2. For devices intended for clinical investigations covered in Annex VII:

- data allowing the devices in question to be identified,
- an investigation plan giving in particular the purpose, scope and number of the devices concerned,
- the name of the doctor and of the institution responsible for the investigations,
- a statement affirming that the device in question complies with the essential requirements apart from the aspects constituting the object of the investigations and that, with regard to these aspects, every precaution has been taken to protect the health and safety of the patient.

3. The manufacturer shall undertake to keep available for the competetent national authorities:

3.1. For custom-made devices, documentation enabling the design, manufacture and performances of the product, including the expected performances, to be understood, so as to allow conformity with the requirement of this Directive to be assessed.

The manufacturer shall take all necessary measures to see that the manufacturing process ensures that the products manufactured conform to the documentation referred to in the first paragraph.

3.2. For devices intended for clinical investigations, the documentation shall also contain:

- a general description of the product,
- design drawings, manufacturing methods, in particular as regards sterilisation, and diagrams of parts, sub-assemblies, circuits, etc.
- the descriptions and explanations necessary for the understanding of the said drawings and diagrams and of the operation of the product.
- a list of the standards laid down in Article 5, applied in full or in part, and a description of the solutions adopted to satisfy the essential requirements of the Directive where the standards in Article 5 have not been applied.
- the results of the design calculations, checks and technical tests carried out, etc.

The manufacturer shall take all necessary measures to see that the manufacturing process ensures that the products manufactured conform to the documentation referred to in 3.1 and in the first paragraph of this section.

The manufacturer may authorise the evaluation, by audit where necessary, of the effectiveness of these measures.

Annex VII
Clinical evaluation

1. General provisions

1.1. Adequacy of the clinical data presented, as referred to in section 4.2 of Annex II, and in section 3 of Annex III, shall be based, account being taken as appropriate of the relevant harmonised standards, on either:
1.1.1. a collection of currently available relevant scientific literature covering the intended use of the device and the technique therefore, as well as, if appropriate, a written report making a critical assessment of this collation; or
1.1.2. the results of all clinical investigations made, including those carried out in accordance with section 2.
1.2. All data must remain confidential unless it is deemed essential that they be divulged.

2. Clinical investigation

2.1. *Purpose*

The purpose of clinical investigation is to:

- verify that, under normal conditions of use, the performances of the device comply with those indicated in section 2 of Annex I,
- determine any undesirable side effects, under normal conditions of use, and assess whether they are acceptable risks having regard to the intended performance of the device.

2.2. *Ethical consideration*

Clinical investigations shall be made in accordance with the Declaration of Helsinki approved by the 18th World Medical Assembly in Helsinki, Finland, in 1964, and amended by the 29th World Medical Assembly in Tokyo, Japan, in 1975 and the 35th World Medical Assembly in Venice, Italy, in 1983. It is mandatory that all measures relating to the protection of human subjects are carried out in the spirit of the Declaration of Helsinki. This includes every step in the clinical investigation from first consideration of need and justification of study to publication of results.

2.3. *Methods*

2.3.1. Clinical investigations shall be performed according to an appropriate state of the art plan of investigation defined in such a way as to confirm or refute the manufacturer's claims for the device; the investigations shall include an adequate number of observations to guarantee the scientific validity of the conclusions.
2.3.2. The procedures utilised to perform the investigations shall be appropriate to the device under examination.
2.3.3. Clinical investigations shall be performed in circumstances equivalent to those which would be found in normal conditions of use of the device.
2.3.4. All appropriate features, including those involving the safety and performances of the device, and its effects on the patients, shall be examined.
2.3.5. All adverse events shall be fully recorded.
2.3.6. The investigations shall be performed under the responsibility of an appropriately qualified medical specialist, in an appropriate environment.

The medical specialist shall have access to the technical data regarding the device.
2.3.7. The written report, signed by the responsible medical specialist, shall comprise a critical evaluation of all the data collected during the clinical investigation.

Annex VIII
Minimum criteria to be met when designing inspection bodies to be notified

1. The body, its director and the staff responsible for carrying out the evaluation and verification operations shall not be the designer, manufacturer, supplier or installer of devices which they control, not the authorised representative of any of those parties. They may not become directly involved in the design, construction, marketing or maintenance of the devices, nor represent the parties engaged in these activities. this does not preclude the possibility of exchanges of technical information between the manufacturer and the body.

2. The body and its staff must carry out the evaluation and verification operations with the highest degree of professional integrity and technical competence and must be free from all pressures and inducements, particularly financial, which might influence their judgement or the results of the inspection, especially from persons or groups of persons with an interest in the results of verifications.

3. The body must be able to carry out all the tasks in one of Annexes II to V assigned to such a body and for which it has been notified, whether those tasks are carried out by the body itself or under its responsibility. In particular, it must have at its disposal the necessary staff and possess the necessary facilities to enable it to perform properly the technical and administrative tasks connected with evaluation and verification; it must also have access to the equipment necessary for the verification required.

4. The staff responsible for control operations must have:
- sound vocational training covering all the evaluation and verification operations for which the body has been designated,
- satisfactory knowledge of the requirements of the controls they carry out and adequate experience of such operations,
- the ability required to draw up the certificates, records and reports to demonstrate that the controls have been carried out.

The impartiality of inspection staff must be guaranteed. Their remuneration must not depend on the number of controls carried out, nor on the results of such controls.

6. The body must take out liability insurance unless liability is assumed by the State in accordance with national law, or the Member State itself is directly responsible for controls.

7. The staff of the body are bound to observe professional secrecy with regard to all information gained in carrying out their tasks (except *vis-à-vis* the competent administrative authorities of the State in which their activities are carried out) under this Directive or any provision of national law giving effect to it.

Annex IX

CE conformity marking

– The CE conformity marking shall consist of the initials "CE" taking the following form:

– If the CE marking is reduced or enlarged the proportions given in the above graduated drawing must be respected.

– The various components of the CE marking must have substantially the same vertical dimensions which may not be less than 5 mm.

This minimum dimension may be waived for small-scale devices.

COUNCIL DIRECTIVE 93/42/EEC

of 14 June 1993

concerning medical devices

THE COUNCIL OF THE EUROPEAN COMMUNITIES,

Having regard to the Treaty establishing the European Economic Community, and in particular Article 100a thereof,

Having regard to the proposal from the Commission,[1]

In cooperation with the European Parliament,[2]

Having regard to the opinion of the Economic and Social Committee,[3]

Whereas measures should be adopted in the context of the internal market; whereas the internal market is an area without internal frontiers in which the free movement of goods, persons, services and capital is ensured;

Whereas the content and scope of the laws, regulations and administrative provisions in force in the Member States with regard to the safety, health protection and performance characteristics of medical devices are different; whereas the certification and inspection procedures for such devices differ from one Member State to another; whereas such disparities constitute barriers to trade within the Community;

Whereas the national provisions for the safety and health protection of patients, users and, where appropriate, other persons, with regard to the use of medical devices should be harmonised in order to guarantee the free movement of such devices within the internal market;

Whereas the harmonised provisions must be distinguished from the measures adopted by the Member States to manage the funding of public health and sickness insurance schemes relating directly or indirectly to such devices; whereas, therefore, these provisions do not affect the ability of the Member States to implement the abovementioned measures provided Community law is complied with;

Whereas medical devices should provide patients, users and third parties with a high level of protection and attain the performance levels attributed to them by the manufacturer; whereas, therefore, the maintenance or improvement of the level of protection attained in the Member States is one of the essential objectives of this Directive;

Whereas certain medical devices are intended to administer medicinal products within the meaning of Council Directive 65/65/EEC of 26 January 1965 on the approximation of provisions laid down by law, regulation or administrative action relating to proprietary medicinal products,[4] whereas, in such cases the placing on the market of the medicinal device as a general rule is governed by the present Directive and the placing on the market of the medicinal product is governed by Directive 65/65/EEC; whereas if, however, such a device is placed on the market in such a way that the device and the medicinal product form a single integral unit which is intended exclusively for use in the given combination and which is not reusable, that single-unit product shall be governed

[1] OJ No C 237, 12. 9. 1991 and OJ No C 251, 28. 9. 1992, p. 40.
[2] OJ No C 150, 31. 5. 1993 and OJ No C 176, 28. 6. 1993.
[3] OJ No C 79, 30. 3. 1992, p. 1.

[4] OJ No 22, 9. 6. 1965, p. 369/65. Directive as last amended by Directive 92/27/EEC (OJ No L 113, 30. 4. 1992, p. 8).

by Directive 65/65/EEC; whereas a distinction must be drawn between the abovementioned devices and medical devices incorporating, *inter alia*, substances which, if used separately, may be considered to be a medicinal substance within the meaning of the Directive 65/65/EEC; whereas in such cases, if the substances incorporated in the medical devices are liable to act upon the body with action ancillary to that of the device, the placing of the devices on the market is governed by this Directive; whereas, in this context, the safety, quality and usefulness of the substances must be verified by analogy with the appropriate methods specified in Council Directive 75/318/EEC of 20 May 1975 on the approximation of the laws of the Member States relating to analytical, pharmaco-toxicological and clinical standards and protocols in respect of the testing of proprietary medicinal products;[5]

Whereas the essential requirements and other requirements set out in the Annexes to this Directive, including any reference to 'minimising' or 'reducing' risk must be interpreted and applied in such a way as to take account of technology and practice existing at the time of design and of technical and economical considerations compatible with a high level of protection of health and safety;

Whereas, in accordance with the principles set out in the Council resolution of 7 May 1985 concerning a new approach to technical harmonisation and standardisation,[6] rules regarding the design and manufacture of medical devices must be confined to the provisions required to meet the essential requirements; whereas, because they are essential, such requirements should replace the corresponding national provisions; whereas the essential requirements should be applied with discretion to take account of the technological level existing at the time of design and of technical and economic considerations compatible with a high level of protection of health and safety;

Whereas Council Directive 90/385/EEC of 20 June 1990 on the approximation of the laws of the Member States relating to active implantable medical devices[7] is the first case of application of the new approach to the field of medical devices; whereas in the interest of uniform Community rules applicable to all medical devices, this Directive is based largely on the provisions of Directive 90/385/EEC; whereas, for the same reasons, Directive 90/385/EEC must be amended to insert the general provisions laid down in this Directive;

Whereas the electromagnetic compatibility aspects form an integral part of the safety of medical devices; whereas this Directive should contain specific rules on this subject with regard to Council Directive 89/336/EEC of 3 May 1989 on the approximation of the laws of the Member States relating to electromagnetic compatibility;[8]

Whereas this Directive should include requirements regarding the design and manufacture of devices emitting ionizing radiation; whereas this Directive does not affect the authorisation required by Council Directive 80/836/Euratom of 15 July 1980 amending the Directives laying down the basic safety standards for the health protection of the general public and workers against the dangers of ionising radiation,[9] nor application of Council Directive 84/446/Euratom of 3 September 1984 laying down basic

[5] OJ No L 147, 9. 6. 1975, p. 1. Directive as last amended by Directive 91/507/EEC (OJ No L 270, 26. 9. 1991, p. 32).

[6] OJ No C 136, 4. 6. 1985, p. 1.

[7] OJ No L 189, 20. 7. 1990, p. 17.

[8] OJ No L 139, 23. 5. 1989, p. 19. Directive as last amended by Directive 92/31/EEC (OJ No L 126, 12. 5. 1992, p. 11).

[9] OJ No L 246, 17. 9. 1980, p. 1. Directive as last amended by Directive 84/467/Eurotom (OJ No 265, 5. 10. 1984, p. 4).

measures for the radiation protection of persons undergoing medical examination or treatment;[10] whereas Council Directive 89/391/EEC of 12 June 1989 on the introduction of measures to encourage improvements in the safety and health of workers at work[11] and the specific directives on the same subject should continue to apply;

Whereas, in order to demonstrate conformity with these essential requirements and to enable conformity to be verified, it is desirable to have harmonised European standards to protect against the risks associated with the design, manufacture and packaging of medical devices; whereas such harmonised European standards are drawn up by private-law bodies and should retain their status as non-mandatory texts; whereas, to this end, the European Committee for Standardisation (CEN) and the European Committee for Electrotechnical Standardisation (CENELEC) are recognised as the competent bodies for the adoption of harmonised standards in accordance with the general guidelines on cooperation between the Commission and these two bodies signed on 13 November 1984;

Whereas, for the purpose of this Directive, a harmonised standard is a technical specification (European standard or harmonisation document) adopted, on a mandate from the Commission, by either or both of these bodies in accordance with Council Directive 83/189/EEC of 28 March 1983 laying down a procedure for the provision of information in the field of technical standards and regulations,[12] and pursuant to the abovementioned general guidelines; whereas with regard to possible amendment of the harmonised standards, the Commission should be assisted by the Committee set up under Directive 83/189/EEC; whereas the measures to be taken must be defined in line with procedure I, as laid down in Council Decision 87/373/EEC;[13] whereas, for specific fields, what already exists in the form of European *Pharmacopoeia* monographs should be incorporated within the framework of this Directive; whereas, therefore, several European *Pharmacopoeia* monographs may be considered equal to the abovementioned harmonised standards;

Whereas, in Decision 90/683/EEC of 13 December 1990 concerning the modules for the various phases of the conformity assessment procedures which are intended to be used in the technical harmonisation directives,[14] the Council has laid down harmonised conformity assessment procedures; whereas the application of these modules to medical devices enables the responsibility of manufacturers and notified bodies to be determined during conformity assessment procedures on the basis of the type of devices concerned; whereas the details added to these modules are justified by the nature of the verification required for medical devices;

Whereas it is necessary, essentially for the purpose of the conformity assessment procedures, to group the devices into four product classes; whereas the classification rules are based on the vulnerability of the human body taking account of the potential risks associated with the technical design and manufacture of the devices; whereas the conformity assessment procedures for Class I devices can be carried out, as a general rule, under the sole responsibility of the manufacturers in view of the low level of vulnerability associated with these products; whereas, for Class IIa devices, the intervention of a notified body should

[10] OJ No L 265, 5. 10. 1984, p. 1.
[11] OJ No L 183, 29. 6. 1989, p. 1.
[12] OJ No L 109, 26. 4. 1983, p. 8. Directive as last amended by Commission Decision 92/400/EEC (OJ No L 221, 6. 8. 1992, p. 55).
[13] OJ No L 197, 18. 7. 1987, p. 33.
[14] OJ No L 380, 31. 12. 1990, p. 13.

be compulsory at the production stage; whereas, for devices falling within Classes IIb and III which constitute a high risk potential, inspection by a notified body is required with regard to the design and manufacture of the devices; whereas Class III is set aside for the most critical devices for which explicit prior authorisation with regard to conformity is required for them to be placed on the market;

Whereas in cases where the conformity of the devices can be assessed under the responsibility of the manufacturer the competent authorities must be able, particularly in emergencies, to contact a person responsible for placing the device on the market and established in the Community, whether the manufacturer or another person established in the Community and designated by the manufacturer for the purpose;

Whereas medical devices should, as a general rule, bear the CE mark to indicate their conformity with the provisions of this Directive to enable them to move freely within the Community and to be put into service in accordance with their intended purpose;

Whereas, in the fight against AIDS and in the light of the conclusions of the Council adopted on 16 May 1989 regarding future activities on AIDS prevention and control at Community level,[15] medical devices used for protection against the HIV virus must afford a high level of protection; whereas the design and manufacture of such products should be verified by a notified body;

Whereas the classification rules generally enable medical devices to be appropriately classified; whereas, in view of the diverse nature of the devices and technological progress in this field, steps must be taken to include amongst the implementing powers conferred on the Commission the decisions to be taken with regard to the proper classification or reclassification of the devices or, where appropriate, the adjustment of the classification rules themselves; whereas since these issues are closely connected with the protection of health, it is appropriate that these decisions should come under procedure IIIa, as provided for in Decision 87/373/EEC;

Whereas the confirmation of compliance with the essential requirements may mean that clinical investigations have to be carried out under the responsibility of the manufacturer; whereas, for the purpose of carrying out the clinical investigations, appropriate means have to be specified for the protection of public health and public order;

Whereas the protection of health and the associated controls may be made more effective by means of a medical device vigilance system at Community level;

Whereas this Directive covers the medical devices referred to in Council Directive 76/764/EEC of 27 July 1976 on the approximation of the laws of the Member States on clinical mercury-in-glass, maximum reading thermometers;[16] whereas the abovementioned Directive must therefore be repealed; whereas for the same reasons Council Directive 84/539/EEC of 17 September 1984 on the approximation of the laws of the Member States relating to electro-medical equipment used in human or veterinary medicine[17] must be amended,

HAS ADOPTED THIS DIRECTIVE:

Article 1

Definitions, scope

1. This Directive shall apply to medical devices and their accessories. For the

[15] OJ No C 185, 22. 7. 1989, p. 8.

[16] OJ No L 262, 27. 9. 1976, p. 139. Directive as last amended by Directive 84/414/EEC (OJ No L 228, 25. 8. 1984, p. 25).

[17] OJ No L 300, 19. 11. 1984, p. 179. Directive as amended by the Act of Accessioon of Spain and Portugal.

purposes of this Directive, accessories shall be treated as medical devices in their own right. Both medical devices and accessories shall hereinafter be termed devices.

2. For the purposes of this Directive, the following definitions shall apply:

(a) "medical device" means any instrument, apparatus, appliance, material or other article, whether used alone or in combination, including the software necessary for its proper application intended by the manufacturer to be used for human beings for the purpose of:
- diagnosis, prevention, monitoring, treatment or alleviation of disease,
- diagnosis, monitoring, treatment or alleviation of or compensation for an injury or handicap,
- investigation, replacement or modification of the anatomy or of a psychological process,
- control of conception,

and which does not achieve its principal intended action in or on the human body by pharmacological, immunological or metabolic means, but which may be assisted in its function by such means;

(b) "accessory" means an article which whilst not being a device is intended specifically by its manufacturer to be used in accordance with the use of the device intended by the manufacturer of the device;

(c) "device used for *in vitro* diagnosis" means any device which is a reagent, reagent product, kit, instrument, equipment or system, whether used alone or in combination, intended by the manufacturer to be used *in vitro* for the examination of samples derived from the human body with a view to providing information on the physiological state, state of health or disease, or congenital abnormality thereof;

(d) "custom-made device" means any device specifically made in accordance with a duly qualified medical practitioner's written prescription which gives, under his responsibility, specific design characteristics and is intended for the sole use of a particular patient.

The abovementioned prescription may also be made out by any other person authorised by virtue of his professional qualifications to do so.

Mass-produced devices which need to be adapted to meet the specific requirements of the medical practitioner or any other professional user are not considered to be custom-made devices;

(e) "device intended for clinical investigation" means any device intended for use by a duly qualified medical practitioner when conducting investigations as referred to in Section 2.1 of Annex X in an adequate human clinical environment.

For the purpose of conducting clinical investigation, any other person who, by virtue of his professional qualifications, is authorised to carry out such investigation shall be accepted as equivalent to a duly qualified medical practitioner;

(f) "manufacturer" means the natural or legal person with responsibility for the design, manufacture, packaging and labelling of a device before it is placed on the market under his own name, regardless of whether these operations are carried out by that person himself or on his behalf by a third party.

The obligations of this Directive to be met by manufacturers also apply to the natural or legal persons who assembles, packages, processes, fully refurbishes and/or labels one or more ready-made products and/or assigns to them their intended purpose as a device with a view to their being placed on the market under his own name. This subparagraph does not apply to the person who, while not a manufacturer within the meaning of the first subparagraph, assembles or adapts devices already

on the market to their intended purpose for an individual patient;
(g) "intended purpose" means the use for which the device is intended according to the data supplied by the manufacturer on the labelling, in the instructions and/or in promotional materials;
(h) "placing on the market" means the first making available in return for payment or free of charge of a device other than a device intended for clinical investigation, with a view to distribution and/or use on the Community market, regardless of whether it is new or fully refurbished;
(i) "putting into service" means the stage at which a device is ready for use on the Community market for the first time for its intended purpose.

3. Where a device is intended to administer a medicinal product within the meaning of Article 1 of Directive 65/65/EEC, that device shall be governed by the present Directive, without prejudice to the provisions of Directive 65/65/EEC with regard to the medicinal product.

If, however, such a device is placed on the market in such a way that the device and the medicinal product form a single integral product which is intended exclusively for use in the given combination and which is not reusable, that single product shall be governed by Directive 65/65/EEC. The relevant essential requirements of Annex I to the present Directive shall apply as far as safety and performance related device features are concerned.

4. Where a device incorporates, as an integral part, a substance which, if used separately, may be considered to be a medicinal product within the meaning of Article 1 of Directive 65/65/EEC and which is liable to act upon the body with action ancillary to that of the device, that device must be assessed and authorised in accordance with this Directive.

5. This Directive does not apply to:
(a) *in vitro* diagnostic devices;
(b) active implantable devices covered by Directive 90/385/EEC.
(c) medicinal products covered by Directive 65/65/EEC;
(d) cosmetic products covered by Directive 76/768/EEC;[18]
(e) human blood, human blood products, human plasma or blood cells of human origin or to devices which incorporate at the time of placing on the market such blood products, plasma or cells;
(f) transplants or tissues or cells of human origin nor to products incorporating or derived from tissue or cells of human origin;
(g) transplants or tissues or cells of animal origin, unless a device is manufactured utilizing animal tissue which is rendered non-viable or non-viable products derived from animal tissue.

6. This Directive does not apply to personal protective equipment covered by Directive 89/686/EEC. In deciding whether a product falls under that Directive or the present Directive, particular account shall be taken of the principal intended purpose of the product.

7. This Directive is a specific Directive within the meaning of Article 2 (2) of Directive 89/336/EEC.

8. This Directive does not affect the application of Directive 80/336/Euratom, nor of Directive 84/466/Euratom.

Article 2

Placing on the market and putting into service

Member States shall take all necessary steps to ensure that devices may be placed on the market and put into service only if they do not compromise the safety and health of patients, users and, where applicable, other persons when properly installed, maintained and used in accordance with their intended purpose.

---

[18] OJ No L 262, 27. 9. 1976, p. 169. Directive as last amended by Commission Directive 92/86/EEC (OJ No L 325, 11. 11. 1992, p. 18).

Article 3

Essential requirements

The devices must meet the essential requirements set out in Annex I which apply to them, taking account of the intended purpose of the devices concerned.

Article 4

Free movement, devices intended for special purposes

1. Member States shall not create any obstacle to the placing on the market or the putting into service within their territory of devices bearing the CE marking provided for in Article 17 which indicate that they have been the subject of an assessment of their conformity in accordance with the provision of Article 11.

2. Member States shall not create any obstacles to:

- devices intended for clinical investigation being made available to medical practitioners or authorised persons for that purpose if they meet the conditions laid down in Article 15 and in Annex VIII,
- custom-made devices being placed on the market and put into service if they meet the conditions laid down in Article 11 in combination with Annex VIII; Class IIa, IIb and III devices shall be accompanied by the statement referred to in Annex VIII.

These devices shall not bear the CE marking.

3. At trade fairs, exhibitions, demonstrations, etc. Member States shall not create any obstacles to the showing of devices which do not conform to this Directive, provided that a visible sign clearly indicates that such devices cannot be marketed or put into service until they have been made to comply.

4. Member States may require the information, which must be made available to the user and the patient in accordance with Annex I, point 13, to be in their national language(s) or in another Community language, when a device reaches the final user, regardless of whether it is for professional or other use.

5. Where the devices are subject to other Directives concerning other aspects and which also provide for the affixing of the CE marking, the latter shall indicate that the devices also fulfil the provisions of the other Directives.

However, should one or more of those directives allow the manufacturer, during a transitional period, to choose which arrangements to apply, the CE marking shall indicate that the devices fulfil the provisions only of those directives applied by the manufacturer. In this case, the particulars of these directives, as published in the *Official Journal of the European Communities*, must be given in the documents, notices or instructions required by the directives and accompanying such devices.

Article 5

Reference to standards

1. Member States shall presume compliance with the essential requirements referred to in Article 3 in respect of devices which are in conformity with the relevant national standards adopted pursuant to the harmonised standards the references of which have been published in the *Official Journal of the European Communities*; Member States shall publish the references of such national standards.

2. For the purposes of this Directive, reference to harmonised standards also includes the monographs of the European *Pharmacopoeia* notably on surgical sutures and on interaction between medicinal products and materials used in devices containing such medicinal products, the references of which have been published in the *Official Journal of the European Communities*.

3. If a Member State or the Commission considers that the harmonised standards do not entirely meet the essential requirements referred to in Article 3, the measures to be taken by the Member States with regard to these standards and the publication referred

to in paragraph 1 of this Article shall be adopted by the procedure defined in Article 6 (2).

Article 6

Committee on Standards and Technical Regulations

1. The Commission shall be assisted by the Committee set up by Article 5 of Directive 83/189/EEC.

2. The representative of the Commission shall submit to the Committee a draft of the measures to be taken. The Committee shall deliver its opinion on the draft, within a time limit which the chairman may lay down according to the urgency of the matter, if necessary by taking a vote.

The opinion shall be recorded in the minutes; in addition, each Member State shall have the right to ask to have its position recorded in the minutes.

The Commission shall take the utmost account of the opinion delivered by the Committee. It shall inform the Committee of the manner in which its opinion has been taken into account.

Article 7

Committee on Medical Devices

1. The Commission shall be assisted by the Committee set up by Article 6 (2) of Directive 90/385/EEC.

2. The representative of the Commission shall submit to the Committee a draft of the measures to be taken. The Committee shall deliver its opinion on the draft within a time limit which the chairman may lay down according to the urgency of the matter. The opinion shall be delivered by the majority laid down in Article 148 (2) of the Treaty in the case of decisions which the Council is required to adopt on a proposal from the Commission. The votes of the representatives of the Member States within the Committee shall be weighted in the manner set out in that Article. The chairman shall not vote.

The Commission shall adopt the measures envisaged if they are in accordance with the opinion of the Committee.

If the measures envisaged are not in accordance with the opinion of the Committee, or if no opinion is delivered, the Commission shall, without delay, submit to the Council a proposal relating to the measures to be taken. The Council shall act by a qualified majority.

If, on the expiry of a period of three months from the date of referral to the Council, the Council has not acted, the proposed measures shall be adopted by the Commission.

3. The Committee may examine any question connected with implementation of this Directive.

Article 8

Safeguard clause

1. Where a Member State ascertains that the devices referred to in Article 4 (1) and (2) second indent, when correctly installed, maintained and used for their intended purpose, may compromise the health and/or safety of patients, users or, where applicable, other persons, it shall take all appropriate interim measures to withdraw such devices from the market or prohibit or restrict their being placed on the market or put into service. The Member State shall immediately inform the Commission of any such measures, indicating the reasons for its decision and, in particular, whether non-compliance with this Directive is due to:

(a) failure to meet the essential requirements referred to in Article 3;
(b) incorrect application of the standards referred to in Article 5, in so far as it is claimed that the standards have been applied;
(c) shortcomings in the standards themselves.

2. The Commission shall enter into consultation with the parties concerned as soon as possible. Where, after such consultation, the Commission finds that:

– measures are justified, it shall

immediately so inform the Member States which took the initiative and the other Member States; where the decision referred to in paragraph 1 is attributed to shortcomings in the standards, the Commission shall, after consulting the parties concerned, bring the matter before the Committee referred to in Article 6 (1) within two months if the Member State which has taken the decision intends to maintain it and shall initiate the procedures referred to in Article 6.

- the measures are unjustified, it shall immediately so inform the Member State which took the initiative and the manufacturer or his authorised representative established within the Community.

3. Where a non-complying device bears the CE marking, the competent Member State shall take appropriate action against whomsoever has affixed the mark and shall inform the Commission and the other Member States thereof.

4. The Commission shall ensure that the Member States are kept informed of the progress and outcome of this procedure.

Article 9

Classification

1. Devices shall be divided into Classes I, IIa, IIb and III. Classification shall be carried out in accordance with the classification rules set out in Annex IX.

2. In the event of a dispute between the manufacturer and the notified body concerned, resulting from the application of the classification rules, the matter shall be referred for decision to the competent authority to which the notified body is subject.

3. The classification rules set out in Annex IX may be adapted in accordance with the procedure referred to in Article 7 (2) in the light of technical progress and any information which becomes available under the information system provided for in Article 10.

Article 10

Information on incidents occurring following placing of devices on the market

1. Member States shall take the necessary steps to ensure that any information brought to their knowledge, in accordance with the provisions of this Directive, regarding the incidents mentioned below involving a Class I, IIa, IIb or III device is recorded and evaluated centrally:

(a) any malfunction or deterioration in the characteristics and/or performance of a device, as well as any inadequacy in the labelling or the instructions for use which might lead to or might have led to the death of a patient or user or to a serious deterioration in his state of health;

(b) any technical or medical reason in relation to the characteristics or performance of a device for the reasons referred to in subparagraph (a) above, leading to systematic recall of devices of the same type by the manufacturer.

2. Where a Member State requires medical practitioners or the medical institutions to inform the competent authorities of any incidents referred to in paragraph 1, it shall take the necessary steps to ensure that the manufacturer of the device concerned, or his authorised representative established in the Community, is also informed of the incident.

3. After carrying out an assessment, if possible together with the manufacturer, Member States shall, without prejudice to Article 8, immediately inform the Commission and the other Member States of the incidents referred to in paragraph 1 for which relevant measures have been taken or are contemplated.

Article 11

Conformity assessment procedures

1. In the case of devices falling within Class III, other than devices which are

custom-made or intended for clinical investigations, the manufacturer shall, in order to affix the CE marking, either:

(a) follow the procedure relating to the EC declaration of conformity set out in Annex II (full quality assurance); or

(b) follow the procedure relating to EC type-examination set out in Annex III, coupled with:

(i) the procedure relating to EC verification set out in Annex IV; or

(ii) the procedure relating to the EC declaration of conformity set out in Annex V (production quality guarantee).

2. In the case of devices falling within Class IIa, other than devices which are custom-made or intended for clinical investigations, the manufacturer shall, in order to affix the CE marking, follow the procedure relating to the EC declaration of conformity set out in Annex VII, coupled with either:

(a) the procedure relating to EC verification set out in Annex IV; or

(b) the procedure relating to the EC declaration of conformity set out in Annex V (production quality guarantee); or

(c) the procedure relating to the EC declaration of conformity set out in Annex VI (product quality assurance).

Instead of applying these procedures, the manufacturer may also follow the procedure referred to in paragraph 3 (a).

3. In the case of devices falling with Class IIb, other than devices which are custom-made or intended for clinical investigations, the manufacturer shall, in order to affix the CE marking, either:

(a) follow the procedure relating to the EC declaration of conformity set out in Annex II (full quality assurance); in this case, point 4 of Annex II is not applicable; or

(b) follow the procedure relating to EC type-examination set out in Annex III, coupled with:

(i) the procedure relating to EC verification set out in Annex IV, or

(ii) the procedure relating to the EC declaration of conformity set out in Annex V (production quality assurance); or

(iii) the procedure relating to the EC declaration of conformity set out in Annex VI (product quality assurance).

4. The Commission shall, no later than five years from the date of implementation of this Directive, submit a report to the Council on the operation of the provisions referred to in Article 10 (1), Article 15 (1), in particular in respect of Class I and Class IIa devices, and on the operation of the provisions referred to in Annex II, Section 4.3 second and third subparagraphs and in Annex III, Section 5 second and third subparagraphs to this Directive, accompanied, if necessary, by appropriate proposals.

5. In the case of devices falling within Class I, other than devices which are custom-made or intended for clinical investigations, the manufacturer shall, in order to affix the CE marking, follow the procedure referred to in Annex VII and draw up the EC declaration of conformity required before placing the device on the market.

6. In the case of custom-made devices, the manufacturer shall follow the procedure referred to in Annex VIII and draw up the statement set out in that Annex before placing each device on the market.

Member States may require that the manufacturer shall submit to the competent authority a list of such devices which have been put into service in their territory.

7. During the conformity assessment procedure for a device, the manufacturer and/or the notified body shall take account of the results of any assessment and verification operations which, where appropriate, have been carried out in accordance with this Directive at an intermediate stage of manufacture.

8. The manufacturer may instruct his authorised representative established in the Community to initiate the pro-

cedures provided for in Annexes III, IV, VII and VIII.

9. Where the conformity assessment procedure involves the intervention of a notified body, the manufacturer, or his authorised representative established in the Community, may apply to a body of his choice within the framework of the tasks for which the body has been notified.

10. The notified body may require, where duly justified, any information or data, which is necessary for establishing and maintaining the attestation of conformity in view of the chosen procedure.

11. Decisions taken by the notified bodies in accordance with Annexes II and III shall be valid for a maximum of five years and may be extended on application, made at a time agreed in the contract signed by both parties, for further periods of five years.

12. The records and correspondence relating to the procedures referred to in paragraphs 1 to 6 shall be in an official language of the Member State in which the procedures are carried out and/or in another Community language acceptable to the notified body.

13. By derogation from paragraphs 1 to 6, the competent authorities may authorise, on duly justified request, the placing on the market and putting into service, within the territory of the Member State concerned, of individual devices for which the procedures referred to in paragraphs 1 to 6 have not been carried out and the use of which is in the interest of protection of health.

Article 12

Particular procedure for systems and procedure packs

1. By way of derogation from Article 11 this Article shall apply to systems and procedure packs.

2. Any natural or legal person who puts devices bearing the CE marking together within their intended purpose and within the limits of use specified by their manufacturers, in order to place them on the market as a system or procedure pack, shall draw up a declaration by which he states that:

(a) he has verified the mutual compatibility of the devices in accordance with the manufacturers' instructions and has carried out his operations in accordance with these instructions; and

(b) he has packaged the system or procedure pack and supplied relevant information to users incorporating relevant instructions from the manufacturers; and

(c) the whole activity is subjected to appropriate methods of internal control and inspection.

Where the conditions above are not met, as in cases where the system or procedure pack incorporate devices which do not bear a CE marking or where the chosen combination of devices is not compatible in view of their original intended use, the system or procedure pack shall be treated as a device in its own right and as such be subjected to the relevant procedure pursuant to Article 11.

3. Any natural or legal person who sterilised, for the purpose of placing on the market, systems or procedure packs referred to in paragraph 2 or other CE-marked medical devices designed by their manufacturers to be sterilised before use, shall, at his choice, follow one of the procedures referred to in Annex IV, V or VI. The application of the abovementioned Annexes and the intervention of the notified body are limited to the aspects of the procedure relating to the obtaining of sterility. The person shall draw up a declaration stating that sterilisation has been carried out in accordance with the manufacturer's instructions.

4. The products referred to in paragraphs 2 and 3 themselves shall not bear an additional CE marking. They shall be accompanied by the information referred to in point 13 of Annex I which includes, where appropriate, the information supplied by the manufacturers of the devices which have been put together. The declaration referred to in paragraphs 2 and 3

above shall be kept at the disposal of competent authorities for a period of five years.

Article 13

Decisions with regard to classification, derogation clause

1. Where a Member State considers that:

(a) application of the classification rules set out in Annex IX requires a decision with regard to the classification of a given device or category of devices; or

(b) a given device or family of devices should be classified, by way of derogation from the provisions of Annex IX, in another class; or

(c) the conformity of a device or family of devices should be established, by way of derogation from the provisions of Article 11, by applying solely one of the given procedures chosen from among those referred to in Article 11,

it shall submit a duly substantiated request to the Commission and ask it to take necessary measures. These measures shall be adopted in accordance with the procedure referred to in Article 7 (2).

2. The Commission shall inform the Member States of the measures taken and, where appropriate, publish the relevant parts of these measures in the *Official Journal of the European Communities*.

Article 14

Registration of persons responsible for placing devices on the market

1. Any manufacturer who, under his own name, places devices on the market in accordance with the procedures referred to in Article 11 (5) and (6) and any other natural or legal person engaged in the activities referred to in Article 12 shall inform the competent authorities of the Member State in which he has his registered place of business of the address of the registered place of business and the description of the devices concerned.

2. Where a manufacturer who places devices referred to in paragraph 1 on the market under his own name does not have a registered place of business in a Member State, he shall designate the person(s) responsible for marketing them who is (are) established in the Community. These persons shall inform the competent authorities of the Member State in which they have their registered place of business of the address of the registered place of business and the category of devices concerned.

3. The Member States shall on request inform the other Member States and the Commission of the details referred to in paragraphs 1 and 2.

Article 15

Clinical investigation

1. In the case of devices intended for clinical investigations, the manufacturer, or his authorised representative established in the Community, shall follow the procedure referred to in Annex VIII and notify the competent authorities of the Member States in which the investigations are to be conducted.

2. In the case of devices falling within Class III and implantable and long-term invasive devices falling within Class IIa or IIb, the manufacturer may commence the relevant clinical investigation at the end of a period of 60 days after notification, unless the competent authorities have notified him within that period of a decision to the contrary based on considerations of public health or public policy.

Member States may however authorise manufacturers to commence the relevant clinical investigations before the expiry of the period of 60 days, in so far as the relevant ethics committee has issued a favourable opinion on the programme of investigation in question.

3. In the case of devices other than those referred to in the second paragraph, Member States may authorise manufacturers to commence clinical investigations, immediately after the

date of notification, provided that the ethics committee concerned has delivered a favourable opinion with regard to the investigational plan.
4. The authorisation referred to in paragraph 2 second subparagraph and paragraph 3, may be made subject to authorisation from the competent authority.
5. The clinical investigations must be conducted in accordance with the provisions of Annex X. The provisions of Annex X may be adjusted in accordance with the procedure laid down in Article 7(2).
6. The Member States shall, if necessary, take the appropriate steps to ensure public health and public policy.
7. The manufacturer or his authorised representative established in the Community shall keep the report referred to in point 2.3.7 of Annex X at the disposal of the competent authorities.
8. The provisions of paragraphs 1 and 2 do not apply where the clinical investigations are conducted using devices which are authorised in accordance with Article 11 to bear the CE marking unless the aim of these investigations is to use the devices for a purpose other than that referred to in the relevant conformity assessment procedure. The relevant provisions of Annex X remain applicable.

Article 16

Notified bodies

1. The Member States shall notify the Commission and other Member States of the bodies which they have designated for carrying out the tasks pertaining to the procedures referred to in Article 11 and the specific tasks for which the bodies have been designated. The Commission shall assign identification numbers to these bodies, hereinafter referred to as 'notified bodies'.

The Commission shall publish a list of the notified bodies, together with the identification numbers it has allocated to them and the tasks for which they have been notified, in the *Official Journal of the European Communities.* It shall ensure that the list is kept up to date.
2. Member States shall apply the criteria set out in Annex XI for the designation of bodies. Bodies that meet the criteria laid down in the national standards which transpose the relevant harmonised standards shall be presumed to meet the relevant criteria.
3. A Member State that has notified a body shall withdraw that notification if it finds that the body no longer meets the criteria referred to in paragraph 2. It shall immediately inform the other Member States and the Commission thereof.
4. The notified body and the manufacturer, or his authorised representative established in the Community, shall lay down, by common accord, the time limits for completion of the assessment and verification operations referred to in Annexes II to VI.

Article 17

CE marking

1. Devices, other than devices which are custom-made or intended for clinical investigations, considered to meet the essential requirements referred to in Article 3 must bear the CE marking of conformity when they are placed on the market.
2. The CE marking of conformity, as shown in Annex XII, must appear in a visible, legible and indelible form on the device or its sterile pack, where practicable and appropriate, and on the instructions for use. Where applicable, the CE marking must also appear on the sales packaging.

It shall be accompanied by the identification number of the notified body responsible for implementation of the procedures set out in Annexes II, IV, V and VI.
3. It is prohibited to affix marks or inscriptions which are likely to mislead third parties with regard to the meaning or the graphics of the CE marking. Any other mark may be affixed to the device, to the packaging or to the

instruction leaflet accompanying the device provided that the visibility and legibility of the CE marking is not thereby reduced.

Article 18

Wrongly affixed CE marking

Without prejudice to Article 8:

(a) where a Member State establishes that the CE marking has been affixed unduly, the manufacturer or his authorised representative established within the Community shall be obliged to end the infringement under conditions imposed by the Member State;

(b) where non-compliance continues, the Member State must take all appropriate measures to restrict or prohibit the placing on the market of the product in question or to ensure that it is withdrawn from the market, in accordance with the procedure in Article 8.

Article 19

Decisions in respect of refusal or restriction

1. Any decision taken pursuant to this Directive:

(a) to refuse or restrict the placing on the market or the putting into service of a device or the carrying out of clinical investigations; or

(b) to withdraw devices from the market,

shall state the exact grounds on which it is based. Such decisions shall be notified without delay to the party concerned, who shall at the same time be informed of the remedies available to him under the national law in force in the Member State in question and of the time limits to which such remedies are subject.

2. In the event of a decision as referred to in paragraph 1, the manufacturer, or his authorised representative established in the Community, shall have an opportunity to put forward his viewpoint in advance, unless such consultation is not possible because of the urgency of the measure to be taken.

Article 20

Confidentiality

Without prejudice to the existing national provisions and practices on medical secrets, Member States shall ensure that all the parties involved in the application of this Directive are bound to observe confidentiality with regard to all information obtained in carrying out their tasks. This does not affect the obligations of Member States and notified bodies with regard to mutual information and the dissemination of warnings, nor the obligations of the persons concerned to provide information under criminal law.

Article 21

Repeal and amendment of Directives

1. Directive 76/764/EEC is hereby repealed with effect from 1 January 1995.

2. In the title and Article 1 of Directive 84/539/EEC, "human or" is deleted.

In Article 2 of Directive 84/539/EEC, the following subparagraph is added to paragraph 1:

"If the appliance is at the same time a medical device within the meaning of Directive 93/42/EEC [19] and if it satisfies the essential requirements laid down therein for that device, the device shall be deemed to be in conformity with the requirements of this Directive.

3. Directive 90/385/EEC is hereby amended as follows:

1. in article 1 (2) the following two subparagraphs are added:

"(h) 'placing on the market' means the first making available in return for payment or free of charge of a device other than a device intended for clinical investigation, with a view to distribution and/or use on the Community market, regardless of whether it is new or fully refurbished;

(i) 'manufacturer' means the natural or legal person with responsibility for the design, manufacture, pack-

[19] OJ No L 169, 12. 7. 1993, p. 1.

aging and labelling of a device before it is placed on the market under his own name, regardless of whether these operations are carried out by that person himself or on his behalf by a third party.

The obligations of this Directive to be met by manufacturers also apply to the natural or legal person who assembles, packages, processes, fully refurbishes and/or labels one or more ready-made products and/or assigns to them their intended purpose as a device with a view to their being placed on the market under his own name. This subparagraph does not apply to the person who, while not a manufacturer within the meaning of the first subparagraph, assembles or adapts devices already on the market to their intended purpose for an individual patient;"

2. in Article 9 the following paragraphs are added:

"5. During the conformity assessment procedure for a device, the manufacturer and/or the notified body shall take account of the results of any assessment and verification operations which, where appropriate, have been carried out in accordance with this Directive at an intermediate stage of manufacture.

6. Where the conformity assessment procedure involves the intervention of a notified body, the manufacturer, or his authorised representative established in the Community, may apply to a body of his choice within the framework of the tasks for which the body has been notified.

7. The notified body may require, where duly justified, any information or data which is necessary for establishing and maintaining the attestation of conformity in view of the chosen procedure.

8. Decisions taken by the notified bodies in accordance with Annexes II and III shall be valid for a maximum of five years and may be extended on application, made at a time agreed in the contract signed by both parties, for further periods of five years.

9. By derogation from paragraphs 1 and 2 the competent authorities may authorise, on duly justified request, the placing on the market and putting into service, within the territory of the Member State concerned, of individual devices for which the procedures referred to in paragraphs 1 and 2 have not been carried out and the use of which is in the interest of protection of health.";

3. the following article 9a is inserted after Article 9:

"*Article 9a*

1. Where a Member State considers that the conformity of a device or family of devices should be established, by way of derogation from the provisions of Article 9, by applying solely one of the given procedures chosen from among those referred to in Article 9, it shall submit a duly substantiated request to the Commission and ask it to take the necessary measures. These measures shall be adopted in accordance with the procedure referred to in Article 7 (2) of Directive 93/42/EEC ([20]).

2. The Commission shall inform the Member States of the measures taken and, where appropriate, publish the relevant parts of these measures in the *Official Journal of the EuropeanCommunities*.

4. Article 10 shall be amended as follows:

– the following subparagraph shall be added to paragraph 2:

"Member States may however authorise manufacturers to start the clinical investigations in question before the expiry of the 60-day period, provided that the Ethical Committee concerned has delivered a favourable opinion with respect to the investigation programme in question ",

[20] OJ No L 169, 12. 7. 1993, p. 1.

– the following paragraph shall be inserted:
"2a. The authorisation referred to in the second subparagraph of paragraph 2 may be subject to approval by the competent authority.";

5. the following is added to Article 14:
"In the event of a decision as referred to in the previous paragraph the manufacturer, or his authorised representative established in the Community, shall have an opportunity to put forward his viewpoint in advance, unless such consultation is not possible because of the urgency of the measures to be taken."

Article 22

Implementation, transitional provisions

1. Member States shall adopt and publish the laws, regulations and administrative provisions necessary to comply with this Directive not later than 1 July 1994. They shall immediately inform the Commission thereof.

The Standing Committee referred to in Article 7 may assume its tasks from the date of notification[21] of this Directive. The Member States may take the measures referred to in Article 16 on notification of this Directive.

When Member States adopt these provisions, these shall contain a reference to this Directive or shall be accompanied by such a reference at the time of their official publication. The procedure for such reference shall be adopted by Member States.

Member States shall apply these provisions with effect from 1 January 1995.

2. Member States shall communicate to the Commission the texts of the provisions of national law which they adopt in the field covered by this Directive.

3. Member States shall take the necessary action to ensure that the notified bodies which are responsible pursuant to Article 11 (1) to (5) for conformity assessment take account of any relevant information regarding the characteristics and performances of such devices, including in particular the results of any relevant tests and verification already carried out under pre-existing national law, regulations or administrative provisions in respect of such devices.

4. Member States shall accept the placing on the market and putting into service of devices which conform to the rules in force in their territory on 31 December 1994 during a period of five years following adoption of this Directive.

In the case of devices which have been subject to EEC pattern approval in accordance with Directive 76/764/EEC, Member States shall accept their being placed on the market and put into service during the period up to 30 June 2004.

Article 23

This Directive is addressed to the Member States.

Done at Luxembourg, 14 June 1993.

For the Council
The President
J. TRØJBORG

---

[21] This Directive was notified to the Member States on 29 June 1993.

Annex I
Essential requirements

I. General requirements

1. The devices must be designed and manufactured in such a way that, when used under the conditions and for the purposes intended, they will not compromise the clinical condition or the safety of patients, or the safety and health of users or, where applicable, other persons, provided that any risks which may be associated with their use constitute acceptable risks when weighed against the benefits to the patient and are compatible with a high level of protection of protection of health and safety.

2. The solutions adopted by the manufacturer for the design and construction of the devices must conform to safety principles, taking account of the generally acknowledged state of the art.

In selecting the most appropriate solutions, the manufacturer must apply the following principles in the following order:

- eliminate or reduce risks as far as possible (inherently safe design and construction),
- where appropriate take adequate protection measures including alarms if necessary, in relation to risks that cannot be eliminated,
- inform users of the residual risks due to any shortcomings of the protection measures adopted.

3. The devices must achieve the performances intended by the manufacturer and be designed, manufactured and packaged in such a way that they are suitable for one or more of the functions referred to in Article 1 (2) (a), as specified by the manufacturer.

4. The characteristics and performances referred to in Sections 1, 2 and 3 must not be adversely affected to such a degree that the clinical condition and safety of the patients and, where applicable, of other persons are compromised during the lifetime of the device as indicated by the manufacturer, when the device is subjected to the stresses which can occur during normal conditions of use.

5. The devices must be designed, manufactured and packed in such a way that their characteristics and performances during their intended use will not be adversely affected during the transport and storage taking account of the instructions and information provided by the manufacturer.

6. Any undesirable side-effects must constitute acceptable risks when weighed against the performances intended.

II. Requirements regarding design and construction

7. Chemical, physical and biological properties

7.1. The devices must be designed and manufactured in such a way as to guarantee the characteristics and performances referred to in Section I on the “General requirements”. Particular attention must be paid to:

- the choice of materials used, particularly as regards toxicity and, where appropriate, flammability,
- the compatibility between the materials used and biological tissues, cells and body fluids, taking account of the intended purpose of the device.

7.2. The devices must be designed, manufactured and packed in such a way as to minimise the risk posed by contaminants and residues to the persons involved in the transport, storage and use of the devices and to the patients, taking account of the intended purpose of the product. Particular attention must be paid to the tissues exposed and to the duration and frequency of exposure.

7.3. The devices must be designed and

manufactured in such a way that they can be used completely safely with the materials, substances and gases with which they enter into contact during their normal use or during routine procedures; if the devices are intended to administer medicinal products they must be designed and manufactured in such a way as to be compatible with the medicinal products concerned according to the provisions and restrictions governing these products and that their performance is maintained in accordance with the intended use.

7.4. Where a device incorporates, as an integral part, a substance which, if used separately, may be considered to be a medicinal product, as defined in Article 1 of Directive 65/65/EEC and which is liable to act upon the body with action ancillary to that of the device, the safety, quality and usefulness of the substance must be verified, taking account of the intended purpose of the device, by analogy with the appropriate methods specified in Directive 75/318/EEC.

7.5. The devices must be designed and manufactured in such a way as to reduce to a minimum the risks posed by substances leaking from the device.

7.6. Devices must be designed and manufactured in such a way as to reduce, as much as possible, risks posed by the unintentional ingress of substances into the device taking into account the device and the nature of the environment in which it is intended to be used.

8. Infection and microbial contamination

8.1. The devices and manufacturing processes must be designed in such a way as to eliminate or reduce as far as possible the risk of infection to the patient, user and third parties. The design must allow easy handling and, where necessary, minimise contamination of the device by the patient or vice versa during use.

8.2. Tissues of animal origin must originate from animals that have been subjected to veterinary controls and surveillance adapted to the intended use of the tissues.

Notified bodies shall retain information on the geographical origin of the animals.

Processing, preservation, testing and handling of tissues, cells and substances of animal origin must be carried out so as to provide optimal security. In particular safety with regard to viruses and other transferable agents must be addressed by implementation of validated methods of elimination or viral inactivation in the course of the manufacturing process.

8.3. Devices delivered in a sterile state must be designed, manufactured and packed in a non-reusable pack and/or according to appropriate procedures to ensure that they are sterile when placed on the market and remain sterile, under the storage and transport conditions laid down, until the protective packaging is damaged or opened.

8.4. Devices delivered in a sterile state must have been manufactured and sterilised by an appropriate, validated method.

8.5. Devices intended to be sterilised must be manufactured in appropriately controlled (e.g. environmental) conditions.

8.6. Packaging systems for non-sterile devices must keep the product without deterioration at the level of cleanliness stipulated and, if the devices are to be sterilised prior to use, minimise the risk of microbial contamination; the packaging system must be suitable taking account of the method of sterilisation indicated by the manufacturer.

8.7 The packaging and/or label of the device must distinguish between identical or similar products sold in both sterile and non-sterile condition.

9. Construction and environmental properties

9.1. If the device is intended for use in combination with other devices or equipment, the whole combination, including the connection system must be safe and must not impair the specified performances of the devices. Any restrictions on use must be indicated on the label or in the instructions for use.

9.2. Devices must be designed and manufactured in such a way as to remove or minimise as far as is possible:

- the risk of injury, in connection with their physical features, including the volume/pressure ratio, dimensional and where appropriate ergonomic features,
- risks connected with reasonably foreseeable environmental conditions, such as magnetic fields, external electrical influences, electrostatic discharge, pressure, temperature or variations in pressure and acceleration,
- the risks of reciprocal interference with other devices normally used in the investigations or for the treatment given,
- risks arising where maintenance or calibration are not possible (as with implants), from ageing of materials used or loss of accuracy of any measuring or control mechanism.

9.3. Devices must be designed and manufactured in such a way as to minimise the risks of fire or explosion during normal use and in single fault condition. Particular attention must be paid to devices whose intended use includes exposure to flammable substances which could cause combustion.

10. Devices with a measuring function

10.1. Devices with a measuring function must be designed and manufactured in such a way as to provide sufficient accuracy and stability within appropriate limits of accuracy and taking account of the intended purpose of the device. The limits of accuracy must be indicated by the manufacturer.

10.2. The measurement, monitoring and display scale must be designed in line with ergonomic principles, taking account of the intended purpose of the device.

10.3. The measurements made by devices with a measuring function must be expressed in legal units conforming to the provisions of Council Directive 80/181/EEC.[1]

11. Protection against radiation

11.1. *General*

11.1.1. Devices shall be designed and manufactured in such a way that exposure of patients, users and other persons to radiation shall be reduced as far as possible compatible with the intended purpose, whilst not restricting the application of appropriate specified levels for therapeutic and diagnostic purposes.

11.2. *Intended radiation*

11.2.1. Where devices are designed to emite hazardous levels of radiation necessary for a specific medical purpose the benefits of which is considered to outweigh the risks inherent in the emission, it must be possible for the user to control the emissions. Such devices shall be designed and manufactured to ensure reproducibility and tolerance of relevant variable parameters.

11.2.2. Where devices are intended to emit potentially hazardous, visible and/or invisible radiation, they must be fitted, where practicable, with visual displays and/or audible warnings of such emissions.

11.3. *Unintended radiation*

11.3.1. Devices shall be designed and manufactured in such a way that exposure of patients, users and other persons to the emission of unintended, stray or scattered radiation is reduced as far as possible.

11.4. *Instructions*

11.4.1. The operating instructions for devices emitting radiation must give detailed information as to the nature of the emitted radiation, means of protecting the patient and the user and on ways of avoiding misuse and of

---

[1] OJ No L 39, 15. 2. 1980, p. 40. Directive as last amended by Directive 896/17/EEC (OJ No L 357, 7. 12. 1989, p. 28).

eliminating the risks inherent in installation.

11.5. *Ionising radiation*

11.5.1. Devices intended to emit ionising radiation must be designed and manufactured in such a way as to ensure that, where practicable, the quantity, geometry and quality of radiation emitted can be varied and controlled taking into account the intended use.

11.5.2. Devices emitting ionising radiation intended for diagnostic radiology shall be designed and manufactured in such a way as to achieve appropriate image and/or output quality for the intended medical purpose whilst minimising radiation exposure of the patient and user.

11.5.3. Devices emitting ionising radiation, intended for therapeutic radiology shall be designed and manufactured in such a way as to enable reliable monitoring and control of the delivered dose, the beam type and energy and where appropriate the quality of radiation.

12. Requirements for medical devices connected to or equipped with an energy source

12.1. Devices incorporating electronic programmable systems must be designed to ensure the repeatability, reliability and performance of these systems according to the intended use. In the event of a single fault condition (in the system) appropriate means should be adopted to eliminate or reduce as far as possible consequent risks.

12.2. Devices where the safety of the patient depends on an internal power supply must be equipped with a means of determining the state of power supply.

12.3. Devices where the safety of the patient depends on the external power supply must include an alarm system to signal any power failure.

12.4. Devices intended to monitor one or more clinical parameters of a patient must be equiupped with appropriate alarm systems to alert the user of situations which could lead to death or severe deterioration of the patient's state of health.

12.5. Devices must be designed and manufactured in such a way as to minimise the risks of creating electromagnetic fields which could impair the operation of other devices or equipment in the usual environment.

12.6. *Protection against electrical risks*

Devices must be designed and manufactured in such a way as to avoid, as far as possible, the risk of accidental electric shocks during normal use and in single fault condition, provided the devices are installed correctly.

12.7. *Protection against mechanical and thermal risks*

12.7.1. Devices must be designed and manufactured in such a way as to protect the patient and user against mechanical risks connected with, for example, resistance, stability and moving parts.

12.7.2. Devices must be designed and manufactured in such a way as to reduce to the lowest possible level the risks arising from vibration generated by the devices, taking account of technical progress and of the means available for limiting vibrations, particularly at source, unless the vibrations are part of the specified performance.

12.7.3. Devices must be designed and manufactured in such a way as to reduce to the lowest possible level the risks arising from the noise emitted, taking account of technical progress and of the means available to reduce noise, particularly at source, unless the noise emitted is part of the specified performance.

12.7.4. Terminals and connectors to the electricity, gas or hydraulic and pneumatic energy supplies which the user has to handle must be designed and constructed in such a way as to minimise all possible risks.

12.7.5. Accessible parts of the devices (excluding the parts or areas intended to supply heat or reach given temperatures) and their surroundings must not attain potentially dangerous temperatures under normal use.

12.8. *Protection against the risks posed to the patient by energy supplies or substances*
12.8.1. Devices for supplying the patient with energy or substances must be designed and constructed in such a way that the flow-rate can be set and maintained accurately enough to guarantee the safety of the patient and of the user.
12.8.2. Devices must be fitted with the means of preventing and/or indicating any inadequacies in the flow-rate which could pose a danger.

Devices must incorporate suitable means to prevent, as far as possible, the accidental release of dangerous levels of energy from an energy and/or substance source.
12.9. The function of the controls and indicators must be clearly specified on the devices.
Where a device bears instructions required for its operation or indicates operating or adjustment parameters by means of a visual system, such information must be understandable to the user and, as appropriate, the patient.

13. Information supplied by the manufacturer

13.1. Each device must be accompanied by the information needed to use it safely and to identify the manufacturer, taking account of the training and knowledge of the potential users.

This information comprises the details on the label and the data in the instructions for use.

As far as practicable and appropriate, the information needed to use the device safely must be set out on the device itself and/or on the packaging for each unit or, where appropriate, on the sales packaging. If individual packaging of each unit is not practicable, the information must be set out in the leaflet supplied with one or more devices.

Instructions for use must be included in the packaging for every device. By way of exception, no such instructions for use are needed for devices in Class I or IIa if they can be used safely without any such instructions.

13.2. Where appropriate, this information should take the form of symbols. Any symbols and identification colours used must conform to the harmonised standards. In areas for which no standards exist, the symbols and colours must be described in the documentation supplied with the device.
13.3. The *label* must bear the following particulars:
(a) the name or trade name and address of the manufacturer. For devices imported into the Community, in view of their distribution in the Community, the label, or the outer packaging, or instructions for use, shall contain in addition the name and address of either the person responsible referred to in Article 14 (2) or of the authorised representative of the manufacturer established within the Community or of the importer established within the Community, as appropriate;
(b) the details strictly necessary for the user to identify the device and the contents of the packaging;
(c) where appropriate, the word "STERILE";
(d) where appropriate, the batch code, preceded by the word "LOT", or the serial number;
(e) where appropriate, an indication of the date by which the device should be used, in safety, expressed as the year and month;
(f) where appropriate, an indication that the device is for single use;
(g) if the device is custom-made, the words "custom-made device";
(h) if the device is intended for clinical investigations, the words "exclusively for clinical investigations";
(i) any special storage and/or handling conditions;
(j) any special operating instructions;
(k) any warnings and/or precautions to take.
(l) year of manufacture for active devices other than those covered by (e). This indication may be included in the batch or serial number;

(m) where applicable, method of sterilisation.

13.4. If the intended purpose of the device is not obvious to the user, the manufacturer must clearly state it on the label and in the instructions for use.

13.5. Wherever reasonable and practicable, the devices and detachable components must be identified, where appropriate in terms of batches, to allow all appropriate action to detect any potential risk posed by the devices and detachable components.

13.6. Where appropriate, the instructions for use must contain the following particulars:

(a) the details referred to in Section 13.3, with the exception of (d) and (e);

(b) the performance referred to in Section 3 and any undesirable side-effects;

(c) if the device must be installed with or connected to other medical devices or equipment in order to operate as required for its intended purpose, sufficient details of its characteristics to identify the correct devices or equipment to use in order to obtain a safe combination;

(d) all the information needed to verify whether the device is properly installed and can operate correctly and safely, plus details of the nature and frequency of the maintenance and calibration needed to ensure that the devices operate properly and safely at all times;

(e) where appropriate, information to avoid certain risks in connection with implantation of the device;

(f) information regarding the risks of reciprocal interference posed by the presence of the device during specific investigations or treatment;

(g) the necessary instructions in the event of damage to the sterile packaging and, where appropriate, details of appropriate methods of resterilisation;

(h) if the device is reusable, information on the appropriate processes to allow reuse, including cleaning, disinfection, packaging and, where appropriate, the method of sterilisation of the device to be resterilised, and any restriction on the number of reuses.

Where the devices are supplied with the intention that they be sterilised before use, the instructions for cleaning and sterilisation must be such that, if correctly followed, the device will still comply with the requirements in Section I;

(i) details of any further treatment or handling needed before the device can be used (for example, sterilisation, final assembly, etc);

(j) in the case of devices emitting radiation for medical purposes, details of the nature, type, intensity and distribution of this radiation.

The instructions for use must also include details allowing the medical staff to brief the patient on any contra-indications and any precautions to be taken. These details should cover in particular:

(k) precautions to be taken in the event of changes in the performance of the device;

(l) precautions to be taken as regards exposure, in reasonably foreseeable environmental conditions, to magnetic fields, external influences, electrostatic discharge, pressure or variations in pressure, acceleration, thermal ignition sources, etc;

(m) adequate information regarding the medicinal product or products which the device in question is designed to administer, including any limitations in the choice of substances to be delivered;

(n) precautions to be taken against any special, unusual risks related to the disposal of the device.

(o) medicinal substances incorporated into the device as an integral part in accordance with Section 7.4;

(p) degree of accuracy claimed for devices with a measuring function.

14. Where conformity with the essential requirements must be based on clinical data, as in Section 1 (6), such data must be established in accordance with Annex X.

Annex II

EC declaration of conformity

(Full quality assurance system)

1. The manufacturer must ensure application of the quality system approved for the design, manufacture and final inspection of the products concerned, as specified in Sections 3.3 and 4, and to Community surveillance as specified in Section 5.

2. The declaration of conformity is the procedure whereby the manufacturer who fulfils the obligations imposed by Section 1 ensures and declares that the products concerned meet the provisions of this Directive which apply to them.

The manufacturer must affix the CE marking in accordance with Article 17 and draw up a written declaration of conformity. This declaration must cover a given number of the products manufactured and be kept by the manufacturer.

3. Quality system

3.1. The manufacturer must lodge an application for assessment of his quality system with a notified body.

The application must include:

- the name and address of the manufacturer and any additional manufacturing site covered by the quality system,
- all the relevant information on the product or product category covered by the procedure,
- a written declaration that no application has been lodged with any other notified body for the same product-related quality system,
- the documentation on the quality system,
- an undertaking by the manufacturer to fulfil the obligations imposed by the quality system approved,
- an undertaking by the manufacturer to keep the approved quality system adequate and efficacious,
- an undertaking by the manufacturer to institute and keep up to date a systematic procedure to review experience gained from devices in the post-production phase and to implement appropriate means to apply any necessary corrective action. This undertaking must include an obligation for the manufacturer to notify the competent authorities of the following incidents immediately on learning of them:
  (i) any malfunction or deterioration in the characteristics and/or performances of a device, as well as any inaccuracies in the instructions for use which might lead to or might have led to the death of a patient or user or to a serious deterioration in his state of health;
  (ii) any technical or medical reason connected with the characteristics or performance of a device leading for the reasons referred to in subparagraph (i) to systematic recall of devices of the same type by the manufacturer.

3.2. Application of the quality system must ensure that the products conform to the provisions of this Directive which apply to them at every stage, from design to final inspection. All the elements, requirements and provisions adopted by the manufacturer for his quality system shall be documented in a systematic and orderly manner in the form of written policies and procedures such as quality programmes, quality plans, quality manuals and quality records.

It shall include in particular an adequate description of:

(a) the manufacturer's quality objectives;

(b) the organisation of the business and in particular:
   - the organisational structures, the responsibilities of the managerial staff and their organisational authority where quality of design and manufacture of the products is concerned,
   - the methods of monitoring the efficient operation of the quality system and in particular its ability to achieve the desired quality of design and of product, including control of products which fail to conform;

(c) the procedures for monitoring and verifying the design of the products and in particular:
   - a general description of the product, including any variants planned,
   - the design specifications, including the standards which will be applied and the results of the risk analysis, and also a description of the solutions adopted to fulfil the essential requirements which apply to the products if the standards referred to in Article 5 are not applied in full,
   - the techniques used to control and verify the design and the processes and systematic measures which will be used when the products are being designed,
   - if the device is to be connected to other device(s) in order to operate as intended, proof must be provided that it conforms to the essential requirements when connected to any such device(s) having the characteristics specified by the manufacturer,
   - a statement indicating whether or not the device incorporates, as an integral part, a substance as referred to in Section 7.4 of Annex I and data on the tests conducted in this connection,
   - the clinical data referred to in Annex X,
   - the draft label and, where appropriate, instructions for use;

(d) the inspection and quality assurance techniques at the manufacturing stage and in particular:
   - the processes and procedures which will be used, particularly as regards sterilisation, purchasing and the relevant documents,
   - the product identification procedures drawn up and kept up to date from drawings, specifications or other relevant documents at every stage of manufacture;

(e) the appropriate tests and trials which will be carried out before, during and after manufacture, the frequency with which they will take place, and the test equipment used; it must be possible to trace back the calibration of the test equipment adequately.

3.3. The notified body must audit the quality system to determine whether it meets the requirements referred to in Section 3.2. It must presume that quality systems which implement the relevant harmonised standards conform to these requirements.

The assessment team must include at least one member with past experience of assessments of the technology concerned. The assessment procedure must include an inspection on the manufacturer's premises and, in duly substantiated cases, on the premises of the manufacturer's suppliers and/or subcontractors to inspect the manufacturing processes.

The decision is notified to the manufacturer. It must contain the conclusions of the inspection and a reasoned assessment.

3.4. The manufacturer must inform the notified body which approved the quality system of any plan for substantial changes to the quality system or the product-range covered. The notified body must assess the changes proposed and shall verify whether after these changes the quality system will still meet the requirements referred to in Section 3.2. It must notify the manufacturer of its decision. This decision must contain the conclusions of the inspection and a reasoned assessment.

4. Examination of the design of the product

4.1. In addition to the obligations imposed by Section 3, the manufacturer must lodge with the notified body an application for examination of the design dossier relating to the product which he plans to manufacture and which falls into the category referred to in Section 3.1.

4.2. The application shall describe the design, manufacture and performances of the product in question. It must include the documents needed to assess whether the product conforms to the requirements of this Directive, as referred to in Section 3.2 (c).

4.3. The notified body must examine the application and, if the product conforms to the relevant provisions of this Directive, shall issue the application with an EC design-examination certificate. The notified body may require the application to be completed by further tests or proof to allow assessment of conformity with the requirements of the Directive. The certificate shall contain the conclusions of the examination, the conditions of validity, the data needed for identification of the approved design, where appropriate, a description of the intended purpose of the product.

In the case of devices referred to in Annex I, paragraph 7.4, the notified body shall, in view of the aspects addressed in that paragraph, consult one of the competent bodies established by the Member States in accordance with Directive 65/65/EEC before taking a decision.

The notified body will give due consideration to the views expressed in this consultation when making its decision. It will convey its final decision to the competent body concerned.

4.4. Changes to the approved design must receive further approval from the notified body which issued the EC design-examination certificate wherever the changes could affect conformity with the essential requirements of the Directive or with the conditions prescribed for use of the product. The applicant shall inform the notified body which issued the EC design-examination certificate of any such changes made to the approved design. This additional approval must take the form of a supplement to the EC design-examination certificate.

5. Surveillance

5.1. The aim of surveillance is to ensure that the manufacturer duly fulfils the obligations imposed by the approved quality system.

5.2. The manufacturer must authorise the notified body to carry out all the necessary inspections and shall supply it with all relevant information, in particular:

- the documentation on the quality system,
- the data stipulated in the part of the quality system relating to design, such as the results of analyses, calculation tests, etc,
- the data stipulated in part of the quality system relating to manufacture, such as inspection reports and test data, calibration data, qualification reports of the personnel concerned, etc.

5.3. The notified body must periodically carry out appropriate inspections and assessments to make sure that the manufacturer applies the approved quality system and must supply the manufacturer with an assessment report.

5.4. In addition, the notified body may pay unannounced visits to the manufacturer. At the time of such visits, the notified body may, where necessary, carry out or ask for tests in order to check that the quality system is working properly. It must provide the manufacturer with an inspection report and, if a test has been carried out, with a test report.

6. Administrative provisions

6.1. The manufacturer must, for a period ending at least five years after the last product has been manufactured, keep at the disposal of the national authorities:

- the declaration of conformity,
- the documentation referred to in the fourth indent of Section 3.1,
- the changes referred to in Section 3.4,
- the documentation referred to in Section 4.2, and
- the decisions and reports from the notified body as referred to in Sections 3.3. 4.3, 4.4., 5.3 and 5.4.

6.2. The notified body must make available to the other notified bodies and the competent authority, on request, all relevant information concerning quality system approvals issued, refused or withdrawn.

6.3. In respect of devices subject to the procedure in Section 4, when neither the manufacturer nor his authorised representative is established in the Community, the obligation to keep available the technical documentation shall fall to the person responsible for placing the device on the Community market or the importer referred to in Annex I, Section 13.3 (a).

7. Application to devices in Classes IIa and IIb

In line with Article 11 (2) and (3), this Annex may apply to products in Classes IIa and IIb. Section 4, however, does not apply.

Annex III
EC type-examination

1. EC type–examination is the procedure whereby a notified body ascertains and certifies that a representative sample of the production covered fulfils the relevant provisions of this Directive.

2. The application includes:

- the name and address of the manufacturer and the name and address of the authorised representative if the application is lodged by the representative,
- the documentation described in Section 3 needed to assess the conformity of the representative sample of the production in question, hereinafter referred to as the "type", with the requirements of this Directive. The applicant shall make a "type" available to the notified body. The notified body may request other samples as necessary,
- a written declaration that no application has been lodged with any other notified body for the same type.

3. The documentation must allow an understanding of the design, the manufacture and the performances of the product and must contain the following items in particular:

- a general description of the type, including any variants planned,
- design drawings, methods of manufacture envisaged, in particular as regards sterilisation, and diagrams of components, sub-assemblies, circuits, etc,
- the descriptions and explanations necessary to understand the above-mentioned drawings and diagrams and the operation of the product,
- a list of the standards referred to in Article 5, applied in full or in part, and descriptions of the solutions adopted to meet the essential requirements if the standards referred to in Article 5 have not been applied in full,
- the results of the design calculations, risk analysis, investigations, technical tests, etc carried out,
- a statement indicating whether or not the device incorporates, as an

integral part, a substance as referred to in Section 7.4 of Annex I and data on the tests conducted in this connection,
- the clinical data referred to in Annex X,
- the draft label and, where appropriate, instructions for use.

4. The notified body must:

4.1. examine and assess the documentation and verify that the type has been manufactured in conformity with that documentation; it must also record the items designed in conformity with the applicable provisions of the standards referred to in Article 5, as well as the items not designed on the basis of the relevant provisions of the abovementioned standards;

4.2. carry out or arrange for the appropriate inspections and the tests necessary to verify whether the solutions adopted by the manufacturer meet the essential requirements of this Directive if the standards referred to in Article 5 have not been applied; if the device is to be connected to other device(s) in order to operate as intended, proof must be provided that it conforms to the essential requirements when connected to any such device(s) having the characteristics specified by the manufacturer;

4.3. carry out or arrange for the appropriate inspections and the tests necessary to verify whether, if the manufacturer has chosen to apply the relevant standards, these have actually been applied;

4.4. agree with the applicant on the place where the necessary inspections and test will be carried out.

5. If the type conforms to the provisions of this Directive, the notified body issues the applicant with an EC type-examination certificate. The certificate must contain the name and address of the manufacturer, the conclusions of the inspection, the conditions of validity and the data needed for identification of the type approved. The relevant parts of the documentation shall be annexed to the certificate and a copy kept by the notified body.

In the case of devices referred to in Annex I, paragraph 7.4, the notified body shall, in view of the aspects addressed in that paragraph, consult one of the competent bodies established by the Member States in accordance with Directive 65/65/EEC before taking a decision.

The notified body will give due consideration to the views expressed in this consultation when making its decision. It will convey its final decision to the competent body concerned.

6. The applicant must inform the notified body which issued the EC type-examination certificate of any significant change made to the approved product.

Changes to the approved product must receive further approval from the notified body which issued the EC type-examination certificate wherever the changes may affect conformity with the essential requirements or with the conditions prescribed for use of the product. This new approval must, where appropriate, take the form of a supplement to the initial EC type-examination certificate.

7. Administrative provisions

7.1. The notified body must make available to the other notified bodies on request, all relevant information on EC type-examination certificates and supplements issued, refused or withdrawn.

7.2. Other notified bodies may obtain a copy of the EC type-examination certificates and/or the supplements thereto. The Annexes to the certificates must be made available to the other notified bodies on reasoned application, after the manufacturer has been informed.

7.3. The manufacturer or his authorised representative must keep with the technical documentation copies of EC type-examination certificates and their additions for a period ending at least five years after the last device has been manufactured.

7.4. When neither the manufacturer nor his authorised representative is established in the Community, the obligation to keep available the technical documentation shall fall to the person responsible for placing the device on the Community market or the importer referred to in Annex I, Section 13.3 (a).

Annex IV
EC verification

1. EC verification is the procedure whereby the manufacturer or his authorised representative established in the Community ensures and declares that the products which have been subject to the procedure set out in Section 4 conform to the type described in the EC type-examination certificate and meet the requirements of this Directive which apply to them.

2. The manufacturer must take all the measures necessary to ensure that the manufacturing process produces products which conform to the type described in the EC type-examination certificate and to the requirements of the Directive which apply to them. Before the start of manufacture, the manufacturer must prepare documents defining the manufacturing process, in particular as regards sterilisation where necessary, together with all the routine, pre-established provisions to be implemented to ensure homogeneous production and, where appropriate, conformity of the products with the type described in the EC type-examination certificate and with the requirements of this Directive which apply to them. The manufacturer must affix the CE marking in accordance with Article 17 and draw up a declaration of conformity.

In addition, for products placed on the market in sterile condition, and only for those aspects of the manufacturing process designed to secure and maintain sterility, the manufacturer must apply the provisions of Annex V, Sections 3 and 4.

3. The manufacturer must undertake to institute and keep up to date a systematic procedure to review experience gained from devices in the post-production phase and to implement appropriate means to apply any necessary corrective action. This undertaking must include an obligation for the manufacturer to notify the competent authorities of the following incidents immediately on learning of them:

(i) any malfunction or deterioration in the characteristics and/or performance of a device, as well as any inadequacy in the labelling or the instructions for use which might lead to or might have led to the death of a patient or user or to a serious deterioration in his state of health;

(ii) any technical or medical reason connected with the characteristics or performance of a device for the reasons referred to in subparagraph (i) leading to systematic recall of devices of the same type by the manufacturer.

4. The notified body must carry out the appropriate examinations and tests in order to verify the conformity of the product with the requirements of the Directive either by examining and testing every product as specified in Section 5 or by examining and testing products on a statistical basis as specified in Section 6, as the manufacturer decides.

The aforementioned checks do not apply to those aspects of the manufacturing process designed to secure sterility.

5. Verification by examination and testing of every product

5.1. Every product is examined individually and the appropriate tests defined in the relevant standard(s) referred to in Article 5 or equivalent

tests must be carried out in order to verify, where appropriate, the conformity of the products with the EC type described in the type-examination certificate and with the requirements of the Directive which apply to them.
5.2. The notified body must affix, or have affixed, its identification number to each approved product and must draw up a written certificate of conformity relating to the tests carried out.

6. Statistical verification

6.1. The manufacturer must present the manufactured products in the form of homogeneous batches.
6.2. A random sample is taken from each batch. The products which make up the sample are examined individually and the appropriate tests defined in the relevant standard(s) referred to in Article 5 or equivalent tests must be carried out to verify, where appropriate, the conformity of the products with the type described in the EC type-examination certificate and with the requirements of the Directive which apply to them in order to determine whether to accept or reject the batch.
6.3. Statistical control of products will be based on attributes, entailing a sampling system ensuring a limit quality corresponding to a probability of acceptance of 5%, with a non-conformity percentage of between 3 and 7%. The sampling method must be established by the harmonised standards referred to in Article 5, taking account of the specific nature of the product categories in question.
6.4. If the batch is accepted, the notified body affixes or has affixed, its identification number to each product and draws up a written certificate of conformity relating to the tests carried out. All products in the batch may be put on the market except any in the sample which failed to conform.

If a batch is rejected the competent notified body must take appropriate measures to prevent the batch from being placed on the market. In the event of frequent rejection of batches, the notified body may suspend the statistical verification.

The manufacturer may, on the responsibility of the notified body, affix the notified body's identification number during the manufacturing process.

7. Administrative provisions

The manufacturer or his authorised representative must, for a period ending at least five years after the last product has been manufactured, keep at the disposal of the national authorities:

- the declaration of conformity,
- the documentation referred to in Section 2,
- the certificates referred to in Sections 5.2 and 6.4.
- where appropriate, the type-examination certificate referred to in Annex III.

8. Application to devices in Class IIa

In line with Article 11 (2), this Annex may apply to products in Class IIa, subject to the following exemptions:
8.1. in derogation from Sections 1 and 2, by virtue of the declaration of conformity the manufacturer ensures and declares that the products in Class IIa are manufactured in conformity with the technical documentation referred to in Section 3 of Annex VII and meet the requirements of this Directive which apply to them;
8.2. in derogation from Sections 1, 2, 5 and 6, the verifications conducted by the notified body are intended to confirm the conformity of the products in Class IIa with the technical documentation referred to in Section 3 of Annex VII.

Annex V
EC declaration of conformity

(Production quality assurance)

1. The manufacturer must ensure application of the quality system approved for the manufacture of the products concerned and carry out the final inspection, as specified in Section 3, and is subject to the Community surveillance referred to in Section 4.

2. The declaration of conformity is the part of the procedure whereby the manufacturer who fulfils the obligations imposed by Section 1 ensures and declares that the products concerned conform to the type described in the EC type-examination certificate and meet the provisions of this Directive which apply to them.

The manufacturer must affix the CE marking in accordance with Article 17 and draw up a written declaration of conformity. This declaration must cover a given number of identical specimens of the products manufactured and must be kept by the manufacturer.

3. Quality system

3.1. The manufacturer must lodge an application for assessment of his quality system with a notified body.

The application must include:

- the name and address of the manufacturer,
- all the relevant information on the product or product category covered by the procedure,
- a written declaration that no application has been lodged with any other notified body for the same products,
- the documentation on the quality system,
- an undertaking to fulfil the obligations imposed by the quality system is approved,
- an undertaking to maintain the practicability and effectiveness of the approved quality system,
- where appropriate, the technical documentation on the types approved and a copy of the EC type-examination certificates,
- an undertaking by the manufacturer to institute and keep up to date a systematic procedure to review the experience gained from devices in the post-production phase and to implement appropriate means to apply any necessary corrective action. This undertaking must include an obligation for the manufacturer to notify the component authorities of the following incidents immediately on learning of them:
  (i) any malfunction or deterioration in the characteristics and/or performance of a device, as well as any inadequacy in the labelling or the instructions for use which might lead to or might have led to the death of a patient or user or to a serious deterioration in their state of health;
  (ii) any technical or medical reason connected with the characteristics or performance of a device for the reasons referred to in subparagraph (i) above leading to a systematic recall of devices of the same type by the manufacturer.

3.2. Application of the quality system must ensure that the products conform to the type described in the EC type-examination certificate.

All the elements, requirements and provisions adopted by the manufacturer for his quality system must be documented in a systematic and orderly manner in the form of written policy statements and procedures. This quality system documentation must permit uniform interpretation of the quality policy and procedures such as quality programmes, plans, manuals and records.

It must include in particular an adequate description of:

(a) the manufacturer's quality objectives;
(b) the organisation of the business and in particular:
   - the organisational structures, the responsibilities of the managerial staff and their organisational authority where manufacture of the products is concerned,
   - the methods of monitoring the efficient operation of the quality system and in particular its ability to achieve the desired quality of product, including control of products which fail to conform;
(c) the inspection and quality assurance techniques at the manufacturing stage and in particular:
   - the processes and procedures which will be used, particularly as regards sterilisation, purchasing and the relevant documents,
   - the product-identification procedures drawn up and kept up to date from drawings, specifications or other relevant documents at every stage of manufacture;
(d) the appropriate tests and trials which will be carried out before, during and after manufacture, the frequency with which they will take place, and the test equipment used; it must be possible adequately to trace back the calibration of the test equipment.

3.3. The notified body must audit the quality system to determine whether it meets the requirements referred to in Section 3.2. It must presume that quality systems which implement the relevant harmonised standards conform to these requirements.

The assessment team must include at least one member with past experience of assessments of the technology concerned. The assessment procedure must include an inspection on the manufacturer's premises and, in duly substantiated cases, on the premises of the manufacturer's suppliers to inspect the manufacturing processes.

The decision must be notified to the manufacturer after the final inspection and contain the conclusions of the inspection and a reasoned assessment.

3.4. The manufacturer must inform the notified body which approved the quality system of any plan for substantial changes to the quality system.

The notified body must assess the changes proposed and verify whether after these changes the quality system still meets the requirements referred to in Section 3.2.

After the abovementioned information has been received the decision is notified to the manufacturer. It must contain the conclusions of the inspection and a reasoned assessment.

4. Surveillance

4.1. The aim of surveillance is to ensure that the manufacturer duly fulfils the obligations imposed by the approved quality system.

4.2. The manufacturer authorises the notified body to carry out all the necessary inspections and must supply it with all relevant information, in particular:
- the documentation on the quality system,
- the data stipulated in the part of the quality system relating to manufacture, such as inspection reports and test data, calibration data, qualification reports of the personnel concerned, etc.

4.3. The notified body must periodically carry out appropriate inspections and assessments to make sure that the manufacturer applies the approved quality system and supply the manufacturer with an assessment report.

4.4. In addition, the notified body may pay unannounced visits to the manufacturer. At the time of such visits, the notified body may, where necessary, carry out or ask for tests in order to check that the quality system is working properly. It must provide the manufacturer with an inspection report and, if a test has been carried out, with a test report.

5. Administrative provisions

5.1. The manufacturer must, for a period ending at least five years after

the last product has been manufactured, make available to the national authorities:
- the declaration of conformity,
- the documentation referred to in the fourth indent of Section 3.1.,
- the changes referred to in Section 3.4,
- the documentation referred to in the seventh indent of Section 3.1,
- the decisions and reports from the notified body as referred to in Section 4.3 and 4.4.
- where appropriate, the type-examination certificate referred to in Annex III.

5.2. The notified body must make available to the other notified bodies, on request, all relevant information concerning the quality system approvals issued, refused or withdrawn.

6. Application to devices in Class IIa

In line with Article 11 (2), this Annex may apply to products in Class IIa, subject to the following exemption:

6.1. in derogation from Sections 2, 3.1 and 3.2, by virtue of the declaration of conformity the manufacturer ensures and declares that the products in Class IIa are manufactured in conformity with the technical documentation referred to in Section 3 of Annex VII and meet the requirements of this Directive which apply to them.

Annex VI
EC declaration of conformity

(Product quality assurance)

1. The manufacturer must ensure application of the quality system approved for the final inspection and testing of the product, as specified in Section 3 and must be subject to the surveillance referred to in Section 4.

In addition, for products placed on the market in sterile condition, and only for those aspects of the manufacturing process designed to secure and maintain sterility, the manufacturer must apply the provisions of Annex V, Sections 3 and 4.

2. The declaration of conformity is the part of the procedure whereby the manufacturer who fulfils the obligations imposed by Section 1 ensures and declares that the products concerned conform to the type described in the EC type-examination certificate and meet the provisions of this Directive which apply to them.

The manufacturer affixes the CE marking in accordance with Article 17 and draws up a written declaration of conformity. This declaration must cover a given number of identified specimens of the products manufactured and be kept by the manufacturer. The CE marking must be accompanied by the identification number of the notified body which performs the tasks referred to in this Annex.

3. Quality system

3.1. The manufacturer lodges an application for assessment of his quality system with a notified body.

The application must include:
- the name and address of the manufacturer,
- all the relevant information on the product or product category covered by the procedure,
- a written declaration specifying that no application has been lodged with any other notified body for the same products,
- the documentation on the quality system,
- an undertaking by the manufacturer to fulfil the obligations imposed by the quality system approved,
- an undertaking by the manufacturer to keep the approved quality system adequate and efficacious,

- where appropriate, the technical documentation on the types approved and a copy of the EC type-examination certificates,
- an undertaking by the manufacturer to institute and keep up to date a systematic procedure to review experience gained from devices in the post-production phase and to implement appropriate means to apply any necessary corrective action. This undertaking must include an obligation for the manufacturer to notify the competent authorities of the following incidents immediately on learning of them:
  - (i) any malfunction or deterioration in the characteristics and/or performances of a device, as well as any inadequacy in the labelling or the instructions for use which might lead to or might have led to the death of a patient or user or to a serious deterioration in his state of health;
  - (ii) any technical or medical reason connected with characteristics or the performance of a device for the reasons referred to in subparagraph (i) leading to a systematic recall of devices of the same type by the manufacturer.

3.2. Under the quality system, each product or a representative sample of each batch is examined and the appropriate tests defined in the relevant standard(s) referred to in Article 5 or equivalent tests are carried out to ensure that the products conform to the type described in the EC type-examination certificate and fulfil the provisions of this Directive which apply to them. All the elements, requirements and provisions adopted by the manufaturer must be documented in a systematic and orderly manner in the form of written measures, procedures and instructions. This quality system documentation must permit uniform interpretation of the quality programmes, quality plans, quality manuals and quality records.

It must include in particular an adequate description of:
- the quality objectives and the organisational structure, responsibilities and powers of the managerial staff with regard to product quality,
- the examinations and tests that will be carried out after manufacture; it must be possible to trace back the calibration of the test equipment adequately,
- the methods of monitoring the efficient operation of the quality system,
- the quality records, such as reports concerning inspections, tests, calibration and the qualifications of the staff concerned, etc.

The aforementioned checks do not apply to those aspects of the manufacturing process designed to secure sterility.

3.3. The notified body audits the quality system to determine whether it meets the requirements referred to in section 3.2. It must presume that quality systems which implement the relevant harmonised standards conform to these requirements.

The assessment team must include at least one member with past experience of assessments of the technology concerned. The assessment procedure must include an inspection on the manufacturer's premises and, in duly substantiated cases, on the premises of the manufacturer's suppliers to inspect the manufacturing processes.

The decision must be notified to the manufacturer. It must contain the conclusions of the inspection and a reasoned assessment.

3.4. The manufacturer must inform the notified body which approved the quality system of any plan for substantial changes to the quality system.

The notified body must assess the changes proposed and verify whether after these changes the quality system will still meet the requirements referred to in Section 3.2.

After receiving the abovementioned information it must notify the manufacturer of its decision. This decision must contain the conclusions of the inspection and a reasoned assessment.

4. Surveillance

4.1. The aim of surveillance is to ensure that the manufacturer duly fulfils the obligations imposed by the approved quality system.

4.2. The manufacturer must allow the notified body access for inspection purposes to the inspection, testing and storage locations and supply it with all relevant information, in particular:

- the documentation on the quality system,
- the technical documentation,
- the quality records, such as inspection reports, test data, calibration data, qualification reports of the staff concerned, etc.

4.3. The notified body must periodically carry out appropriate inspections and assessments to make sure that the manufacturer applies the quality system and must supply the manufacturer with an assessment report.

4.4. In addition, the notified body may pay unannounced visits to the manufacturer. At the time of such visits, the notified body may, where necessary, carry out or ask for tests in order to check that the quality system is working properly and that the production conforms to the requirements of the Directive which apply to it. To this end, an adequate sample of the final products, taken on site by the notified body, must be examined and the appropriate tests defined in the relevant standard(s) referred to in Article 5 or equivalent tests must be carried out. Where one or more of the samples fails to conform, the notified body must take the appropriate measures.

It must provide the manufacturer with an inspection report and, if a test has been carried out, with a test report.

5. Administrative provisions

5.1. The manufacturer must, for a period ending at least five years after the last product has been manufactured, make available to the national authorities:

- the declaration of conformity,
- the documentation referred to in the seventh indent of Section 3.1,
- the changes referred to in Section 3.4,
- the decisions and reports from the notified body as referred to in the final indent of Section 3.4. and in Sections 4.3 and 4.4.
- where appropriate, the certificate of conformity referred to in Annex III.

5.2. The notified body must make available to the other notified bodies, on request, all relevant information concerning the quality system approvals issued, refused or withdrawn.

6. Application to devices in Class IIa

In line with Article 11 (2), this Annex may apply to products in Class IIa, subject to this derogation:

6.1. by derogation from Sections 2, 3.1 and 3.2 by virtue of the declaration of conformity the manufacturer ensures and declares that the products in Class IIa are manufactured in conformity with the technical documentation referred to in Section 3 of Annex VII and meet the requirements of this Directive which apply to them.

Annex VII
EC declaration of conformity

1. The EC declaration of conformity is the procedure whereby the manufacturer or his authorised representative established in the Community who fulfils the obligations imposed by Section 2 and, in the case of products placed on the market in a sterile condition and devices with a measuring function, the obligations imposed by Section 5 ensures and declares that the products concerned meet the provisions of this Directive which apply to them.

2. The manufacturer must prepare the technical documentation described in Section 3. The manufacturer or his authorised representative established in the Community must make this documentation, including the declaration of conformity, available to the national authorities for inspection purposes for a period ending at least five years after the last product has been manufactured.

Where neither the manufacturer nor his authorised representative are established in the Community, this obligation to keep the technical documentation available must fall to the person(s) who place(s) the product on the Community market.

3. The technical documentation must allow assessment of the conformity of the product with the requirements of the Directive. It must include in particular:

- a general description of the product, including any variants planned,
- design drawings, methods of manufacture envisaged and diagrams of components, sub-assemblies, circuits, etc,
- the descriptions and explanations necessary to understand the abovementioned drawings and diagrams and the operation of the product,
- the results of the risk analysis and a list of the standards referred to in Article 5, applied in full or in part, and descriptions of the solutions adopted to meet the essential requirements of the Directive if the standards referred to in Article 5 have not been applied in full,
- in the case of products placed on the market in a sterile condition, description of the methods used,
- the results of the design calculations and of the inspections carried out, etc.; if the device is to be connected to other device(s) in order to operate as intended, proof must be provided that it conforms to the essential requirements when connected to any such device(s) having the characteristics specified by the manufacturer,
- the test reports and, where appropriate, clinical data in accordance with Annex X,
- the label and instructions for use.

4. The manufacturer shall institute and keep up to date a systematic procedure to review experience gained from devices in the post-production phase and to implement appropriate means to apply any necessary corrective actions, taking account of the nature and risks in relation to the product. He shall notify the competent authorities of the following incidents immediately on learning of them:

(i) any malfunction or deterioration in the characteristics and/or performance of a device, as well as any inadequacy in the labelling or the instructions for use which might lead to or might have led to the death of a patient or user or to a serious deterioration in his state of health;

(ii) any technical or medical reason connected with characteristics or the performance of a device for the reasons referred to in subparagraph (i) leading to a systematic recall of devices of the same type by the manufacturer.

5. With products placed on the market in sterile condition and Class I devices

with a measuring function, the manufacturer must observe not only the provisions laid down in this Annex but also one of the procedures referred to in Annex IV, V or VI. Application of the abovementioned Annexes and the intervention by the notified body is limited to:

- in the case of products placed on the market in sterile condition, only the aspects of manufacture concerned with securing and maintaining sterile conditions,
- in the case of devices with a measuring function, only the aspects of manufacture concerned with the conformity of the products with the metrological requirements.

Section 6.1 of this Annex is applicable.

6. Application to devices in Class IIa

In line with Article 11 (2), this Annex may apply to products in Class IIa, subject to this derogation:

6.1. where this Annex is applied in conjunction with the procedure referred to in Annex IV, V or VI, the declaration of conformity referred to in the abovementioned Annexes forms a single declaration. As regards the declaration based on this Annex, the manufacturer must ensure and declare that the product design meets the provisions of this Directive which apply to it.

Annex VIII

Statement concerning devices for special purposes

1. For custom-made devices or for devices intended for clinical investigations the manufacturer or his authorised representative established in the Community must draw up the statement containing the information stipulated in Section 2.

2. The statement must contain the following information:

2.1. for custom-made devices:

- data allowing identification of the device in question,
- a statement that the device is intended for exclusive use by a particular patient, together with the name of the patient,
- the name of the medical practitioner or other authorised person who made out the prescription and, where applicable, the name of the clinic concerned,
- the particular features of the device as specified in the relevant medical prescription,
- a statement that the device in question conforms to the essential requirements set out in Annex I and, where applicable, indicating which essential requirements have not been fully met, together with the grounds;

2.2. for devices intended for the clinical investigations covered by Annex X:

- data allowing identification of the device in question,
- an investigation plan stating in particular the purpose, scientific, technical or medical grounds, scope and number of devices concerned,
- the opinion of the ethics committee concerned and details of the aspects covered by its opinion,
- the name of the medical practitioner or other authorised person and of the institution responsible for the investigations,
- the place, starting date and scheduled duration for the investigations,
- a statement that the device in question conforms to the essential requirements apart from the aspects

covered by the investigations and that, with regard to these aspects, every precaution has been taken to protect the health and safety of the patient.

3. The manufacturer must also undertake to keep available for the competent national authorities:

3.1. for custom-made devices, documentation allowing an understanding of the design, manufacture and performances of the product, including the expected performances, so as to allow assessment of conformity with the requirements of this Directive.

The manufacturer shall take all the measures necessary to ensure that the manufacturing process produces products which are manufactured in accordance with the documentation mentioned in the first paragraph;

3.2. for devices intended for clinical investigations, the documentation must contain:

- a general description of the product,
- design drawings, methods of manufacture envisaged, in particular as regards sterilisation, and diagrams of components, sub-assemblies, circuits, etc,
- the descriptions and explanations necessary to understand the above-mentioned drawings and diagrams and the operation of the product,
- the results of the risk analysis and a list of the standards referred to in Article 5, applied in full or in part, and descriptions of the solutions adopted to meet the essential requirements of this Directive if the standards referred to in Article 5 have not been applied,
- the results of the design calculations, and of the inspections and technical tests carried out, etc.

The manufacturers must take all the measures necessary to ensure that the manufacturing process produces products which are manufactured in accordance with the documentation referred to in the first paragraph of this Section.

The manufacturer must authorise the assessment, or audit where necessary, of the effectiveness of these measures.

4. The information contained in the declarations concerned by this Annex should be kept for a period of time of at least five years.

Annex IX
Classification criteria

I. DEFINITIONS

1. Definitions for the classification rules

1.1. *Duration*

Transient
Normally intended for continuous use for less than 60 minutes.

Short term
Normally intended for continuous use for not more than 30 days.

Long term
Normally intended for continuous use for more than 30 days.

1.2. *Invasive devices*

Invasive device
A device which, in whole or in part, penetrates inside the body, either through a body orifice or through the surface of the body.

Body orifice
Any natural opening in the body, including the external surface of the eyeball, or any permanent artificial opening, such as a stoma.

Surgically invasive device
An invasive device which penetrates inside the body through the surface of the body, with the aid or in the context of a surgical operation.

For the purposes of this Directive devices other than those referred to in the previous subparagraph and which produce penetration other than through an established body orifice, shall be treated as surgically invasive devices.
Implantable device
Any device which is intended:
- to be totally introduced into the human body or,
- to replace an epithelial surface or the surface of the eye,
  by surgical intervention which is intended to remain in place after the procedure:

  Any device intended to be partially introduced into the human body through surgical intervention and intended to remain in place after the procedure for at least 30 days is also considered an implantable device.

1.3. *Reusable surgical instrument*

Instrument intended for surgical use by cutting, drilling, sawing, scratching, scraping, clamping, retracting, clipping or similar procedures, without connection to any other active medical device and which can be reused after appropriate procedures have been carried out.

1.4. *Active medical device*

Any medical device operation of which depends on a source of electrical energy or any source of power other than that directly generated by the human body or gravity and which acts by converting this energy. Medical devices intended to transmit energy, substances or other elements between an active medical device and the patient, without any significant change, are not considered to be active medical devices.

1.5. *Active therapeutical device*

Any active medical device, whether used alone or in combination with other medical devices, to support, modify, replace or restore biological functions or structures with a view to treatment or alleviation of an illness, injury or handicap.

1.6. *Active device for diagnosis*

Any active medical device, whether used alone or in combination with other medical devices, to supply information for detecting, diagnosing, monitoring or treating physiological conditions, states of health, illnesses or congenital deformities.

1.7. *Central circulatory system*

For the purposes of this Directive, 'central circulatory system' means the following vessels:
arteriae pulmonales, aorta ascendens, arteriae coronariae, arteria carotis communis, arteria carotis externa, arteria carotis interna, arteriae cerebrales, truncus brachiocephalicus, venae cordis, venae pulmonales, vena cava superior, vena cava inferior.

1.8. *Central nervous system*

For the purposes of this Directive, 'central nervous system' means brain, meninges and spinal cord.

II. IMPLEMENTING RULES

2. Implementing rules

2.1. Application of the classification rules shall be governed by the intended purpose of the devices.
2.1. If the device is intended to be used in combination with another device, the classification rules shall apply separately to each of the devices. Accessories are classified in their own right separately from the device with which they are used.
2.3. Software, which drives a device or influences the use of a device, falls automatically in the same class.
2.4. If the device is not intended to be used solely or principally in a specific part of the body, it must be considered and classified on the basis of the most critical specified use.
2.5. If several rules apply to the same device, based on the performance specified for the device by the manufac-

turer, the strictest rules resulting in the higher classification shall apply.

III. CLASSIFICATION

1. Non-invasive devices

1.1. *Rule 1*

All non-invasive devices are in Class I, unless one of the rules set out hereinafter applies.

1.2 *Rule 2*

All non-invasive devices intended for channelling or storing blood, body liquids or tissues, liquids or gases for the purpose of eventual infusion, administration or introduction into the body are in Class IIa:

- if they may be connected to an active medical device in Class IIa or a higher class,
- if they are intended for use for storing or channelling blood or other body liquids or for storing organs, parts of organs or body tissues,

in all other cases they are in class I.

1.3. *Rule 3*

All non-invasive devices intended for modifying the biological or chemical composition of blood, other body liquids or other liquids intended for infusion into the body are in Class IIb, unless the treatment consists of filtration, centrifugation or exchanges of gas, heat, in which case they are in Class IIa.

1.4. *Rule 4*

All non-invasive devices which come into contact with injured skin:

- are in Class I if they are intended to be used as a mechanical barrier, for compression or for absorption of exudates,
- are in Class IIb if they are intended to be used principally with wounds which have breached the dermis and can only heal by secondary intent,
- are in Class IIa in all other cases, including devices principally intended to manage the micro-environment of a wound.

2. Invasive devices

2.1. *Rule 5*

All invasive devices with respect to body orifices, other than surgically invasive devices and which are not intended for connection to an active medical device:

- are in Class I if they are intended for transient use,
- are in Class IIa if they are intended for short-term use, except if they are used in the oral cavity as far as the pharynx, in an ear canal up to the ear drum or in a nasal cavity, in which case they are in Class I,
- are in Class IIb if they are intended for long-term use, except if they are used in the oral cavity as far as the pharynx, in an ear canal up to the ear drum or in a nasal cavity and are not liable to be absorbed by the mucous membrane, in which case they are in Class IIa.

All invasive devices with respect to body orifices, other than surgically invasive devices, intended for connection to an active medical device in Class IIa or a higher class, are in Class IIa.

2.2. *Rule 6*

All surgically invasive devices intended for transient use are in Class IIa unless they are:

- intended specifically to diagnose, monitor or correct a defect of the heart or of the central circulatory system through direct contact with these parts of the body, in which case they are in Class III.
- reusable surgical instruments, in which case they are in Class I,
- intended to supply energy in the form of ionizing radiation in which case they are in Class IIb,
- intended to have a biological effect or to be wholly or mainly absorbed in which case they are in Class IIb,
- intended to administer medicines by means of a delivery system, if this is done in a manner that is potentially

hazardous taking account of the mode of application, in which case they are in Class IIb.

2.3. *Rule 7*

All surgically invasive devices intended for short-term use are in Class IIa unless they are intended:

- either specifically to diagnose, monitor or correct a defect of the heart or of the central circulatory system through direct contact with these parts of the body, in which case they are in Class III,
- or specifically for use in direct contact with the central nervous system, in which case they are in Class III,
- or to supply energy in the form of ionizing radiation in which case they are in Class IIb,
- or to have a biological effect or to be wholly or mainly absorbed in which case they are in Class III,
- or to undergo chemical change in the body, except if the devices are placed in the teeth, or to administer medicines, in which case they are in Class IIb.

2.4. *Rule 8*

All implantable devices and long-term surgically invasive devices are in Class IIb unless they are intended:

- to be placed in the teeth, in which case they are in Class IIa,
- to be used in direct contact with the heart, the central circulatory system or the central nervous system, in which case they are in Class III,
- to have a biological effect or to be wholly or mainly absorbed, in which case they are in Class III,
- or to undergo chemical change in the body, except if the devices are placed in the teeth, or to administer medicines, in which case they are in Class III.

3. Additional rules applicable to active devices

3.1. *Rule 9*

All active therapeutic devices intended to administer or exchange energy are in Class IIa unless their characteristics are such that they may administer or exchange energy to or from the human body in a potentially hazardous way, taking account of the nature, the density and site of application of the energy, in which case they are in Class IIb.

All active devices intended to control or monitor the performance of active therapeutic devices in Class IIb, or intended directly to influence the performance of such devices are in Class IIb.

3.2. *Rule 10*

Active devices intended for diagnosis are in Class IIa:

- if they are intended to supply energy which will be absorbed by the human body, except for devices used to illuminate the patient's body, in the visible spectrum.
- if they are intended to image *in vivo* distribution of radiopharmaceuticals,
- if they are intended to allow direct diagnosis or monitoring of vital physiological processes, unless they are specifically intended for monitoring of vital physiological parameters, where the nature of variations is such that it could result in immediate danger to the patient, for instance variations in cardiac performance, respiration, activity of CNS in which case they are in Class IIb.

Active devices intended to emit ionizing radiation and intended for diagnostic and therapeutic intervention radiology including devices which control or monitor such devices, or which directly influence their performance, are in Class IIb.

Rule 11

All active devices intended to administer and/or remove medicines, body liquids or other substances to or from the body are in Class IIa, unless this is done in a manner:

- that is potentially hazardous, taking account of the nature of the substances involved, of the part of the body concerned and of the mode of application in which case they are in class IIb.

3.3. *Rule 12*

All other devices are in Class I.

4. Special Rules

4.1. *Rule 13*

All devices incorporating, as an integral part, a substance which, if used separately, can be considered to be a medicinal product, as defined in Article 1 of Directive 65/65/EEC, and which is liable to act on the human body with action ancillary to that of the devices, are in Class III.

4.2. *Rule 14*

All devices used for contraception or the prevention of the transmission of sexually transmitted diseases are in Class IIb, unless they are implantable or long term invasive devices, in which case they are in Class III.

4.3. *Rule 15*

All devices intended specifically to be used for disinfecting, cleaning, rinsing or, when appropriate, hydrating contact lenses are in Class IIb.

All devices intended specifically to be used for disinfecting medical devices are in Class IIa.

This rule does not apply to products that are intended to clean medical devices other than contact lenses by means of physical action.

4.4. *Rule 16*

Non-active devices specifically intended for recording of X-ray diagnostic images are in Class IIa.

4.5. *Rule 17*

All devices manufactured utilising animal tissues or derivatives rendered non-viable are Class III except where such devices are intended to come into contact with intact skin only.

5. *Rule 18*

By derogation from other rules, blood bags are in Class IIb.

Annex X
Clinical evaluation

1. General provisions

1.1. As a general rule, confirmation of conformity with the requirements concerning the characteristics and performances referred to in Section 1 and 3 of Annex I under the normal conditions of use of the device and the evaluation of the undesirable side-effects must be based on clinical data in the case of implantable devices and devices in Class III. Taking account of any relevant harmonised standards, where approporiate, the adequacy of the clinical data must be based on:

1.1.1. either a compilation of the relevant scientific literature currently available on the intended purpose of the device and the techniques employed as well as, if appropriate, a written report containing a critical evaluation of this compilation;

1.1.2. or the results of all the clinical investigations made, including those carried out in conformity with Section 2.

1.2. All the data must remain confidential, in accordance with the provisions of Article 20.

2. Clinical investigation

2.1. *Objectives*

The objectives of clinical investigation are:

- to verify that, under normal conditions of use, the performances of the devices conform to those referred to in Section 3 of Annex I and
- to determine any undesirable side-effects, under normal conditions of use, and assess whether they constitute risks when weighed against the intended performance of the device.

2.2. *Ethical considerations*

Clinical investigations must be carried out in accordance with the Helsinki Declaration adopted by the 18th World Medical Assembly in Helsinki, Finland, in 1964, as last amended by the 41st World Medical Assembly in Hong Kong in 1989. It is mandatory that all measures relating to the protection of human subjects are carried out in the spirit of the Helsinki Declaration. This includes every step in the clinical investigation from first consideration of the need and justification of the study to publication of the results.

2.3. *Methods*

2.3.1. Clinical investigations shall be performed on the basis of an appropriate plan of investigation reflecting the latest scientific and technical knowledge and defined in such a way as to confirm or refute the manufacturer's claims for the device; these investigations must include an adequate number of observations to guarantee the scientific validity of the conclusions.

2.3.2. The procedures used to perform the investigations must be appropriate to the device under examination.

2.3.3. Clinical investigations must be performed in circumstances similar to the normal conditions of use of the device.

2.3.4. All the appropriate features, including those involving the safety and performances of the device, and its effect on patients must be examined.

2.3.5. All adverse incidents such as those specified in Article 10 must be fully recorded and notified to the competent authority.

2.3.6. The investigations must be performed under the responsibility of a medical practitioner or another authorised qualified person in an appropriate environment.

The medical practitioner or other authorised person shall have access to the technical and clinical data regarding the device.

2.3.7. The written report, signed by the medical practitioner or other authorised person responsible, must contain a critical evaluation of all the data collected during the clinical investigation.

Annex XI

Criteria to be met for the designation of notified bodies

1. The notified body, its Director and the assessment and verification staff shall not be the designer, manufacturer, supplier, installer or user of the devices which they inspect, nor the authorised representative of any of these persons. They may not be directly involved in the design, construction, marketing or maintenance of the devices, nor represent the parties engaged in these activities. This in no way precludes the possibility of exchanges of technical information between the manufacturer and the body.

2. The notified body and its staff must carry out the assessment and verification operations with the highest degree of professional integrity and the requisite competence in the field of medical devices and must be free from all pressures and inducements, particularly financial, which might influence their judgment or the results of the inspection, especially from persons or groups of persons with an

interest in the results of the verifications.

Should the notified body subcontract specific tasks connected with the establishment and verification of the facts, it must first ensure that the subcontractor meets the provisions of the Directive and, in particular, of this Annex. The notified body shall keep at the disposal of the national authorities the relevant documents assessing the subcontractor's qualifications and the work carried out by the subcontractor under this Directive.

3. The notified body must be able to carry out all the tasks assigned to such bodies by one of Annexes II to VI and for which it has been notified, whether these tasks are carried out by the body itself or on its responsibility. In particular, it must have the necessary staff and possess the facilities needed to perform properly the technical and administrative tasks entailed in assessment and verification. It must also have access to the equipment necessary for the verifications required.

4. The notified body must have:

- sound vocational training covering all the assessment and verification operations for which the body has been designated,
- satisfactory knowledge of the rules on the inspections which they carry out and adequate experience of such inspections,
- the ability required to draw up the certificates, records and reports to demonstrate that the inspections have been carried out.

5. The impartiality of the notified body must be guaranteed. Their remuneration must not depend on the number of inspections carried out, nor on the results of the inspections.

6. The body must take out civil liability insurance, unless liability is assumed by the State under domestic legislation or the Member State itself carries out the inspections directly.

7. The staff of the notified body are bound to observe professional secrecy with regard to all information gained in the course of their duties (except *vis-à-vis* the competent administrative authorities of the State in which their activities are carried out) pursuant to this Directive or any provision of national law putting it into effect.

Annex XII
CE marking of conformity

The CE conformity marking shall consist of the initials "CE" taking the following form

- If the marking is reduced or enlarged the proportions given in the above graduated drawing must be respected.
- The various components of the CE marking must have substantially the same vertical dimension, which may not be less than 5 mm.

This minimum dimension may be waived for small-scale devices.

COUNCIL DIRECTIVE 90/396/EEC

of 29 June 1990

on the approximation of the laws of the Member States relating to appliances burning gaseous fuels*†

THE COUNCIL OF THE EUROPEAN COMMUNITIES,

Having regard to the Treaty establishing the European Economic Community, and in particular Article 100A thereof,

Having regard to the proposal from the Commission,

In cooperation with the European Parliament,

Having regard to the opinion of the Economic and Social Committee,

Whereas Member States are responsible for ensuring the health and safety on their territory of their people and, where appropriate, of domestic animals and goods in relation to the hazards arising out of the use of appliances burning gaseous fuels;

Whereas, in certain Member States, mandatory provisions define in particular the safety level required of appliances burning gaseous fuels by specifying design, operating characteristics and inspection procedures; whereas these mandatory provisions do not necessarily lead to different safety levels from one Member State to another but do, by their disparity, hinder trade within the Community;

Whereas different conditions as regards types of gas and supply pressures are in force in the Member States; whereas these conditions are not harmonised because each Member State's energy supply and distribution situation is peculiar to it;

Whereas paragraphs 65 and 68 of the White Paper on the completion of the internal market, approved by the European Council in June 1985, provide for a new approach to legislative harmonisation.

Whereas Community law provides – by way of derogation from one of the fundamental rules of the Community, namely the free movement of goods – that obstacles to movement within the Community resulting from disparities in national legislation relating to the marketing of products must be accepted in so far as such obstacles can be recognised as being necessary to satisfy mandatory requirements; whereas, therefore, the harmonisation of legislation in the present case must be limited to the provisions necessary to satisfy both the mandatory and essential requirements regarding safety, health and energy conservation in relation to gas appliances; whereas these requirements must replace the national provisions in this matter because they are essential requirements;

Whereas the maintenance or improvement of the level of safety attained in Member States constitutes one of the essential aims of this Directive and of safety as defined by the essential requirements;

Whereas the essential safety and health requirements must be observed in order to ensure that appliances burning gaseous fuels are safe; whereas energy conservation is considered essential; whereas these requirements must be applied with discernment to take account of the state of the art at the time of construction;

*OJ 1990 L196/15.
†Amended by Dir 93/68.

Whereas this Directive therefore contains only essential requirements; whereas, to facilitate proof of conformity with the essential requirements, it is necessary to have harmonised standards at European level in particular as to the construction, operation and installation of appliances burning gaseous fuels so that products complying with them may be assumed to conform to the essential requirements; whereas these standards, harmonised at European level, are drawn up by private bodies and must remain non-mandatory texts; whereas for that purpose the European Committee for Standardisation (CEN) and the European Committee for Electrotechnical Standardisation (CENELEC) are recognised as the competent bodies for the adoption of harmonised standards in accordance with the general guidelines for cooperation between the Commission and those two bodies signed on 13 November 1984; whereas, for the purposes of this Directive, a harmonised standard is a technical specification (European standard or harmonisation document) adopted by one or both of those bodies upon a remit from the Commission in accordance with Council Directive 83/189/EEC of 28 March 1983 (OJ 1983 L109/8) laying down a procedure for the provision of information in the field of technical standards and regulations, as last amended by Directive 88/182/EEC (OJ 1988 L81/75), and the abovementioned general guidelines;

Whereas, pending the adoption of harmonised standards within the meaning of this Directive, conformity with the essential requirements and the free movement of appliances burning gaseous fuels should be facilitated by the acceptance, at Community level, of products conforming to the national standards whose conformity with the essential requirements has been confirmed by a Community checking procedure;

Whereas a check on compliance with the relevant technical requirements is necessary in order to provide effective protection for users and third parties; whereas the existing certification procedures differ from one Member State to another; whereas, in order to avoid multiple inspections, which are in effect barriers to the free movement of appliances burning gaseous fuels, arrangements should be made for the mutual recognition of certification procedures by the Member States; whereas, in order to facilitate mutual recognition of certification procedures, harmonised Community procedures should be set up and the criteria for appointing the bodies responsible for carrying out these procedures should be harmonised;

Whereas the Member States' responsibility on their territory for safety, health and energy conservation covered by the essential requirements must be recognised in a safeguard clause providing for an adequate Community procedure;

Whereas the addressees of any decision taken under this Directive must be informed of the reasons for such a decision and the legal remedies available to them;

Whereas the Council adopted, on 17 September 1984, a framework Directive on gas appliances (84/530/EEC) (OJ 1986 L196/56), as last amended by Directive 86/312/EEC (OJ 1984 L300/106), and a separate Directive on gas water heaters (84/531/EEC) (OJ 1984 L300/106) as last amended by Directive 88/665/EEC (OJ 1988 L382/42); whereas those Directives cover the same area as this Directive and should therefore be repealed;

Whereas the measures aimed at the gradual establishment of the internal market must be adopted by 31 December 1992; whereas the internal market consists of an area without internal frontiers in which the free movement of goods, persons, services and capital is ensured,

HAS ADOPTED THIS DIRECTIVE:

Chapter 1

Scope, placing on the market and free movement

Article 1

1. This Directive shall apply to:
- appliances burning gaseous fuels used for cooking, heating, hot water production, refrigeration, lighting or washing and having, where applicable, a normal water temperature not exceeding 105 °C, hereinafter referred to as "appliances". Forced draught burners and heating bodies to be equipped with such burners will also be considered as appliances,
- safety devices, controlling devices or regulating devices and sub-assemblies, other than forced draught burners and heating bodies to be equipped with such burners separately marketed for trade use and designed to be incorporated into an appliance burning gaseous fuel or assembled to constitute such an appliance, hereinafter referred to as "fittings".

2. Appliances specifically designed for use in industrial processes carried out on industrial premises are excluded from the scope defined in paragraph 1.

3. For the purposes of this Directive, "gaseous fuel" means any fuel which is in a gaseous state at a temperature of 15 °C under a pressure of 1 bar.

4. For the purposes of this Directive, an appliance is said to be "normally used" when it is:
- correctly installed and regularly serviced in accordance with the manufacturer's instructions,
- used with a normal variation in the gas quality and a normal fluctuation in the supply pressure, and
- used in accordance with its intended purpose or in a way which can be reasonably foreseen.

Article 2

1. Member States shall take all necessary steps to ensure that the appliances referred to in Article 1 may be placed on the market and put into service only if, when normally used, they do not compromise the safety of persons, domestic animals and property.

2. Member States shall communicate the types of gas and corresponding supply pressures used on their territory to the other Member States and the Commission before 1 January 1991. They shall also communicate all changes in good time. The Commission shall ensure that this information is published in the *Official Journal of the European Communities*.

Article 3

Appliances and fittings as referred to in Article 1 shall satisfy the essential requirements applicable to them set out in Annex I.

Article 4

1. Member States may not prohibit, restrict or impede the placing on the market and the putting into service of appliances which comply with all the provisions of this Directive, including the conformity assessment procedures laid down in Chapter II, and which bear the CE marking provided for in Article 10.

2. Member States may not prohibit, restrict or impede the placing on the market of fittings as referred to in Article 1 accompanied by a certificate as referred to in Article 8 (4).

Article 5

1. Member States shall presume compliance with the essential requirements referred to in Article 3 of appliances and fittings when they conform to:

(a) national standards applicable to them implementing the harmonised standard whose reference numbers have been published in the *Official Journal of the European Communities*.

Member States shall publish the reference numbers of these national standards;

(b) the national standards applicable to them referred to in paragraph 2 in so far as, in the areas covered by such standards, no harmonised standards exist.

2. Member States shall communicate to the Commission the texts of their national standards as referred to in paragraph 1 (b) which they regard as complying with the essential requirements referred to in Article 3. The Commission shall forward these national standards to the other Member States. In accordance with the procedure provided for in Article 6 (2), it shall notify the Member States of those national standards which are presumed to conform with the essential requirements referred to in Article 3.

Article 6

1. Where a Member State or the Commission considers that the standards referred to in Article 5 (1) do not entirely meet the essential requirements referred to in Article 3, the Commission or the Member State concerned shall bring the matter before the standing committee set up under Directive 83/189/EEC, hereinafter referred to as "the committee", giving the reasons therefore. The committee shall deliver an opinion without delay.

In the light of the committee's opinion, the Commission shall inform the Member States whether or not it is necessary to withdraw those standards from the publications referred to in Article 5 (1).

2. After receipt of the communication referred to in Article 5 (2), the Commission shall consult the committee. Upon receipt of the latter's opinion, the Commission shall, within one month, inform the Member States whether or not the national standard(s) in question are to enjoy the presumption of conformity; if they are, the Member States shall publish the reference numbers of those standards. The Commission shall also publish them in the *Official Journal of the European Communities*.

Article 7

1. Where a Member State finds that normally used appliances bearing the CE marking might compromise the safety of persons, domestic animals or property, it shall take all appropriate measures to withdraw those appliances from the market and prohibit or restrict their being placed on the market.

The Member State concerned shall immediately inform the Commission of any such measure, indicating the reasons for its decision and, in particular, whether non-compliance is due to:

(a) failure to meet the essential requirements referred to in Article 3, where the appliance does not correspond to the standards referred to in Article 5 (1);
(b) incorrect application of the standards referred to in Article 5 (1);
(c) shortcomings in the standards referred to in Article 5 (1) themselves.

2. The Commission shall enter into consultation with the parties concerned as soon as possible. Where, after such consultation, the Commission finds that any measure as referred to in paragraph 1 is justified, it shall immediately so inform the Member State that took the measure and the other Member States.

Where the decision referred to in paragraph 1 is attributed to shortcomings in the standards, the Commission, after consulting the parties concerned, shall bring the matter before the committee within two months if the Member State which has taken the measures intends to maintain them, and shall initiate the procedures referred to in Article 6.

3. Where an appliance which does not comply bears the CE marking, the competent Member State shall take appropriate action against whomsoever has affixed the mark and shall inform the Commission and the other Member States thereof.

4. The Commission shall ensure that the Member States are kept informed of the progress and outcome of the procedures.

Chapter II
Means of certification of conformity

Article 8

1. The means of certification of conformity of series-manufactured appliances shall be:
(a) the EC type-examination as referred to in section 1 of Annex II, and
(b) prior to their being placed on the market, at the choice of the manufacturer:
   - the EC declaration of conformity to type referred to in section 2 of Annex II, or
   - the EC declaration of conformity to type (guarantee of production quality) referred to in section 3 of Annex II, or
   - the EC declaration of conformity to type (guarantee of product quality) referred to in section 4 of Annex II, or
   - EC verification as referred to in section 5 of Annex II.

2. In the case of production of an appliance as a single unit or in small quantities, EC verification by single unit, as referred to in section 6 of Annex II, may be chosen by the manufacturer.
3. After completion of the procedures referred to in paragraphs 1 (b) and 2, the CE marking of conformity shall be affixed to conforming appliances in accordance with Article 10.
4. The procedures referred to in paragraph 1 shall be applied in respect of fittings as referred to in Article 1 with the exception of the affixing of the CE marking of conformity and, where appropriate, the drawing-up of the declaration of conformity. A certificate shall be issued declaring the conformity of the fittings with the provisions of this Directive which apply to them and stating their characteristics and how they must be incorporated into an appliance or assembled to assist compliance with the essential requirements applicable to finished appliances.

The certificate shall be supplied with the fitting.
5. (a) Where the appliances are covered by other Directives dealing with other aspects and specifying the affixing of the CE marking, the latter shall indicate that the appliances are also presumed to conform to the provisions of those other Directives.
(b) However, where one or more of these Directives allow the manufacturer, during a transitional period, to choose which arrangements to apply, the CE marking shall indicate conformity to the provisions only of those Directives applied by the manufacturer. In this case, particulars of the Directives applied, as published in the *Official Journal of the Eropean Communities*, must be given in the documents, notices or instructions required by the Directives and accompanying such devices.
6. Records and correspondence relating to the means of certification of conformity shall be drawn up in the official language(s) of the Member State where the body responsible for carrying out these procedures is established or in a language accepted by it.

Article 9

1. Member States shall notify the Commission and the other Member States of the bodies, which they have appointed to carry out the procedures referred to in Article 8 together with the specific tasks which these bodies have been appointed to carry out and the identification numbers assigned to them beforehand by the Commission.

The Commission shall, for information, publish in the *Official Journal of the European Communities*, a list of these bodies, and the identification numbers it has assigned to them and shall ensure that the list is kept up to date.
2. Member States shall apply the criteria set out in Annex V for assessing the bodies to be notified.

Bodies which satisfy the assessment criteria laid down in the applicable harmonised standards shall be presumed to satisfy the criteria set out in that Annex.

3. A Member State which has notified a body must withdraw approval if it finds that the body no longer meets the criteria referred to in paragraph 2. It shall immediately inform the Commission and the other Member States accordingly.

Chapter III
EC mark of conformity

Article 10

1. The EC mark of conformity and the inscriptions set out in Annex III shall be affixed in a visible, easily legible and indelible form to the appliance or to a data plate attached to it. The data plate shall be so designed that it cannot be re-used.
2. The affixing of markings on the appliances which are likely to deceive third parties as to the meaning and form of the CE marking shall be prohibited. Any other marking may be affixed to the appliance or to the data plate provided that the visibility and legibility of the CE marking is not hereby reduced.

Article 11

Without prejudice to Article 7:
(a) where a Member State establishes that the CE marking has been affixed unduly, the manufacturer or his authorised representative established within the Community shall be obliged to make the product comply as regards the provisions concerning the CE markings and so end the infringement under conditions imposed by the Member State;
(b) where non-compliance continues, the Member State must take all appropriate measures to restrict or prohibit the placing on the market of the appliance in question or to ensure that it is withdrawn from the market in accordance with the procedures laid down in Article 7.

Chapter IV
Final provisions

Article 12

Any decision taken pursuant to this Directive which includes restriction on the placing on the market and/or putting into service of an appliance shall state the precise grounds on which it is based. It shall be notified without delay to the party concerned, who shall at the same time be informed of the legal remedies available to him under the laws in force in the Member State in question and of the time limits to which such remedies are subject.

Article 13

Directives 84/530/EEC and 84/531/EEC are hereby repealed.

Article 14

1. Before 1 July 1991, Member States shall adopt and publish the laws, regulations and administrative provisions necessary to comply with this Directive. They shall inform the Commission thereof.
They shall apply such provisions from 1 January 1992.
2. However, in derogation from Article 2, Member States may, during the period up to 31 December 1995 and without prejudice to Articles 30 to 36 of the Treaty, permit the placing on the market and/or the putting into service of appliances and fittings which comply with the regulations in force in the Member States before 1 January 1992.
3. Member States shall communicate to the Commission the texts of the provisions of national law which they adopt in the field covered by this Directive.

Article 15

This Directive is addressed to the Member States.

Annex I
Essential requirements

Preliminary remark

The obligations resulting from the essential requirements for appliances in this Annex also apply to fittings where the corresponding risk exists.

1. General conditions

1.1. Appliances must be so designed and built as to operate safely and present no danger to persons, domestic animals or property when normally used as defined in Article 1 (4) of this Directive.

1.2. When placed on the market, all appliances must:

- be accompanied by technical instructions intended for the installer,
- be accompanied by instructions for use and servicing, intended for the user,
- bear appropriate warning notices, which must also appear on the packaging.

The instructions and warning notices must be in the official language or languages of the Member States of destination.

1.2.1. The technical instructions intended for the installer must contain all the instructions for installation, adjustment and servicings required to ensure that those operations are correctly performed and that the appliance may be used safely. In particular, the instructions must specify:

- the type of gas used,
- the gas supply pressure used,
- the flow of fresh air required:
- for the combustion air supply,
- to avoid the formation of dangerous unburned gas mixtures for appliances not fitted with the device referred to in point 3.2.3,
- the conditions for the dispersal of combustion products,
- for forced draught burners and heating bodies intended to be equipped with such burners, their characteristics, the requirements for assembly, to assist compliance with the essential requirements applicable to finished appliances and, where appropriate, the list of combinations recommended by the manufacturer.

1.2.2. The instructions for use and servicing intended for the user must contain all the information required for safe use, and must in particular draw the user's attention to any restrictions on use.

1.2.3. The warning notices on the appliance and its packaging must clearly state the type of gas used, the gas supply pressure and any restrictions on use, in particular the restriction whereby the appliance must be installed only in areas where there is sufficient ventilation.

1.3. Fittings intended to be part of an appliance must be so designed and built as to fulfil correctly their intended purpose when incorporated in accordance with the instructions for installation.

The instructions for installation, adjustment, operation and maintenance must be provided with the fittings concerned.

2. Materials

2.1. Materials must be appropriate for their intended purpose and must withstand the technical, chemical and thermal conditions to which they will foreseeably be subjected.

2.2. The properties of materials that are important for safety must be guaranteed by the manufacturer or the supplier of the appliance.

3. Design and Construction

3.1. General

3.1.1. Appliances must be so constructed that, when used normally, no instability, distortion, breakage or wear likely to impair their safety can occur.

3.1.2. Condensation produced at the start-up and/or during use must not affect the safety of appliances.

3.1.3. Appliances must be so designed and constructed as to minimise the risk of explosion in the event of a fire of external origin.

3.1.4. Appliances must be so constructed that water and inappropriate air penetration into the gas circuit does not occur.

3.1.5. In the event of a normal fluctuation of auxiliary energy, appliances must continue to operate safely.

3.1.6. Abnormal fluctuation or failure of auxiliary energy or its restoration must not lead to an unsafe situation.

3.1.7. Appliances must be so designed and constructed as to obviate hazards of electrical origin. In the area in which it applies, compliance with the safety objectives in respect of electrical hazards laid down in Directive 73/23/EEC (OJ 1973 L77/29) shall be equivalent to fulfilment of this requirements.

3.1.8. All pressurised parts of an appliance must withstand the mechanical and thermal stresses to which they are subjected without any deformation affecting safety.

3.1.9. Appliances must be so designed and constructed that failure of a safety, controlling or regulating device may not lead to an unsafe situation.

3.1.10. If an appliance is equipped with safety and controlling devices, the functioning of the safety devices must not be overruled by that of the controlling devices.

3.1.11. All parts of appliances which are set or adjusted at the stage of manufacture and which should not be manipulated by the user or the installer must be appropriately protected.

3.1.12. Levers and other controlling and setting devices must be clearly marked and give appropriate instructions so as to prevent any error in handling. Their design must be such as to preclude accidental manipulation.

3.2. Unburned gas release

3.2.1. Appliances must be so constructed that the gas leakage rate is not dangerous.

3.2.2. Appliances must be so constructed that gas release during ignition and re-ignition and after flame extinction is limited in order to avoid a dangerous accumulation of unburned gas in the appliance.

3.2.3. Appliances intended to be used in indoor spaces and rooms must be fitted with a special device which avoids a dangerous accumulation of unburned gas in such spaces or rooms.

Appliances which are not fitted with such devices must be used only in areas where there is sufficient ventilation to avoid a dangerous accumulation of unburned gas.

Member States may define on their territory adequate space ventilation conditions for the installation of such appliances, bearing in mind the features peculiar to them.

Large-scale kitchen appliances and appliances powered by gas containing toxic components must be equipped with the aforesaid device.

3.3. Ignition

Appliances must be so constructed that, when used normally:

– ignition and re-ignition is smooth,
– cross-lighting is assured.

3.4. Combustion

3.4.1. Appliances must be so constructed that, when used normally, flame stability is assured and combustion products do not contain unacceptable concentrations of substances harmful to health.

3.4.2. Appliances must be so constructed that, when used normally, there will be no accidental release of combustion products.

3.4.3. Appliances connected to a flue for the dispersal of combustion products must be so constructed that in abnormal draught conditions there is no release of combustion products in a dangerous quantity into the room concerned.

3.4.4. Independent flueless domestic heating appliances and flueless instantaneous water heaters must not cause, in the room or space concerned, a carbon monoxide concentration likely to present a danger to the health of persons exposed, bearing in mind the foreseeable duration of their exposure.

3.5. Rational use of energy

Appliances must be so constructed as to ensure rational use of energy, reflecting the state of the art and taking into account safety aspects.

3.6. Temperatures

3.6.1. Parts of appliances which are intended to be placed in close proximity to the floor or other surfaces must not reach temperatures which present a danger in the surrounding area.
3.6.2. The surface temperature of knobs and levers of appliances intended to be manipulated must not present a danger to the user.
3.6.3. The surface temperatures of external parts of appliances intended for domestic use, with the exception of surfaces or parts which are associated with the transmission of heat, must not under operating conditions present a danger to the user and in particular to children, for whom an appropriate reaction time must be taken into account.

3.7. Foodstuffs and water used for sanitary purposes

Without prejudice to the Community rules in this area, materials and components used in the construction of an appliance, which may come into contact with food or water used for sanitary purposes, must not impair their quality.

Annex II
Procedure for certification of conformity

1. EC type-examination

1.1. The EC type-examination is that part of the procedure by which a notified body checks and certifies that an appliance, representative of the production envisaged, meets the provisions of this Directive which apply to it.
1.2. The application for type-examination must be lodged by the manufacturer or his authorised representative established within the Community with a single notified body.
1.2.1. The application must include:
- the name and address of the manufacturer and, if the application is lodged by the authorised representative, his name and address,
- a written declaration that the application has not been lodged with any other notified body,
- the design documentation, as described in Annex IV.

1.2.2. The manufacturer must place at the disposal of the notified body an appliance, representative of the production envisaged, hereinafter called "type". The notified body may request further samples of the type if needed for the test programme.

The type may additionally cover variants of the product provided that those variants do not have different characteristics with respect to types of risk.
1.3. The notified body must:
1.3.1. examine the design documentation and verify that the type has been manufactured in conformity with the design documentation and identify the elements which have been designed in accordance with the applicable provisions of the standards referred to in Article 5 and the essential requirements of this Directive;
1.3.2. perform, or have performed, the appropriate examinations and/or tests to check whether the solutions

adopted by the manufacturer meet the essential requirements where the standards referred to in Article 5 have not been applied;

1.3.3. perform, or have performed, the appropriate examinations and/or tests to check whether the applicable standards have effectively been applied where the manufacturer has chosen to do so, thereby assuring conformity with the essential requirements.

1.4. Where the type satisfies the provisions of this Directive, the notified body must issue an EC type-examination certificate to the applicant. The certificate must contain the conclusions of the examination, the conditions, if any, for its validity and the necessary data for identification of the approved type and, if relevant, descriptions of its functioning. Relevant technical elements such as drawings and diagrams must be annexed to the certificate.

1.5. The notified body must inform the other notified bodies forthwith of the issuing of the EC type-examination certificate and any additions to the said type as referred to in point 1.7. They may obtain a copy of the EC type-examination certificate and/or its additions and on a reasoned request may obtain a copy of the Annexes to the certificate and the reports on the examinations and tests carried out.

1.6. A notified body which refuses to issue or withdraws an EC type-examination certificate must inform the Member State which notified it and the other notified bodies accordingly, giving the reasons for its decision.

1.7. The applicant must keep the notified body that has issued the EC type-examination certificate informed of all modifications to the approved type which might affect conformity with the essential requirements.

Modifications to the approved type must receive additional approval from the notified body that issued the EC type-examination certificate where such changes affect conformity with the essential requirements or the prescribed conditions for use of the appliance. This additional approval is to be given in the form of an addition to the original EC type-examination certificate.

2. EC declaration of conformity to type

2.1. The EC declaration of conformity to type is that part of the procedure whereby the manufacturer declares that the appliances concerned are in conformity with the type as described in the EC type-examination certificate and satisfy the essential requirements of this Directive which apply to them. The manufacturer or his authorised representative established within the Community must affix the CE marking on each appliance and draw up a written declaration of conformity. The declaration of conformity may cover one or more appliances and must be kept by the manufacturer. The CE marking must be followed by the identification number of the notified body responsible for the random checks set out in point 2.3.

2.2. The manufacturer must take all necessary measures to ensure that the manufacturing process, including final product inspection and testing, results in homogeneity of production and conformity of the appliances with the type as described in the EC type-examination certificate and with the requirements of this Directive which apply to them. A notified body, chosen by the manufacturer, must carry out random checks on the appliances as set out in point 2.3.

2.3. On-site checks of appliances must be undertaken at random by the notified body at intervals of one year or less. An adequate number of appliances must be examined and appropriate tests as set out in the applicable standards referred to in Article 5 or equivalent tests must be carried out in order to ensure conformity with the corresponding essential requirements of this Directive. The notified body shall in each case determine whether these tests need to be carried out in full or in part. Where one or more appliances are rejected, the notified body shall take the appropriate measures to prevent the marketing thereof.

3. EC declaration of conformity to type (guarantee of production quality)

3.1. The EC declaration of conformity to type (guarantee of production quality) is the procedure whereby a manufacturer who fulfils the obligations in point 3.2 declares that the appliances concerned are in conformity with the type as described in the EC type-examination certificate and satisfy the essential requirements of this Directive which applies to them. The manufacturer or his authorised representative established within the Community must affix the CE marking to each appliance and draw up a written declaration of conformity. This declaration may cover one or more appliances and must be kept by the manufacturer. The CE marking must be followed by the identification number of the notified body responsible for EC surveillance.

3.2. The manufacturer shall apply a quality system that ensures conformity of the appliances with the type as described in the EC type-examination certificate and with the essential requirements of this Directive which apply to them. The manufacturer is subject to EC surveillance as specified in point 3.4.

3.3. Quality system

3.3.1. The manufacturer must lodge an application for approval of his quality system with a notified body of his choice for the appliances in question.

The application must include:

- the quality system documentation,
- an undertaking to carry out the obligations arising from the quality system as approved,
- an undertaking to maintain the approved quality system to ensure its continuing suitability and effectiveness,
- documentation relating to the approved type and a copy of the EC type-examination certificate.

3.3.2. All the elements, requirements and provisions adopted by the manufacturer must be documented in a systematic and logical manner in the form of measures, procedures and written instructions. This quality system documentation must permit a uniform interpretation of the quality programmes, plans, manuals and records. It shall contain, in particular, an adequate description of:

- the quality objectives, the organisational structure and responsibilities of management and of their powers with regard to appliance quality,
- the manufacturing processes, quality control and quality assurance techniques and systematic actions that will be used,
- the examinations and tests that will be carried out before, during and after manufacture and the frequency with which they will be carried out,
- the method of monitoring attainment of the required appliance quality and the effective operation of the quality system.

3.3.3. The notified body shall examine and evaluate the quality system to determine whether it satisfies the requirements referred to in point 3.3.2. It will presume conformity with these requirements in respect of quality systems that implement the corresponding harmonised standard.

It must notify its decision to the manufacturer and inform the other notified bodies thereof. The notification to the manufacturer must contain the conclusions of the examination, the name and address of the notified body and the reasoned assessment decision in respect of the appliances concerned.

3.3.4. The manufacturer must keep the notified body that has approved the quality system informed of any updating of the quality system in relation to changes brought about by, for example, new technologies and quality concepts.

The notified body must examine the proposed modifications and decide whether the modified quality system complies with the relevant provisions or whether reappraisal is necessary. It must notify the manufacturer of its decision. The notification must include the conclusions of the inspection and the reasoned assessment decision.

3.3.5. A notified body that withdraws approval of a quality system must so inform the other notified bodies, giving the reasons for the decision.

3.4. EC surveillance

3.4.1. The purpose of EC surveillance is to ensure that the manufacturer duly fulfils the obligations arising out of the approved quality system.

3.4.2. The manufacturer must allow the notified body access for inspection purposes to the place of manufacture, inspection, testing and storage and must provide it with all necessary information, in particular:

- the quality system documentation,
- the quality records, such as inspection reports and test data, calibration data, reports on qualifications of the staff concerned, etc.

3.4.3. The notified body must carry out a check at least once every two years to ensure that the manufacturer is maintaining and applying the approved quality system and must supply a report of the check to the manufacturer.

3.4.4. Furthermore, the notified body may make unannounced visits to the manufacturer. During these visits, the notified body may carry out tests on appliances or have them carried out. It must supply the manufacturer with an inspection report and, if appropriate, a test report.

3.4.5. The manufacturer may supply the notified body's report on request.

4. EC declaration of type conformity (guarantee of product quality)

4.1. The EC declaration of type conformity (guarantee of product quality) is that part of the procedure whereby a manufacturer who fulfils the obligations in point 4.2 declares that the appliances concerned are in conformity with the type as described in the EC type-examination certificate and satisfy the essential requirements of this Directive which apply to them. The manufacturer or his authorised representative established within the Community must affix the CE marking to each appliance and draw up a written declaration of conformity. This declaration may cover one or more appliances and must be kept by the manufacturer. The CE marking must be followed by the identification number of the notified body responsible for EC surveillance.

4.2. The manufacturer shall apply an approved quality system for the final inspection of the appliances and the tests, as specified in point 4.3, and is subject to EC surveillance as specified in point 4.4.

4.3. Quality system

4.3.1. Under this procedure, the manufacturer must lodge an application for approval of his quality system with a notified body of his choice for the appliances in question.

The application must include:

- the quality system documentation,
- an undertaking to carry out the obligations arising from the quality system as approved,
- an undertaking to maintain the approved quality system to ensure its continuing suitability and effectiveness,
- the documentation relating to the approved type and a copy of the EC type-examination certificate.

4.3.2. As part of the quality system, each appliance must be examined and appropriate tests as laid down in the applicable standard(s) referred to in

Article 5 or equivalent tests carried out to check its conformity with the essential requirements relating to it in this Directive.

All the elements, requirements and provisions adopted by the manufacturer must be documented in a systematic and logical manner in the form of measures, procedures and written instructions. This quality system documentation must permit a uniform interpretation of the quality programmes, plans, manuals and records.

The quality system documentation shall contain, in particular, an adequate description of:

- the quality objectives, the organisational structure and responsibilities of management and of their powers with regard to appliance quality,
- the checks and tests to be carried out after manufacture,
- the method of verifying the effective operation of the quality system.

4.3.3. The notified body shall examine and evaluate the quality system to determine whether it satisfies the requirements referred to in point 4.3.2. It will presume conformity with these requirements in respect of quality systems that implement the corresponding harmonised standard. It must notify the manufacturer of its decision and inform the other notified bodies thereof. The notification to the manufacturer must contain the conclusions of the examination, the name and address of the notified body and the reasoned assessment decision for the appliances concerned.

4.3.4. The manufacturer must keep the notified body which approved the quality system informed of any adaptation of the quality system made necessary, e.g. by new technology and quality concepts.

The notified body must examine the proposed changes and decide whether the amended quality system satisfies the relevant provisions or whether a reassessment is necessary. It must notify the manufacturer of its decision. The notification must contain the conclusions of the inspection and the reasoned assessment decision.

4.3.5. A notified body which withdraws approval of a quality system must inform the other notified bodies that it has done so and give reasons for its decision.

4.4. EC surveillance

4.4.1. The purpose of EC surveillance is to ensure that the manufacturer duly fulfils the obligations arising out of the approved quality system.

4.4.2. The manufacturer must allow the notified body access for inspection to the place of inspection, testing and storage and must provide it with all necessary information, in particular:

- the quality system documentation,
- the quality files such as inspection reports and test data, calibration data, report on qualifications of the staff concerned, etc.

4.4.3. The notified body must carry out a check at least once every two years to ensure that the manufacturer is maintaining and applying the approved quality system and must supply a report on the check to the manufacturer.

4.4.4. Furthermore, the notified body may make unannounced visits to the manufacturer. During these visits, the body may carry out tests on appliances or have them carried out. It must supply the manufacturer with an inspection report and, if appropriate, a test report.

4.4.5. The manufacturer may supply the notified body's report on request.

5. EC verification

5.1. EC verification is the procedure whereby the manufacturer or his authorised representative established within the Community ensures and declares that the appliances subject to the provisions of point 3 are in conformity to the type as described in the EC type-examination certification and satisfy the requirements of this Directive that apply to them.

5.2. The manufacturer or his authorised representative established within the Community must take all measures necessary in order that the manufacturing process ensures conformity of the appliances to the type as described in the EC type-examination certification and to the requirements of this Directive that apply to them. The manufacturer or his authorised representative established within the Community must affix the CE marking to each appliance and draw up a written declaration of conformity. The declaration of conformity may cover one or more appliances and must be kept by the manufacturer or his authorised representative established within the Community.

5.3. The notified body must carry out the appropriate examinations and tests in order to check the conformity of the appliance to the requirements of this Directive by examination and testing of every appliance, as specified in point 5.4, or by examination and testing of appliances on a statistical basis, as specified in point 5.5, at the choice of the manufacturer.

5.4. Verification by checking and testing of each appliance.

5.4.1. All appliances must be individually examined and appropriate tests, as set out in the relevant standard(s) referred to in Article 5, or equivalent tests, must be carried out in order to verify their conformity with the type as described in the EC type-examination certificate and the requirements of this Directive that apply to them.

5.4.2. The notified body must affix, or cause to be affixed, its identification number on each appliance and draw up a written certificate of conformity relating to the tests carried out. The certificate of conformity may cover one or more appliances.

5.4.3. The manufacturer or his authorised representative must ensure that he is able to supply the notified body's certificates of conformity on request.

5.5. Statistical verification

5.5.1. Manufacturers must present the appliances manufactured in the form of uniform batches and must take all necessary measures in order that the manufacturing process ensures the uniformity of each batch produced.

5.5.2. Statistical control is as follows:

Appliances are subject to statistical control by attributes. They should be grouped into identifiable batches consisting of units of a single model manufactured under the same conditions. A batch is examined at random intervals. The appliances constituting a sample are examined individually and appropriate tests, as laid down in the respective standard(s) referred to in Article 5, or equivalent tests are carried out to determine whether the batch is to be accepted or rejected.

A sampling system with the following characteristics is applied:

– a level of quality corresponding to a probability of acceptance of 95%, with a non-conformity percentage of between 0.5 and 1.5%,

– a limit quality corresponding to a probability of acceptance of 5%, with a percentage of non-conformity of between 5 and 10%.

5.5.3. Where batches are accepted, the notified body must affix, or cause to be affixed, its identification number to each appliance and draw up a written certificate of conformity relating to the tests carried out. All appliances in the batch may be placed on the market except for those products from the sample which were found not to be in conformity.

Where a batch is rejected, the notified body must take appropriate measures to prevent the placing on the market of that batch. In the event of frequent rejection of batches the notified body may suspend the statistical verification.

The manufacturer may, under the responsibility of the notified body, affix the latter's identification number during the manufacturing process.

5.5.4. The manufacturer or his authorised representative must ensure that he is able to supply the notified body's certificates of conformity on request.

6. EC unit verification

6.1. EC unit verification is the procedure whereby the manufacturer or his authorised representative established within the Community ensures and declares that the appliance concerned, which has been issued with the certificate referred to in point 2, conforms to the requirements of this Directive that apply to it. The manufacturer or his authorised representative must affix the CE marking to the appliance and draw up a written declaration of conformity which he must keep.

6.2. The notified body must examine the appliance and carry out the appropriate tests, taking account of the design documentation in order to ensure its conformity with the essential requirements of this Directive.

Then notified body must affix, or cause to be affixed, its identification number to the approved appliance and must draw up a written certificate of conformity concerning the tests carried out.

6.3. The aim of the technical documentation relating to the design of the instrument, as referred to in Annex IV, is to enable conformity to the requirements of this Directive to be assessed and the design, manufacture and operation of the appliance to be understood.

The design documentation referred to in Annex IV must be made available to the notified body.

6.4. If deemed necessary by the notified body, the examinations and tests may be carried out after installation of the appliance.

6.5. The manufacturer or his authorised representative must ensure that he is able to supply the notified body's certificates of conformity on request.

Annex III

CE conformity marking and inscriptions

1. The CE conformity marking consists of the initials "CE" as shown below:

The CE marking must be followed by the identification number of the notified body involved in the production control phase.

2. The appliance or its data plate must bear the CE marking together with the following inscriptions:

- the manufacturer's name or identification symbol,
- the trade name of the appliance.
- the type of electrical supply used, if applicable,
- the appliance category,
- the last two digits of the year in which the CE marking was affixed.

Information needed for installation purposes may be added to the nature of the appliance.

3. If the CE marking is reduced or enlarged the proportions given in the above graduated drawing must be respected.

The various components of the CE marking must have substantially the same vertical dimension, which may not be less than 5 mm.

Annex IV
Design documentation

The design documentation must contain the following information, in so far as is required by the notified body for assessment:
- a general description of the appliance,
- conceptual designs and manufacturing drawings and diagrams of components, sub-assemblies, circuits, etc,
- descriptions and explanations necessary for the understanding of the above, including the operation of the appliances,
- a list of the standards referred to in Article 5, applied in full or in part, and descriptions of the solutions adopted to meet the essential requirements where the standards referred to in Article 5 have not been applied.
- test reports,
- manuals for installation and use,

Where appropriate, the design documentation must contain the following elements:
- attestations relating to the equipment incorporated in the appliance,
- attestations and certificates relating to the methods of manufacture and/or inspection and/or monitoring of the appliance,
- any other document making iut possible for the notified body to improve its assessment.

Annex V
Minimum criteria for assessment of notified bodies

The notified bodies designated by the Member States must fulfil the following minimum conditions:
- availability of personnel and of the necessary means and equipment,
- technical competence and professional integrity of personnel.
- independence in carrying out tests, preparing reports, issuing certificates and performing the surveillance provided for this Directive, of management and technical staff in relation to all circles, groups or persons directly or indirectly involved in the field of the appliances.
- maintenance of professional secrecy by staff,
- possession of civil liability insurance unless that liability is covered by the State under national law.

Fulfilment of the conditions in the first two indents must be peridocally verified by the component authorities of the Member States or by bodies designated by the Member States.

COUNCIL DIRECTIVE 89/686/EEC

of 21 December 1989

on the approximation of the laws of the Member States relating to personal protective equipment*†

THE COUNCIL OF THE EUROPEAN COMMUNITIES,

Having regard to the Treaty establishing the European Economic Community, and in particular Article 100A thereof,

Having regard to the proposal from the Commission,

In cooperation with the European Parliament,

Having regard to the opinion of the Economic and Social Committee,

Whereas it is necessary to adopt measures with the aim of progressively establishing the internal market over a period expiring on 31 December 1992; whereas the internal market comprises an area without internal frontiers in which the free movement of goods, persons, services and capital is guaranteed;

Whereas various Member States have, over recent years, adopted provisions covering numerous items of personal protective equipment with a view in particular to safeguarding public health, improving safety at work and ensuring user protection;

Whereas these national provisions are often very detailed as regards the requirements relating to the design, manufacture, quality level, testing and certification of personal protective equipment with a view to the protection of individuals against injury and illness;

Whereas, in particular, the national provisions relating to safety at work make the use of personal protective equipment compulsory; whereas many requirements oblige employers to make appropriate personal protective equipment available to their staff in the absence or inadequacy of priority public protection measures;

Whereas national provisions relating to personal protective equipment differ significantly from one Member State to another; whereas they may thus constitute a barrier to trade with direct consequences for the creation and operation of the common market;

Whereas it is necessary to harmonise these different national provisions in order to ensure the free movement of these products, without in any way reducing the valid levels of protection already required in the Member States, and to provide for any necessary increase therein;

Whereas the provisions governing the design and manufacture of personal protective equipment laid down in this Directive which are fundamental, in particular, to attempts to ensure a safer working environment are without prejudice to provisions relating to the use of such equipment and the organisation of the health and safety of workers at the workplace;

Whereas this Directive defines only the basic requirements to be satisfied by personal protective equipment; whereas, in order to facilitate proof of conformity with those basic requirements, it is essential that harmonised European standards be available relating, in particular, to the design and manufacture of, and the specifications and test methods applicable to, personal protective equipment, since compliance therewith confers on these products a presumption of conformity with the abovementioned basic requirements;

---

*OJ 1989 L399/18.

†Amended by Dir 93/95 and Dir 93/68.

whereas such harmonised European standards are drawn up by private bodies and must retain the status of non-mandatory texts; whereas, to this end, the European Committee for Standardisation (CEN) and the European Committee for Electrotechnical Standardisation (CENELEC) are the competent bodies which have been authorised to adopt harmonised standards in accordance with the general guidelines governing cooperation between the Commission and those two institutions ratified on 13 November 1984; whereas, for the purposes of this Directive, a harmonised standard is a text containing technical specifications (a European standard or a harmonisation document) which has been adopted by one or both of the abovementioned bodies at the instigation of the Commission in accordance with Council Directive 83/189/EEC of 28 March 1983 (OJ 1983 L109/8) laying down a procedure for the provision of information in the field of technical standards and regulations, as amended by Directive 88/182/EEC (OJ 1988, L81/75), and pursuant to the abovementioned general guidelines;

Whereas, pending the adoption of harmonised standards, which will be very numerous because of the broad scope of application and the preparation of which within the deadline set for the creation of the internal market will involve a great deal of work, it would be advisable to maintain, on a transitional basis and subject to the requirements of the Treaty, the status quo as regards conformity with existing national standards for personal protective equipment not covered by a harmonised standard at the date of adoption of this Directive;

Whereas, given the general and horizontal nature of the role played by the Standing Committee set up pursuant to Article 5 of Directive 83/189/EEC in Community standardisation policy and, more particularly, its part in the preparation of standardisation applications and the operation of the existing European standardisation agreements, this Standing Committee is especially suited to the task of assisting the Commission in monitoring the conformity of harmonised standards throughout the Community;

Whereas compliance with these technical requirements must be monitored in order to ensure adequate user and third-party protection; whereas existing monitoring procedures may differ appreciably from one Member State to another; whereas, in order to avoid numerous checks which merely impede the free movement of personal protective equipment, provision should be made for the mutual recognition of inspections conducted by the Member States; whereas, in order to facilitate such recognition, it is necessary, in particular, to lay down harmonised Community procedures and to harmonise the criteria to be taken into account in selecting the bodies responsible for examination, monitoring and verification;

Whereas the legislative framework should be improved so that both sides of industry will make an effective and appropriate contribution to the process of standardisation,

HAS ADOPTED THIS DIRECTIVE:

Chapter 1
Scope, placing on the market and free movement

Article 1

1. This Directive applies to personal protective equipment, hereinafter referred to as "PPE".

It lays down the conditions governing its placing on the market and free movement within the Community and the basic safety requirements which PPE must satisfy in order to ensure the health protection and safety of users.

2. For the purposes of this Directive, PPE shall mean any device or appliance designed to be worn or held by an individual for protection against one or more health and safety hazards.

PPE shall also cover:

(a) a unit constituted by several devices or appliances which have been integrally combined by the manufacturer for the protection of an individual against one or more potentially simultaneous risks;
(b) a protective device or appliance combined, separably or inseparably, with personal non-protective equipment worn or held by an individual for the execution of a specific activity;
(c) interchangeable PPE components which are essential to its satisfactory functioning and used exclusively for such equipment.

3. Any system placed on the market in conjunction with PPE for its connection to another external, additional device shall be regarded as an integral part of that equipment even if the system is not intended to be worn or held permanently by the user for the entire period of risk exposure.

4. This Directive does not apply to:

- PPE covered by another directive designed to achieve the same objectives as this Directive with regard to placing on the market, free movement of goods and safety,
- the PPE classes specified in the list of excluded products in Annex I, independently of the reason for exclusion mentioned in the first indent.

Article 2

1. Member States shall take all appropriate measures to ensure that the PPE referred to in Article 1 may be placed on the market and brought into service only if it preserves the health and ensures the safety of users without prejudice to the health or safety of other individuals, domestic animals or goods, when properly maintained and used for its intended purpose.

2. This Directive shall be without prejudice to the right of Member States to lay down – in conformity with the Treaty – any requirements which they consider necessary to ensure user protection, provided that this does not give rise to modifications to PPE which could result in its non-conformity with the provisions of this Directive.

3. Member States shall not prevent the presentation at trade fairs, exhibitions and the like of PPE which is not in conformity with the provisions of this Directive, provided that an appropriate notice is displayed drawing attention to this fact and the prohibition on its acquisition and/or use for any purpose whatsoever until it has been brought into conformity by the manufacturer or his representative established in the Community.

Article 3

The PPE referred to in Article 1 must satisfy the basic health and safety requirements laid down in Annex II.

Article 4

1. Member States may not prohibit, restrict or hinder the placing on the market of PPE or PPE components which comply with the provisions of this Directive and which bear the CE marking attesting their conformity to all the provisions of this Directive, including the certification procedures in Chapter II.

2. Member States shall not prohibit, restrict or impede the placing on the market of PPE conditions which do not bear the CE marking, and which are intended to be incorporated in PPE, provided that they are not essential to its satisfactory functioning.

Article 5

1. Member States shall regard as in conformity with the basic requirements referred to in Article 3 the PPE referred to in Article 8 (3) bearing the CE marking with respect to which the manufacturer is able to produce, on demand, the declaration of conformity referred to in Article 12.

2. Member States shall presume that the PPE referred to in Article 8 (2) satisfies the basic requirements referred to in Article 3 if it bears the CE marking with respect to which the manufacturer is able to produce, on demand, not only the declaration referred to in Article 12 but also the certificate issued by the body of which

notification has been given in accordance with Article 9 attesting to their conformity to the relevant national standards, transposing the harmonised standards, assessed at the EC type examination level in accordance with the first indent of Article 10 (4) (a) and (b).

Where a manufacturer has not applied or has only partly applied the harmonised standards or where there are no such standards the certificate issued by the body of which notification has been given must state the conformity to the basic requirements in accordance with the second indent of Article 10 (4) (a) and (b).

3. . . .

4. Member States shall ensure that by 30 June 1991 appropriate steps are taken to enable both sides of industry to have an influence at national level on the process of formulating the harmonised standards and keeping them under review.

5. (a) Where the PPE is subject to other Directives concerning other aspects and which also provide for the affixing of the CE marking, the latter shall indicate that the PPE is also presumed to conform to the provisions of the other Directives.

(b) However, where one or more of these Directives allow the manufacturer, during a transitional period, to choose which arrangements to apply, the CE marking shall indicate conformity to the provisions only of those Directives applied by the manufacturer. In this case, particulars of the Directives applied, as published in the *Official Journal of the European Communities*, must be given in the documents, notices or instructions required by the Directives and accompanying such PPE.

Article 6

1. Should a Member State or the Commission consider that the harmonised standards referred to in Article 5 do not completely satisfy the relevant basic requirements referred to in Article 3, the Commission or the Member State concerned shall refer the matter to the committee created pursuant to Directive 83/189/EEC (OJ 1983 L109/8), setting out its reasons. The committee shall deliver an opinion without delay.

In the light of the committee's opinion, the Commission shall notify Member States of whether or not it is necessary to withdraw the standards concerned from publications made pursuant to Article 5.

2. The Standing Committee set up by Article 6(2) of Directive 89/392/EEC (OJ 1989 L183/9) may be apprised, in accordance with the procedure described below, of any matter to which the implementation and practical application of this Directive give rise.

The representative of the Commission shall submit to the committee a draft of the measures to be taken. The committee shall deliver its opinion on the draft, within a time limit which the chairman may lay down according to the urgency of the matter, if necessary by taking a vote.

The opinion shall be recorded in the minutes; in addition, each Member State shall have the right to ask to have its position recorded in the minutes.

The Commission shall take the utmost account of the opinion delivered by the committee. It shall inform the committee of the manner in which its opinion has been taken into account.

Article 7

1. If a Member State discovers that PPE bearing the CE marking and used in accordance with its intended purpose could compromise the safety of individuals, domestic animals or property, it shall take all necessary measures to remove that equipment from the market and prohibit the marketing or free movement thereof.

The Member State concerned shall immediately inform the Commission of such action, indicating the reasons for its decision and, in particular, stating whether non-conformity is due to:

(a) failure to comply with the basic

requirements referred to in Article 3;
(b) the unsatisfactory application of the standards referred to in Article 5;
(c) a shortcoming in the standards referred to in Article 5.

2. The Commission shall initiate discussions with the parties concerned as soon as possible. If, after such consultation, the Commission decides that the action taken was justified, it shall immediately inform the Member State concerned and all the other Member States to that effect. If, after such consultation, the Commission decides that the action taken was not justified, it shall immediately inform the Member State concerned and the manufacturer or his authorised representative established in the Community to that effect. If the decision referred to in paragraph 1 is in response to a shortcoming in the standards, the Commission shall refer the matter to the Committee referred to in Article 6 (1) if the Member State concerned intends to adhere to its decision and shall initiate the procedure referred to in Article 6 (2).

3. If PPE which is not in conformity with the relevant requirements bears the CE marking, the Member State concerned shall take the appropriate measures with regard to those responsible for affixing the mark and shall inform the Commission and the other Member States accordingly.

4. The Commission shall ensure that the Member States are kept informed of the progress and results of the procedure provided for in this article.

Chapter II
Certification procedures

Article 8

1. Before placing a PPE model on the market, the manufacturer or his authorised representative established in the Community shall assemble the technical documentation referred to in Annex III so that this can, if necessary, be submitted to the competent authorities.

2. Prior to the series production of PPE other than those referred to in paragraph 3, the manufacturer or his authorised representative established in the Community shall submit a model for EC type-examination as referred to in Article 10.

3. EC type-examination shall not be required in the case of PPE models of simple design where the designer assumes the user can himself assess the level of protection provided against the minimal risks concerned the effects of which, when they are gradual, can be safely identified by the user in good time.

This category shall cover exclusively PPE intended to protect the wearer against:

- mechanical action whose effects are superficial (gardening gloves, thimbles, etc),
- cleaning materials of weak action and easily reversible effects (gloves affording protection against diluted detergent solutions, etc),
- risks encountered in the handling of hot components which do not expose the user to a temperature exceeding 50 °C or to dangerous impacts (gloves, aprons for professional use, etc),
- atmospheric agents of a neither exceptional nor extreme nature (headgear, seasonal clothing, footwear, etc),
- minor impacts and vibrations which do not affect vital areas of the body and whose effects cannot cause irreversible lesions (light anti-scalping helmets, gloves, light footwear, etc),
- sunlight (sunglasses).

4. Production of PPE shall be subject:
(a) according to the manufacturer's choice, to one of the two procedures referred to in Article 11 in the case of PPE of complex design intended to protect against mortal danger or against dangers that may seriously and irreversibly harm the health, the immediate effects of which the designer assumes the user cannot identify in

sufficient time. This category shall cover exclusively:

- filtering respiratory devices for protection against solid and liquid aerosols or irritant, dangerous, taxic or radiotoxic gases,
- respiratory protection devices providing full insulation from the atmosphere, including those for use in diving,
- PPE providing only limited protection against chemical attack or against ionizing radiation,
- emergency equipment for use in high-temperature environments the effects of which are comparable to those of an air temperature of 100 °C or more and which may or may not be characterised by the presence of infrared radiation, flames or the projection of large amounts of molten material,
- emergency equipment for use in low-temperature environments the effects of which are comparable to those of an air temperature of −50 °C or less,
- PPE to protect against falls from a height,
- PPE against electrical risks and dangerous voltages or that used as insulation in high-tension work,

(b) the EC declaration of conformity referred to in Article 12 for all PPE.

Article 9

1. Member States shall notify the Commission and the other Member States of the bodies which they have appointed to carry out the procedures referred to in Article 8 together with the specific tasks which these bodies have been appointed to carry out and the identification numbers assigned to them beforehand by the Commission.

The Commission shall publish in the *Official Journal of the European Communities* a list of the notified bodies and their identification numbers and the tasks for which they have been notified. The Commission shall ensure that this list is kept up to date.

2. Member States shall apply the criteria laid down in Annex V in assessing the bodies to be indicated in such notification. Bodies meeting the assessment criteria laid down in the relevant harmonised standards shall be presumed to fulfil those criteria.

3. A Member State shall withdraw its approval from such a body if it establishes that the latter no longer satisfies the criteria referred to in Annex V. It shall inform the Commission and the other Member States of its action forthwith.

EC type-examination

Article 10

1. EC type-examination is the procedure whereby the approved inspection body establishes and certifies that the PPE model in question satisfies the relevant provisions of this Directive.

2. Application for EC type-examination shall be made by the manufacturer or his authorised representative to a single approved inspection body in respect of the model in question. The authorised representative shall be established in the Community.

3. The application shall comprise:

- the name and address of the manufacturer or his authorised representative and of the PPE production plant in question,
- the manufacturer's technical file referred to in Annex III.

It shall be accompanied by the appropriate number of specimens of the model to be approved.

4. The inspection body of which notification has been given shall conduct the EC type-examination in accordance with the undermentioned procedures:

(a) Examination of the manufacturer's technical file

- It shall examine the manufacturer's technical file to establish its suitability with respect to the harmonised standards referred to in Article 5.
- Where a manufacturer has not applied, or has only partly applied, the harmonised standards or where

there are no such standards, the body of which notification has been given must check the suitability of the technical specifications used by the manufacturer with respect to the basic requirements before examining the manufacturer's technical file to establish its suitability with respect to these technical specifications.

(b) Examination of the model

- When examining the model, the inspection body shall verify that it has been produced in accordance with the manufacturer's technical file and can be used in complete safety for its intended purpose.
- It shall conduct the necessary examinations and tests to establish the conformity of the model with the harmonised standards.
- Where a manufacturer has not applied or has only partly applied the harmonised standards or where there are no such standards the body of which notification has been given shall conduct the necessary examinations and tests to establish the conformity of the model with the technical specifications used by the manufacturer, subject to their being suitable with respect to these basic requirements.

5. If the model satisfies the relevant provisions, the inspection body shall draw up an EC type-examination certificate and shall notify the applicant to this effect. This certificate shall reproduce the findings of the examination, indicate any conditions attaching to its issue and incorporate the descriptions and drawings necessary for the identification of the approved model.

The Commission, the other approved inspection bodies and the other Member States may obtain a copy of the certificate and, in response to a reasoned request, a copy of the manufacturer's technical file and the reports of the examinations and tests conducted.

The file shall be held at the disposal of the competent authorities for 10 years following the placing of the PPE on the market.

6. Any inspection body which refuses to issue an EC type-examination certificate shall inform the other approved inspection bodies of this fact. An inspection body withdrawing an EC type-examination certificate shall inform the Member State which approved it, to this effect. That Member State shall then inform the other Member States and the Commission, setting out the reasons for the decision.

Checking of PPE manufactured

Article 11

A. "EC" quality control system for the final product

1. A manufacturer shall take all steps necessary to ensure that the manufacturing process, including the final inspection of PPE and tests, ensures the homogeneity of production and the conformity of PPE with the type described in the EC type-approval certificate and with the relevant basic requirements of this Directive.
2. A body of which notification has been given, chosen by a manufacturer, shall carry out the necessary checks. Those checks shall be carried out at random, normally at intervals of at least one year.
3. An adequate sample of PPE taken by the body of which notification has been given shall be examined and appropriate tests defined in the harmonised standards or necessary to show conformity to the basic requirements of this Directive shall be carried out to check the conformity of PPE.
4. Where a body is not the body that issued the relevant EC type-approval certificate it shall contact the body of which notification has been given in the event of difficulties in connection with the assessment of the conformity of samples.
5. The body of which notification has been given shall provide the manufacturer with a test report. If the report concludes that production is not homogeneous or that the PPE examined do not conform to the type described in the EC type-approval cer-

tificate or the relevant basic requirements, the body shall take measures appropriate to the nature of the fault or faults recorded and inform the Member States which gave notification thereof accordingly.

6. The manufacturer must be able to present, on request, the report of the body of which notification has been given.

B. System for ensuring EC quality of production by means of monitoring

1. *The system*

(a) Under this procedure the manufacturer submits an application for the approval of his quality-control system to a body of which notification has been given, of his choice.

That application shall include:

- all the information relating to the category of PPE concerned, including, where appropriate, documentation relating to the model approved,
- documentation on the quality-control system,
- the undertaking to maintain the obligations arising from the quality-control system and to maintain its adequacy and efficiency.

(b) Under the quality-control system, each PPE shall be examined and the appropriate tests referred to in Section A paragraph 3 shall be carried out to check their conformity to the relevant basic requirements of this Directive.

The documentation on the quality-control system shall in particular include an adequate description of:

- the quality objectives, the organisation chart, the responsibilities of executives and their powers in respect of product quality,
- the checks and tests which must be carried out after manufacture,
- the means to be employed to check the efficient operation of the quality-control system.

(c) The body shall assess the quality-control system to determine whether it satisfies the provisions referred to in paragraph 1 (b). It shall assume that quality-control systems applying the relevant harmonised standard satisfy those provisions.

The body carrying out audits shall make all necessary objective evaluations of the components of the quality-control system and shall check in particular whether the system ensures conformity of PPE manufactured with the approved model.

The decision shall be communicated to the manufacturer. It shall include the conclusions of the check and the reasoned assessment decision.

(d) The manufacturer shall inform the body which approved the quality-control system of any plan to alter the quality-control system.

The body shall examine the proposed changes and decide whether the altered quality-control system satisfies the relevant provisions. It shall communicate its decision to the manufacturer. The communication shall include the conclusions of the check and the reasoned assessment decision.

2. *Supervision*

(a) The purpose of supervision is to ensure that a manufacturer correctly fulfils the obligations arising from the approved quality-control system.

(b) The manufacturer shall authorise the body to have access, for purposes of inspection, to PPE inspection, testing and storage sites and shall provide the body with all requisite information, in particular:

- documentation on the quality-control system,
- technical documentation,
- quality control manuals.

(c) The body shall periodically carry out audits to ensure that the manufacturer is maintaining and applying the approved quality-control system and shall provide the manufacturer with a copy of the audit report.

(d) In addition, the body may make unannounced visits to the manufacturer. In the course of such visits the body shall provide the manufacturer with a report of the visit and, if appropriate, with an audit report.

(e) The manufacturer must be able to present, on request, the report of the body of which notification has been given.

EC declaration of production conformity

Article 12

The EC declaration of conformity is the procedure whereby the manufacturer or his authorised representative established in the EC:

1. draws up a declaration using the form laid down in Annex VI certifying that the PPE placed on the market are in conformity with the provisions of this Directive with a view to its submission to the competent authorities;
2. affixes the CE marking of conformity provided for by Article 13 to each PPE.

Chapter III CE marking

Article 13

1. The CE conformity marking shall consist of the initials "CE" in the form shown in the specimen in Annex IV. In the event of the involvement of a notified body in the production control phase as indicated in Article 11, its identification number shall be added.

2. The CE marking must be affixed to each piece of manufactured PPE so as to be visible, legible and indelible throughout the expected life of the PPE; however, if this is not possible in view of the characteristics of the product, the CE marking may be affixed to the packaging.

3. The affixing of markings on the PPE which are likely to deceive third parties as to the meaning and form of the CE marking shall be prohibited. Any other marking may be affixed to the PPE or its packaging provided that the visibility and legibility of the CE marking is not thereby reduced.

4. Without prejudice to Article 7:

(a) where a Member State establishes that the CE marking has been affixed unduly, the manufacturer or his authorised representative established within the Community shall be obliged to make the product conform as regards the provision concerning the CE marking and to end the infringement under the conditions imposed by the Member State;

(b) where non-conformity continues, the Member State must take all appropriate measures to restrict or prohibit the placing on the market of the product in question or to ensure that it is withdrawn from the market in accordance with the procedures laid down in Article 7.

Chapter IV Final provisions

Article 14

Any decision taken in implementation of this Directive and leading to restrictions on the marketing of PPE shall be accompanied by a detailed explanation of the grounds on which it is based. The interested party shall be notified of the decision without delay and informed of the possibilities for appeal under the legislation in force in the Member State concerned and of the deadlines for lodging such appeals.

Article 15

The Commission shall take the necessary steps to ensure that data concerning all the relevant decisions in connection with the management of this Directive are made available.

Article 16

1. Before 31 December 1991, Member States shall adopt and publish the laws, regulations and administrative provisions necessary in order to comply with this Directive. They shall forthwith inform the Commission thereof.

They shall apply the measures in question with effect from 1 July 1992.

2. Furthermore, Member States shall allow, for the period until 30 June 1995, the placing on the market and putting into service on PPE in conformity with the national regulations in force in their territory on 30 June 1992.

3. Member States shall communicate to the Commission the texts of the provisions of national law which they adopt in the field governed by this Directive.

Article 17

This Directive is addressed to the Member States.

Annex I

Exhaustive list of PPE classes not covered by this Directive

1. PPE designed and manufactured specifically for use by the armed forces or in the maintenance of law and order (helmets, shields, etc).
2. PPE for self-defence (aerosol canisters, personal deterrent weapons, etc).
3. PPE designed and manufactured for private use against:
   - adverse atmospheric conditions (headgear, seasonal clothing, footwear, umbrellas, etc),
   - damp and water (dish-washing gloves, etc),
   - heat (gloves etc).
4. PPE intended for the protection or rescue of persons on vessels or aircraft, not worn all the time.
5. Helmets and visors intended for users of two or three-wheeled motor vehicles.

Annex II

Basic health and safety requirements

1. General requirements applicable to all PPE

PPE must provide adequate protection against all risks encountered.

1.1. Design principles

1.1.1. *Ergonomics*

PPE must be so designed and manufactured that in the foreseeable conditions of use for which it is intended the user can perform the risk-related activity normally whilst enjoying appropriate protection of the highest possible level.

1.1.2. *Levels and classes of protection*

1.1.2.1. *Highest level of protection possible.* The optimum level of protection to be taken into account in the design is that beyond which the constraints imposed by the wearing of the PPE would prevent its effective use during the period of exposure to the risk or normal performance of the activity.

1.1.2.2. *Classes of protection appropriate to different levels of risk.* Where differing foreseeable conditions of use are such that several levels of the same risk can be distinguished, appropriate classes of protection must be taken into account in the design of the PPE.

1.2. Innocuousness of PPE

1.2.1. *Absence of risks and other "inherent" nuisance factors*

PPE must be so designed and manufactured as to preclude risks and other nuisance factors under foreseeable conditions of use.

1.2.1.1. *Suitable constituent materials.* PPE materials and parts, including any of their decomposition products, must not adversely affect user hygiene or health.

1.2.1.2. *Satisfactory surface condition of all PPE parts in contact with the user.* Any PPE part in contact or in potential contact with the user when such equipment is worn must be free of roughness, sharp edges, projections and the like which could cause excessive irritation or injuries.

1.2.1.3. *Maximum permissible user impediment.* Any impediment caused by PPE to movements to be made, postures to be adopted and sensory perception must be minimised; nor must PPE cause movements which endanger the user or other persons.

1.3. Comfort and efficiency

1.3.1. *Adaptation of PPE to user morphology*

PPE must be so designed and manufactured as to facilitate correct positioning on the user and to remain in place for the foreseeable period of use, bearing in mind ambient factors, movements to be made and postures to be adopted. For this purpose, it must be possible to optimise PPE adaptation to user morphology by all appropriate means, such as adequate adjustment and attachment systems or the provision of an adequate size range.

1.3.2. *Lightness and design strength*

PPE must be as light as possible without prejudicing design strength and efficiency.

Apart from the specific additional requirements which they must satisfy in order to provide adequate protection against the risks in question (see 3), PPE must be capable of withstanding the effects of ambient phenomena inherent under the foreseeable conditions of use.

1.3.3. *Compatibility of different classes or types of PPE designed for simultaneous use*

If the same manufacturer markets several PPE models of different classes or types in order to ensure the simultaneous protection of adjacent parts of the body against combined risks, these must be compatible.

1.4. Information supplied by the manufacturer

In addition to the name and address of the manufacturer and/or his authorised representative established in the Community, the notes that must be drawn up by the former and supplied when PPE is placed on the market must contain all relevant information on:

(a) storage, use, cleaning, maintenance, servicing and disinfection. Cleaning, maintenance or disinfectant products recommended by manufacturers must have no adverse effect on PPE or users when applied in accordance with the relevant instructions;
(b) performance as recorded during technical tests to check the levels or classes of protection provided by the PPE in question;
(c) suitable PPE accessories and the charateristics of appropriate spare parts;
(d) the classes of protection appropriate to different levels of risk and the corresponding limits of use;
(e) the obsolescence deadline or period of obsolescence of PPE or certain of its components;
(f) the type of packaging suitable for transport;
(g) the significance of any markings (see 2.12).
(h) where appropriate, the references of the Directives applied in accordance with Article 5 (6) (b);
(i) the name, address and identification number of the notified body

involved in the design stage of the PPE.

These notes, which must be precise and comprehensible, must be provided at least in the official language(s) of the Member State of destination.

2. Additional requirements common to several classes or types of PPE

2.1. PPE incorporating adjustment systems

If PPE incorporates adjustment systems, the latter must be so designed and manufactured as not to become incorrectly adjusted without the user's knowledge under the foreseeable conditions of use.

2.2. PPE "enclosing" the parts of the body to be protected

As far as possible, PPE "enclosing" the parts of the body to be protected must be sufficiently ventilated to limit perspiration resulting from use; if this is not the case, it must if possible be equipped with devices which absorb perspiration.

2.3. PPE for the face, eyes and respiratory tracts

Any restriction of the user's field of vision or sight by PPE for the face, eyes or respiratory trace must be minimised.

The degree of optical neutrality of the vision systems of these PPE classes must be compatible with the type of relatively meticulous and/or prolonged activites of the user.

If necessary, they must be treated or provided with facilities to prevent moisture formation.

PPE models intended for users requiring sight correction must be compatible with the wearing of spectacles or contact lenses.

2.4. PPE subject to ageing

If it is known that the design performances of new PPE may be significantly affected by ageing, the date of manufacture and/or, if possible, the date of obsolescence, must be indelibly inscribed on every PPE item or interchangeable component placed on the market in such a way as to preclude any misinterpretation; this information must also be indelibly inscribed on the packaging.

If a manufacturer is unable to give an undertaking with regard to the useful life of PPE, his notes must provide all the information necessary to enable the purchaser or user to establish a reasonable obsolescence date, bearing in mind the quality level of the model and the effective conditions of storage, use, cleaning, servicing and maintenance.

Where appreciable and rapid deterioration in PPE performance is likely to be caused by ageing resulting from the periodic use of a cleaning process recommended by the manufacturer, the latter must, if possible, affix a mark to each item of PPE placed on the market indicating the maximum number of cleaning operations that may be carried out before the equipment needs to be inspected or discarded; failing that, the manufacturer must give this information in his notes.

2.5. PPE which may be caught up during use

Where the foreseeable conditions of use include in particular the risk of the PPE being caught up by a moving object thereby creating a danger for the user, the PPE must possess an appropriate resistance threshold above which a constituent part will break and eliminate the danger.

2.6. PPE for use in explosive atmospheres

PPE intended for use in explosive atmospheres must be so designed and manufactured that it cannot be the source of an electric, electrostatic or impact-induced arc or spark likely to cause an explosive mixture to ignite.

2.7. PPE intended for emergency use or rapid installation and/or removal

These PPE classes must be so designed and manufactured as to minimise the time required for attachment and (or) removal.

Any integral systems permitting correct positioning on, or removal from, the user must be susceptible of rapid and easy operation.

2.8. PPE for use in very dangerous situations

The information notes supplied by the manufacturer together with PPE for use in the very dangerous situations referred to in Article 8 (4) (a) must include, in particular, data intended for the exclusive use of competent trained individuals who are qualified to interpret them and ensure their application by the user.

They must also describe the procedure to be adopted in order to verify that PPE is correctly adjusted and functional when worn by the user.

If PPE incorporates an alarm which is activated in the absence of the level of protection normally provided, this must be so designed and accommodated as to be perceived by the user in the conditions of use for which the PPE is marketed.

2.9. PPE incorporating components which can be adjusted or removed by the user

Any PPE components which can be adjusted or removed by the user for the purpose of replacement must be so designed and manufactured as to facilitate adjustment, attachment and removal without tools.

2.10. PPE for connection to another, external complementary device

If PPE incorporates a system permitting connection to another, complementary, device, the attachment mechanism must be so designed and manufactured as to enable it to be mounted only on appropriate equipment.

2.11. PPE incorporating a fluid circulation system

If PPE incorporates a fluid circulation system, the latter must be so chosen, or designed, and incorporated as to permit adequate fluid renewal in the vicinity of the entire part of the body to be protected, irrespective of user gestures, posture or movement under the foreseeable conditions of use.

2.12. PPE bearing one or more identification or recognition marks directly or indirectly relating to health and safety

The identification or recognition marks directly or indirectly relating to health and safety affixed to these types or classes of PPE must preferably take the form of harmonised pictograms or ideograms and must remain perfectly legible throughout the foreseeable useful life of the PPE. In addition, these marks must be complete, precise and comprehensible so as to prevent any misinterpretation; in particular, when such marks incorporate words or sentences, the latter must appear in the official language(s) of the Member State where the equipment is to be used.

If PPE (or a PPE component) is too small to allow all or part of the necessary marking to be affixed, the relevant information must be mentioned on the packing and in the manufacturer's notes.

2.13. PPE in the form of clothing capable of signalling the user's presence visually

PPE in the form of clothing intended for foreseeable conditions of use in which the user's presence must be visibly and individually signalled must have one (or more) judiciously positioned means of or devices for emitting direct or reflected visible radiation of appropriate luminous intensity and photometric and colorimetric properties.

2.14. "Multi-risk" PPE

All PPE designed to protect the user against several potentially simultaneous risks must be so designed and manufactured as to satisfy, in particular, the basic requirements specific to each of those risks (see 3).

3. Additional requirements specific to particular risks

3.1. Protection against mechanical impact

3.1.1. *Impact caused by falling or projecting objects and collision of parts of the body with an obstacle*

Suitable PPE for this type of risk must be sufficiently shock-absorbent to prevent injury resulting, in particular, from the crushing or penetration of the protected part, at least up to an impact-energy level above which the excessive dimensions or mass of the absorbing device would preclude effective use of the PPE for the foreseeable period of wear.

3.1.2. *Falls*

3.1.2.1. *Prevention of falls due to slipping.* The outsoles for footwear designed to prevent slipping must be so designed, manufactured or equipped with added elements as to ensure satisfactory adhesion by grip and friction having regard to the nature or state of the surface.

3.1.2.2. *Prevention of falls from a height.* PPE designed to prevent falls from a height or their effects must incorporate a body harness and an attachment system which can be connected to a reliable anchorage point. It must be designed so that under the foreseeable conditions of use the vertical drop of the user is minimised to prevent collision with obstacles and the braking force does not, however, attain the threshold value at which physical injury or the tearing or rupture of any PPE component which might cause the user to fall can be expected to occur.

It must also ensure that after braking the user is maintained in a correct position in which he may await help if necessary.

The manufacturer's notes must specify in particular all relevant information relating to:

- the characteristics required for the reliable anchorage point and the necessary minimum clearance below the user,
- the proper way of putting on the body harness and of connecting the attachment system to the reliable anchorage point.

3.1.3. *Mechanical vibration*

PPE designed to prevent the effects of mechanical vibrations must be capable of ensuring adequate attenuation of harmful vibration components for the part of the body at risk.

Under no circumstances must the effective value of the accelerations transmitted to the user by those vibrations exceed the limit values recommended in the light of the maximum foreseeable daily exposure of the part of the body at risk.

3.2. Protection against (static) compression of part of the body

PPE designed to protect part of the body against (static) compressive stress must be sufficiently capable of attenuating its effects to prevent serious injury or chronic complaints.

3.3. Protection against injury (abrasion, perforation, cuts, bites)

PPE constituent materials and other components designed to protect all or part of the body against superficial injury caused by machinery, such as abrasion, perforation, cuts or bites, must be so chosen or designed and incorporated as to ensure that these PPE classes provide sufficient resistance to abrasion, perforation and gashing (see also 3.1) under the foreseeable conditions of use.

3.4. Prevention of drowning (lifejackets, armbands and lifesaving suits)

PPE designed to prevent drowning must be capable of returning to the surface as quickly as possible, without danger to his health, a user who may be exhausted or unconscious after falling into a liquid medium, and of keeping him afloat in a position which permits breathing while awaiting help.

PPE may be wholly or partially inherently buoyant or may be inflated either by gas which can be manually or automatically released or orally.

Under the foreseeable conditions of use:

- PPE must, without prejudice to its satisfactory operation, be capable of withstanding the effects of impact with the liquid medium and the environmental factors inherent in that medium,
- inflatable PPE must be capable of inflating rapidly and fully.

Where particular foreseeable conditions of use so require, certain types of PPE must also satisfy one or more of the following additional requirements:

- it must have all the inflation devices referred to in the second subparagraph, and/or a light or sound-signalling device,
- it must have a device for hitching and attaching the body so that the user may be lifted out of the liquid medium,
- it must be suitable for prolonged use throughout the period of activity exposing the user, possibly dressed, to the risk of falling into the liquid medium or requiring his immersion in it.

3.4.1. *Buoyancy aids*

Clothing which will ensure an effective degree of buoyancy, depending on its foreseeable use, which is safe when worn and which affords positive support in water. In foreseeable conditions of use, this PPE must not restrict the user's freedom of movement but must enable him, in particular, to swim or take action to escape from danger or rescue other persons.

3.5. Protection against the harmful effects of noise

PPE designed to prevent the harmful effects of noise must be capable of attenuating the latter to such an extent that the equivalent sound levels perceived by the user do not under any circumstances exceed the daily limit values laid down by Council Directive 86/188/EEC of 12 May 1986 (OJ 1986 L137/28) on the protection of workers from the risks related to exposure to noise at work.

All PPE must bear labelling indicating the noise attenuation level and the value of the comfort index provided by the PPE; should this not be possible, the labelling must be fixed to the packaging.

3.6. Protection against heat and/or fire

PPE designed to protect all or part of the body against the effects of heat and/or fire must possess thermal insulation capacity and mechanical strength appropriate to foreseeable conditions of use.

3.6.1. *PPE constituent materials and other components*

Constituent materials and other components suitable for protection against radiant and convective heat must possess an appropriate coefficient of transmission of incident heat flux and be sufficiently incombustible to preclude any risk of spontaneous ignition under the foreseeable conditions of use.

Where the outside of these materials and components must be reflective, its reflective power must be appropriate to the intensity of the heat flux due to radiation in the infra-red range.

Materials and other components of equipment intended for brief use in high-temperature environments and of PPE which may be splashed by hot products such as large quantities of molten material must also possess sufficient thermal capacity to retain most of the stored heat until after the user has left the danger area and removed his PPE.

PPE materials and other components which may be splashed by large amounts of hot products must also possess sufficient mechanical-impact absorbency (see 3.1).

PPE materials and other components which may accidentally come into contact with flame and those used in the manufacture of fire-fighting equipment must also possess a degree of non-flammability corresponding to the risk class associated with the foreseeable conditions of use. They must not melt when exposed to flames nor contribute to flame propagation.

3.6.2. *Complete PPE ready for use*

Under the foreseeable conditions of use:

1. the quantity of heat transmitted by PPE to the user must be sufficiently low to prevent the heat accumulated during wear in the part of the body at risk from attaining, under any circumstances, the pain or health impairment threshold;
2. PPE must if necessary prevent liquid or steam penetration and must not cause burns resulting from contact between its protective integument and the user.

If PPE incorporates refrigeration devices for the absorption of incident heat by means of liquid evaporation or solid sublimation, their design must be such that any volatile substances released are discharged beyond the outer protective integument and not towards the user.

If PPE incorporates a breathing device, the latter must adequately fulfil the protective function assigned to it under the foreseeable conditions of use.

The manufacturer's notes accompanying each PPE model intended for brief use in high-temperature environments must in particular provide all relevant data for the determination of the maximum permissible user exposure to the heat transmitted by the equipment when used in accordance with its intended purpose.

3.7. Protection against cold

PPE designed to protect all or part of the body against the effects of cold must possess thermal insulating capacity and mechanical strength appropriate to the foreseeable conditions of use for which it is marketed.

3.7.1. *PPE constituent materials and other components*

Constituent materials and other components suitable for protection against cold must possess a coefficient of transmission of incident thermal flux as low as required under the foreseeable conditions of use. Flexible materials and other components of PPE intended for use in a low-temperature environment must retain the degree of flexibility required for the necessary gestures and postures.

PPE materials and other components which may be splashed by large amounts of cold products must also possess sufficient mechanical-impact absorbency (see 3.1).

3.7.2. *Complete PPE ready for use*

Under the foreseeable conditions of use:

1. the flux transmitted by PPE to the user must be sufficiently low to prevent the cold accumulated during wear at any point on the part of the body being protected, including the tips of fingers and toes in the case of hands or feet, from attaining, under any circumstances, the pain or health-impairment threshold;
2. PPE must as far as possible prevent the penetration of such liquids as rain water and must not cause injuries resulting from contact between its cold protective integument and the user.

If PPE incorporates a breathing device, this must adequately fulfil the protective function assigned to it under the foreseeable conditions of use.

The manufacturer's notes accompanying each PPE model intended for brief use in low-temperature environments must provide all relevant data concerning the maximum permissible user exposure to the cold transmitted by the equipment.

3.8. Protection against electric shock

PPE designed to protect all or part of the body against the effects of electric current must be sufficiently insulated against the voltages to which the user is likely to be exposed under the most unfavourable foreseeable conditions.

To this end, the constituent materials and other components of these PPE classes must be so chosen or designed and incorporated as to ensure that the leakage current measured through the protective integument under test conditions at voltages correlated with those likely to be encountered *in situ* is minimised and, at all events, below a maximum conventional permissible value which correlates with the tolerance threshold.

Together with their packaging, PPE types intended exclusively for use during work or activities in electrical installations which are or may be under tension must bear markings indicating, in particular, their protection class and (or) corresponding operating voltage, their serial number and their date of manufacture; a space must also be provided outside the protective integument of such PPE for the subsequent inscription of the date of entry into service and those of the periodic tests or inspections to be conducted.

The manufaturer's notes must indicate, in particular, the exclusive use for which these PPE types are intended and the nature and frequency of the dielectric tests to which they are to be subjected during their useful life.

3.9. Radiation protection

3.9.1. *Non-ionising radiation*

PPE designed to prevent acute or chronic eye-damage from sources of non-ionising radiation must be capable of absorbing or reflecting the majority of the energy radiated in the harmful wavelengths without unduly affecting the transmission of the innocuous part of the visible spectrum, the perception of contrasts and the ability to distinguish colours where required by the foreseeable conditions of use.

To this end, protective glass must be so designed and manufactured as to possess, for each harmful wave, a spectral transmission factor such that the radiant-energy illumination density capable of reaching the user's eye through the filter is minimised and, under no circumstances, exceeds the maximum permissible exposure value.

Furthermore, the glasses must not deteriorate or lose their properties as a result of the effects of radiation emitted under the foreseeable conditions of use and all marketed specimens must bear the protection-factor number corresponding to the spectral distribution curve of their transmission factor.

Glasses suitable for radiation sources of the same type must be classified in the ascending order of their protection factors and the manufacturer's notes must indicate, in particular, the transmission curves which make it possible to select the most appropriate PPE bearing in mind such inherent factors of the effective conditions of use as distance to source and the spectral distribution of the energy radiated at that distance.

The relevant protection-factor number must be marked on all specimens of filtering glasses by the manufacturer.

3.9.2. *Ionising radiation*

3.9.2.1. *Protection against external radioactive contamination.* PPE constituent materials and other components designed to protect all or part of the body against radioactive dust, gases, liquids or mixtures thereof must be so chosen or designed and incorporated as to ensure that this equipment effectively prevents the penetration of the contaminants under the foreseeable conditions of use.

Depending on the nature or condition of these contaminants, the necessary leak-tightness can be provided by the impermeability of the protective integument and/or by any other appropriate means, such as ventilation and pressurisation systems designed to prevent the back-scattering of these contaminants.

Any decontamination measures to which PPE is subject must not prejudice its possible re-use during the foreseeable useful life of these classes of equipment.

3.9.2.2. *Limited protection against external irradiation.* PPE intended to provide complete user protection against external irradiation or, failing this, adequate attenuation thereof, must be designed to counter only weak electron (e.g. beta) or weak photon (e.g. X, gamma) radiation.

The constituent materials and other components of these PPE classes must be so chosen or designed and incorporated as to provide the degree of user protection required by the foreseeable conditions of use without leading to an increase in exposure time as a result of the impedance of user gestures, posture or movement (see 1.3.2.).

PPE must bear a mark indicating the type and thickness of the constituent material(s) suitable for the foreseeable conditions of use.

3.10. Protection against dangerous substances and infective agents

3.10.1. *Respiratory protection*

PPE intended for the protection of the respiratory tract must make it possible to supply the user with breathable air when the latter is exposed to a polluted atmosphere and/or an atmosphere having inadequate oxygen concentration.

The breathable air supplied to the user by the PPE must be obtained by appropriate means, for example after filtration of the polluted air through the protective device or appliance or by a piped supply from an unpolluted source.

The constituent materials and other components of these PPE classes must be so chosen or designed and incorporated as to ensure appropriate user respiration and respiratory hygiene for the period of wear concerned under the foreseeable conditions of use.

The leak-tightness of the facepiece and the pressure drop on inspiration and, in the case of the filtering devices, purification capacity must be such as to keep contaminant penetration from a polluted atmosphere low enough not to be prejudicial to the health or hygiene of the user.

The PPE must bear the manufacturer's identification mark and details of the specific characteristics of that type of equipment which, in conjunction with the instructions for use, will enable a trained and qualified user to employ the PPE correctly.

The manufacturer's notes must also in the case of filtering devices, indicate the deadline for the storage of filters as new and kept in their original packaging.

3.10.2. *Protection against cutaneous and ocular contact*

PPE intended to prevent the surface contact of all or part of the body with dangerous substances and infective agents must be capable of preventing the penetration or diffusion of such substances through the protective integument under the foreseeable conditions of use for which the PPE is placed on the market.

To this end, the constituent materials and other components of these PPE classes must be so chosen, or designed and incorporated as to ensure, as far as possible, complete leak-tightness, which will allow where necessary prolonged daily use or, failing this, limited leak-tightness necessitating a restriction of the period of wear.

Where, by virtue of their nature and the foreseeable conditions of their use, certain dangerous substances or infective agents possess high penetrative power which limits the duration of the protection provided by the PPE in question, the latter must be subjected to standard tests with a view to their classification on the basis of efficiency. PPE which is considered to be in conformity with the test specifications must bear a mark indicating, in particular, the names or, failing this, the codes of the substances used in the tests and the corresponding standard period of protection. The manufacturer's notes must also contain, in particular, an explanation of the codes (if necessary), a detailed description of the standard tests and all appropriate information for the determination of the maximum permissible period of wear under the different foreseeable conditions of use.

3.11. Safety devices for diving equipment

1. Breathing equipment

The breathing equipment must make it possible to supply the user with a breathable gaseous mixture, under foreseeable conditions of use and taking account in particular of the maximum depth of immersion.

2. Where the foreseeable conditions of use so require, the equipment must comprise:

(a) a suit which protects the user against the pressure resulting from the depth of immersion (see 3.2.) and/or against cold (see 3.7);

(b) an alarm designed to give the user prompt warning of an approaching failure in the supply of breathable gaseous mixture (see 2.8);

(c) a life-saving suit enabling the user to return to the surface (see 3.4.1.).

Annex III
Technical documentation supplied by the manufacturer

The documentation referred to in Article 8 (1) must comprise all relevant data on the means used by the manufacturer to ensure that a PPE complies with the basic requirements relating to it.

In the case of PPE models referred to in Article 8 (2), the documentation must comprise in particular:

1. the manufacturer's technical file consisting of:

(a) overall and detailed plans of the PPE accompanied, where appropriate, by calculation notes and the results of prototype tests in so far as necessary for the verification of compliance with the basic requirements;

(b) an exhaustive list of the basic safety requirements and of the harmonised standards or other technical specifications referred to in Articles 3 and 5, taken into account in the design of the model;

2. a description of the control and test facilities to be used in the manufacturer's plant to check compliance of production PPE with the harmonised standards or other technical specifications and to maintain quality level;

3. a copy of the information notice referred to in Annex II, 1.4.

Annex IV

CE conformity marking and information

– The CE conformity marking shall consist of the intials "CE" taking the following form:

– If the CE marking is reduced or enlarged the proportions given in the above graduated drawing must be respected.

– The various components of the CE marking must have substantially the same vertical dimension, which may not be less than 5 mm. This minimum dimension may be waived for small-scale PPE.

Additional information

– The last two digits of the year in which the CE marking was affixed; this information is not required in the case of the PPE referred to in Article 8 (3).

Annex V

Conditions to be fulfilled by the bodies of which notification has been given

(Article 9 (2))

The bodies designated by the Member States must fulfil the following minimum conditions:

1. availability of personnel and of the necessary means and equipment;
2. technical competence and professional integrity of personnel;
3. independence, in carrying out the tests, preparing the reports, issuing the certificates and performing the surveillance provided for in the Directive, of staff and technical personnel in relation to all circles, groups or persons directly or indirectly concerned with PPE;
4. maintenance of professional secrecy by personnel;
5. subscription of a civil liability insurance unless that liability is covered by the State under national law.

Fulfilment of the conditions under 1 and 2 shall be verified at intervals by the competent authorities of the Member States.

Annex VI

Model EC declaration of conformity

The manufacturer or his authorised representative established in the Community:[1]

....

....

declares that the new PPE described hereafter[2]

....

....

....

is in conformity with the provisions of Council Directive 89/686/EEC and, where such is the case, with the national standard tranposing harmonised standard No .... (for the PPE referred to in Article 8 (3))

[1]Business name and full address; authorised representatives must also give the business name and address of the manufacturer.

[2]Description of the PPE (make, type, serial number, etc).

is identical to the PPE which is the subject of EC certificate of conformity No .... issued by[3] ....

....

....

is subject to the procedure set out in Article 11 point A or point B[4] of Directive 89/686/EEC under the supervision of the notified body[3] ....

....

....

Done at ...., on ....

....

Signature[5]

[3]Name and address of the approved body.

[4] Delete whichever is inapplicable.

[5] Name and position of the person empowered to sign on behalf of the manufacturer or his authorised representative.

COUNCIL DIRECTIVE 91/263/EEC

of 29 April 1991

on the approximation of the laws of the Member States concerning telecommunications terminal equipment, including the mutual recognition of their conformity

THE COUNCIL OF THE EUROPEAN COMMUNITIES,

Having regard to the Treaty establishing the European Economic Community, and in particular Article 100A thereof,

Having regard to the proposal from the Commission,[1]

In cooperation with the European Parliament,[2]

Having regard to the opinion of the Economic and Social Committee,[3]

Whereas Directive 86/361/EEC[4] introduced the initial stage of the mutual recognition of type approval for telecommunications terminal equipment and in particular in its Article 9 envisaged a further stage for full mutual recognition of type approval for terminal equipment;

Where Decision 87/95/EEC[5] sets out the measures to be implemented for the promotion of standardisation in Europe and the preparation and implementation of standards in the field of information technology and telecommunications;

Whereas the Commission has issued a Green Paper on the development of the common market for telecommunications services and equipment proposing to accelerate the introduction of the full mutual recognition of type approval as the measure vital for the development of a competitive Community-wide terminal market;

Whereas the Council, in its resolution of 30 June 1988 on the development of the common market for telecommunications services and equipment up to 1992,[6] considers as a major goal in the telecommunications policy the full mutual recognition of type approval for terminal equipment on the basis of the rapid development of common European conformity specifications;

Whereas the terminal equipment sector is a vital part of the telecommunications industry, which is one of the industrial mainstays of the economy in the Community;

Whereas harmonising conditions for the placing on the market of telecommunications terminal equipment will create the conditions for an open and unified market;

Whereas real, comparable access to third country markets for European manufacturers should preferably the achieved through multilateral negotiations within GATT, although bilateral talks between the Community and third countries may also contribute to this process;

Whereas the Council Resolution of 7 May 1985 provides for a new approach to technical harmonisation and standards;[7]

Whereas the scope of the Directive must be based on a general definition of the term "terminal equipment" so as

---

[1] OJ No C 211, 17. 8. 1989, p. 12.
[2] OJ No C 113, 7. 5. 1990; and OJ No C 19, 28. 1. 1991, p. 88.
[3] OJ No C 329, 30. 12. 1989, p. 1.
[4] OJ No L 217, 5. 8. 1986, p. 21.
[5] OJ No L 36, 7. 2. 1987, p. 31.
[6] OJ No C 257, 4. 10. 1988, p. 1.
[7] OJ No C 136, 4. 6. 1985, p. 1.

to allow the technical development of products;

Whereas Community law in its present form provides – notwithstanding one of the fundamental rules of the Community, namely the free movement of goods – that obstacles to movement within the Community, resulting from disparaties in national legislation relating to the marketing of products, must be accepted in so far as such requirements can be recognised as being necessary to satisfy imperative requirements; whereas, therefore, the harmonisation of laws in this case must be limited only to those requirements necessary to satisfy the essential requirements relating to terminal equipment; whereas these requirements must replace the relevant national requirements because they are essential;

Whereas the essential requirements must be satisfied in order to safeguard the general interest; whereas these requirements must be applied with discernment to take account of the state of the art at the time of manufacture and economic requirements;

Whereas Council Directive 73/23/EEC of 19 February 1973 on the harmonisation of the laws of the Member States relating to electrical equipment designed for use within certain voltage limits[8] and Council Directive 83/189/EEC of 28 March 1983 laying down a procedure for the provision of information in the field of technical standards and regulations,[9] as amended by Directive 88/182/EEC,[10] are applicable, *inter alia*, to the fields of telecommunications and information technology;

Whereas Council Directive 89/336/EEC of 3 May 1989 on the approximation of the laws of Member States relating to electromagnetic compatibility[11] is applicable, *inter alia*, to the fields of telecommunications and information technology; whereas it is, however, appropriate to delete the provisions of Directive 89/336/EEC in so far as they refer to the definition of telecommunications terminal equipment and to the conformity assessment procedures to be applied for such equipment;

Whereas in respect of the essential requirements and in order to help manufacturers to prove conformity to those requirements, it is desirable to have standards harmonised at European level to safeguard the general interest in the design and manufacture of terminal equipment and in order to allow checks of conformity to those requirements; whereas these standards harmonised at European level are drawn up by private-law bodies and must retain their non-binding status; whereas for this purpose the European Committee for Standardisation (CEN), the European Committee for Electrotechnical Standardisation (CENELEC) and the European Telecommunications Standards Institute (ETSI), are the bodies recognised as competent to adopt harmonised standards; whereas, within the meaning of this Directive, a harmonised standard is a technical specification (European standard or harmonisation document) adopted by one of these bodies, on the basis of a remit from the Commission in accordance with the provision of Directive 83/189/EEC, and in accordance with the general guidelines referred to above;

Whereas in respect of the essential requirements related to interworking with public telecommunications networks, and in cases where it is justified, through such networks, it is in general not possible to comply with such requirements other than by the application of unique technical solutions; whereas such solutions shall therefore be mandatory;

Whereas the proposals for common technical regulations are, as a general rule, drawn up on the basis of harmo-

[8] OJ No L 77, 26. 3. 1973, p. 29.
[9] OJ No L 109, 26. 4. 1983, p. 8.
[10] OJ No L 81, 26. 3. 1988, p. 75.
[11] OJ No L 139, 23. 5. 1989, p. 19.

nised standards, and, in order to ensure an appropriate technical coordination on a broad European basis, of additional consultations, in particular with the Telecommunications Regulations Application Committee (TRAC) set up by members of the European Conference of Postal and Telecommunications Administrations (CEPT) in a memorandum of understanding signed in 1991;

Whereas it is essential to ensure that notified bodies are of a high standard throughout the Community and meet minimum criteria of competence, impartiality and financial and other independence from clients;

Whereas it is appropriate to set up a committee bringing together parties directly concerned with the implementation of this Directive, in particular the national bodies designated for certifying conformity, to assist the Commission in executing the tasks entrusted to it by this Directive; whereas representatives from the telecommunication organisations, users, consumers, manufacturers, service providers and the trade unions should have the right to be consulted;

Whereas the Member States' responsibility for safety, health and the other aspects covered by the essential requirements on their territory must be recognised in a safeguard clause providing for adequate Community protection procedures;

Whereas the addressees of any decision taken under this Directive must be informed of the reasons for such a decision and the means of appeal open to them;

Whereas measures must be adopted with the aim of progressively establishing the internal market over a period expiring on 31 December 1992; whereas the internal market comprises an area without internal frontiers in which the free movement of goods, persons, services and capital is ensured,

HAS ADOPTED THIS DIRECTIVE:

Chapter 1
Scope, placing on the market and free circulation

Article 1

1. This Directive shall apply to terminal equipment.

2. For the purpose of this Directive:

- "public telecommunications network" means the public telecommunications infrastructure which permits the conveyance of signals between defined network termination points by wire, by microwave, by optical means or by other electromagnetic means,
- "terminal equipment" means equipment intended to be connected to the public telecommunications network, *i.e.*:

(a) to be connected directly to the termination of a public telecommunications network;
or
(b) to interwork with a public telecommunications network being connected directly or indirectly to the termination of a public telecommunications network

in order to send, process or receive information.

The system of connection may be wire, radio, optical or other electromagnetic system,

- "technical specification" means a specification contained in a document which lays down the characteristics required of a product such as levels of quality, performance, safety or dimensions, including the requirements applicable to the product as regards terminology, symbols, testing and test methods, packaging, marking and labelling,
- "standard" means a technical specification adopted by a recognised standards body for repeated or continuous application, compliance with which is not compulsory.

3. The intended purpose of the equip-

ment, shall be declared by the manufacturer or supplier of the equipment. However, terminal equipment within the meaning of paragraph 2 which makes use of a system of communication employing the radio frequency spectrum is presumed to be intended for connection to the public telecommunications network.

Article 2

1. Notwithstanding Article 1, equipment which is capable of being connected to the public telecommunications network, but is not intended for such a purpose, shall be accompanied by a manufacturer's or supplier's declaration, the model of which is to be found in Annex VIII and by the operating manual. At the time of placing the equipment on the market for the first time, a copy of such documentation shall be transmitted to the notified body referred to in Article 10 (1) in the Member State where this first placing on the market takes place. In addition, such equipment shall be subject to the provisions of Article 11 (4).
2. The manufacturer or supplier shall be prepared to justify once, at the request of any notified body referred to in Article 10 (1), the intended purpose of such equipment on the basis of its relevant technical characteristics, its functions and indications of the market segment it is intended for.

Article 3

1. Member States shall take all appropriate measures to ensure that terminal equipment may be placed on the market and put into service only if it bears the CE marking provided for in Article 11 attesting its conformity to the requirements of this Directive, including the conformity assessment procedures laid down in Chapter II and where it is properly installed and maintained and used for its intended purpose.
2. Member States shall also take all appropriate measures to ensure that equipment referred to in Article 2 may be placed and allowed to remain on the market only if it complies with the requirements laid down by this Directive for this equipment and may not be connected to the public telecommunications network within the meaning of Article 1 (2).
3. Member States shall also take all appropriate measures to ensure that terminal equipment or equipment referred to in Article 2 is disconnected from the public telecommunications network if it is not used for its intended purpose. Member States may moreover take all appropriate measures, according to their national laws, to prevent connection to the public telecommunications network of terminal equipment that is not used in conformity with its intended purpose.
4. (a) Where the terminal equipment is subject to other Directives concerning other aspects and which also provide for the affixing of the CE marking, the latter shall indicate that the equipment is also presumed to conform to the provision of those other Directives.
(b) However, where one or more of these Directives allow the manufacturer, during a transitional period, to choose which arrangements to apply, the CE marking shall indicate conformity to the provisions only of those Directives applied by the manufacturer. In this case, particulars of the Directives applied, as published in the *Official Journal of the European Community*, must be given in the documents, notices or instructions required by the Directives and accompanying the terminal equipment.

Article 4

Terminal equipment shall satisfy the following essential requirements:

(a) user safety, in so far as this requirement is not covered by Directive 73/23/EEC;
(b) safety of employees of public telecommunications networks operators, in so far as this requirement is not covered by Directive 73/23/EEC;
(c) electromagnetic compatibility requirements in so far as they are specific to terminal equipment;

(d) protection of the public telecommunications network from harm;
(e) effective use of the radio frequency spectrum, where appropriate;
(f) interworking of terminal equipment with public telecommunications network equipment for the purpose of establishing, modifying, charging for, holding and clearing real or virtual connection;
(g) interworking of terminal equipment via the public telecommunications network, in justified cases.

The cases where terminal equipment supports:

(i) reserved service according to Community law;
(ii) a service which the Council has decided that there should be Community-wide availability,
are considered as justified cases and the requirements concerning this interworking are determined in accordance with the procedure provided for in Article 14.

In addition, after consultation of representatives of the bodies referred to in Article 13 (3) and taking due account of the result of these consultations, the Commission may propose that this essential requirement is recognised as being justified for other terminal equipment in accordance with the procedure provided for in Article 14.

Article 5

Member States shall not impede the placing on the market and the free circulation and use on their territory of terminal equipment which complies with the provisions of this Directive.

Article 6

1. Member States shall presume compliance with the essential requirements referred to in Article 4 (a) and (b) in respect of terminal equipment which is in conformity with the national standards implementing the relevant harmonised standards, the references of which have been published in the *Official Journal of the European Communities*. Member States shall publish the references of such national standards.
2. The Commission shall, in accordance with the procedure laid down in Article 14, adopt:

- as a first step, the measure identifying the type of terminal equipment for which a common technical regulation is required, as well as the associated scope statement for that regulation, with a view to its transmission to the relevant standardisation bodies,
- as a second step, once they have been prepared by the relevant standardisation bodies, the corresponding harmonised standards, or parts thereof, implementing the essential requirements referred to in Article 4 (c) to (g) which shall be transformed into common technical regulations, compliance with which shall be mandatory and the reference of which shall be published in the *Official Journal of the European Communities*.

Article 7

Where a Member State or the Commission considers that the harmonised standards referred to in Article 6 exceed or do not entirely meet the essential requirements referred to in Article 4, the Commission or the Member State concerned shall bring the matter before the Committee referred to in Article 13, hereinafter referred to as 'the Committee', giving the reasons therefore. The Committee shall deliver an opinion as soon as possible.

In the light of the Committee's opinion and after consultation of the standing Committee set up by Directive 83/189/EEC, the Commission shall inform the Member States whether or not it is necessary to withdraw reference to those standards and any related technical regulations from the *Official Journal of the European Communities* and shall take the necessary steps to correct the shortcomings noted in the standards.

Article 8

1. Where a Member State finds that terminal equipment bearing the markings under the provision laid down in Chapter III does not comply with the relevant essential requirements when properly used in accordance with the purpose intended by the manufacturer, it shall take all appropriate measures to withdraw such products from the market or to prohibit or restrict their being placed on the market.

The Member State concerned shall immediately inform the Commission of any such measure indicating the reasons for its decision, and in particular whether non-compliance is due to:

(a) incorrect application of the harmonised standards or common technical regulations referred to in Article 6;

(b) shortcomings in the harmonised standards or common technical regulations referred to in Article 6 themselves.

2. The Commission shall enter into consultation with the parties concerned as soon as possible. Where, after such consultation, the Commission finds that any measure as referred to in paragraph 1 is justified it shall immediately so inform the Member State that took the action and the other Member States. Where the decision referred to in paragraph 1 is attributed to shortcomings in the harmonised standards or common technical regulations, the Commission, after consulting the parties concerned, shall bring the matter before the Committee within two months if the Member State which has taken the measure intends to maintain them, and shall initiate the procedure referred to in Article 7.

3. Where terminal equipment which does not comply with the relevant essential requirements bears the CE marking the competent Member State shall take appropriate action against whomsoever has affixed the mark and shall inform the Commission and the other Member States thereof.

4. The Commission shall keep the Member State informed of the progress and outcome of this procedure.

Chapter II
Conformity assessment

Article 9

1. According to the choice of the manufacturer or his authorised representative established within the Community, terminal equipment shall be subject to either the EC type-examination, as described in Annex I, or to the EC declaration of conformity, as described in Annex IV.

2. An EC type-examination as described in Annex I shall be accompanied by a declaration issued according to the EC declaration of conformity to type procedure as described in Annex II or Annex III.

3. The records and correspondence relating to the procedure referred to in this Article shall be in an official language of the Member State where the said procedure will be carried out, or in a language acceptable to the notified body involved.

4. Article 10 (4) of Directive 89/336/EEC is hereby deleted.

Article 10

1. Member States shall notify the Commission and the other Member States of the bodies established within the Community which they have designated for carrying out the certification, product checks and associated surveillance tasks pertaining to the procedures referred to in Article 9, together with the identification numbers assigned to them beforehand by the Commission.

Member States shall apply the minimum criteria set out in Annex V for the designation of such bodies. Bodies that satisfy the criteria fixed by the relevant harmonised standards shall be presumed to satisfy the criteria set out in Annex V.

2. Member States shall inform the Commission of test laboratories established in the Community which they have designated for carrying out tests

pertaining to the procedures referred to in Article 9. Notified bodies shall apply the criteria fixed by the appropriate parts of the relevant harmonised standards for the designation of such laboratories.

3. The Commission shall publish in the *Official Journal of the European Communities* a list of notified bodies together with their identification numbers and a list of test laboratories together with the tasks for which they have been designated and shall ensure that these lists are kept up to date.

4. A Member State has designated a notified body or a test laboratory under paragraph 1 or 2 shall annul the designation if the notified body or the test laboratory no longer meets the relevant criteria for designation. It shall immediately inform the other Member States and the Commission accordingly and withdraw the notification. Where a Member State or the Commission considers that a notified body or a test laboratory designated by a Member State does not meet the relevant criteria the matter shall be brought before the Committee referred to in Article 13, which shall give its opinion within three months; in the light of the Committee's opinion the Commission shall inform the Member State concerned of any changes needed if that notified body or test laboratory is to retain its recognised status.

5. In order to facilitate the determination of conformity of terminal equipment with technical regulations and standards, the notified bodies shall recognise documentation issued by third country relevant bodies, when agreements between the Community and the third country concerned have been concluded on the basis of a mutually satisfactory understanding.

6. The notified bodies referred to in Article 10 (1), when issuing an EC type-examination certificate as referred to in Annex I, followed by the appropriate document referred to in Annex II or III, or a decision on quality assurance assessment as referred to in Annex IV, issue at the same time an administrative approval for the connection of the concerned terminal equipment to the public telecommunications network.

Chapter III
CE marking of conformity and inscriptions

Article 11

1. The marking of terminal equipment complying with this Directive shall consist of the CE marking consisting of the initials CE, followed by the identification number of the notified body involved in the production control stage and a symbol indicating that the equipment is intended and is suitable to be connected to the public telecommunications network. The form of CE marking to be used, together with the other information, is shown in Annex VI.

2. The affixing of markings on the equipment which are likely to deceive third parties as to the meaning and form of the CE marking specified in Annexes VI and VII shall be prohibited. Any other marking may be affixed to the eqipment provided that the visibility and legibility of the CE marking is not thereby reduced.

3. Terminal equipment shall be identified by the manufacturer by means of type, batch and/or serial numbers and by the name of the manufacturer and/or supplier responsible for placing it on the market.

4. Equipment manufacturers or suppliers who place on the market equipment as referred to in Article 2 shall affix the symbol specified in Annex VII in such a way that it follows the CE marking and visually forms an integral part of the total marking. Article 11 (2) of Directive 93/68 states, "in Article 11 (4) the term 'the EC mark' is replaced by 'the initials CE as shown in Annex VI'."

Article 12

Without prejudice to Article 8:

(a) Where a Member State establishes

that the CE marking has been affixed unduly, the manufacturer or his authorised representative established within the Community shall be obliged to make the equipment conform as regards the provisions concerning the CE marking and to end the infringement under the conditions imposed by the Member State.

(b) where non-conformity continues, the Member State must take all appropriate measures to restrict or prohibit the placing on the market of the equipment in question or to ensure that it is withdrawn from the market in accordance with the procedures laid down in Article 8.

Chapter IV Committee

Article 13

1. The Commission shall be assisted by a Committee of an advisory nature composed of the representatives of the Member States and chaired by the representative of the Commission. The Committee shall be called the Approvals Committee for Terminal Equipment (ACTE).
2. The representative of the Commission shall submit to the Committee a draft of the measure to be taken. The Committee shall deliver its opinion on the draft, within a time limit which the chairman may lay down according to the urgency of the matter, if necessary by taking a vote.

The opinion shall be recorded in the minutes; in addition, each Member State shall have the right to ask to have its position recorded in the minutes.

The Commission shall take the utmost account of the opinion delivered by the Committee. It shall inform the Committee of the manner in which its opinion has been taken into account.

3. The Commission will periodically consult the representative of the telecommunications organisations, the consumers, the manufacturers, the service providers and trade unions and will inform the Committee on the outcome of such consultations, with a view to taking due account of the outcome.

Article 14

1. Notwithstanding Article 13 (1) and (2), the following procedure shall apply for matters covered by Article 4 (g) and 6 (2).
2. The representative of the Commission shall submit to the Committee established in Article 13 a draft of the measures to be taken as referred to in Article 4 (g) and 6 (2). The Committee shall deliver its opinion on the draft within a time limit which the chairman may lay down according to the urgency of the matter. The opinion shall be delivered by the majority laid down in Article 148 (2) of the Treaty in the case of decisions which the Council is required to adopt on a proposal from the Commission. The votes of the representatives of the Member States within the Committee shall be weighted in the manner set out in that Article. The chairman shall not vote.
3. The Commission shall adopt the measures envisaged if they are in accordance with the opinion of the Committee.
4. If the measures envisaged are not in accordance with the opinion of the Committee, or if no opinion is delivered, the Commission shall, without delay, submit to the Council a proposal relating to the measure to be taken. The Council shall act by qualified majority. If, within three months from the date of referral to it, the Council has not acted, the proposed measure shall be adopted by the Commission.

Chapter V Final and transitional provisions

Article 15

The Commission shall draw up every second year a report on the implementation of this Directive, including progress on drawing up the relevant harmonised standards and on transforming them into technical regulations, as well as any problems that have arisen in the course of implementation. The report will also outline the activities of the Committee, and assess progress in achieving an open competitive market for terminal equipment at Community level consistent with the essential requirements referred to in Article 4.

Article 16

1. Directive 86/361/EEC is hereby repealed, with effect from 6 November 1992. References made to the repealed Directive shall be construed as being made to this Directive.
2. Notwithstanding paragraph 1 and Article 10 (2), Member States may designate as test laboratories such bodies which have been notified under Directive 86/361/EEC, without applying the criteria of Article 10 (2) for a period of 18 months after the effective date of repeal of Directive 86/361/EEC, it being understood that these laboratories will continue to observe the criteria for which they were notified.
3. Notwithstanding paragraph 1, any type approval granted by Member States in accordance with Directive 86/361/EEC may remain valid under the legislation of the Member States within the criteria of validity appropriate to the original approval.
4. Notwithstanding paragraph 1, measures adopted under Directive 86/361/EEC shall be submitted to the Committee under the procedure of Article 14 for possible transposition into common technical regulations.

Article 17

1. Member States shall take the measure necessary to comply with this Directive not later than 6 November 1992. They shall forthwith inform the Commission thereof.

When Member States adopt these measures, they shall contain a reference to this Directive or shall be accompanied by such reference on the occasion of their official publication. The methods of making such a reference shall be laid down by the Member States.

2. Member States shall inform the Commission of the main provisions of domestic law which they adopt in the field governed by this Directive.

Article 18

This Directive is addressed to the Member States.

Done at Luxembourg, 29 April 1991.

For the Council
The President
R. GOEBBELS

Annex I
EC type-examination

1. EC type-examination is that part of the procedure whereby a notified body ascertains and attests that a specimen, representative of the production envisaged, meets the provisions of the Directive that apply to it.

2. The application for the EC type-examination shall be lodged by the manufacturer or his authorised representative established within the Community with a notified body of his choice.

The application shall include:

- the name and address of the manufacturer and, if the application is lodged by the authorised representative, his name and address in addition,
- a written declaration that the same application has not been lodged with any other notified body,
- the technical documentation, as described in point 3.

The applicant shall place at the disposal of the notified body a specimen, representative of the production envisaged and hereinafter called "type".[1] The notified body may request further specimens if needed for carrying out the test programme.

3. The technical documentation shall enable the conformity of the product with the essential requirements of the Directive to be assessed. It shall, as far as relevant for such assessment, cover the design, manufacture and operation of the product.

For example, the documentation shall contain as far as is relevant for assessment:

- a general type-description sufficient to identify the product preferably by provision of photographs,
- design and manufacturing drawings and lists of components, sub-assemblies, circuits, etc,
- descriptions and explanations necessary for the understanding of said drawings and lists and the operation of the product,
- a list of the standards referred to in Article 6, applied in full or in part, and descriptions of the solutions adopted to meet the essential requirements of the Directive when the standards referred to in Article 6 have not been applied,
- results of examinations carried out, etc,
- test reports,
- proposed user information or handbook.

4. The notified body shall:

4.1. examine the technical documentation, verify that the type has been manufactured in conformity with it and identify the elements which have been designed in accordance with the relevant provisions of the standards referred to in Article 6 (1), as well as the components of those standards;

4.2. perform, or have performed, the appropriate examinations and necessary tests to check whether the solutions adopted by the manufacturer meet the essential requirements of the Directive which are specified in Article 4 (a) and (b);

4.3. perform, or have performed, the appropriate examinations and necessary tests to check that the type meets the relevant common technical regulations specified in Article 6 (2);

4.4. agree with the applicant on the locations where the examinations and necessary tests are to be carried out.

5. Where the type meets the provisions of the Directive, the notified body shall issue an EC type-examination certificate to the applicant. The certificate shall contain the name and address of the manufacturer, conclusions of the

[1] A type may cover several versions of the product provided that the differences between the versions do not affect the level of safety and the other requirements concerning the performance of the product.

examination, conditions for its validity and the necessary data for identification of the approved type.

A list of the relevant parts of the technical documentation shall be annexed to the certificate and a copy kept by the notified body.

6. The applicant shall inform the notified body that holds the technical documentation concerning the EC type-examination certificate of all modifications to the approved product which must receive additional approval where such changes may affect the conformity with the essential requirements or the prescribed conditions for use of the product. This additional approval is given in the form of an addition to the original EC type-examination certificate.

7. Each notified body shall communicate to the other notified bodies the relevant information concerning the EC type-examination certificates and additions issued and withdrawn.

8. The other notified bodies may request copies of the EC type-examination certificates and/or their additions. The Annexes to the certificate shall be kept at the disposal of the other notified bodies.

9. The manufacturer or his authorised representative shall keep with the technical documentation copies of EC type-examination certificates and their additions for a period ending at least 10 years after the last product has been manufactured.

Where neither the manufacturer nor his authorised representative is established within the Community, the obligation to keep the technical documentation available shall be the responsibility of the person who places the product on the Community market.

Annex II

Conformity to type

1. Conformity to type is that part of the procedure whereby the manufacturer or his authorised representative established within the Community ensures and declares that the products concerned are in conformity with the type as described in the EC type-examination certificate and satisfy the requirements of the Directive that applies to them. The manufacturer or his authorised representative established within the Community shall affix the markings provided for in Article 11 (1) to each product and draw up a written declaration of conformity to type.

2. The manufacturer shall take all measures necessary to ensure that the manufacturing process assures compliance of the manufactured products with the type as described in the EC type-examination certificate and with the requirements of the Directive that apply to them.

3. The manufacturer or his authorised representative shall keep a copy of the declaration of conformity for a period ending at least 10 years after the last product has been manufactured.

Where neither the manufacturer nor his authorised representative is established within the Community, the obligation to keep the declaration of conforming to type available shall be the responsibility of the person who places the product on the Community market.

4. A notified body chosen by the manufacturer shall carry out, or have carried out, product checks at random intervals. An adequate sample of the final products, which may be taken on site by the notified body or on its behalf, shall be examined and appropriate tests shall be carried out to check the conformity of products with the relevant requirements of the Directive. In those cases where one or more of the products checked do not conform, the notified body shall take appropriate measures.

Annex III
Production quality assurance

1. Production quality assurance is the procedure whereby the manufacturer who satisfies the obligations of point 2 ensures and declares that the products concerned are in conformity with the type as described in the EC type-examination certificate and satisfy the requirements of the Directive that apply to them. The manufacturer or his authorised representative established within the Community shall affix the markings provided for in Article 11 (1) to each product and draw up a written declaration of conformity to type.

2. The manufacturer shall operate an approved quality system for production, final product inspection and testing as specified in point 3 and shall be subject to monitoring as specified in point 4.

3. Quality system

3.1. The manufacturer shall lodge an application for assessment of his quality system with a notified body of his choice, for the products concerned.

The application shall include:
- all relevant information for the product category envisaged,
- the documentation concerning the quality system,
- if applicable, the technical documentation of the approved type and a copy of the EC type-examination certificate.

3.2. The quality system shall ensure compliance of the products with the type as described in the EC type-examination certificate and with the requirements of the Directive that apply to them.

All the element, requirements and provisions adopted by the manufacturer shall be documented in a systematic and orderly manner in the form of written policies, procedures and instructions. The quality system documentation must permit a consistent interpretation of the quality programmes, plan, manuals and records.

It shall contain in particular an adequate description of:
- the quality objectives and the organisational structure, responsibilities and powers of the management with regard to product quality,
- the manufacturing, quality control and quality assurance techniques, processes and systematic actions that will be used,
- the examinations and tests that will be carried out before, during and after manufacture, and the frequency with which they will be carried out,
- the quality records, such as inspection reports and test data, calibration data, qualification reports of the personnel concerned, etc,
- the means to monitor the achievement of the required product quality and the effective operation of the quality system.

3.3. The notified body shall assess the quality system to determine whether it satisfies the requirements referred to in point 3.2. It shall presume conformity with these requirements in respect of quality systems that implement the relevant harmonised standard.[1]

The auditing team shall have at least one member with experience of evaluation in the product technology concerned. The evaluation procedure shall include an inspection visit to the manufacturer's premises.

The decision shall be notified to the manufacturer. The notification shall contain the conclusions of the examination and the reasoned assessment decision.

[1] This harmonised standard will be EN 29002, supplemented, if necessary, to take into account the specific nature of the procedure for which it is implemented.

3.4. The manufacturer shall undertake to fulfil the obligations arising out of the quality system as approved and to uphold it so that it remains adequate and efficient.

The manufacturer or his authorised representative shall keep the notified body that has approved the quality system informed of any intended updating of the quality system.

The notified body shall evaluate the modifications proposed and decide whether the amended quality system will still satisfy the requirements referred to in point 3.2 or whether a re-assessment is required.

It shall notify its decision to the manufacturer. The notification shall contain the conclusions of the examination and the reasoned assessment decision.

4. Surveillance under the responsibility of the notified body

4.1. The purpose of surveillance is to make sure that the manufacturer duly fulfils the obligations arising out of the approved quality system.

4.2. The manufacturer shall allow the notified body access for inspection purpose to the locations of manufacture, inspection and testing, and storage and shall provide it with all necessary information, in particular:

- the quality system documentation,
- the quality records, such as inspection reports and test data, calibration data, qualification reports of the personnel concerned, etc.

4.3. The notified body shall carry out audits at reasonable intervals to make sure that the manufacturer maintains and applies the quality system and shall provide an audit report to the manufacturer.

4.4. Additionally, the notified body may pay unexpected visits to the manufacturer. During such visits the notified body may carry out, or cause to be carried out, tests to verify that the quality system is functioning correctly, if necessary. The notified body shall provide the manufacturer with a visit report and, if a test has taken place, with a report.

5. The manufacturer shall, for a period ending at least 10 years after the last product has been manufactured, keep at the disposal of the national authorities:

- the documentation referred to in the second indent of point 3.1.,
- the updating referred to in the second paragraph of point 3.4.,
- the decisions and reports from the notified body which are referred to in the final paragraph of points 3.4., 4.3. and 4.4.

6.Each notified body referred to in Article 10 (1) shall make available to the other notified bodies referred to in that article the relevant information concerning the quality system approvals issued and withdrawn.

Annex IV
Full quality assurance

1. Full quality assurance is the procedure whereby the manufacturer who satisfies the obligations of point 2 ensures and declares that the products concerned satisfy the requirements of the Directive that apply to them. The manufacturer or his authorised representative shall affix the markings provided for in Article 11(1) to each product and draw up a written declaration of conformity.

2. The manufacturer shall operate an approved quality system for design, manufacture and final product inspection and testing as specified in point 3 and shall be subject to surveillance as specified in point 4.

3. Quality system

3.1. The manufacturer shall lodge an application for assessment of his quality system with a notified body.

The application shall include:

- all relevant information for the products envisaged,
- the quality system's documentation.

3.2. The quality system shall ensure compliance of the products with the requirements of the Directive that apply to them.

All the elements, requirements and provisions adopted by the manufacturer shall be documented in a systematic and orderly manner in the form of written policies, procedures and instructions. This quality system documentation shall ensure a common understanding of the quality policies and procedures such as a quality programmes, plans, manuals and records.

It shall contain in particular an adequate description of:

- the quality objectives and the organisational structure, responsibilities and powers of the management with regard to design and product quality,
- the technical specifications, including the harmonised standards and technical regulations as well as relevant test specifications that will be applied and, where the standards referred to in Article 6 (1) will not be applied in full, the means will be used to ensure that the essential requirements of the Directive that apply to the products will be met,
- the design control and design verification techniques, processes and systematic actions that will be used when designing the products pertaining to the product category covered,
- the corresponding manufacturing, quality control and quality assurance techniques, processes and systematic actions that will be used,
- the examinations and tests that will be carried out before, during and after manufacture, and the frequency with which they will be carried out; as well as the results of the tests carried out before manufacture where appropriate,
- the means by which it is ensured that the test and examination facilities respect the appropriate requirements for the performance of the necessary test,
- the quality records, such as inspection reports and test data, calibration data, qualification reports of the personnel concerned, etc,
- the means to monitor the achievement of the required design and product quality and the effective operation of the quality system.

3.3. The notified body shall assess the quality system to determine whether it satisfies the requirements referred to in point 3.2. It shall presume compliance with these requirements in respect of quality systems that implement the relevant harmonised standard.[1]

The notified body shall assess in particular whether the quality control system ensures conformity of the products with the requirements of the Directive in the light of the relevant documentation supplied in respect of points 3.1. and 3.2. including, where relevant, test results supplied by the manufacturer.

The auditing team shall have at least one member experienced as an assessor in the product technology concerned. The evaluation procedure shall include an assessment visit to the manufacturer's premises.

The decision shall be notified to the manufacturer. The notification shall contain the conclusions of the examination and the reasoned assessment decision.

3.4. The manufacturer shall undertake to fulfil the obligations arising out of the quality system as approved and to uphold it so that it remains adequate and efficient.

The manufacturer or his authorised representative shall keep the notified body that has approved the quality system informed of any intended updating of the quality system.

The notified body shall evaluate the modifications proposed and decide

---

[1] This harmonised standard shall be EN 29001, supplemented, if necessary, to take into account the specific nature of the products for which it is implemented.

whether the amended quality system will still satisfy the requirements referred to in point 3.2. or whether a re-assessment is required.

It shall notify its decision to the manufacturer. The notification shall contain the conclusions of the examination and the reasoned assessment decision.

4. EC surveillance under the responsibility of the notified body

4.1. The purpose of surveillance is to make sure that the manufacturer duly fulfils the obligations arising out of the approved quality system.

4.2. The manufacturer shall allow the notified body access for inspection purposes to the locations of design, manufacture, inspection and testing, and storage and shall provide it with all necessary information, in particular:

- the quality system documentation,
- the quality records as foreseen by the design part of the quality system, such as results of analyses, calculations, tests, etc,
- the quality records as foreseen by the manufacturing part of the quality system, such as inspection reports and test data, calibration data, qualification reports of the personnel concerned, etc.

4.3. The notified body shall carry out audits at reasonable intervals to make sure that the manufacturer maintains and applies the quality system and shall provide an audit report to the manufacture.

4.4. Additionally, the notified body may pay unexpected visits to the manufacturer. At the time of such visits, the notified body may carry out tests or have them carried out in order to check the proper functioning of the quality system where necessary; it shall provide the manufacturer with a visit report and, if a test has been carried out, with a test report.

5. The manufacturer shall, for a period ending at least 10 years after the last product has been manufactured, keep at the disposal of the national authorities:

- the documentation referred to in the second indent of point 3.1.,
- the updating referred to in the second paragraph of point 3.4.,
- the decisions and reports from the notified body which are referred to in the final paragraph of points 3.4., 4.3. and 4.4.

6. Each notified body referred to in Article 10 (1) shall make available to the other notified bodies referred to in that article the relevant information concerning quality system approvals including references to the product(s) concerned, issued and withdrawn.

Annex V

Minimum criteria to be taken into account by Member States when designating notified bodies in accordance with Article 10 (1)

1. The notified body, its director and the staff responsible for carrying out the tasks for which the notified body has been designated shall not be a designer, manufacturer, supplier or installer of terminal equipment, or a network operator or a service provider, nor the authorised representative of any of such parties. They shall not become directly involved in the design, construction, marketing or maintenance of terminal equipment, nor represent the parties engaged in these activities. This does not preclude the possibility of exchanges of technical information between the manufacturer and the notified body.

2. The notified body and its staff must

carry out the tasks for which the notified body has been designated with the highest degree of professional integrity and technical competence and must be free from all pressures and inducements, particularly financial, which might influence their judgment or the results of any inspection, especially from persons or groups of persons with an interest in such results.
3. The notified body must have at its disposal the necessary staff and facilities to enable it to perform properly the administrative and technical work associated with the tasks for which it has been designated.
4. The staff responsible for inspections must have:
- sound technical and professional training,
- satisfactory knowledge of the requirements of the tests or inspections that are carried out and adequate experience of such tests or inspections,
- the ability to draw up the certificates, records and reports required to authenticate the performance of the inspections.

5. The impartiality of inspection staff must be guaranteed. Their remuneration must not depend on the number of tests or inspections carried out nor on the results of such inspections.
6. The notified body must take out liability insurance unless its liability is assumed by the State in accordance with national law, or the Member State itself is directly responsible.
7. The staff of the notified body is bound to observe professional secrecy with regard to all information gained in carrying out its tasks (except *vis-à-vis* the competent administrative authorities of the State in which its activities are carried out) under this Directive or any provision of national law giving effect thereto.

Annex VI

Marking for terminal equipment referred to in Article 11 (1)

- The CE conformity marking shall consist of the initials "CE" taking the following form, followed by the additional information referred to in Article 11 (1);

(See *Official Journal of the European Communities* for character font)

- If the CE marking is reduced or enlarged the proportions given in the above graduated drawing must be respected.
- The various components of the CE marking must have substantially the same vertical dimension, which may not be less than 5 mm.

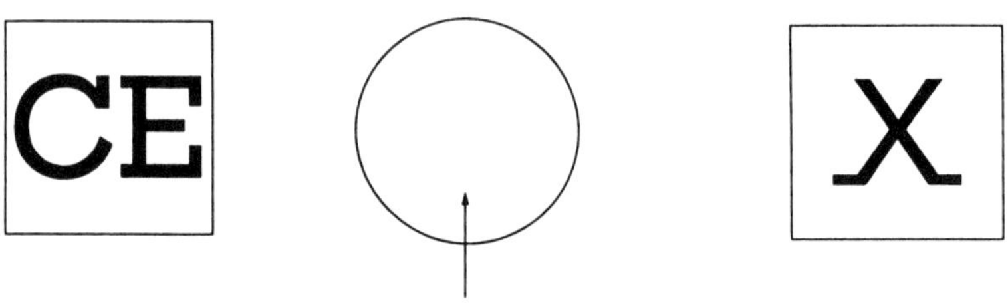

CE initials | Identification number of the notified body | Symbol of suitability for connection to the public telecommunications network

Annex VII

Marking for equipment referred to in Article 11 (4)

- If the CE marking is reduced or enlarged the proportions given in the above graduated drawing must be respected.
- The various components of the CE marking must have substantially the same vertical dimensions, which may not be less than 5 mm.

Annex VIII

Model of a declaration referred to in Article 2 (1)

The manufacturer/supplier[1] ....
....
....
Declares that[2] ....
....
is not intended to be connected to a public telecommunications network.

The connection of such equipment to a public telecommunications network in the Community Member State will be in violation of the national law implementing Directive 91/263/EEC on the approximation of the laws of the Member States concerning telecommunication terminal equipment, including the mutual recognition of their conformity.

DATE, PLACE AND SIGNATURE

---

[1] Name and address.
[2] Equipment identification.

COUNCIL DIRECTIVE 89/106/EEC

of 21 December 1988

on the approximation of laws, regulations and administrative provisions of the Member States relating to construction products[1, 2]

THE COUNCIL OF THE EUROPEAN COMMUNITIES,

Having regard to the Treaty establishing the European Economic Community, and in particular Article 100A thereof,

Having regard to the proposal from the Commission,

In cooperation with the European Parliament,

Having regard to the opinion of the Economic and Social Committee,

Whereas Member States are responsible for ensuring that building and civil engineering works on their territory are designed and executed in a way that does not endanger the safety of persons, domestic animals and property, while respecting other essential requirements in the interests of general well-being;

Whereas Member States have provisions, including requirements, relating not only to building safety but also to health, durability, energy economy, protection of the environment, aspects of economy, and other aspects important in the public interest;

Whereas these requirements, which are often the subject of national provisions laid down by law, regulation or administrative action, have a direct influence on the nature of construction products employed and are reflected in national product standards, technical approvals and other technical specifications and provisions which, by their disparity, hinder trade within the Community;

Whereas paragraph 71 of the White Paper on completing the internal market, approved by the European Council in June 1985, states that, within the general policy, particular emphasis will be placed on certain sectors, including construction; whereas the removal of technical barriers in the construction field, to the extent that they cannot be removed by mutual recognition of equivalence among all the Member States, should follow the new approach set out in the Council Resolution of 7 May 1985 (OJ 1985 C136/1) which calls for the definition of essential requirements on safety and other aspects which are important for the general well-being, without reducing the existing and justified levels of protection in the Member States;

Whereas the essential requirements constitute both the general and specific criteria with which construction works must comply; whereas such requirements are to be understood as requiring that the said works conform with an appropriate degree of reliability with one, some or all of these requirements when and where this is laid down in regulations;

Whereas, as a basis for the harmonised standards or other technical specifications at European level and for the drawing up or granting of European technical approval, interpretative documents will be established in order to give concrete form to the essential requirements at a technical level;

Whereas these essential requirements provide the basis for the preparation of harmonised standards at European

[1] OJ 1989 L40/12.
[2] Amended by Dir 93/68.

level for construction products; whereas, in order to achieve the greatest possible advantage for a single internal market, to afford access to that market for as many manufacturers as possible, to ensure the greatest possible degree of market transparency and to create the conditions for a harmonised system of general rules in the construction industry, harmonised standards should be established as far as, and as quickly as, possible; whereas these standards are drawn up by private bodies and must remain non-mandatory texts; whereas, for that purpose, the European Committee for Standardisation (CEN) and the European Committee for Electrotechnical Standardisation (CENELEC) are recognised as the competent bodies for the adoption of harmonised standards in accordance with the general guidelines for cooperation between the Commission and those two bodies signed on 13 November 1984; whereas, for the purposes of this Directive, a harmonised standard is a technical specification (European standard or harmonised document) adopted by one or both of those bodies upon a mandate given by the Commission in accordance with the provisions of Council Directive 83/189/EEC of 28 March 1983 (OJ 1983 L109/8) laying down a procedure for the provision of information in the field of technical standards and regulations;

Whereas the special nature of construction products requires the precise formulation of these harmonised standards; whereas it is therefore necessary to draw up interpretative documents in order to establish links between mandates for standards and the essential requirements; whereas harmonised standards, expressed as far as possible in terms of product performance, take account of these interpretative documents, which shall be drawn up in cooperation with the Member States;

Whereas performance levels and requirements to be fulfilled by products in future in the Member States shall be laid down in classes in the interpretative documents and in the harmonised technical specifications in order to take account of different levels of essential requirements for certain works and of different conditions prevailing in the Member States;

Whereas harmonised standards should include classifications that allow construction products which meet the essential requirements and which are produced and used lawfully in accordance with technical traditions warranted by local climatological and other conditions to continue to be placed on the market;

Whereas a product is presumed fit for use if it conforms to a harmonised standard, a European technical approval or a non-harmonised technical specification recognised at Community level; whereas, in cases where products are of little importance with respect to the essential requirements and where they deviate from existing technical specifications, their fitness for use can be certified by recourse to an approved body;

Whereas products thus considered fit for use are easily recognisable by the EC mark; whereas they must be allowed free movement and free use for their intended purpose throughout the Community;

Whereas, in the case of products where European standards cannot be produced or foreseen within a reasonable period of time or of products which deviate substantially from a standard, the fitness for use of such products may be proved by recourse to European technical approvals on the basis of common guidelines; whereas the common guidelines for the granting of European technical approvals will be adopted on the basis of the interpretative documents;

Whereas, in the absence of harmonised standards and European technical approvals, national or other non-harmonised technical specifications may be recognised as providing a suitable basis for a presumption that the essential requirements are met;

Whereas it is necessary to ensure the conformity of products with harmonised standards and with non-harmonised technical specifications recognised at European level by means of procedures of production control by manufacturers and of supervision, testing assessment and certification by independent qualified third parties, or by the manufacturer himself;

Whereas a special procedure should be provided as an interim measure for products where standards or technical approvals recognised at European level do not yet exist; whereas this procedure should facilitate recognition of the results of tests performed in another Member State according to the technical requirements of the Member State of destination;

Whereas a Standing Committee on Construction should be set up comprising experts designated by Member States to assist the Commission on questions arising from the implementation and practical application of this Directive;

Whereas the responsibility of Member States for safety, health and other matters covered by the essential requirements on their territory should be recognised in a safeguard clause providing for appropriate protective measures,

HAS ADOPTED THIS DIRECTIVE:

Chapter I
Field of application – definitions – requirements – technical specifications – free movement of goods

Article 1

1. This Directive shall apply to construction products in so far as the essential requirements in respect of construction works under Article 3 (1) relate to them.
2. For the purposes of this Directive, "construction product" means any product which is produced for incorporation in a permanent manner in construction works, including both buildings and civil engineering works.

"Construction products" are hereinafter referred to as "products"; construction works including both buildings and civil engineering works are hereinafter referred to as "works".

Article 2

1. Member States shall take all necessary measures to ensure that the products referred to in Article 1, which are intended for use in works, may be placed on the market only if they are fit for this intended use, that is to say they have such characteristics that the works in which they are to be incorporated, assembled, applied or installed, can, if properly designed and built, satisfy the essential requirements referred to in Article 3 when and where such works are subject to regulations containing such requirements.
2. (a) When products are subject to other Directives with regard to other aspects and which also provide for the affixing of the CE conformity marking, referred to in Article 4 (2), the latter shall indicate that the products are also presumed to conform to the provisions of those other Directives.
(b) However, where one or more of these Directives allow the manufacturer, during a transitional period, to choose which arrangements to apply, the CE marking shall indicate conformity only to the Directives applied by the manufacturer. In this case, particulars of the Directives applied, as published in the *Official Journal of the European Communities*, must be given in the documents, notices or instructions required by the Directives and accompanying such products.
3. When a further Directive concerns mainly other aspects and only to a minor extent the essential requirements of this Directive, that subsequent Directive shall contain provisions ensuring that it also covers the requirements of this Directive.

4. This Directive shall not affect the right of Member States to specify – with due observance of the provisions of the Treaty – the requirements they deem necessary to ensure that workers are protected when using products, provided it does not mean the products are modified in a way unspecified in this Directive.

Article 3

1. The essential requirements applicable to works which may influence the technical characteristics of a product are set out in terms of objectives in Annex I. One, some or all of these requirements may apply; they shall be satisfied during an economically reasonable working life.

2. In order to take account of possible differences in geographical or climatic conditions or in ways of life as well as different levels of protection that may prevail at national, regional or local level, each essential requirement may give rise to the establishment of classes in the documents referred to in paragraph 3 and the technical specifications referred to in Article 4 for the requirement to be respected.

3. The essential requirements shall be given concrete form in documents (interpretative documents) for the creation of the necessary links between the essential requirements laid down in paragraph 1 and the standardisation mandates, mandates for guidelines for European technical approval or the recognition of other technical specifications within the meaning of Articles 4 and 5.

Article 4

1. Standards and technical approvals shall, for the purposes of this Directive, be referred to as "technical specifications".

For the purposes of this Directive, harmonised standards shall be the technical specifications adopted by CEN, CENELEC or both, on mandates given by the Commission in conformity with Directive 83/189/EEC on the basis of an opinion given by the Committee referred to in Article 19 and in accordance with the general provisions concerning cooperation between the Commission and these two bodies signed on 13 November 1984.

2. Member States shall presume that the products are fit for use if they enable works in which they are employed, provided the latter are properly designed and built, to satisfy the essential requirements referred to in Article 3 where such products bear the CE marking indicating that they satisfy all the provisions of this Directive, including the conformity assessment procedures laid down in Chapter V and the procedure laid down in Chapter III. The CE marking shall indicate:

(a) that they comply with the relevant national standards transposing the harmonised standards, references to which have been published in the *Official Journal of the European Communities*. Member States shall publish the references of these national standards:

(b) that they comply with a European technical approval, delivered according to the procedure of Chapter III, or

(c) that they comply with the national technical specifications referred to in paragraph 3 in as much as harmonised specifications do not exist; a list of these national specifications shall be drawn up according to the procedure in Article 5 (2).

3. Member States may communicate to the Commission the texts of their national technical specifications which they regard as complying with the essential requirements referred to in Article 3. The Commission shall forward these national technical specifications forthwith to the other Member States. In accordance with the procedure provided for in Article 5 (2), it shall notify the Member States of those national technical specifications in respect of which there is presumption of conformity with the essential requirements referred to in Article 3.

This procedure will be initiated and

managed by the Commission in consultation with the committee referred to in Article 19.

Member States shall publish the references to these technical specifications. The Commission shall also publish them in the *Official Journal of the European Communities*.

4. Where a manufacturer, or his agent, established in the Community, has not applied, or has applied only in part, the existing technical specifications referred to in paragraph 2, which require, according to the criteria set out in Article 13 (4), the product to be submitted for a declaration of conformity as defined in Annex III (2) (ii), second and third possibilities, the corresponding decisions under Article 13 (4) and Annex III shall apply and such a product's fitness for use within the meaning of Article 2 (1) shall be established in accordance with the procedure set out in Annex III (2) (ii), second possibility.

5. The Commission, in consultation with the committee referred to in Article 19, shall draw up, manage and revise periodically a list of products which play a minor part with respect to health and safety and in respect of which a declaration of compliance with the "acknowledged rule of technology", issued by the manufacturer, will authorise such products to be placed on the market.

6. The CE marking signifies that products satisfy the requirements of paragraphs 2 and 4 of this Article. It is for the manufacturer or his authorised representative established within the Community to take responsibility for affixing the CE marking on the product itself, on a label attached to it, on its packaging, or on the accompanying commercial documents.

The model of the CE marking and conditions of its use are given in Annex III.

Products referred to in paragraph 5 shall not bear the CE marking.

Article 5

1. Where a Member State or the Commission is of the opinion that the harmonised standards or European technical approvals referred to in Article 4 (2), points (a) and (b), or the mandates referred to in Chapter II, do not satisfy the provisions of Articles 2 and 3, that Member State or the Commission shall notify the committee referred to in Article 19, setting out its reasons. The committee shall deliver an urgent opinion.

In the light of the opinion of the committee, and after consultation with the committee set up under Directive 83/189/EEC where it concerns harmonised standards, the Commission shall inform Member States if the standards or approvals concerned should be withdrawn in the publications referred to in Article 7 (3).

2. On reception of the communication referred to in Article 4 (3), the Commission shall consult the committee referred to in Article 19. In the light of the opinion of the committee, the Commission shall notify Member States whether the technical specification in question should benefit from the presumption of conformity and, if so, publish a reference to it in the *Official Journal of the European Communities*.

If the Commission or a Member State believes that a technical specification no longer fulfills the conditions necessary for presumption of conformity with the provisions of Articles 2 and 3, the Commission shall consult the committee referred to in Article 19. In the light of the opinion of the said committee, the Commission shall notify the Member States whether the national technical specification in question should continue to benefit from presumption of conformity, and, if not, whether the reference to it referred to in Article 4 (3) should be withdrawn.

Article 6

1. Member States shall not impede the free movement, placing on the market or use in their territory of products which satisfy the provisions of this Directive.

Member States shall ensure that the

use of such products, for the purpose for which they were intended, shall not be impeded by rules or conditions imposed by public bodies or private bodies acting as a public undertaking or acting as a public body on the basis of a monopoly position.

2. Member States shall, however, allow products not covered by Article 4 (2) to be placed on the market in their territory if they satisfy national provisions consistent with the Treaty until the European technical specifications referred to in Chapters II and III provide otherwise. The Commission and the committee referred to in Article 19 will monitor and review the development of the European technical specifications on a regular basis.

3. If the relevant European technical specifications, either themselves or on the basis of the interpretative documents referred to in Article 3 (3), distinguish between different classes corresponding to different performance levels, Member States may determine the performance levels also to be observed in their territory only within the classifications adopted at Community level and only subject to the use of all or some classes or one class.

Chapter II
Harmonised standards

Article 7

1. In order to ensure the quality of harmonised standards for products, the standards shall be established by the European standards organisations on the basis of mandates given by the Commission in accordance with the procedure laid down in Directive 83/189/EEC and, after consulting the committee referred to in Article 19, in accordance with the general provisions concerning cooperation between the Commission and these bodies signed on 13 November 1984.

2. The resulting standards shall be expressed as far as practicable in product performance terms, having regard to the interpretative documents.

3. Once the standards have been established by the European standards organisations, the Commission shall publish the references of the standards in the "C" series of the *Official Journal of the European Communities*.

Chapter III
European technical approval

Article 8

1. European technical approval is a favourable technical assessment of the fitness for use of a product for an intended use, based on fulfilment of the essential requirements for building works for which the product is used.

2. European technical approval may be granted to:

(a) products for which there is neither a harmonised standard, nor a recognised national standard, nor a mandate for a harmonised standard, and for which the Commission, after consulting the committee referred to in Article 19, considers that a standard could not, or not yet, be elaborated; and

(b) products which differ significantly from harmonised or recognised national standards.

Even in the case where a mandate for a harmonised standard has been issued, the provisions referred to in (a) do not exclude the granting of European technical approval for products for which guidelines for such approval exist. This shall apply until the entry into force of the harmonised standard in the Member States.

3. In special cases, the Commission may, as a derogation from paragraph 2 (a), authorise the issue of European technical approval, after consulting the committee referred to in Article 19, for products for which there is a mandate for a harmonised standard, or for which the Commission has established that a harmonised standard can be elaborated. The authorisation shall be valid for a fixed period.

4. European technical approval shall in general be issued for a five-year period. This period may be extended.

Article 9

1. European technical approval for a product shall be based on examinations, tests and an assessment on the basis of the interpretative documents referred to in Article 3 (3) and of the guidelines referred to in Article 11 for this product or the corresponding family of products.

2. Where guidelines referred to in Article 11 do not or not yet exist, European technical approval may be issued by reference to the relevant essential requirements and the interpretative documents where the assessment of the product is adopted by the approval bodies acting jointly in the organisation referred to in Annex II. If the approval bodies cannot agree, the matter shall be referred to the committee referred to in Article 19.

3. The European technical approval for a product shall be issued in a Member State in accordance with the procedure laid down in Annex II at the request of the manufacturer or his agent established in the Community.

Article 10

1. Each Member State shall notify the other Member States and the Commission of the names and addresses of the bodies which it has authorised to issue European technical approvals.

2. The approval bodies must satisfy the requirements of this Directive and in particular must be able:

- to assess the fitness for use of new products on the basis of scientific and practical knowledge,
- to take impartial decisions in relation to the interests of the manufacturers concerned or their agents, and
- to collate the contributions of all the interested parties in a balanced assessment.

3. The list of approval bodies which are competent to issue European technical approvals, as well as any amendments to that list, shall be published in the "C" series of the *Official Journal of the European Communities*.

Article 11

1. The Commission shall, after consulting the committee referred to in Article 19, issue mandates for establishing guidelines for European technical approval for a product or family of products to the organisation of approval bodies designated by the Member States.

2. The guidelines for European technical approval for a product or family of products should contain the following, in particular:

(a) a list of the relevant interpretative documents referred to in Article 3 (3);
(b) specific requirements for the products within the meaning of the essential requirements referred to in Article 3 (1);
(c) the test procedures;
(d) method of assessing and judging the results of the tests;
(e) the inspection and conformity procedures which must correspond to Article 13, 14 and 15;
(f) the period of validity of the European technical approval.

3. The guidelines for European technical approval shall, after consultation with the committee referred to in Article 19, be published by the Member States in their official language or languages.

Chapter IV
Interpretative documents

Article 12

1. The Commission shall, after consulting the committee referred to in Article 19, instruct technical committees in which the Member States participate to draw up the interpretative documents referred to in Article 3 (3).

2. The interpretative documents shall:

(a) give concrete form to the essential requirements laid down in Article 3 and in Annex I by harmonising the terminology and the technical bases and indicating classes or levels for each requirement where necessary and where the state of

scientific and technical knowledge so permits;
(b) indicate methods of correlating these classes or levels of requirement with the technical specifications referred to in Article 4, for example, methods of calculation and of proof, technical rules for project design, etc;
(c) serve as a reference for the establishment of harmonised standards and guidelines for European technical approval and for recognition of national technical specifications in accordance with Article 4 (3).

3. The Commission shall publish the interpretative documents in the "C" series of the *Official Journal of the European Communities* after soliciting the opinion of the committee referred to in Article 19.

Chapter V
Attestation of conformity

Article 13

1. The manufacturer, or his agent established in the Community, shall be responsible for the attestation that products are in conformity with the requirements of a technical specification within the meaning of Article 4.
2. Products that are the subject of an attestation of conformity shall benefit from the presumption of conformity with technical specifications within the meaning of Article 4. Conformity shall be established by means of testing or other evidence on the basis of the technical specifications in accordance with Annex III.
3. The attestation of conformity of a product is dependent on:
(a) the manufacturer having a factory production control system to ensure that production conforms with the relevant technical specifications; or
(b) for particular products indicated in the relevant technical specifications, in addition to a factory production control system, an approved certification body being involved in assessment and surveillance of the production control or of the product itself.

4. The choice of the procedure within the meaning of paragraph 3 for a given product or family of products shall be specified by the Commission, after consultation of the committee referred to in Article 19, according to:
(a) the importance of the part played by the product with respect to the essential requirements, in particular those relating to health and safety;
(b) the nature of the product;
(c) the effect of the variability of the product's characteristics on its serviceability;
(d) the susceptibility to defects in the product manufacture;

in accordance with the particulars set out in Annex III.

In each case, the least onerous possible procedure consistent with safety shall be chosen.

The procedure thus determined shall be indicated in the mandates and in the technical specifications or in the publication thereof.

5. In the case of individual (and non-series) production, a declaration of conformity in accordance with Annex III (2) (ii), third possibility, shall suffice, unless otherwise provided by the technical specifications for products which have particularly important implications for health and safety.

Article 14

1. In accordance with Annex III, the procedures described shall lead:
(a) in the case of Article 13 (3) (a), to the production of a declaration of conformity for a product by the manufacturer, or his agent established in the Community; or
(b) in the case of Article 13 (3) (b), to the issue by an approved certification body of a certificate of conformity for a system of production control and surveillance or for the product itself.

Detailed rules for the implementation of the procedures of attestation of conformity are given in Annex III.

2. The manufacturer's declaration of

conformity or the certificate of conformity shall entitle the manufacturer, or his agent established in the Community, to affix the corresponding CE marking on the product itself, on a label attached to it, on its packaging or on the accompanying commercial documents. The model of the EC mark and the rules for its use in respect of each of the procedures of attestation of conformity are given in Annex III.

Article 15

1. Member States shall ensure that the CE marking is correctly used.
2. Without prejudice to Article 21:
(a) where a Member State establishes that the CE marking has been affixed unduly, the manufacturer or his agent established within the Community shall be obliged to make the product conform as regards the provisions concerning the CE marking and to end the infringement under conditions imposed by the Member State;
(b) where non-conformity continues, the Member State must take all appropriate measures to restrict or prohibit the placing on the market of the product in question or to ensure that it is withdrawn from the market in accordance with the procedures laid down in article 21.
3. Member States shall take the measures necessary to prohibit the affixing to products or their packaging of markings which are likely to deceive third parties as to the meaning and form of the CE marking. Any other marking may be affixed to the construction products on a label fixed to the product packaging or on the accompanying commercial documents provided that the visibility and legibility of the CE marking is not thereby reduced.

Chapter VI
Special procedures

Article 16

1. In the absence of technical specifications, as defined in Article 4, for any given product, the Member State of destination shall, on request in individual cases, consider the product to be in conformity with the national provisions in force if they have satisfied tests and inspections carried out by an approved body in the producing Member State according to the methods in force in the Member State of destination or recognised as equivalent by that Member State.
2. The producing Member State shall inform the Member State of destination, in accordance with whose provisions the tests and inspections are to be carried out, of the body it intends to approve for this purpose. The Member State of destination and the producing Member State shall provide each other with all necessary information. On conclusion of this exchange of information the producing Member State shall approve the body thus designated. If a Member State has misgivings, it shall substantiate its position and inform the Commission.
3. Member States shall ensure that the designated bodies afford one another all necessary assistance.
4. Where a Member State establishes that an approved body is not carrying out the tests and inspections properly in conformity with its national provisions, it shall notify the Member State in which the body is approved thereof. That Member State shall inform the notifying Member State within a reasonable time limit of what action has been taken. If the notifying Member State does not consider the action taken to be sufficient, it may prohibit the placing on the market and use of the product in question or make it subject to special conditions. It shall inform the other Member State and the Commission thereof.

Article 17

Member States of destination shall attach the same value to reports and attestations of conformity issued in the producing Member State in accordance with the procedure referred to in

Article 16, as they do to their own corresponding national documents.

Chapter VII
Approved bodies

Article 18

1. Member States shall notify the Commission and the other Member States of the certification and inspection bodies and the testing laboratories which they have designated for the tasks which must be carried out for the purposes of technical approval, certificates of conformity, inspections and tests, in accordance with this Directive, together with their names and addresses and the identification numbers assigned to them beforehand by the Commission.

The Commission shall publish in the *Official Journal of the European Communities* a list of the notified bodies and laboratories with their identification numbers and the tasks and products for which they have been notified. The Commission shall ensure that this list is kept up to date.

2. Certification bodies, inspection bodies and testing laboratories shall comply with the criteria laid down in Annex IV.

3. Member States shall indicate the products which fall within the competence of the bodies and laboratories referred to in paragraph 1 and the nature of the tasks to be assigned to them.

Chapter VIII
Standing Committee on Construction

Article 19

1. A Standing Committee on Construction is hereby set up.

2. The committee shall be made up of representatives appointed by the Member States. It shall be chaired by a representative of the Commission. Each Member State shall appoint two representatives. The representatives may be accompanied by experts.

3. The committee shall draw up its own rules of procedure.

Article 20

1. The committee referred to in Article 19 may, at the request of its chairman or a Member State, examine any question posed by the implementation and the practical application of this Directive.

2. The provisions necessary for:

(a) the establishment of classes of requirements in so far as they are not included in the interpretative documents and the establishment of the procedure for attesting conformity in mandates for standards pursuant to Article 7 (1) and guidelines for approvals pursuant to Article 11 (1);

(b) the giving of instructions for the drawing-up of interpretative documents pursuant to Article 12 (1) and decisions on interpretative documents pursuant to Article 12 (3);

(c) the recognition of national technical specifications in accordance with Article 4 (3);

shall be adopted in accordance with the procedure laid down in paragraphs 3 and 4.

3. The representative of the Commission shall submit to the committee a draft of the measures to be taken. The committee shall deliver its opinion on the draft within a time limit which the chairman may lay down according to the urgency of the matter. The opinion shall be delivered by the majority laid down in Article 148 (2) of the Treaty in the case of decisions which the Council is required to adopt on a proposal from the Commission. The votes of the representatives of the Member State within the committee shall be weighed in the manner set out in that Article. The chairman shall not vote.

4. The Commission shall adopt the measures envisaged if they are in accordance with the opinion of the committee.

If the measures envisaged are not in accordance with the opinion of the committee, or if no opinion is delivered, the Commission shall, without delay, submit to the Council a proposal relating to the measures to be taken. The Council shall act by qualified majority.

If, within three months of the proposal being submitted to it, the Council has not acted, the proposed measures shall be adopted by the Commission.

Chapter IX Safeguard clause

Article 21

1. Where a Member State ascertains that a product declared to be in conformity with the terms of this Directive does not comply with Articles 2 and 3, it shall take all appropriate measures to withdraw those products from the market, prohibit the placing thereof on the market or restrict free movement thereof.

The Member State concerned shall immediately inform the Commission of any such measure, indicating the reasons for its decision, and in particular whether non-conformity is due to:

(a) failure to comply with Articles 2 and 3, where the product does not meet the technical specifications referred to in Article 4;
(b) incorrect application of the technical specifications referred to in Article 4;
(c) shortcomings in the technical specifications referred to in Article 4 themselves.

2. The Commission shall carry out a consultation of the parties concerned as soon as possible. Where the Commission finds, after this consultation, that the action is justified, it shall immediately so inform the Member State that took the action as well as the other Member States.

3. Where the decision referred to in paragraph 1 is attributed to shortcomings in the standards or technical specifications, the Commission, after consulting the parties concerned, shall bring the matter before the committee referred to in Article 19, as well as the committee set up under Directive 83/189/EEC in the case of shortcomings in a harmonised standard, within two months if the Member State which has taken the measures intends to uphold them, and shall start the procedures referred to in Article 5 (2).

4. The Member State concerned shall take appropriate action against whomsoever made the declaration of conformity and shall inform the Commission and the other Member States thereof.

5. The Commission shall ensure that the Member States are kept informed of the progress and outcome of this procedure.

Chapter X Final provisions

Article 22

1. Member States shall bring into force the laws, regulations and administrative provisions necessary to comply with the provisions of this Directive within 30 months of its notification. (This Directive was notified to the Member States on 27 December 1988.) They shall forthwith inform the Commission thereof.

2. Member States shall communicate to the Commission the texts of the provisions of national law which they adopt in the field governed by this Directive.

Article 23

At the latest by 31 December 1993, the Commission, in consultation with the committee referred to in Article 19, shall re-examine the practicability of the procedures laid down by this Directive and, where necessary, submit proposals for appropriate amendments.

Article 24

This Directive is addressed to the Member States.

Annex I

Essential requirements

The products must be suitable for construction works which (as a whole and in their separate parts) are fit for their intended use, account being taken of economy, and in this connection satisfy the following essential requirements where the works are subject to regulations containing such requirements. Such requirements must, subject to normal maintenance, be satisfied for an economically reasonable working life. The requirements generally concern actions which are forseeable.

1. Mechanical resistance and stability

The construction works must be designed and built in such a way that the loadings that are liable to act on it during its constructions and use will not lead to any of the following:

(a) collapse of the whole or part of the work;

(b) major deformations to an inadmissible degree;

(c) damage to other parts of the works or to fittings or installed equipment as a result of major deformation of the load-bearing construction;

(d) damage by an event to an extent disproportionate to the original cause.

2. Safety in case of fire

The construction works must be designed and built in such a way that in the event of an outbreak of fire:

- the load-bearing capacity of the construction can be assumed for a specific period of time,
- the generation and spread of fire and smoke within the works are limited,
- the spread of the fire to neighbouring construction works is limited,
- occupants can leave the works or be rescued by other means,
- the safety of rescue teams is taken into consideration.

3. Hygiene, health and the environment

The construction work must be designed and built in such a way that it will not be a threat to the hygiene or health of the occupants or neighbours, in particular as a result of any of the following:

- the giving-off of toxic gas,
- the presence of dangerous particles or gases in the air,
- the emission of dangerous radiation,
- pollution or poisoning of the water or soil,
- faulty elimination of waste water, smoke, solid or liquid wastes,
- the presence of damp in parts of the works or on surfaces within the works.

4. Safety in use

The construction work must be designed and built in such a way that it does not present unacceptable risks of accidents in service or in operation such as slipping, falling, collision, burns, electrocution, injury from explosion.

5. Protection against noise

The construction works must be designed and built in such a way that noise perceived by the occupants or people nearby is kept down to a level that will not threaten their health and will allow them to sleep, rest and work in satisfactory conditions.

6. Energy economy and heat retention

The construction works and its heating, cooling and ventilation installations must be designed and built in such a way that the amount of energy required in use shall be low, having regard to the climatic conditions of the location and the occupants.

Annex II

European technical approval

1. A request for approval may be made by a manufacturer, or his agent established in the Community, only to a single body authorised for this purpose.

2. The approval bodies designated by the Member States form an organisation. In the performance of its duties, this organisation is obliged to work in close coordination with the Commission, which shall consult the committee referred to in Article 19 of the Directive on important matters. Where a Member State has designated more than one approval body, the Member State shall be responsible for coordinating such bodies; it shall also designate the body which shall be spokesman in the organisation.

3. The common procedural rules for making the request, the preparation and the granting of approvals are drawn up by the organisation comprising the designated approval bodies. The common procedural rules are adopted by the Commission on the basis of the opinion of the committee in accordance with Article 20.

4. In the framework of the organisation comprising them, the approval bodies shall afford each other all necessary support. This organisation is also responsible for coordination on specific questions of technical approval. If necessary, the organisation shall establish sub-groups for this purpose.

5. The European technical approvals are published by the approval bodies, which notify all other approved bodies. At the request of an authorised approval body, a complete set of supporting documents for an approval which has been granted is to be forwarded to the latter for information.

6. The costs arising from European technical approval procedure shall be paid by the applicant in accordance with national rules.

Annex III

Attestation of conformity with technical specifications

1. Methods of control of conformity

When the procedures for attestation of conformity of a product with technical specifications pursuant to Article 13 are being determined, the following methods of control of conformity shall be used; the choice and combination of methods for any given system shall depend on requirements for the particular product or group of products according to the criteria indicated in Article 13 (3) and (4):

(a) initial type-testing of the product by the manufacturer or an approved body;

(b) testing of samples taken at the factory in accordance with a prescribed test plan by the manufacturer or an approved body;

(c) audit-testing of samples taken at

the factory, on the open market or on a construction site by the manufacturer or an approved body;
(d) testing of samples from a batch which is ready for delivery, or has been delivered, by the manufacturer or an approved body;
(e) factory production control;
(f) initial inspection of factory and of factory production control by an approved body;
(g) continuous surveillance, judgement and assessment of factory production control by an approved body.

In the Directive, factory production control means the permanent internal control of production exercised by the manufacturer. All the elements, requirements and provisions adopted by the manufacturer shall be documented in a systematic manner in the form of written policies and procedures. This production control system documentation shall ensure a common understanding of quality assurance and enable the achievement of the required product characteristics and the effective operation of the production control system to be checked.

2. *Systems of conformity attestation*

Preference is given to application of the following systems of conformity attestation.

(i) Certification of the conformity of the product by an approved certification body on the basis of:
   (a) *(tasks for the manufacturer)*
      (1) factory production control;
      (2) further testing of samples taken at the factory by the manufacturer in accordance with a prescribed test plan;
   (b) *(tasks for the approved body)*
      (3) initial type-testing of the product;
      (4) initial inspection of factory and of factory production control;
      (5) continuous surveillance, assessment and approval of factory production control;
      (6) possibly, audit-testing of samples taken at the factory, on the market or on the construction site.

(ii) Declaration of conformity of the product by the manufacturer on the basis of:
   First possibility:
   (a) *(tasks for the manufacturer)*
      (1) initial type-testing of the product;
      (2) factory production control;
      (3) possibly, testing of samples taken at the factory in accordance with a prescribed test plan;
   (b) *(tasks for the approved body)*
      (4) certification of factory production control on the basis of:
      – initial inspection of factory and of factory production control,
      – possibly, continuous surveillance, assessment and approval of factory production control.

   Second possibility:
      (1) initial type-testing of the product by an approved laboratory;
      (2) factory production control.

   Third possibility:
   (a) initial type-testing by the manufacturer;
   (b) factory production control.

3. *Bodies involved in the attestation of conformity*

With respect to the function of the bodies involved in the attestation of conformity, distinction shall be made between

(i) *certification body*, which means an impartial body, governmental or non-governmental, possessing the necessary competence and responsibility to carry out conformity certification according to given rules of procedure and management;

(ii) *inspection body*, which means an impartial body having the organisation, staffing, competence and integrity to perform according to specified criteria functions such as assessing, recommending for acceptance and subsequent audit of manufacturers' quality control operations, and selection and evaluation of products on site or in factories or elsewhere, according to specific criteria;

(iii) *testing laboratory*, which means a laboratory which measures, examines, tests, calibrates or otherwise determines the characteristics or performance of materials or products.

In case (i) and (ii) (first possibility) of paragraph 2, the three functions 3 (i) to (iii) may be performed by one and the same body or by different bodies, in which case the inspection body and/or the testing laboratory involved in the attestation of conformity carries out its function on behalf of the certification body.

For the criteria concerning the competence, impartiality and integrity of certification bodies, inspection bodies and testing laboratories, see Annex IV.

4.1. CE conformity mark

- The CE conformity marking shall consist of the initials "CE" as given below.

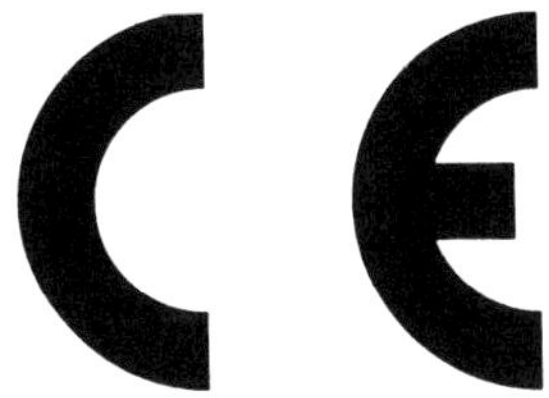

- If the CE marking is reduced or enlarged the proportions given in the above graduated drawing must be respected.
- The various components of the CE marking must have substantially the same vertical dimension, which may not be less than 5 mm.
- The CE marking shall be followed by the identificaiton number of the body involved in the production control stage.

Additional information

- The CE marking shall be accompanied by the name or identifying mark of the producer, the last two digits of the year in which the marking was affixed, and where appropriate, the number of the EC certificate of conformity and, where appropriate, indications to identify the characteristics of the product on the basis of the technical specifications.

Annex IV

Approval of testing laboratories, inspection bodies and certification bodies

The testing laboratories, the inspection bodies and the certification bodies designated by the Member States must fulfil the following minimum conditions:

1. availability of personnel and of the necessary means and equipment;
2. technical competence and professional integrity of personnel;
3. impartiality, in carrying out the tests, preparing the reports, issuing the certificates and performing the surveillance provided for in the Directive, of staff and technical personnel in

relation to all circles, groups or persons directly or indirectly concerned with construction products;

4. maintenance of professional secrecy by personnel;

5. subscription of a civil liability insurance unless that liability is covered by the State under national law.

Fulfilment of the conditions under 1 and 2 shall be verified at intervals by the competent authorities of Member States.

COUNCIL DIRECTIVE 92/42/EEC

of 21 May 1992

on efficiency requirements for new hot-water boilers fired with liquid or gaseous fuels*†

THE COUNCIL OF THE EUROPEAN COMMUNITIES,

Having regard to the Treaty establishing the European Economic Community, and in particular Article 100A thereof,

Having regard to the proposal from the Commission,

In cooperation with the European Parliament,

Having regard to the opinion of the Economic and Social Committee,

Whereas Decision 91/565/EEC (OJ 1991 L307/34) provides for the promotion of energy efficiency in the Community under the SAVE programme;

Whereas it is important to promote measures aimed at the progressive establishment of the internal market in the period up to 31 December 1992; whereas the internal market encompasses an area without internal frontiers, in which the free circulation of goods, persons, services and capital is assured;

Whereas the Council Resolution of 15 January 1985 (OJ 1985 C20/1) on the improvement of energy-saving programmes in the Member States invites Member States to pursue and where necessary increase their efforts to promote the more rational use of energy by the further development of intergrated energy-saving policies;

Whereas the Council Resolution of 16 September 1986 (OJ 1986 C241/1) concerns new Community energy-policy objectives for 1995 and convergence of the policies of the Member States, and in particular the objective of improving the efficiency of final energy demand by at least 20%;

Whereas Article 130R of the Treaty provides that action by the Community relating to the environment shall have the objective of ensuring a prudent and rational utilisation of natural resources;

Whereas it is appropriate to take as a base a high level of protection in proposals for the approximation of the provisions laid down by law, regulation or administrative action in Member States and concerning health, safety, environmental protection and consumer protection;

Whereas the Council Resolution of 21 June 1989 (OJ 1989 C182/4) declares "that the Community should take proper account of potential climatic change linked to the greenhouse effect" and the Council's conclusions of 29 October 1990 state that CO_2 emissions in the year 2000 should be stabilised throughout the Community at their 1990 level;

Whereas the importance of the domestic and tertiary sector, which absorbs a major proportion of the final consumption of energy in the Community, is considerable;

Whereas this sector will become even more important through trends towards more central heating and a general increase in thermal comfort;

Whereas better boiler efficiency is in the consumer's interest; whereas energy saving will be reflected in fewer

---

*OJ 1992 L167/17.
†Amended by Dir 93/68.

imports of hydrocarbons; whereas reduction in the Community's energy dependence will have a positive impact on its trade balance;

Whereas Council Directive 78/170/EEC of 13 February 1978 (OJ 1978 L52/32) on the performance of heat generators for space heating and the production of hot water in new or existing non-industrial buildings and on the insulation of heat and domestic hot-water distribution in new non-industrial buildings, has given rise to the establishment of substantially different efficiency levels between one Member State and another;

Whereas the requirement of high efficiency for hot-water boilers will reduce the range of technical properties of equipment placed on the market, thus facilitating series production and making for economies of scale; whereas the absence of a measure laying down energy requirements at a sufficiently high level may result, with the completion of the internal market, in a significant drop in the efficiency levels of heating installations through the spread on the market of low-efficiency boilers;

Whereas local climatic conditions and the energy and occupancy characteristics of buildings differ greatly within the Community; whereas Member States must take this diversity into account when determining the conditions for putting boilers into service in implementation of this Directive; whereas these circumstances justify the fact that Member States where back-boilers and boilers designed to be installed in the living space are widely installed at the date of the adoption of this Directive should continue to authorise, within specific limits, the placing on their markets and the putting into service of such boilers; whereas these arrangements should be subject to particular surveillance by the Commission;

Whereas this Directive, which is aimed at eliminating technical barriers with regard to boiler efficiency, must follow the new approach established by the Council Resolution of 7 May 1985 (OJ 1985 C136/1) which specifically lays down that legislative harmonisation is limited to the adoption, by means of Directives based on Article 100 of the EEC Treaty, of the essential requirements with which products put on the market must conform and that "these essential requirements shall be worded precisely enough in order to create legally binding obligations which can be enforced and to enable the certification bodies to certify products as being in conformity, having regard to those requirements in the absence of standards";

Having regard to Directive 83/189/EEC (OJ 1983 L109/8) laying down a procedure for the provision of information in the field of technical standards and regulations;

Having regard to Decision 90/683/EEC (OJ 1990 L380/13) concerning the modules for the various phases of the conformity assessment procedures which are intended to be used in the technical harmonisation Directives;

Whereas boilers complying with the efficiency requirements should bear the CE mark and, where appropriate, signs in order to enable them to move freely and to be put into service in accordance with their intended purpose within the Community;

Having regard to Directive 89/106/EEC (OJ 1989 L40/12) on the approximation of laws, regulations and administrative provisions of the Member States relating to construction products;

Whereas efficiency requirements to encourage the rational use of energy as laid down in Council Directive 90/396/EEC of 29 June 1990 (OJ 1990 L196/15) on the approximation of the laws of the Member States relating to appliances burning gaseous fuels should be established for the gas boilers referred to in this Directive,

HAS ADOPTED THIS DIRECTIVE:

Article 1

This Directive, which comes under the SAVE programme concerning the promotion of energy efficiency in the Community, determines the efficiency requirements applicable to new hot-water boilers fired by liquid or gaseous fuels with a rated output of no less than 4 kW and no more than 400 kW, hereinafter called "boilers".

Article 2

For the purposes of this Directive:

- *boiler:* the combined boiler body-burner unit, designed to transmit to water the heat released from burning,
- *appliance:*
  - the boiler body designed to have a burner fitted,
  - the burner designed to be fitted to a boiler body,
- *effective rated output (expressed in kW):* the maximum calorific output laid down and guaranteed by the manufacturer as being deliverable during continuous operation while complying with the useful efficiency indicated by the manufacturer,
- *useful efficiency (expressed in %):* the ratio between the heat output transmitted to the boiler water and the product of the net calorific value at constant fuel pressure and the consumption expressed as a quantity of fuel per unit time,
- *part load (expressed in %):* the ratio between the effective output of a boiler operating intermittently or at an output lower than the effective rated output and the same effective rated output;
- *average temperature of the boiler water:* the average of the water temperatures at the entry and exit of the boiler,
- *standard boiler:* a boiler for which the average water temperature can be restricted by design,
- *back-boiler:* a boiler designed to supply a central-heating system and to be installed in a fireplace recess as part of a back boiler/gas fire combination,
- *low-temperature boiler:* a boiler which can work continuously with a water supply temperature of 35 to 40 °C, possibly producing condensation in certain circumstances, including condensing boilers using liquid fuel,
- *gas condensing boiler:* a boiler designed to condense permanently a larger part of the water vapour contained in the combustion gases,
- *boiler to be installed in the living space:* a boiler with an effective rated output of less than 37 kW, designed to provide heat to the part of the living space in which it is installed by means of the emission of heat from the casing having an open expansion chamber, supplying hot water using gravity circulation; such boilers shall bear on their casings the explicit indication that they must be installed in living space.

Article 3

1. The following shall be excluded from this Directive:

- hot-water boilers capable of being fired by different fuels including solid fuels,
- equipment for the instantaneous preparation of hot water,
- boilers designed to be fired by fuels the properties of which differ appreciably from the properties of the liquid and gaseous fuels commonly marketed (industrial waste gas, biogas, etc),
- cookers and appliances designed mainly to heat the premises in which they are installed and, as a subsidiary function, to supply hot water for central heating and sanitary hot water,
- appliances with rated outputs of less than 6 kW using gravity circulation and designed solely for the production of stored sanitary hot water,
- boilers manufactured on a one-off basis.

2. In the case of boilers with a dual function, that of heating premises and also providing sanitary hot water, the efficiency requirements referred to in Article 5 (1) concern the heating function only.

Article 4

1. Member States may not prohibit, restrict or impede the placing on the market or entry into service within their territories of appliances and boilers which comply with the provisions of this Directive and which bear the CE marking provided for in Article 7 attesting to the conformity to all the provisions of this Directive, including the conformity assessment procedures laid down in Articles 7 and 8, save as otherwise laid down in the Treaty or other Directives or Community provisions.
2. Member States shall take all necessary measures to ensure that boilers cannot be put into service unless they satisfy the efficiency requirements set out in Article 5 (1) and the conditions for entry into service which the Member States lay down on the basis of local climatic conditions and the energy and occupancy characteristics of the buildings.
3. However, Member States where back-boilers and/or boilers that are to be installed in the living space, are widely installed at the date of the adoption of the present Directive, shall continue to authorise their entry into service, provided that their efficiency both at rated output and a 30% part load is not more than 4% below the requirements laid down in Article 5 (1) for standard boilers.
4. The effects of the provisions in paragraphs 2 and 3 shall be constantly monitored by the Commission and analysed in the report to be submitted under Article 10. To this end the Member States shall forward to the Commission any information it requires to submit to the Council the proposed amendments, provided for in that Article, designed to ensure at all events the energy efficiency and free movement of boilers in the Community.
5. (a) Where the boilers are subject to other Directives concerning other aspects which also provide for the affixing of the CE marking, the latter shall indicate that the boilers in question are also presumed to conform to the provisions of those other Directives.
(b) However, where one or more of these Directives allow the manufacturer, during a transitional period, to choose which arrangements to apply, the CE marking shall indicate conformity to the provisions only of those Directives applied by the manufacturer. In this case, particulars of the Directives applied, as published in the *Official Journal of the European Communities*, must be given in the documents, notices or instructions required by the Directives and accompanying the boilers.

Article 5

1. Boilers must comply with the following useful efficiency requirements:
- at rated output, *i.e.* operating at rated output Pn expressed in kW, at an average boiler-water temperature of 70 °C,
  and
- a part load, *i.e.* operating at 30% part load, at an average boiler-water temperature which varies according to the type of boiler.

The useful efficiency requirements to be complied with are set out in the following table:

Type of boiler	*Range of power output*	*Efficiency at rated output*		*Efficiency at partload*	
	kW	*Average boiler-water temperature (in °C)*	*Efficiency requirement expressed (in %)*	*Average boiler-water temperature in °C*	*Efficiency requirement expressed (in %)*
Standard boilers	4 to 400	70	⩾ 84 + 2 logPn	⩾ 50	⩾ 80 + 3 logPn
Low-temperature boilers (*)	4 to 400	70	⩾ 87.5 + 1.5 logPn	40	⩾ 87.5 + 1.5 logPn
Gas condensing boilers	4 to 400	70	⩾ 91 + 1 logPn	30 (**)	⩾ 97 + 1 logPn

2. The harmonised standards relating to the requirements of this Directive drawn up under mandate from the Commission in accordance with Directive 83/189/EEC and 88/182/EEC (OJ 1988 L81/75) shall determine, *inter alia*, the verification methods valid for production and measurements. Appropriate tolerances must be incorporated in the efficiency levels.

Article 6

1. Under the procedures laid down in Article 7, Member States may decide to apply a specific system of labels enabling the energy performance of boilers to be clearly ascertained. This system shall apply to boilers the efficiency of which is superior to the requirements for standard boilers set out in Article 5 (1).

If its efficiency at rated output and its efficiency at part load are equal to or greater than the relevant values for standard boilers, a boiler shall be awarded an "★" as set out in Annex I, section 2.

If its efficiency at rated output and its efficiency at part load are three or more points higher than the relevant values for standard boilers a boiler shall be awarded "★★".

Every extra step of efficiency of three points at rated output and at part load will allow the attribution of an extra "★" as set out in Annex II.

2. Member States may not authorise any other label likely to be confused with those referred to in paragraph 1.

Article 7

1. Member States shall deem that boilers which comply with the harmonised standards, the reference numbers of which have been published in the *Official Journal of the European Communities* and for which the Member States have published the reference numbers of the national standards transposing those harmonised standards, to be in conformity with the essential efficiency requirements stipulated in Article 5 (1). Such boilers must bear the CE marking referred to in Annex I section 1, and be accompanied by the EC declaration of conformity.

2. The conformity of series-produced boilers shall be certified by:

- examination of the efficiency of a boiler type in accordance with module B as described in Annex III,
- a declaration of conformity to the approved type in accordance with module C, D or E as described in Annex IV.

For boilers burning gaseous fuels, the procedures for assessing the conformity of their efficiency shall be those used to assess conformity to the safety requirements laid down in Directive 90/396/EEC on the approximation of the laws of the Member States relating to appliances burning gaseous fuels.

3. When appliances marketed separately are placed on the market, they must bear the CE marking and be accompanied by the EC declaration of conformity, which defines the parameters enabling them after assembly to achieve the useful efficiency levels laid down in Article 5 (1).

4. The CE mark of conformity to the requirements of this Directive and to the other provisions concerning the granting of the CE marking, and also the inscriptions specified in Annex I, shall be affixed on boilers in a visible, easily legible and indelible manner. The affixing on such products of any markings likely to deceive third parties as to the meaning and form of the CE marking shall be prohibited. Any other marking may be affixed to the boilers and appliances provided that the visibility and legibility of the CE marking is not thereby reduced.

5. (a) Where a Member State or a notified body establishes that the CE marking has been affixed unduly, the manufacturer or his agent established within the Community shall be obliged to make the boiler comply as regards the provisions concerning the CE marking and to end the infringement under conditions imposed by the Member State;

(b) where non-compliance continues, the Member State must take all appropriate measures to restrict or prohibit the placing on the market of the product in question or so ensure that it is withdrawn from the market, and shall inform the Commission and the other Member States accordingly.

Article 8

1. Member States shall inform the Commission and the other Member States of the bodies which they have appointed to carry out the procedures referred to in Article 7, together with specific tasks which these bodies have been appointed to carry out and the identification numbers assigned to them beforehand by the Commission.

The Commission shall publish in the *Official Journal of the European Communties* a list of the notified bodies with their identification numbers and the tasks for which they have been notified. The Commission shall ensure that this list is kept up to date.

2. Member States shall implement the minimum criteria laid down in Annex V for the appointment of such bodies. Bodies which satisfy the criteria laid down in the corresponding harmonised standards shall be deemed to comply with the criteria laid down in that Annex.

3. A Member State which has notified a particular body must withdraw that notification if it finds that the body concerned no longer satisfies the criteria referred to in paragraph 2. It shall immediately inform the other Member States and the Commission accordingly and shall withdraw the notification.

Article 9

1. By 1 January 1993, Member States shall adopt and publish the provisions necessary to comply with this Directive. They shall forthwith inform the Commission thereof.

They shall apply those provisions from 1 January 1994.

When Member States adopt those provisions, they shall contain a reference to this Directive or shall be accompanied by such a reference on the occasion of their official publication. The methods of making such a reference shall be laid down by the Member States.

2. Until 31 December 1997, Member States shall permit the placing on the market and putting into service of appliances complying with the national rules and schemes in force within their territories on the date of the adoption of this Directive.

Article 10

Three years after the implementation of this Directive the Commission shall submit a report to the European Parliament and to the Council on the results achieved. That report shall be accompanied by proposals for any changes to be made to this Directive in the light of those results and of advances in technology.

Article 11

This Directive is addressed to the Member States.

Annex I

CE Conformity marks and additional specific markings

1. Conformity marking

The CE conformity marking consists of the letters "CE" taking the following form:

- If the CE marking is reduced or enlarged the proportions given in the above graduated drawing must be respected.
- The various components of the CE marking must have substantially the same vertical dimension, which may not be less than 5 mm.

2. Additional specific markings

- The last two digits of the year in which the CE marking was affixed.
- The energy performance label awarded under Article 6 of this Directive consists of the following symbol:
  The energy performance label awarded under Article 6 of this Directive consists of the following symbol:

Annex II

Award of energy-performance labels

Efficiency requirements to be met both at nominal ouput and at part-load of 0.3 Pn

Label	*Efficiency requirement at nominal output Pn and at an average boiler-water temperature of 70 °C* %	*Efficiency requirement at part-load of 0.3 Pn and at an average boiler-water temperature of ≥ 50 °C* %
★	≥ 84 + 2 log Pn	≥ 80 + 3 log Pn
★★	≥ 87 + 2 log Pn	≥ 83 + 3 log Pn
★★★	≥ 90 + 2 log Pn	≥ 86 + 3 log Pn
★★★★	≥ 93 + 2 log Pn	≥ 89 + 3 log Pn

Annex III

Module B: EC type-examination

1. This module describes that part of the procedure by which a notified body ascertains and attests that an example, representative of the production envisaged, meets the relevant provisions of the Directive.

2. The application for EC type-examination is lodged by the manufacturer or his authorised representative established within the Community with a notified body of his choice.

The application must include:

- the name and address of the manufacturer and, if the application is lodged by the authorised representative, the name and address in addition,
- a written declaration that the same application has not been lodged with any other notified body,
- the technical documents, as described in section 3.

The applicant must place at the disposal of the notified body an example representative of the production envisaged, hereinafter called "type". The notified body may request further examples if needed for carrying out the test programme.

3. The technical documents must enable the conformity of the appliance with the requirements of the Directive to be assessed. They must, as far as is relevant for such assessment, cover the design, manufacture and operation of the appliance and contain as far as is relevant for assessment:

- a general type-description,
- conceptual design and manufacturing drawings and diagrams of components, sub-assemblies, circuits, etc,
- descriptions and explanations necessary for the understanding of the drawings and diagrams and the operation of the product,
- a list of the standards referred to in Article 5 (2), applied in full or in part, and descriptions of the solutions adopted to meet the essential requirements of the Directive where the standards referred to in Article 5 have not been applied,
- results of design calculations made, examinations carried out, etc,
- test reports.

4. The notified body must:

4.1. examine the technical documents, verify that the type has been manufactured in conformity with those documents and identify the elements which have been designed in accordance with the relevant provisions of the standards referred to in Article 5 (2) as well as the components which have been designed without applying the relevant provisions of those standards;

4.2. perform or have performed the appropriate examinations and necessary tests to check whether, where the standards referred to in Article 5 (2)

have not been applied, the solutions adopted by the manufacturer meet the essential requirements of the Directive;
4.3. perform or have performed the appropriate examinations and necessary tests to check whether, where the manufacturer has chosen to apply the relevant standards, these have actually been applied;
4.4. agree with the applicant the location where the examinations and necessary tests are to be carried out.
5. Where the type meets the relevant provisions of this Directive, the notified body issues an EC type-examination certificate to the applicant. The certificate contains the name and address of the manufacturer, the conclusion of the examination and necessary data for identification of the approved type.

A list of the relevant parts of the technical documents is annexed to the certificate and a copy kept by the notified body.

If the manufacturer or his authorised representative established in the Community is refused a type certificate, the notified body must provide detailed reasons for such refusal.

Provision must be made for an appeals procedure.
6. The applicant informs the notified body that holds the technical documents concerning the EC type-examination certificate of all modifications to the approved appliance which must receive additional approval where such changes may affect the conformity with the essential requirements or the prescribed conditions for use of the product. This additional approval is given in the form of an addition to the original EC type-examination certificate.
7. Each notified body must communicate to the other notified bodies the relevant information concerning the EC type-examination certificates and additions issued and withdrawn.
8. The other notified bodies may receive copies of the EC type-examination certificates and/or their additions. The Annexes to the certificates must be kept at the disposal of the other notified bodies.
9. The manufacturer or his authorised representative established within the Community must keep with the technical documents copies of EC type-examination certificates and their additions for a period of at least 10 years after the last date of manufacture of the product concerned.

Where neither the manufacturer nor his authorised representative is established within the Community, the obligation to keep the technical documents available is the responsibility of the person who places the product on the Community market.

Annex IV

Module C: Conformity to type

1. This module describes that part of the procedure whereby the manufacturer or his authorised representative established within the Community ensures and declares that the appliances concerned are in conformity with the type as described in the EC type-examination certificate and satisfy the requirements of this Directive that apply to them. The manufacturer or his authorised representative established within the Community shall affix the CE marking to each appliance and draw up a written declaration of conformity.
2. The manufacturer must take all measures necessary to ensure that the manufacturing process assures the conformity of the manufactured appliances with the type as described in the EC type-examination certificate and with the efficiency requirements of the Directive.
3. The manufacturer or his authorised representative must keep a copy of the

declaration of conformity for a period of at least 10 years after the last date of manufacture of the product concerned.

Where neither the manufacturer nor his authorised representative is established within the Community, the obligation to keep the technical documents available is the responsibility of the person who places the product on the Community market.

4. A notified body chosen by the manufacturer must perform or have performed examinations of the product at random intervals. A suitable sample of the finished products, taken on the spot by the notified body, is examined and appropriate tests, defined in the applicable standard or standards referred to in Article 5 (2) or equivalent tests are carried out to check the conformity of the product with the requirements of the corresponding Directive. In the event of one or more samples of the products examined not conforming, the notified body must take the appropriate measures.

Module D: Production quality assurance

1. This module describes the procedure whereby the manufacturer who satisfies the obligations of section 2 ensures and declares that the appliances concerned are in conformity with the type as described in the EC type-examination certificate and satisfy the requirements of this Directive. The manufacturer or his authorised representative established within the Community shall affix the CE marking to each appliance and draw up a written declaration of conformity. The CE mark is accompanied by the identification number of the notified body responsible for the checks referred to in section 4.

2. The manufacturer must operate an approved quality system for production, final appliance inspection and testing as specified in section 3. He is subject to the checks referred to in section 4.

3. *Quality system*

3.1. The manufacturer lodges an application for assessment of his quality system with a notified body of his choice, for the appliances concerned.

The application must include:
- all relevant information for the appliance category envisaged,
- the documents concerning the quality system,
- the technical documents pertaining to the approved type and a copy of the EC type-examination certificate.

3.2. The quality system must ensure conformity of appliances with the type as described in the EC type-examination certificate and with the requirements of this Directive that apply to them.

All the elements, requirements and provisions adopted by the manufacturer must be documented in a systematic and orderly manner in the form of written policies, procedures and instructions. The quality system documents must permit a consistent interpretation of the quality programmes, plans, manuals and quality records.

It must contain in particular an adequate description of:
- the quality objectives and the organisational structure, responsibilities and powers of the management with regard to appliance quality,
- the manufacturing, quality control and quality assurance techniques, processes and systematic actions that will be used,
- the examinations and tests that will be carried out before, during and after manufacture, and the frequency with which they will be carried out,
- the quality records, such as inspection reports and test data, calibration data, qualification reports of the personnel concerned, etc,
- the means of monitoring the achievement of the required appliance quality and the effective operation of the quality system.

3.3. The notified body must assess the quality system to determine whether it satisfies the requirements referred to in section 3.2. It must presume conformity with those requirements in respect of quality systems that implement the relevant harmonised standard. The auditing team must have at least one member with experience of assessing the relevant product technology. The assessment procedure includes an inspection visit to the manufacturer's premises.

The decision is notified to the manufacturer. The notification must contain the conclusions of the examination and the duly substantiated assessment decision.

3.4. The manufacturer must undertake to fulfil the obligations arising out of the quality system as approved and maintain it at an adequate and efficient level.

The manufacturer or his authorised representative must keep the notified body that has approved the quality system informed of any proposed change in the quality system.

The notified body must assess the changes proposed and decide whether the altered quality system will still satisfy the requirements referred to in 3.2 or whether reassessment is required.

It must notify the manufacturer of its decision. The notification must contain the conclusions of the examination and the substantiated assessment decision.

4. *Monitoring under the responsibility of the notified body*

4.1. The purpose of monitoring is to make sure that the manufacturer duly fulfils the obligations arising out of the approved quality system.

4.2. The manufacturer must allow the notified body access for inspection purposes to the manufacturing, inspection, testing and storage premises and provide it with all necessary information, in particular:

- the quality system documents,
- the quality records, such as inspection reports and test data, calibration data, qualification reports of the personnel concerned, etc.

4.3. The notified body must periodically carry out audits to make sure that the manufacturer maintains and applies the quality system and provides an audit report to the manufacturer.

4.4. Additionally the notified body may pay unannounced visits to the manufacturer. During such visits the notified body may carry out tests or have them carried out to verify that the quality system is functioning correctly; if necessary, the notified body must provide the manufacturer with a visit report and, if a test has taken place, with a test report.

5. The manufacturer must, for a period of at least 10 years after the last date of manufacture of the product, keep at the disposal of the national authorities:

- the document referred to in the second indent of 3.1,
- the updating referred to in the second paragraph of 3.4,
- the decisions and reports from the notified body which are referred to in the final paragraph of 3.4, and in 4.3 and 4.4.

6. Each notified body must give the other notified bodies the relevant information concerning the quality system approvals issued and withdrawn.

Module E: Product quality assurance

1. This module describes the procedure whereby the manufacturer who satisfies the obligations of section 2 ensures and declares that the boilers and appliances are in conformity with the type as described in the EC type-examination certificate. The manufacturer or his authorised representative established within the Community shall affix the CE marking to each boiler and appliance and draw up a written declaration of conformity. The CE marking must be accompanied by the identification number of the notified body responsible for the checks referred to in section 4.

2. The manufacturer must operate an approved quality system for final boiler and appliance inspection and

testing as specified in section 3. He must be subject to the checks referred to in section 4.

3. *Quality system*

3.1. The manufacturer lodges an application with a notified body of his choice for the assessment of the quality system for his boilers and appliances.

The application must include:

- all relevant information for the boiler or appliance category envisaged,
- the quality system's documentation,
- the technical documents pertaining to the approved type and a copy of the EC type-examination certificate.

3.2. Under the quality system, each boiler or appliance is examined and appropriate tests as defined in the relevant standard(s) referred to in Article 5 or equivalent tests are carried out in order to verify its conformity with the relevant requirements of the Directive. All the elements, requirements and provisions adopted by the manufacturer must be documented in a systematic and orderly manner in the form of written policies, procedures and instructions. This quality system documentation must enable the quality programmes, plans, manuals and records to be interpreted in a uniform manner.

It must in particular contain an adequate description of:

- the quality objectives and the organisational structure, responsibilities and powers of the management with regard to product quality,
- the examination and tests that will be carried out after manufacture,
- the means of monitoring the effective operation of the quality system,
- quality records, such as inspection reports and test data, calibration data, qualification reports of the personnel concerned, etc.

3.3. The notified body must assess the quality system to determine whether it satisfies the requirements referred to in 3.2. It must presume conformity with these requirements in respect of quality systems that implement the relevant harmonised standard.

The auditing team must have at least one member with experience of assessing the relevant product technology. The assessment procedure must include an inspection visit to the manufacturer's premises.

The manufacturer must be notified of the decision. The notification must contain the conclusions of the examination and the substantiated assessment decision.

3.4. The manufacturer must undertake to fulfil the obligations arising out of the quality system as approved and maintain it at an adequate and efficient level.

The manufacturer or his authorised representative must keep the notified body which has approved the quality system informed of any proposed change in the quality system.

The notified body must assess the changes proposed and decide whether the altered quality system will still satisfy the requirements referred to in 3.2 or whether a reassessment is required.

It must notify the manufacturer of its decision. The notification must contain the conclusions of the examination and the substantiated assessment decision.

4. *Monitoring under the responsibility of the notified body*

4.1. The purpose of monitoring is to make sure that the manufacturer duly fulfils the obligations arising out of the approved quality system.

4.2. The manufacturer must allow the notified body access for inspection purposes to the inspection, testing and storage premises and provide it with all necessary information, in particular:

- the quality system documentation,
- the technical documents,
- the quality records, such as inspection reports and test data, calibration data, qualification reports of the personnel concerned, etc.

4.3. The notified body must periodically carry out audits to ensure that the manufacturer maintains and applies the quality system and must

provide an audit report to the manufacturer.
4.4. Additionally, the notified body may pay unannounced visits to the manufacturer. During such visits the notified body may carry out tests or have them carried out to verify that the quality system is functioning correctly; if necessary, the notified body must provide the manufacturer with a visit report and, if a test has been carried out, with a test report.
5. The manufacturer must, for a period of at least 10 years after the last date of manufacture of the boiler or appliance, keep at the disposal of the national authorities:
- the documents referred to in the third indent of 3.1,
- the changes referred to in the second paragraph of 3.4,
- the decisions and reports from the notified body which are referred to in the final paragraph of 3.4, and in 4.3 and 4.4.

6. Each notified body must forward to the other notified bodies the relevant information concerning the quality system approvals issued and withdrawn.

Annex V

Minimum criteria to be taken into account by Member States for the notification of bodies

1. The body, its director and the staff responsible for carrying out the verification tests may not be the designer, manufacturer, supplier or installer of appliances which they inspect, nor the authorised representative of any of those parties. They may not become either involved directly or as authorised representatives in the design, construction, marketing or maintenance of such boilers and appliances. This does not preclude the possibility of exchanges of technical information between the manufacturer and the body.
2. The body and its staff must carry out the verification tests with the highest degree of professional integrity and technical competence and must be free from all pressures and inducements, particularly financial, which might influence their judgment of the results of the inspection, especially from persons or groups of persons with an interest in the result of verifications.
3. The body must have at its disposal the necessary staff and possess the necessary facilities to enable it to perform properly the administrative and technical tasks connected with verification; it must also have access to the equipment required for special verification.
4. The staff responsible for inspection must have:
- sound technical and professional training,
- satisfactory knowledge of the requirements of the tests they carry out and adequate experience of such tests,
- the ability to draw up the certificates, records and reports required to authenticate the performance of the tests.

5. The impartiality of inspection staff must be guaranteed. Their remuneration must not depend on the number of tests carried out or on the results of such tests.
6. The body must take out liability insurance unless its liability is assumed by the State in accordance with national law, or the Member State itself is directly responsible for the tests.
7. The staff of the body must be bound to observe professional secrecy (except

vis-à-vis the competent administrative authorities of the State in which its activities are carried out) under this Directive or any provision of national law giving effect to it.

COUNCIL DIRECTIVE 73/23/EEC

of 19 February 1973

on the harmonisation of the laws of Member States relating to electrical equipment designed for use within certain voltage limits*†

Council Directive of 19 February 1973 on the harmonisation of the laws of Member States relating to electrical equipment designed for use within certain voltage limits (73/23/EEC).

THE COUNCIL OF THE EUROPEAN COMMUNITIES,

Having regard to the Treaty establishing the European Economic Community, and in particular Article 100 thereof,

Having regard to the proposal from the Commission,

Having regard to the opinion of the European Parliament,

Having regard to the opinion of the Economic and Social Committee,

Whereas the provisions in force in the Member States designed to ensure safety in the use of electrical equipment used within certain voltage limits may differ, thus impeding trade;

Whereas in certain Member States in respect of certain electrical equipment, the safety legislation takes the form of preventive and repressive measures by means of binding provisions;

Whereas in other Member States in order to achieve the same objective, the safety legislation provides for reference to technical standards laid down by standards bodies;

Whereas such a system offers the advantage of rapid adjustment to technical progress without neglecting safety requirements;

Whereas certain Member States carry out administrative operations to approve standards;

Whereas such approval neither affects the technical content of the standards in any way nor limits their conditions of use;

Whereas such approval cannot therefore alter the effects, from a Community point of view, of harmonised and published standards;

Whereas within the Community the free movement of electrical equipment should follow when this equipment complies with certain safety requirements recognised in all Member States;

Whereas without prejudice to any other form of proof, the proof of compliance with these requirements may be established by reference to harmonised standards which incorporate these conditions;

Whereas those harmonised standards should be established by common agreement by bodies to be notified by each Member State to the other Member States and to the Commission and should be publicised as widely as possible;

Whereas such harmonisation should for the purpose of trade eliminate the inconveniences, resulting from differences between national standards;

Whereas, without prejudice to any other form of proof, the compliance of electrical equipment with the harmonised standards may be presumed from the affixing or issue of marks or certificates by the competent organisations or, in the absence thereof, from a

---

*OJ No 1077 p. 29, 1973/03/26.
†Amended by Dir 93/68.

manufacturer's Declaration of compliance;

Whereas in order to facilitate the removal of barriers to trade the member states should recognise such marks or certificates or such Declaration as elements of proof;

Whereas, with this end in view, the said marks or certificates should be publicised in particular by their publication in the *Official Journal of the European Communities*;

Whereas as a transitional measure, the free movement of electrical equipment for which harmonised standards do not yet exist may be achieved by applying the safety provisions or standards already laid down by other international bodies or by one of the bodies which establish harmonised standards;

Whereas it is possible that electrical equipment may be placed in free circulation even though it does not comply with the safety requirements, and whereas it is therefore desirable to lay down suitable provisions to minimise this danger;

Whereas Decision 90/683/EEC[1] establishes the modules for the various phases of the conformity assessment procedures which are intended to be used in the technical harmonisation Directives;

Whereas the choice of procedures must not lead to a lowering of safety standards of electrical equipment, which have already been established throughout the Communituy;

HAS ADOPTED THIS DIRECTIVE:

Article 1

For the purposes of this Directive "electrical equipment" means any equipment designed for use with a voltage rating of between 50 and 1,000 V for alternating current and between 75 and 1,500 V for direct current, other than the equipment and phenomena listed in Annex II.

[1] OJ No L 380, 31. 12. 1990, p. 13.

Article 2

1. The Member States shall take all appropriate measures to ensure that electrical equipment may be placed on the market only if, having been constructed in accordance with good engineering practice in safety matters in force in the Community, it does not endanger the safety of persons, domestic animals or property when properly installed and maintained and used in applications for which it was made.

2. The principal elements of the safety objectives referred to in paragraph 1 are listed in Annex I.

Article 3

The Member States shall take all appropriate measures to ensure that if electrical equipment is of such a nature as to comply with the provisions of Article 2, subject to the conditions laid down in Articles 5, 6, 7 and 8, the free movement thereof within the Community shall not be impeded for reasons of safety.

Article 4

In relation to electrical equipment the Member States shall ensure that stricter safety requirements than those laid down in Article 2 are not imposed by electricity supply bodies for connection to the grid, or for the supply of electricity to users of electrical equipment.

Article 5

The Member States shall take all appropriate measures to ensure that, in particular, electrical equipment which complies with the safety provisions of harmonised standards shall be regarded by their competent administrative authorities as complying with the provisions of Article 2, for the purposes of placing on the market and free movement as referred to in Article 2 and 3 respectively.

Standards shall be regarded as harmonised once they are drawn up by common agreement between the bodies notified by the Member States in accordance with the procedure laid down in Article 11, and published under national procedures. The standards shall be kept up to date in the light of technological progress and the developments in good engineering practice in safety matters.

For purposes of information the list of harmonised standards and their references shall be published in the Official Journal of the European Communities.

Article 6

1. Where harmonised standards as defined in Article 5 have not yet been drawn up and published, the Member States shall take all appropriate measures to ensure that, for the purposes of placing on the market of free movement as referred to in Articles 2 and 3 respectively, their competent administrative authorities shall also regard, as complying with the provisions of Article 2, electrical equipment which complies with the safety provisions of the international Commission on the Rules for the approval of electrical equipment (CEE) or of the international electrotechnical Commission (IEC) in respect of which the publication procedure laid down in paragraphs 2 and 3 has been applied.

2. The safety provisions referred to in paragraph 1 shall be notified to the Member States by the Commission as from the entry into force of this Directive, and thereafter as and when they are published. The Commission, after consulting the Member States, shall state the provisions and in particular the variants which it recommends to be published.

3. The Member States shall inform the Commission within a period of three months of such objections as they may have to the provisions thus notified, stating the safety grounds on account of which the provisions should not be recognised.

For purposes of information those safety provisions against which no objection has been raised shall be published in the *Official Journal of the European Communities.*

Article 7

Where harmonised standards within the meaning of Article 5 or safety provisions published in accordance with Article 6 are not yet in existence, the Member States shall take all appropriate measures to ensure that, for the purpose of placing on the market or free movement as referred to in Articles 2 and 3 respectively, their competent administrative authorities shall also regard as complying with the provisions of Article 2, electrical equipment manufactured in accordance with the safety provisions of the standards in force in the Member State of manufacture, if it ensures a safety level equivalent to that required in their own territory.

Article 8

1. Before being placed on the market, the electrical equipment referred to in Article 1 must have affixed to it the CE marking provided for in Article 10 attesting to its conformity to the provisions of this Directive, including the conformity assessment procedure described in Annex IV.

2. In the event of a challenge the manufacturer or importer may submit a report, drawn up by a body, which is notified in accordance with the procedure set out in Article 11, on the conformity of the electrical equipment with the provisions of Article 2.

3. (a) Where electrical equipment is subject to other Directives concerning other aspects which also provide for the affixing of the CE marking, the latter shall indicate that the equipment in question is also presumed to conform to the provisions of those other Directives.

3. (b) However, where one or more of these Directives allow the manufacturer, during a transitional period, to choose which arrangements to apply, the CE marking shall indicate conformity to the provisions only of those Directives applied by the manufacturer. In this case, particulars of the Directives applied, as published in the *Official Journal of the European Commu-*

nities, must be given in the documents, notices or instructions required by the Directives and accompanying the electrical equipment.

Article 9

1. If, for safety reasons, a Member State prohibits the placing on the market of any electrical equipment or impedes its free movement, it shall immediately inform the other Member States concerned and the Commission, indicating the grounds for its Decision and stating in particular:

Whether its non-conformity with Article 2 is attributable to a shortcoming in the harmonised standards referred to in Article 5, the provisions referred to in Article 6 or the standards referred to in Article 7.

Whether its non-conformity is attributable to faulty application of such standards or publications or to failure to comply with good engineering practice as referred to in Article 2.

2. If other Member States raise objections to the Decision referred to in paragraph 1, the Commission shall immediately consult the Member States concerned.

3. If an agreement has not been reached within three months from the date of notification as laid down in paragraph 1, the Commission shall obtain the opinion of one of the bodies notified in accordance with the procedure laid down in Article 11 having its registered office outside the territory of the Member States concerned and which has not been involved in the procedure provided for in Article 8. The opinion shall state the extent to which the provisions of Article 2 have not been complied with.

4. The Commission shall communicate the opinion of this body to all the Member States which may, within a period of one month, make their observations known to the Commission. The Commission shall at the same time note any observations by the parties concerned on the abovementioned opinion.

5. Having taken note of these observations the Commission shall, if necessary, formulate the appropriate recommendations or opinions.

Article 10

1. The CE conformity marking referred to in Annex III shall be affixed by the manufacturer or his authorised representative established within the Community to the electrical equipment or, failing that, to the packaging, the introduction sheet or the guarantee certificate so as to be visible, easily legible and indelible.

2. The affixing on electrical equipment of any markings liable to deceive third parties as to the meaning and form of the CE markings shall be prohibited. However, any other marking may be affixed to the electrical equipment, its packaging, the instruction sheet or the guarantee certificate provided that the visibility and legibility of the CE marking is not thereby reduced.

3. Without prejudice to Article 9:

(a) where a Member State establishes that the CE marking has been affixed unduly, the manufacturer or his agent established within the Community market shall be obliged to make the electrical equipment comply as regards the provisions concerning the CE marking and to end the infringement under conditions imposed by the Member State;

(b) where non-compliance continues, the Member State must take all appropriate measures to restrict or prohibit the placing on the market of the electrical equipment in question or to ensure that it is withdrawn from the market in accordance with the procedure laid down in Article 9.

Article 11

Each Member State shall inform the other Member States and the Commission of the following:

- The bodies referred to in Article 5.
- The bodies which may make a report in accordance with the provisions of Article 8 or give an opinion in accordance with the provisions of Article 9.

– The place of publication referred to in Article 5 (2).

Any amendment to the above shall be notified by each Member State to the other Member States and to the Commission.

Article 12

This Directive shall not apply to electrical equipment intended for export to third countries.

Article 13

1. The Member States shall put into force the laws, Regulations and administrative provisions necessary to comply with the requirements of this Directive within eighteen months of its notification and shall forthwith inform the Commission thereof.

However, in the case of Denmark, the time limit shall be extended to five years.

2. The Member States shall communicate to the Commission the texts of the main provisions of national laws which they adopt in the field covered by this Directive.

Article 14

This directive is addressed to the Member States.

Done at Brussels, 19 February 1973.

For the Council
The President
A. LAVENS

Annex I

Principal elements of the safety objectives for electrical equipment designed for use within certain voltage limits

1. General conditions

(a) the essential characteristics, the recognition and observance of which will ensure that electrical equipment will be used safely and in applications for which it was made, shall be marked on the equipment, or, if this is not possible, on an accompanying notice.

(b) the manufacturers or brand name or trade mark should be clearly printed on the electrical equipment or, where that is not possible, on the packaging.

(c) the electrical equipment, together with its component parts should be made in such a way as to ensure that it can be safely and properly assembled and connected.

(d) the electrical equipment should be so designed and manufactured as to ensure that protection against the hazards set out in points 2 and 3 of this Annex is assured providing that the equipment is used in application for which it was made and is adequately maintained.

2. Protection against hazards arising from the electrical equipment

Measures of a technical nature should be prescribed in accordance with point 1, in order to ensure:

(a) that persons and domestic animals are adequately protected against danger of physical injury or other harm which might be caused by electrical contact direct or indirect

(b) that temperatures, arcs or radiation which would cause a danger, are not produced

(c) that persons, domestic animals and property are adequately protected against non-electrical dangers caused by the electrical equipment which are revealed by experience

(d) that the insulation must be suitable for foreseeable conditions.

3. Protection against hazards which may be caused by external influences on the electrical equipment

Technical measures are to be laid down in accordance with point 1, in order to ensure:

(a) that the electrical equipment meets the expected mechanical requirements in such a way that persons, domestic animals and property are not endangered

(b) that the electrical equipment shall be resistant to non-mechanical influences in expected environmental conditions, in such a way that persons, domestic animals and property are not endangered

(c) that the electrical equipment shall not endanger persons, domestic animals and property in foreseeable conditions of overload.

Annex II

Equipment and phenomena outside the scope of the Directive

- Electrical equipment for use in an explosive atmosphere
- Electrical equipment for radiology and medical purposes
- Electrical parts for goods and passenger lifts
- Electricity meters
- Plugs and socket outlets for domestic use
- Electric fence controllers
- Radio-electrical interference
- Specialised electrical equipment, for use on ships, aircraft or railways, which complies with the safety provisions drawn up by international bodies in which the Member States participate.

Annex III

CE conformity marking and EC declaration of conformity

A. CE conformity marking

The CE conformity marking shall consist of the initials "CE" taking the following form:

- If the CE marking is reduced or enlarged the proportions given in the above graduated drawing must be respected.
- The various components of the CE marking must have substantially the same vertical dimension, which may not be less than 5 mm.

B. EC declaration of conformity

The EC declaration of conformity must contain the following elements:

- nane and address of the manufacturer or his authorised representative established within the Community,
- a description of the electrical equipment,

- reference to the harmonised standards,
- where appropriate, references to the specifications with which conformity is declared,
- identification of the signatory who has been empowered to enter into commitments on behalf of the manufacturer or his authorised representative established within the Community,
- the last two digits of the year in which the CE marking was affixed.

Annex IV

Internal production control

1. Internal production control is the procedure whereby the manufacturer or his authorised representative established within the Community, who carries out the obligations laid down in point 2, ensures and declares that the electrical equipment satisfies the requirements of this Directive that apply to it. The manufacturer or his authorised representative established within the Community must affix the CE marking to each product and draw up a written declaration of conformity.

2. The manufacturer must establish the technical documentation described in point 3 and he or his authorised representative established within the Community must keep it on Community territory at the disposal of the relevant national authorities for inspection purposes for a period ending at least 10 years after the last product has been manufactured.

Where neither the manufacturer nor his authorised representative is established within the Community, this obligation is the responsibility of the person who places the electrical equipmnent on the Community market.

3. Technical documentation must enable the conformity of the electrical equipment to the requirements of this Directive to be assessed. It must, as far as relevant for such assessment, cover the design, manufacture and operation of the electrical equipment. It must include:

- a general description of the electrical equipment,
- conceptual design and manufacturing drawings and schemes of components, sub-assemblies, circuits, etc.,
- descriptions and explanations necessary for the understanding of said drawings and schemes and the operation of the electrical equipment,
- a list of the standards applied in full or in part, and descriptions of the solutions adopted to satisfy the safety aspects of this Directive where standards have not been applied.
- results of design calculations made, examinations carried out, etc.
- test reports.

4. The manufacturer or his authorised representative must keep a copy of the declaration of conformity with the technical documentation.

5. The manufacturer must take all measures necessary in order that the manufacturing process shall ensure compliance of the manufactured products with the technical documentation referred to in point 2 and with the requirements of this Directive that apply to them.

Proposal for a European Parliament and Council Directive on *in vitro* diagnostic medical devices

(95/C 172/02)

COM(95) 130 final – 95/0013(COD)

(Submitted by the Commission on 19 April 1995)

THE COUNCIL OF THE EUROPEAN COMMUNITIES,

Having regard to the Treaty establishing the European Economic Community, and in particular Article 100a thereof,

Having regard to the proposal from the Commission,

Having regard to the opinion of the Economic and Social Committee,

Acting in accordance with the procedure referred to in Article 189b of the Treaty establishing the European Community.

Whereas measures should be adopted for the progressive establishment of the internal market; whereas the internal market is an area without internal frontiers in which the free movement of goods, persons, services and capital is ensured;

Whereas the content and scope of the laws, regulations and administrative provisions in force in the Member States with regard to the safety, health protection and performance characteristics and authorisation procedures for *in vitro* diagnostic medical devices are different; whereas the existence of such disparities constitutes barriers to trade and the need to establish harmonised rules was confirmed by a comparative survey of national legislation carried out on behalf of the Commission

Whereas the harmonisation of national legislations is the only means of removing these barriers to free trade and preventing new barriers; whereas this objective cannot be achieved in a satisfactory manner at another level by the individual Member States; whereas this Directive only lays down necessary and sufficient requirements for the free circulation of the *in vitro* diagnostic medical devices to which it is applicable;

Whereas the harmonised provisions must be distinguished from the measures adopted by the Member States to manage the funding of public health and sickness insurance schemes relating directly or indirectly to such devices; whereas, therefore, the provisions do not affect the ability of the Member States to implement the abovementioned measures provided Community law is complied with;

Whereas *in vitro* diagnostic medical devices should provide patients, users and third parties with a high level of protection and attain the performance levels attributed to them by the manufacturer; whereas, therefore, the maintenance or improvement of the level of protection attained in the Member States is one of the essential objectives of this Directive;

Whereas, in accordance with the principles set out in the Council resolution of 7 May 1985 concerning a new approach to technical harmonisation and standardisation,[1] rules regarding the design and manufacture of relevant products must be confined to the

[1] OJ No C 136, 4. 6. 1985, p. 1.

provisions required to meet the essential requirements; whereas, because they are essential, such requirements should replace the corresponding national provisions; whereas the essential requirements, including requirements to minimise and reduce risks, should be applied with discretion, taking into account the technology and practice at the time of design and of technical and economic considerations compatible with a high level of protection of health and safety;

Whereas the major part of medical devices are covered by Council Directive 90/385/EEC relating to active implantable medical devices[2] and Council Directive 93/42/EEC relating to all other medical devices[3] with the exclusion of *in vitro* diagnostic medical devices, the present Directive shall extend the harmonisation to *in vitro* diagnostic medical devices; whereas, in the interest of uniform Community rules, this Directive is based largely on the provisions of Directives 90/385/EEC and 93/42/EEC;

Whereas certified international reference materials and materials used for external quality assessment schemes are not covered by this Directive, calibrators and control materials needed by the user to establish or verify performances of devices are *in vitro* diagnostic medical devices:

Whereas reagents produced within user laboratories and which are not subject to commercial transactions, taking account of the principle of subsidiary, are not included in this Directive;

Whereas the electromagnetic compatibility aspects form an integral part of the essential requirements of this Directive, Council Directive 89/336/EEC of 3 May 1989 relating to electromagnetic compatibility[4] shall not apply;

Whereas, in order to ease the task of proving conformity with the essential requirements and to enable conformity to be verified, it is desirable to have harmonised standards in respect of the prevention of risks associated with the design, manufacture and packaging of medical devices; whereas such harmonised standards are drawn up by private-law bodies and should retain their status as non-mandatory texts; whereas, to this end, the European Committee for Standardisation (CEN) and the European Committee for Electrotechnical Standardisation (CENELEC) are recognised as the competent bodies for the adoption of harmonised standards in accordance with the general guidelines on cooperation between the Commission and these two bodies signed on 13 November 1984;

Whereas, for the purpose of this Directive, a harmonised standard is a technical specification (European standard or harmonisation document) adopted, on a mandate from the Commission, by either or both of these bodies in accordance with council Directive 83/189/EEC of 28 March 1983 laying down a procedure for the provision of information in the field of technical standards and regulations,[5] and pursuant to the abovementioned general guidelines;

Wheras, in Decision 90/683/EEC of 13 December 1990 concerning the modules for the various phases of the conformity assessment procedures which are intended to be used in the technical harmonisation directives,[6] the Council has laid down harmonised conformity assessment procedures; whereas the details added to these modules are justified by the nature of

[2] OJ No L 189, 20. 7. 1990, p. 17.
[3] OJ No L 169, 12. 7. 1993, p. 1.
[4] OJ No. L 139, 23. 5. 1989, p. 19. Directive as last amended by Directive 93/68/EEC (OJ No L 220, 30. 8. 1993, p. 1).
[5] OJ No L 109, 26. 4. 1983, p. 8. Directive as last amended by Directive 94/10/EC of the European parliament and the Council (OJ No L 100, 19. 4. 1994, p. 30).
[6] OJ No L 380, 31. 12. 1990, p. 13. Decision as last amended by Decision 93/465/EEC (OJ No L 220, 30. 8. 1993, p. 23).

the verification required for *in vitro* diagnostic medical devices and by the need for consistency with the previous directives on medical devices;

Whereas it is necessasry, essentially for the purpose of the conformity assessment procedures, to group *in vitro* diagnostic medical devices into two product classes; whereas, since the large majority of such devices do not constitute a direct risk to patients and are used by competently trained professionals, and the results obtained can often be confirmed by other means, the conformity assessment procedures can be carried out, as a general rule, under the sole responsibility of the manufacturer; whereas, taking account of existing national regulations and of notifications received following the procedure of Directive 83/189/EEC, the intervention of notified bodies is needed only for defined devices, the correct performance of which is essential to medical practice and the failure of which can cause a serious risk to health;

Whereas the list of *in vitro* diagnostic medical devices to be subjected to third-party conformity assessment needs updating, taking account of technological progress and of evolution in the domain of protection of health; whereas such updating measures must be taken in line with the procedure IIIa as laid down in Council Decision 87/373/EEC;[7] whereas a system of adverse incident reporting (vigilance) constitutes a useful tool for the surveillance of the market, including the performance of new devices; whereas information obtained from vigilance as well as from external quality assessment schemes becomes useful for decision-making on classification of devices;

Whereas medical devices should, as a general rule, bear the CE marking to indicate their conformity with the provisions of this Directive to enable them to move freely within the Community and to be put into service in accordance with their intended purpose;

Whereas manufacturers shall have the possibility, when the intervention of a notified body is required, to choose amongst those bodies published by the Commission; whereas the Member States do not have an obligation to designate such notified bodies, they must ensure however that bodies designated as notified bodies comply with the assessment criteria laid down in this Directive;

Whereas the competent authorities in charge of market surveillance must be able, particularly in emergencies, to contract the manufacturer or his authorised representative established in the Community; whereas cooperation and exchange of information between Member States are necessary in view of uniform application of this Directive, in particular for the purpose of market surveillance;

Whereas this Directive includes *in vitro* diagnostic medical devices incorporating substances derived from the human body, it does not affect national regulations relating to the ethics of use of such substances; whereas for the purpose of overall consistency between medical-devices directives, Directive 93/42/EEC needs to be amended accordingly.

HAVE ADOPTED THIS DIRECTIVE:

Article 1

Scope, Definitions

1. The Directive shall apply to *in vitro* diagnostic medical devices and their accessories. For the purposes of this Directive, accessories shall be treated as *in vitro* diagnostic medical devices in their own right. Both *in vitro* diagnostic medical devices and accessories shall hereinafter be termed devices.

2. For the purpose of this Directive, the following definitions shall apply:

(a) *"medical device"* means any instrument, apparatus, appliance,

[7] OJ No L 197, 18. 7. 1987, p. 23.

material or other article, whether used alone or in combination, including the software necessary for its proper application, intended by the manufacturer to be used for human beings for the purpose of:

- diagnosis, prevention, monitoring, treatment or alleviation of disease,
- diagnosis, monitoring, treatment, alleviation or compensation for an injury or handicap.
- investigation, replacement or modification of the anatomy or of a physiological process,
- control of conception.

and which does not achieve its principal intended action in or on the human body by pharmacological, immunological or metabolic means, but which may be assisted in its function by such means;

(b) *"in vitro diagnostic medical device"* means any medical device which is a reagent, reagent product, calibrator, control material, kit, instrument, apparatus, equipment or system whether used alone or in combination, intended by the manufacturer to be used *in vitro* for the examination of specimens including blood and tissue donations, derived from the human body, solely or principally for the purpose of providing information concerning a physiological state, state of health or disease or congenital abnormality or to determine the safety and compatibility with potential recipients. For the purpose of this Directive, a specimen receptacle, whether evacuated or not, specifically intended by its manufacturer to contain a specimen for the purpose of *in vitro* diagnostic examination is considered to be a device.

Products for general laboratory use are not devices unless such products, in view of their characteristics, are specifically intended by their manufacturer to be used for *in vitro* diagnostic examination;

(c) *"accessory"* means an article which, whilst not being a device in the meaning of paragraph 2 (b), is intended specifically by its manufacture to be used together with a device to enable this device to be used in accordance with its intended purpose;

(d) *"device for self-testing"* means any device intended by the manufacturer to be used in the home environment;

(e) *"device for performance evaluation"* means any device intended by the manufacturer to be subject to one or more performance evaluation studies in clinical laboratories, or in other appropriate environments outside his own premises;

(f) *"manufacturer"* means the natural or legal person with responsibility for the design, manufacture, packaging and labelling of a device before it is placed on the market under his own name, regardless of whether these operations are carried out by that person himself or on his behalf by a third party.

The obligations of this Directive to be met by manufacturers also apply to the natural or legal person who assembles, packages, processes, fully refurbishes and/or labels one or more ready-made products and/or assigns to them their intended purpose as devices with a view to their being placed on the market under his own name;

(g) *"authorised representative"* means the natural or legal person established within the Community who, being explicitly designated by the manufacturer, acts and may be addressed by authorities and bodies in the Community in substitution for the manufacturer with regard to the latter's obligations in accordance with this Directive.

(h) *"intended purpose"* means the use for which the device is intended according to the data supplied by the manufacturer on the labelling, in the instructions for use and/or in promotional materials;

(i) *"placing on the market"* means the

first making available in return for payment or free of charge of a device with a view to distribution and/or use on the Community market, regardless of whether it is new or fully refurbished;

(j) *"putting into service"* means the stage at which a device is ready for use on the Community market for the first time for its intended purpose;

(k) *"calibrator"*, "control material" means any substance, material or article intended by its manufacturer to establish and/or verify performance characteristics of a device in conjunction with the use of that device.

3. Whether a device incorporates tissues or substances of human origin, this Directive shall not affect national regulations relating to the ethics of the collection of tissues or substances of human origin as well as any regulations relating to the ethics governing distribution of given types of devices of such origin.

4. This Directive shall not apply to devices manufactured and used only within the same institution and on the premises of their manufacture. This does not affect the right of Member States to subject such activities to appropriate protection requirements.

5. This Directive is a specific directive within the meaning of Article 2 (2) of Directive 89/336/EEC.[8]

Article 2

Placing on the market and putting into service

Member States shall take all necessary steps to ensure that devices may be placed on the market and put into service only if they comply with the requirements laid down in this Directive when properly installed, maintained and used in accordance with their intended purpose.

Article 3

Essential requirements

Devices must meet the essential requirements set out in Annex I which apply to them, taking account of the intended purpose of the devices concerned.

Article 4

Free movement

1. Member States shall not create any obstacle to the placing on the market or the putting into service within their territory of devices bearing the CE marking provided for in Article 14 or labelled as devices for performance evaluation if these devices have been the subject of an assessment of their conformity in accordance with the provisions of Article 9.

2. At trade fairs, exhibitions, demonstrations, Member States shall not create any obstacle to the showing of devices which do not conform to this Directive, provided that a visible sign clearly indicates that such devices cannot be marketed or put into service until they have been made to comply.

3. Member States may require, when a device reaches the final user, information, as referred to in Annex 1, point 13 to be in their national language(s) or in other Community language(s) to the extent that it is needed for safe and correct use of the device. In the application of this provision Member States shall take into account the principle of proportionality and in particular whether information can be provided by recognised symbols or codes and whether the device is intended for use by trained professionals.

4. Where the devices are subject to other directives concerning other aspects and which also provide for the affixing of the CE marking, the latter shall indicate that the devices also fulfil the provisions of the other directives.

However, should one or more of

[8] Council Directive 89/336/EEC relating to electromagnetic compatibility (OJ No L 139, 23. 5. 1989, p. 19), as last amended by Directive 93/68/EEC (OJ No L 220, 30. 8. 1993, p. 1).

these directives allow the manufacturer, during a transitional period, to choose which arrangements to apply, the CE marking shall indicate that the devices fulfil the provisions only of those directives applied by the manufacturer. In this case, the particulars of these directives, as published in the *Official Journal of the European Communities*, must be given in the documents, notices or instructions required by the directives and accompanying such devices.

Article 5

Reference to standards

1. Member States shall presume compliance with the essential requirements referred to in Article 3 in respect of devices which are in conformity with the relevant national standards which transpose the harmonised standards the reference of which have been published in the *Official Journal of the European Community*; Member States shall publish the references of such national standards.

2. If the Member State or the Commission considers that the harmonised standards do not entirely meet the essential requirements referred to in Article 3, the measures to be taken with regard to these standards and the publication referred to in paragraph 1 of this Article shall be adopted by the procedure defined in Article 6 (2).

Article 6

Committee on Standards and Technical Regulations

1. The Commission shall be assisted by the Committee set up by Article 5 of Directive 83/189/EEC.

2. The representative of the Commission shall submit to the Committee a draft of the measures to be taken.

The Committee shall deliver its opinion on the draft within a time limit which the chairman may lay down according to the urgency of the matter, if necessary by taking a vote.

The opinion shall be recorded in the minutes; in addition, each Member State shall have the right to ask to have its position recorded in the minutes.

The Commission shall take the utmost account of the opinion delivered by the Committee. It shall inform the Committee of the manner in which its opinion has been taken into account.

Article 7

Committee on Medical Devices

1. The Commission shall be assisted by the Committee set up by Article 6 (2) of Directive 90/385/EEC.

2. The representative of the Commission shall submit to the Committee a draft of the measures to be taken. The Committee shall deliver its opinion on the draft within a time limit which the chairman may lay down according to the urgency of the matter. The opinion shall be delivered by the majority laid down in Article 148 (2) of the Treaty in the case of decisions which the Council is required to adopt on a proposal from the Commission. The votes of the representatives of the Member States within the Committee shall be weighted in the manner set out in that Article. The chairman shall not vote.

The Commission shall adopt the measures envisaged if they are in accordance with the opinion of the Committee.

If the measures envisaged are not in accordance with the opinion of the Committee, or if no opinion is delivered, the Commission shall, without delay, submit to the Council a proposal relating to the measures to be taken. The Council shall act by a qualified majority.

If, on the expiry of a period of three months from the date of referal to the Council, the Council has not acted, the proposed measures shall be adopted by the Commission.

3. The Committee referred to in paragraph 1 may examine any question connected with the implementation of this Directive.

Article 8

Safeguard clause

1. Where a Member State ascertains that the devices referred to in Article 4 (1), when correctly installed, maintained and used for their intended purpose may compromise the health and/or safety of patients, users or, where applicable, other persons, and the safety of property, it shall take all appropriate interim measures to withdraw such devices from the market or prohibit or restrict their being placed on the market or put into service. The Member State shall immediately inform the Commission of any such measures, indicating the reasons for its decision and, in particular, whether non-compliance with this Directive is due to:

(a) failure to meet the essential requirements referred to in Article 3;
(b) incorrect application of the standards referred to in Article 5, in so far as it is claimed that the standards have been applied;
(c) shortcomings in the standards themselves.

2. The Commission shall enter into consultation with the parties concerned as soon as possible. Where, after such consultation, the Commission finds that:

– the measures are justified, it shall immediately so inform the Member State which took the initiative and the other Member States; where the decision referred to in paragraph 1 is attributed to shortcomings in the standards, the Commission shall, after consulting the parties concerned, bring the matter before the committee referred to in Article 6 (1) within two months if the Member State which has taken the decision intends to maintain it and shall initiate the procedures referred to in Article 6,
– the measures are unjustified, it shall immediately so inform the Member State which took the initiative and the manufacturer or his authorised representative established within the Community.

3. Where a non-complying device bears the CE marking, the competent Member State shall take appropriate action against whomsoever has affixed the marking and shall inform the Commission and the other Member States thereof.

4. The Commission shall ensure that the Member States are kept informed of the progress and outcome of this procedure.

Article 9

Conformity assessment procedures

1. For all devices other than those covered by Annex 2 and devices for performance evaluation, the manufacturer shall, in order to affix the CE marking, follow the procedure referred to in Annex 3 and draw up the EC declaration of conformity required before placing the device on the market.

For all devices for self-testing the manufacturer shall, prior to the drawing up of the aforementioned declaration of conformity, fulfil the supplementary requirements set out in Annex 3 point 6. Instead of applying this procedure, the manufacturer may also follow the procedure referred to in paragraph 2.

2. For all devices covered by Annex 2 other than those intended for performance evaluation, the manufacturer shall, in order to affix the CE marking, either:

(a) follow the procedure relating to the EC declaration of conformity set out in Annex 4 (full quality assurance); or
(b) follow the procedure relating to EC type examination set out in Annex 5 coupled with:
    (i) the procedure relating to EC verification set out in Annex 6; or
    (ii) the procedure relating to the EC declaration of conformity set out in Annex 7 (production quality assurance).

3. In the case of devices for performance evaluation, the manufacturer shall follow the procedure referred to

in Annex 8 and draw up the statement set out in that Annex before placing such devices on the market.
4. During the conformity assessment procedure for a device, the manufacturer and, if involved, the notified body shall take account of the results of any assessment and verification operations which, where appropriate, have been carried out in accordance with this Directive at an intermediate stage of manufacture.
5. The manufacturer may instruct his authorised representative established in the Community to initiate the procedures provided for in Annexes 3, 5, 6 and 8.
6. The manufacturer must keep the declaration of conformity, the technical documentation referred to in Annexes 3 to 8, as well as the decisions, reports and certificates established by notified bodies, and make it available to the national authorities for inspection purposes for a period ending five years after the last product has been manufactured. Where the manufacturer is not established in the Community, the obligations to make the aforementioned documentation available on request applies to his authorised representative established within the Community.
7. Where the conformity assessment procedure involves the intervention of a notified body, the manufacture, or his authorised representative established in the Community, may apply to a body of his choice within the framework of the tasks for which the body has been notified.
8. The notified body may require, where duly justified, any information or data, which is necessary for establishing and maintaining the attestation of conformity in view of the chosen procedure.
9. Decisions taken by the notified bodies in accordance with Annexes 3, 4 and 5 shall be valid for a maximum of five years and may be extended on application, made at a time agreed in the contract signed by both parties, for further periods of five years.
10. The records and correspondence relating to the procedures referred to in paragraphs 1 to 3 shall be in one of the official languages of the European Community and in the case of involvement of a notified body in a Community language acceptable to the notified body.
11. By derogation from paragraphs 1 to 3, the competent authorities may authorise, on duly justified request, the placing on the market and putting into service, within the territory of the Member State concerned, of individual devices for which the procedures referred to in paragraphs 1 to 3 have not been carried out and the use of which is in the interest of protection of health.

Article 10

Registration of manufacturers

1. Any manufacturer, who places devices on the market under his own name, shall notify the competent authorities of the Member State in which he has his registered place of business of the address of the registered place of business, the categories of devices as defined in terms of common characteristics of technology and/or analytes and of any significant change thereto. The abovementioned notification shall also include any new device as referred to in Article 11 (4).
2. Where a manufacturer, who places devices on the market under his own name, does not have a registered place of business in a Member State, he shall designate an authorised representative established in the Community. The authorised representative shall inform the competent authorities of the Member State, in which he has his registered place of business, of all particulars as referred to in paragraph 1.
3. The Member States shall on request inform the other Member States and the Commission of the details referred to in paragraphs 1 and 2. The procedures implementing the present Article shall be adopted in accordance with the procedure referred to in Article 7 (2).

Article 11

Vigilance

1. Member States shall take the necessary steps to ensure that any information brought to their knowledge, in accordance with the provisions of this Directive, regarding the incidents mentioned below involving devices bearing the CE marking is recorded and evaluated centrally:

(a) any malfunction, failure or deterioration in the characteristics and/or performance of a device, as well as any inadequacy in the labelling or the instructions for use which, directly or indirectly, might lead to or might have led to the death of a patient or user or to a serious deterioration in his state of health;

(b) any technical or medical reason in relation to the characteristics or performance of a device for the reasons referred to in subparagraph (a), leading to systematic recall of devices of the same type by the manufacturer.

2. Where a Member State requires medical practitioners, the medical institutions or the organisers of external quality assessment schemes to inform the competent authorities of any incidents referred to in paragraph 1, it shall take the necessary steps to ensure that the manufacturer of the device concerned, or his authorised representative established in the Community, is also informed of the incident.

3. After carrying out an assessment, if possible together with the manufacturer, Member States shall, without prejudice to Article 8, immediately inform the Commission and the other Member States of the incidents referred to in paragraph 1 for which relevant measures have been taken or are contemplated.

4. Where, in the context of notification referred to in Article 10, a device notified, bearing a CE marking, is a 'new' product, the manufacturer shall indicate this fact on his notification. The competent authority so notified may at any time within the following two years and on justified grounds, require the manufacturer to submit a report relating to the experience with the device subsequent to its being placed on the market. For this purpose, a device is 'new' if:

(a) such a device, for the relevant analyte or other parameter, has not been continuously available on the Community market during the previous three years;

(b) the procedure involves analytical technology not used in connection with a given analyte or other parameter on the Community market continuously during the previous three years.

5. The Member States shall on request inform the other Member States of the details referred to in paragraphs 1 and 4. The procedures implementing the present Article shall be adopted in accordance with the procedure referred to in Article 7 (2).

Article 12

Modification of Annex 2, derogation clause

1. Where a Member State considers that:

(a) the list of devices covered by Annex 2 should be modified or extended; or

(b) the conformity of a device or a category of devices should be established by way of derogation from the provisions of Article 9, by applying one of several alternative procedures taken from amongst those referred to in Article 9,

it shall submit a duly substantiated request to the Commission and ask it to take the necessary measures. The measures shall be adopted in accordance with Article 7 (2).

2. When a measure is to be taken in accordance with paragraph 1, due consideration shall be given:

(a) to any relevant information available from the vigilance procedures and from external quality assessment schemes as referred to in Article 11;

(b) to the following criteria:
- (i) whether total reliance has to be put on the result obtained with a given device, having a direct impact on the subsequent medical action; and
- (ii) whether an action taken on the basis of an incorrect result obtained using a given device could prove to be hazardous to the patient or third party as a consequence of false positive or false negative results; and
- (iii) whether the involvement of a notified body is appropriate and justified.

3. The Commission shall inform the Member States of the measures taken and, where appropriate, publish these measures in the *Official Journal of the European Communities.*

Article 13

Notified bodies

1. Member States shall notify the Commission and other Member States of the bodies which they have designated for carrying out the tasks pertaining to the procedures referred to in Article 9 and the specific tasks for which the bodies have been designated. The Commission shall assign identification numbers to these bodies, hereinafter referred to as "notified bodies".

The Commission shall publish a list of the notified bodies, together with the identification numbers it has allocated to them and the tasks for which they have been notified, in the *Official Journal of the European Communities.* It shall ensure that the list is kept up to date.

2. Member States shall apply the criteria set out in Annex 9 for the designation of bodies. Bodies that meet the criteria laid down in the national standards which transpose the relevant harmonised standards shall be presumed to meet the relevant criteria.

3. A Member State that has notified a body shall withdraw that notification if it finds that the body no longer meets the criteria referred to in paragraph 2. It shall immediately inform the other Member States and the Commission of any withdrawal of notification for the aforementioned or any other reason.

4. The notified body and the manufacturer, or its authorised representative established in the Community, shall lay down, by common accord, the time limits for completion of the assessment and verification operations referred to in Annexes 3 to 7.

5. The notified body must make available to the other notified bodies and the competent authority, on request, all relevant information concerning approvals and certificates issued, refused or withdrawn.

Article 14

CE marking

1. Devices, other than devices for performance evaluation, considered to meet the essential requirements referred to in Article 3 must bear the CE marking of conformity when they are placed on the market.

2. The CE marking of conformity, as shown in Annex 10, must appear in a visible, legible and indelible form on the device, where practicable and appropriate, and on the instructions for use. Where applicable the CE marking of conformity must also appear on the sales packaging. The CE marking shall be accompanied by the identification number of the notified body responsible for implementation of the procedures set out in Annexes 4, 6 and 7.

3. It is prohibited to affix marks or inscriptions which are likely to mislead third parties with regard to the meaning or the graphics of the CE marking. Any other mark may be affixed to the device, to the packaging or to the instruction leaflet accompanying the device provided that the visibility and legibility of the CE marking is not thereby reduced.

Article 15

Unduly affixed CE marking

Without prejudice to Article 8:

(a) where a Member State establishes

that the CE marking has been affixed unduly, the manufacturer or his authorised representative established within the Community shall be obliged to end the infringement under conditions imposed by the Member State;

(b) where non-compliance continues, the Member State must take all appropriate measures to restrict or prohibit the placing on the market of the product in question or to ensure that it is withdrawn from the market, in accordance with the procedure in Article 8.

Article 16

Decision in respect of refusal or restriction

1. Any decision taken pursuant to this Directive:
   (a) to refuse or restrict the placing on the market or the putting into service of a device; or
   (b) to withdraw devices from the market,

shall state the exact grounds on which it is based. Such decisions shall be notified without delay to the party concerned, who shall at the same time be informed of the remedies available to him under the national law in force in the Member State in question and of the time limits to which such remedies are subject.

2. In the event of a decision as referred to in paragraph 1, the manufacturer, or his authorised representative established in the Community, shall have an opportunity to put forward his viewpoint in advance, unless such consultation is not possible because of the urgency of the measure to be taken.

Article 17

Confidentiality

Without prejudice to the existing national provisions and practices on medical secrets. Member States shall ensure that all the parties involved in the application of this Directive are bound to observe confidentiality with regard to information obtained in carrying out their tasks. This does not affect the obligations of Member States and notified bodies with regard to mutual information and the dissemination of warnings, nor the obligations of the persons concerned to provide information under criminal law.

Article 18

Cooperation between Member States

Member States shall take appropriate measures to ensure that competent authorities charged with the implementation of the present Directive cooperate with each other and convey to each other the information necessary to ensure compliance with the provisions of this Directive.

Article 19

Amendment of Directives

1. Directive 93/42/EEC is hereby amended as follows:

(a) Article 1 (2) (c) shall read as follows:

"(c) *in vitro* diagnostic medical device means any medical device which is a reagent, reagent product, calibrator, control material, kit, instrument, apparatus, equipment or system whether used alone or in combination, intended by the manufacturer to be used *in vitro* for the examination of specimens including blood and tissue donations, derived from the human body, solely or principally for the purpose of providing information concerning a physiological state, state of health or disease or congenital abnormality to or determine the safety and compatibility with potential recipients."

(b) Article 1 (5) (f) shall read as follows:

"(f) transplants or tissues or cells of human origin, unless a device is manufactured utilising tissues or substances derived from such tissues which are non viable or rendered non viable. In this case, the Directive shall not affect national regulations relating to the ethics of the collection of tissues or substances of human origin, as well as any regulations relating to the ethics governing distribution of given types of devices of such origin.";

(c) Annex I, paragraph 8.2 shall read as follows:

"8.2 Where a device is manufactured utilising human tissues or substances derived from human tissues, the use of such tissues or substances must be subject to the relevant validated selection and screening procedures, including traceability as appropriate in relation to the inherent risk.

Tissues of animal origin must originate from animals that have been subjected to veterinary controls and surveillance adapted to the intended use of tissues.

Processing, preservation, testing and handling of tissues, cells and substances of human or animal origin must be carried out so as to provide optimal security. In particular safety with regard to viruses and other transmissible agents must be addressed as appropriate by implementation of validated methods of elimination or inactivation of such viruses or other transmissible agents in the course of the manufacturing process, taking into account the sourcing and control methods applied.";

(d) Annex I, paragraph 13.3 shall be completed by the following subparagraph:

"(n) in the case of devices incorporating tissues of human origin or substances derived from such tissues, a statement indicating that the device incorporates tissue or substances derived from tissue of human origin as appropriate";

(e) in Annex II, paragraph 3.2 (c) and Annex III, paragraph 3 the following indent shall be included as third-last indent;

"in the case of devices incorporating tissues of human or animal origin, information on the selection and origin";

(f) Annex IX, Section III, paragraph 4.5: the following subparagraph shall be added:

"All devices manufactured utilising human tissues or substances derived from such tissues are in class III."

2. In Directive 89/392/EEC, the text of Article 1 (3), second indent, "machinery for medical use, used in direct contact with patients", is replaced by the following text:

"– medical devices".

Article 20

Implementation, transitional provisions

1. Member States shall adopt and publish the laws, regulations and administrative provisions necessary to comply with this Directive not later than 1 April 1998. They shall immediately inform the Commission thereof.

The Standing Committee referred to in Article 7 may assume its tasks from the date of notification of this Directive. The Member States may take the measures referred to in Article 13 on notification of this Directive.

When Member States adopt these provisions, these shall contain a reference to this Directive or shall be accompanied by such a reference at the time of their official publication. The procedures for such reference shall be adopted by Member States.

Member States shall apply these provisions with effect from 1 July 1998.

2. Member States shall communicate to the Commission the texts of the provisions of national law which they adopt in the field covered by this Directive.

3. Member States shall take the necessary action to ensure that the notified bodies which are responsible pursuant to Article 9 for conformity assessment take account of any relevant information regarding the characteristics and performance of such devices, including in particular the results of any relevant test and verification already carried out under pre-existing national law, regulations and administrative provisions in respect of such devices.
4. Member States shall accept during a period of four years following adoption of the present Directive the placing on the market and putting into service of devices which conform to the rules in force in their territory on the date of adoption of this Directive.

Article 21

This Directive shall enter into force on the 20th day after its publication.

Article 22

This directive is addressed to the Member States.

Annex I
Essential requirements

I. General requirements

1. The devices must be designed and manufactured in such a way that, when used under the conditions and for the purposes intended, they will not compromise, directly or indirectly, the clinical condition or the safety of the patients, the safety and health of users and, where applicable, other persons, and the safety of property. Any risks which may be associated with their use must be acceptable when weighed against the benefits to the patient and be compatible with a high level of protection of health and safety.
2. The solutions adopted by the manufacturer for the design and construction of the devices must conform to safety principles, taking account of the generally acknowledged state of the art.

In selecting the most appropriate solutions, the manufacturer must apply the following principles in the following order:

- eliminate or reduce risks as far as possible (inherently safe design and construction),
- where appropriate, take adequate protection measures in relation to risks that cannot be eliminated,
- inform users of the residual risks due to any shortcomings of the protection measures.

3. The devices must be designed and manufactured in such a way that they are suitable for the purposes referred to in Article 1 (2) (b), as specified by the manufacturer, taking account of the generally acknowledged state of the art. They must achieve the performance in particular in terms of analytical sensitivity, specificity, accuracy, repeatability, reproducibility and limits of detection stated by the manufacturer.

The traceability of values assigned to calibrators and/or control materials must be assured through available reference measurement procedures and/or available reference materials of higher order.

4. The characteristics and performances referred to in Sections 1 and 3 must not be adversely affected to such a degree that the health and the safety of the patient or the user and, where applicable, of other persons, are compromised during the lifetime of the device, as indicated by the manufacturer, when the device is subject to the stresses which can occur during

normal conditions of use. When no lifetime is stated, the same applies for the lifetime reasonably to be expected of a device of that kind, having regard to the intended purpose and the anticipated use of the device.

5. The devices must be designed, manufactured and packed in such a way that their characteristics and performances during their intended use will not be adversely affected in the storage and transport conditions (temperature, humidity, etc.), taking account of the instruction and information provided by the manufacturer.

II. Requirements regarding design and manufacturing

6. Chemical and physical properties

6.1. The devices must be designed and manufactured in such a way as to achieve the characteristics and performances referred to in Section I, "General Requirements". Attention must be paid to the possibility of impairment of analytical performance due to incompatibility between the materials used and the samples (such as biological tissues, cells, body fluids and micro-organisms) intended to be used with the device, taking account of its intended purpose.

6.2. The devices must be designed, manufactured and packed in such a way as to reduce as far as possible the risk posed by leakage products, contaminants and residues to the persons involved in the transport, storage and use of the device, taking account of the intended purpose of the products.

7. Infection and microbial contamination

7.1. The devices must be designed in such a way as to eliminate or reduce as far as possible the risk of infection to the user or other persons. The design must allow easy handling and, where necessary, reduce as far as possible contamination and leakage from the device during use and, in the case of specimen receptacles, the risk of contamination of the specimen. The manufacturing processes must be appropriate for these purposes.

7.2. Where a device incorporates biological substances, the risks of infection must be reduced as far as possible by selecting appropriate donors, appropriate substances and using appropriate inactivation, conservation, test and control procedures.

7.3. Devices labelled either as "STERILE" or as having a special microbiological state must be designed, manufactured and packed in an appropriate pack, according to procedures suitable to ensure that they remain in the microbiological state indicated on the label when placed on the market, under the storage and transport conditions as laid down by the manufacturer, until the protective packaging is damaged or opened.

7.4. Devices labelled either as "STERILE" or as having a special microbiological state must have been processed by an appropriate, validated method.

7.5. Packaging systems for devices other than those referred to in section 7.3 must keep the product without deterioration at the level of cleanliness, if any, as indicated by the manufacturer and, if the devices are to be sterilised prior to use, reduce as far as possible the risk of microbial contamination.

Steps shall be taken to reduce microbial contamination during selection and handling of raw materials, manufacture, storage and distribution where the performance of the device can be adversely affected by such contamination.

8. Construction and environmental properties

8.1. If the device is intended for use in combination with other devices or equipment, the whole combination, including the connection system must be safe and must not impair the specified performances of the devices. Any restrictions on use must be indicated on the label and/or in the instructions for use.

8.2. Devices must be designed and manufactured in such a way as to remove or reduce as far as possible:

- the risk of injury, in connection with their physical features, including aspects of volume × pressure, dimensional and where appropriate ergonomic features,
- risks connected with reasonably foreseeable external influences, such as magnetic fields, external electrical effects, electrostatic discharge, pressure, temperature or variations in pressure or acceleration.

Devices must be designed and manufactured in such a way as to provide an adequare level of intrinsic immunity from electromagnetic disturbance to enable them to operate in accordance with their intended purpose.

8.3. Devices must be designed and manufactured in such a way as to reduce as far as possible the risks of fire or explosion during normal use. Particular attention must be paid to devices whose intended use includes exposure to or use in association with flammable substances or substances which could cause combustion.

8.4. Devices must be designed and manufactured in such a way as to facilitate the management of safe waste disposal.

8.5. The measuring, monitoring or display scale (including colour change and other visual indicators) must be designed and manufactured in line with ergonomic principles, taking account of the intended purpose of the dcvicc.

9. Devices which are instruments or apparatus with a measuring function

9.1. Devices which are instruments or apparatus having a primary analytical measuring function must be designed and manufactured in such a way as to provide adequate stability and accuracy of measurement within appropriate accuracy limits, taking into account the intended purpose of the device and of available and appropriate reference measurement procedures and materials. The accuracy limits have to be specified by the manufacturer.

9.2. When values are expressed numerically, they must be given in legal units conforming to the provisions of Council Directive 80/181/EEC.[1]

10. Protection against radiation

10.1. Devices shall be designed and manufactured in such a way that exposure of users and other persons to the emitted radiation is reduced as far as possible.

10.2. When devices are intended to emit potentially hazardous, visible and/or invisible radiation, they must be:

- designed and manufactured in such a way as to ensure that the quality and quantity of radiation emitted can be adjusted and controlled,
- fitted, where practicable, with visual displays and/or audible warnings of such emissions.

10.3. The operating instructions for devices emitting radiation must give detailed information as to the nature of the emitted radiation, means of protecting the user, and on ways of avoiding misuse and of eliminating the risks inherent in installation.

11. Requirements for medical devices connected to or equipped with an energy source

11.1. Devices incorporating electronic programmable systems must be designed to ensure the repeatability, reliability and performance of these systems according to the intended use.

11.2. Devices must be designed and manufactured in such a way as to minimise the risks of creating electromagnetic fields which could impair the operation of other devices or equipment in the usual environment.

11.3. Devices must be designed and manufactured in such a way as to

---

[1] OJ No L 39, 15. 2. 1980, p. 40. Directive as last amended by Directive 89/617/EEC (OJ No L 357, 7. 12. 1989, p. 28).

avoid, as far as possible, the risk of accidental electric shocks during normal use and in single fault condition, provided the devices are installed and maintained correctly.

11.4. *Protection against mechanical and thermal risks*

11.4.1. Devices must be designed and manufactured in such a way as to protect the user against mechanical risks connected with, for example, resistance, stability and moving parts. Annex I, Sections 1.3 and 1.4 of Directive 89/392/EEC[2] shall apply, where appropriate.

11.4.2. Devices must be designed and manufactured in such a way as to reduce to the lowest possible level the risks arising from vibration generated by the devices, taking account of technical progress and of the means available for limiting vibrations, particularly at source, unless the vibrations are part of the specified performance.

11.4.3. Devices must be designed and manufactured in such a way as to reduce to the lowest possible level the risks arising from the noise emitted, taking account of technical progress and of the means available to reduce noise, particularly at source, unless the noise emitted is part of the specified performance.

11.4.4. Terminals and connectors to electricity, gas or hydraulic and pneumatic energy supplies which the user has to handle must be designed and manufactured in such a way as to minimise all possible risk.

11.4.5. Accessible parts of the devices (excluding the parts or areas intended to supply heat or reach given temperatures) and their surroundings must not attain potentially dangerous temperatures under normal use.

[2] Council Directive 89/392/EEC (OJ No L 183, 29. 6. 1989), as last amended by Council Directive 93/68/EEC (OJ No L 220, 30. 8. 1993, p. 1).

12. Requirements for devices for self-testing

Devices for self-testing must be designed and manufactured in such a way that they perform appropriately for their intended purpose taking into account the skills and the means available to users and the influence resulting from variation that can reasonably be anticipated in users' technique and environment. The information and instructions provided by the manufacturer should be easily understood and applied by the user.

12.1. Devices for self-testing must be designed and manufactured in such a way as to reduce as far as practicable the risk of user's error in handling and in interpretation of the result.

12.2. Devices for self-testing must, where reasonably possible, include a method of user control, i.e. a procedure by which the user can reasonably verify that, at the time of use, the product will perform as intended.

13. Information supplied by the manufacturer

13.1 Each device must be accompanied by the information needed to use it properly, taking account of the training and knowledge of the potential users, and to identify the manufacturer.

This information comprises the data on the label and in the instructions for use.

As far as practicable and appropriate, the information needed to use the device properly must be set out on the device itself and/or on the packaging or, where appropriate, on the sales packaging. If individual full labelling of each unit is not practicable, the information must be set out on the packaging and/or in the instructions for use supplied with one or more devices.

Instructions for use must accompany or be included in the packaging of one or more devices.

By way of exception, no such instructions for use are needed for a device if it can be used properly and safely without them.

13.2. Where appropriate, this information to be supplied should take the form of symbols. Any symbol and identification colour used must conform to the harmonised standards. In areas for which no standards exist, the symbols and colours as used must be described in the documentation supplied with the device.

13.3. In the case of a device containing a substance or being a preparation which may be considered as being dangerous, taking account of the nature and quantity of its constituents and the form under which they are present, relevant danger symbols and labelling requirements of Directive 67/548/EEC and Directive 88/379/EEC shall apply. Where there is insufficient space to put all information on the device itself or on its label, the relevant danger symbols shall be put on the label and the other information required by those Directives shall be given in the instructions for use.

The provisions of the abovementioned Directives on the safety data sheet shall apply, unless all relevant information as appropriate is already made available by the instructions for use.

13.4. The label must bear the following particulars which may take the form of symbols as appropriate:

(a) the name or trade name and address of the manufacturer. For devices imported into the Community, in view of their distribution in the Community, the label, or the outer packaging, or instructions for use, shall contain in addition the name and address of the authorised representative of the manufacturer established within the Community;
(b) the details strictly necessary for the user to identify the device and the contents of the packaging;
(c) where appropriate, the word "STERILE" or a statement indicating any special microbiological state or state of cleanliness;
(d) where appropriate, the batch code, preceded by the word "LOT", or the serial number;
(e) where appropriate, an indication of the date by which the device or part of it should be used, expressed as the year, the month and, where relevant, the day;
(f) in cases of devices for performance evaluation, the words "for performance evaluation only";
(g) where appropriate, a statement indicating the *in vitro* nature of the device;
(h) any special storage and/or handling conditions;
(i) where applicable, any special operating instructions;
(j) appropriate warnings and/or precautions to take;
(k) in the case of a device intended for self-testing, this fact should be clearly stated;
(l) in the case of devices incorporating tissues of human origin or substances derived from such tissues, a statement indicating that the device incorporates tissues or substances derived from tissue of human origin.

13.5. If the purpose of the device is not obvious to the user, the manufacturer must clearly state the intended purpose in the instructions for use and, if appropriate, on the label.

13.6. Wherever reasonable and practicable, the devices and detachable components must be identified, where appropriate in terms of batches, to allow all appropriate action to detect any potential risk posed by the devices and detachable components.

13.7. Where appropriate, the instructions for use must contain the following particulars:

(a) the details referred to in Section 13.4, with exception of points (d) and (e);
(b) composition of the reagent product by nature and amount or concentration of the active ingredient(s) of the reagent(s) or kit as well as a statement that the device contains other ingredients influencing the measurement;
(c) the storage conditions and shelf

life following the first opening of the immediate container, together with the storage conditions and stability of working reagents;

(d) the performances referred to in Section 3;

(e) a statement of any special materials required including information necessary to enable those special materials to be identified for proper use;

(f) the type of specimen to be used, any special conditions of collection, pre-treatment and, if necessary, storage conditions;

(g) a detailed description of the procedure to be followed in using the device;

(h) the measurement procedure to be followed with the device, including as appropriate:
- the principle of the method,
- the specific analytical performance characteristics (e.g. analytical sensitivity, specificity, accuracy, repeatability, reproducibility and limits of detection), limitations of the method and information about the use of available reference measurement procedures and materials by the user,
- the details of any further procedure or handling needed before the device can be used (for example, reconstitution, incubation, dilution, instrument checks, etc.),
- the indication whether any particular training is required;

(i) the mathematical approach upon which the calculation of the analytical result is made;

(j) measures to be taken in the event of changes in the analytical performance of the device;

(k) information appropriate to users on:
- internal quality control, including specific validation procedures,
- the traceability of the calibrartion of the devices;

(l) the reference intervals for the quantities being determined;

(m) if the device must be used in combination with or installed with or connected to other medical devices or equipment in order to operate as required for its intended purpose, sufficient details of its characteristics to identify the correct devices or equipment to use in order to obtain a safe and proper combination;

(n) all the information needed to verify whether the device is properly installed and can operate correctly and safely, plus details of the nature and frequency of the maintenance and calibration needed to ensure that the device operates properly and safely; information about safe waste disposal;

(o) the necessary instructions in the event of damage to the sterile protective packaging and details of appropriate methods of resterilisation or decontamination;

(p) if the device is reusable, information on the appropriate processes to allow reuse, including cleaning, disinfection, packaging and resterilisation or decontamination, and any restriction on the number of reuses;

(q) precautions to be taken as regards exposure, in reasonably foreseeable environmental conditions, to magnetic fields, external influences, electrostatic discharge, pressure or variations in pressure, acceleration, thermal ignition sources, etc.;

(r) precautions to be taken against any special, unusual risks related to the use or disposal of the device including special protective measures; where the device includes substances of human or animal origin, attention shall be drawn to their potential infectious nature;

(s) specifications for devices for self-testing:
- the results need to be expressed and presented in a way that is readily understood by a non-professional person; information needs to be provided with advice to the user on action

to be taken (in case of positive, negative or indeterminate result) and on the possibility of false positive or false negative result,
- specific particulars may be omitted provided that the other information supplied by the manufacturer is sufficient to enable the user to know how to use the device and to understand the result(s) produced by the device.
- the information provided must include a statement clearly directing that the user should not take any decision of medical relevance without first consulting his or her medical practitioner;

(t) the date of issue or last revision of the instructions for use.

Annex II

List of devices referred to in Article 9 (2)

1. Reagents and reagent products for blood grouping (A, B, O and Rho/D).
2. Reagents and reagent products for the detection in human specimens of markers of HIV infection and hepatitis B and C.

Annex III

EC declaration of conformity

1. The EC declaration of conformity is the procedure whereby the manufacturer or his authorised representative established in the Community who fulfills the obligations imposed by Sections 2 to 5 and additionally, in the case of devices for self-testing, the obligations imposed by Section 6, ensures and declares that the products concerned meet the provisions of this Directive which apply to them. the manufacturer must affix the CE marking in accordance with Article 14.

2. The manufacturer must prepare the technical documentation described in Section 3 and ensure that the manufacturing process follows the principles of quality assurance as set out in Section 4.

3. The technical documentation must allow assessment of the conformity of the product with the requirements of the Directive. It must include in particular:

- a general description of the product, including any variants planned,
- the documentation of the quality system.
- design information, including performance characteristics and limitation, methods of manufacture and, in the case of instruments, design drawings, diagrams of components, sub-assemblies, circuits, etc.,
- the descriptions and explanations necessary to understand the above-mentioned characteristics, drawings and diagrams and the operation of the product,
- the results of the risk analysis, where appropriate, and a list of the standards referred to in Article 5, applied in full or in part, and descriptions of the solutions adopted to meet the essential requirements of the Directive if the standards referred to in Article 5 have not been applied in full,
- in the case of sterile products or products with a special microbiological state or state of cleanliness, a

description of the procedures used,
- the results of the design calculations and of the inspections carried out, etc.,
- if the device is to be combined with other device(s) in order to operate as intended, proof must be provided that it conforms to the essential requirements when combined with any such device(s) having the characteristics specified by the manufacturer,
- the test reports,
- adequate performance evaluation data, supported by a reference system (when available), originated from studies in a clinical or other appropriate environment or resulting from available literature; this applies in particular to biological materials, culture media and devices using a new technology,
- the labels and instructions for use,

4. The manufacturer shall take necessary measures to ensure that the manufacturing process follows the principles of quality assurance as appropriate for the products manufactured.

The system shall address:
- the organisational structure and responsibilities,
- the manufacturing processes and systematic quality control of production,
- the means to monitor the performance of the quality system.

5. The manufacturer shall institute and keep up to date a systematic procedure to review experience gained from devices in the post-production phase and to implement appropriate means to apply any necessary corrective actions, taking account of the nature and risks in relation to the product. He shall notify the competent authorities of the following incidents immediately on learning of them:

(i) any malfunction, failure or deterioration in the characteristics and/or performance of a device, as well as any inadequacy in the labelling or the instructions for use which, directly or indirectly, might lead to or might have led to the death of a patient or user or to a serious deterioration in his state of health;

(ii) any technical or medical reason connected with the characteristics or the performance of a device for the reasons referred to in subparagraph (i) above leading to systematic recall of devices of the same type by the manufacturer.

6. For devices for self-testing the manufacturer shall lodge an application for examination of the design with a notified body.

6.1. The application shall enable the design of the device to be understood and shall enable conformity with the design-related requirements of the Directive to be assessed.

It shall include:
- test reports including, where appropriate, results of studies carried out with lay persons,
- data showing the handling suitability of the device in view of its intended purpose for self-testing,
- the information to be provided with the device on its label and its instructions for use.

6.2. The notified body shall examine the application and, if the design conforms to the relevant provisions of this Directive, shall issue the applicant with an EC design-examination certificate. The notified body may require the application to be completed by further tests or proof to allow assessment of conformity with the design-related requirements of the Directive. The certificate shall contain the conclusions of the examination, the conditions of validity, the data needed for identification of the approved design and, where appropriate, a description of the intended purpose of the product.

6.3 The applicant shall inform the notified body which issued the EC design-examination certificate of any significant change made to the approved design. Changes to the approved design must receive further approval from the notified body which issued the EC design-examination certificate wherever the changes could affect conformity with the essential requirements of the Directive or with the conditions prescribed for use of the product. This additional approval shall take the form of a supplement to the EC design-examination certificate.

Annex IV

EC declaration of conformity (Full quality assurance system)

1. The manufacturer must ensure application of the quality system approved for the design, manufacture and final inspection of the products concerned, as specified in Section 3, and is subject to audit as laid down in Section 3.3 and to the EC surveillance as specified in Section 4.

2. The declaration of conformity is the procedure whereby the manufacturer who fulfils the obligations imposed by Section 1 ensures and declares that the products concerned meet the provisions of this Directive which apply to them.

The manufacturer shall affix the CE marking in accordance with Article 14 and shall draw up a declaration of conformity covering the products concerned.

3. Quality system

3.1. the manufacturer must lodge an application for assessment of his quality system with a notified body.

The application must include:

- the name and address of the manufacturer and any additional manufacturing site covered by the quality system,
- adequate information on the product or product category covered by the procedure,
- a written declaration that no such application has been lodged with any other notified body for the same product-related quality system.
- the documentation on the quality system,
- an undertaking by the manufacturer to fulfil the obligations imposed by the quality system approved.
- an undertaking by the manufacturer to keep the approved quality system adequate and efficacious,
- an undertaking by the manufacturer to institute and keep up to date a systematic procedure to review experience gained from devices in the post-production phase and to implement appropriate means to apply any necessary corrective action and notification as referred to in Annex 3, Section 5.

3.2. Application of the quality system must ensure that the products conform to the provisions of this Directive which apply to them at every stage, from design to final inspection. All the elements, requirements and provisions adopted by the manufacturer for his quality system must be documented in a systematic and orderly manner in the form of written policies and procedures, such as quality programmes, quality plans, quality manuals and quality records.

It shall include in particular an adequate description of:

(a) the manufacturer's quality objectives;

(b) the organisation of the business and in particular:
  - the organisational structures, the responsibilities of the managerial staff and their organisational authority where quality of design and manufacture of the products is concerned,
  - the methods of monitoring the efficient operation of the quality system and in particular its ability to achieve the desired quality of design and of product, including control of products which fail to conform;

(c) the procedures for monitoring and verifying the design of the products and in particular:
- a general description of the product, including any variants planned,
- all documentation referred to in Annex 3, Section 3, indents 3 to 11,
- in the case of devices for self-testing, the information referred to in Annex 3, Section 6.1,
- the techniques used to control and verify the design and the processes and systematic measures which will be used when the products are being designed;

(d) the inspection and quality assurance techniques at the manufacturing stage and in particular:
- the processes and procedures which will be used, particularly as regards sterilisation and purchasing,
- the product identification procedures drawn up and kept up to date from drawings, specifications or other relevant documents at every stage of manufacture..

(e) the appropriate tests and trials which will be carried out before, during and after manufacture, the frequency with which they will take place, and the test equipment used; it must be possible to trace back the calibration of the test equipment adequately.

3.3.The notified body must audit the quality system to determine whether it meets the requirements referred to in Section 3.2. It must presume that quality systems which implement the relevant harmonised standards conform to these requirements.

The assessment team must include at least one member with past experience of assessments of the technology concerned. The assessment procedure must include an inspection on the manufacturer's premises and, in duly substantiated cases, on the premises of the manufacturer's suppliers and/or subcontractors to inspect the manufacturing processes.

The decision shall be notified to the manufacturer. It must contain the conclusions of the inspection and a reasoned assessment.

3.4. The manufacturer must inform the notified body which approved the quality system of any plan for substantial changes to the quality system of the product range covered.

The notified body must assess the changes proposed and verify whether after these changes the quality system still meets the requirements referred to in Section 3.2. It must notify the manufacturer of its decision. This decision must contain the conclusions of the inspection and a reasoned assessment.

4. Surveillance

4.1. The aim is to ensure that the manufacturer duly fulfils the obligations imposed by the approved quality system.

4.2. The manufacturer must authorise the notified body to carry out all the necessary inspections and supply it with all relevant information, in particular:
- the documentation on the quality system.
- the data stipulated in the part of the quality system relating to design, such as the results of analyses, calculation, tests, etc.,
- the data stipulated in the part of the quality system relating to manufactur, such as inspection reports and test data, qualification reports of the personnel concerned, etc.

4.3. The notified body must periodically carry out appropriate inspections and assessments to make sure that the manufacturer applies the approved quality system and must supply the manufacturer with an assessment report.

4.4. In addition, the notified body may pay unannounced visits to the manufacturer. At the time of such visits, the notified body may, where necessary, carry out or ask for tests in order to check that the quality system is working properly. It must provide the manufacturer with an inspection report and, if a test has been carried out, with a test report.

Annex V
EC type-examination

1. EC type-examination is the part of the procedure whereby a notified body ascertains and certifies that a representative sample of the production envisaged fulfils the relevant provisions of this Directive.
2. The application for EC type-examination shall be lodged by the manufacturer or by his authorised representative established in the Community with a notified body.

The application shall include:
- the name and address of the manufacturer and the name and address of the authorised representative if the application is lodged by the representative,
- the documentation described in Section 3 needed to assess the conformity of the representative sample of the production in question, hereinafter referred to as the "type", with the requirements of this Directive. The applicant shall make a "type" available to the notified body. The notified body may request other samples as necessary,
- a written declaration that no application has been lodged with any other notified body for the same type.

The documentation must allow an understanding of the design, the manufacture and the performances of the product. The documentation shall contain the following items in particular:
- a general description of the type, including any variants planned,
- all documentation referred to in Annex 3, Section 23, indents 3 to 5, 7, 10 and 11.

4.The notified body shall:
4.1. examine and assess the documentation and verify that the type has been manufactured in conformity with that documentation; it shall also record the items designed in conformity with the applicable provision of the standards referred to in Article 5, as well as the items not designed on the basis of the relevant provisions of the abovementioned standards;
4.2. perform or have performed appropriate examinations and the tests necessary to verify whether the solutions adopted by the manufacturer meet the essential requirements of this Directive if the standards referred to in Article 5 have not been applied; if the device is to be combined with other device(s) in order to operate as intended, proof must be provided that it conforms to the essential requirements when combined with any such device(s) having the characteristics specified by the manufacturer;
4.3. carry out or ask for the appropriate examinations and the tests necessary to verify whether, if the manufacturer has chosen to apply the relevant standards, these have actually been applied;
4.4. agree with the applicant on the place where the necessary examinations and tests will be carried out.
5. If the type conforms to the provisions of this Directive, the notified body shall issue the applicant with an EC type-examination certificate. The certificate shall contain the name and address of the manufacturer, the conclusions of the examination, the conditions of validity and the data needed for identification of the type approved. The relevant parts of the documentation shall be annexed to the certificate and a copy shall be kept by the notified body.
6. The applicant shall inform the noti-

fied body which issued the EC type-examination certificate of any significant change made to the approved product.

Changes to the approved product must receive further approval from the notified body which issued the EC type-examination certificate wherever the changes may affect conformity with the essential requirements or with the conditions prescribed for use of the product. This new approval shall, where appropriate, take the form of a supplement to the initial EC type-examination certificate.

7. Administrative provisions

Other notified bodies may obtain a copy of the EC type-examination certificates and/or the supplements thereto. The annexes to the certificates must be made available to the other notified bodies on reasoned application, after the manufacturer has been informed.

Annex VI
EC verification

1. EC verification is the procedure whereby the manufacturer or his authorised representative established in the Community ensures and declares that the products which have been subject to the procedure set out in Section 4 conform to the type described in the EC type-examination certificate and meet the requirements of this Directive which apply to them.

2.1. The manufacturer must take all the measures necessary to ensure that the manufacturing process produces products which conform to the type described in the EC type-examination certificate and to the requirements of the Directive which apply to them. Before the start of manufacture, the manufacturer must prepare documents defining the manufacturing process, in particular as regards sterilisation and the suitability of starting materials, where necessary, and define the necessary testing procedures according to the state of the art. All the routine, pre-established provisions must be implemented to ensure homogeneous production and conformity of the products with the type described in the EC type-examination certificate and with the requirements of the Directive which apply to them. The manufacturer must affix the CE marking in accordance with Article 14 and draw up a declaration of conformity covering the products concerned.

2.2. To the extent that for certain aspects the final testing according to Section 6.3 is not appropriate, adequate in-process testing, monitoring and control methods shall be established by the manufacturer with the approval of the notified body. The provisions of Annex 4, Section 4 shall apply accordingly in relation to the abovementioned approved procedures.

3. The manufacturer must undertake to institute and keep up to date a systematic procedure to review experience gained from devices in the post-production phase and to implement appropriate means to apply any necessary corrective and notification action as referred to in Annex 3, Section 5.

4. The notified body must carry out the appropriate examinations and tests in order to verify the conformity of the product with the requirements of the Directive either by examining and testing every product as specified in Section 5 or by examining and testing products on a statistical basis as specified in Section 6, as the manufacturer decides.

In as far as the conduct of examination and tests on a statistical basis is not appropriate, examination and tests may be carried out on a random basis provided that such procedure in conjunction with the measures taken in accordance with Section 2.2 ensures an equivalent level of conformity.

5. Verification by examination and testing of every product

5.1. Every product is examined individually and the appropriate test defined in the relevant standard(s) referred to in Article 5 or equivalent tests must be carried out in order to verify the conformity of the products with the EC type described in the type-examination certificate and with the requirements of the Directive which apply to them.

5.2. The notified body must affix, or have affixed, its identification number to each approved product and must draw up a written certificate of conformity relating to the tests carried out.

6. Statistical verification

6.1. The manufacturer must present the manufactured products in the form of homogeneous batches.

6.2. A random sample is taken from each batch. The products which make up the sample are examined individually and the appropriate tests defined in the relevant standard(s) referred to in Article 5 or equivalent tests must be carried out to verify, where appropriate, the conformity of the products with the type described in the EC type-examination certificate and with the requirements of the Directive which apply to them in order to determine whether to accept or reject the batch.

6.3. Statistical control of products will be based on attributes, entailing a sampling system ensuring a limit quality corresponding to a probability of acceptance of 5%, with a non-conformity percentage of between 3 and 7%. The sampling method will be established by the harmonised standards referred to in Article 5, taking account of the specific nature of the product categories in question.

6.4. If the batch is accepted, the notified body affixes, or has affixed its identification number to each product and draws up a written certificate of conformity relating to the test carried out. All products in the batch may be put on the market except any in the sample which failed to conform.

If a batch is rejected the competent notified body must take appropriate measures to prevent the batch from being placed on the market. In the event of frequent rejection of batches, the notified body may suspect the statistical verification.

The manufacturer may, on the responsibility of the notified body, affix the notified body's identification number during the manufacturing process.

Annex VII

EC declaration of conformity (Product quality assurance)

1. The manufacturer must ensure application of the quality system approved for the manufacture of the products concerned and carry out the final inspection, as specified in Section 3, and is subject to the EC surveillance referred to in Section 4.

2. The declaration of conformity is the part of the procedure whereby the manufacturer who fulfils the obligations imposed by Section 1 ensures and declares that the products concerned conform to the type described in the EC type-examination certificate and meet the provisions of this Directive which apply to them.

The manufacturer must lodge an application for assessment of his quality system with a notified body.

3. Quality system

3.1. The manufacturer must lodge an application for assessment of his quality system with a notified body.

The application must include:

- all documentation and undertakings referred to in Annex IV, Section 3.1, and
- the technical documentation on the types approved and a copy of the EC type-examination certificates.

3.2. Application of the quality system must ensure that the products conform to the type described in the EC type-examination certificate.

All the elements, requirements and provisions adopted by the manufacturer for his quality system must be documented in a systematic and orderly manner in the form of written policy statements and procedures. This quality system documentation must permit uniform interpretation of the quality policy and procedures such as quality programmes, plans, manuals and records.

It must include in particular an adequate description of:

(a) the manufacturer's quality objectives;

(b) the organisation of the business and in particular:
  - the organisational structures, the responsibilities of the managerial staff and their organisational authority where manufacture of the products is concerned,
  - the methods of monitoring the efficient operation of the quality system and in particular its ability to achieve the desired quality of design and of product, including control of products which fail to conform;

(c) the inspection and quality assurance techniques at the manufacturing stage and in particular:
  - the process and procedures which will be used, particularly as regards sterilisation and purchasing,
  - the product identification procedures drawn up and kept up to date from drawings, specifications or other relevant documents at every stage of manufacture;

(d) the appropriate tests and trials to be carried out before, during and after manufacture, the frequency with which they will take place, and the test equipment used; it must be possible adequately to trace back the calibration of the test equipment.

3.3. The notified body must audit the quality system to determine whether it meets the requirements referred to in Section 3.2. It must presume that quality systems which implement the relevant harmonised standards conform to these requirements.

The assessment team must include at least one member with past experience of assessment of the technology concerned. The assessment procedure must include an inspection of the manufacturer's premises and, in duly substantiated cases, on the premises of the manufacturer's suppliers and/or subcontractors to inspect the manufacturing processes.

The decision must be notified to the manufacturer after the final inspection and contain the conclusions of the inspection and a reasoned assessment.

3.4. The manufacturer shall inform the notified body which approved the quality system of any plan for substantial changes to the quality system.

The notified body must assess the changes proposed and verify whether after these changes the quality system still meets the requirements referred to in Section 3.2.

It must notify the manufacturer of its decision. This decision must contain the conclusions of the inspection and a reasoned assessment.

4. Surveillance

The provision of Annex 4, Section 4, shall apply.

Annex VIII

Statement and procedures concerning devices for performance evaluation

1. For devices for performance evaluation the manufacturer or his authorised representative established in the Community shall draw up the statement containing the information stipulated in Section 2 of this Annex and ensure that the relevant provisions of this Directive are met.

2. The statement shall contain the following information:

- the list of laboratories or other institutions taking part in the performance evaluation study,
- in the case of devices for self-testing, the place, starting date and scheduled duration for the investigations and the number of laypersons involved,
- a statement that the device in question conforms to the requirement of the Directive, apart from the aspects covered by the investigation and apart from those specifically itemised in the statement, and that every precaution has been taken to protect the health and safety of the patient, user and third parties.

3. The manufacturer shall also undertake to keep available for the competent national authorities the documentation allowing an understanding of the design, manufacture and performance of the product, including the expected performance, so as to allow assessment of conformity with the requirements of this Directive. This documentation must be kept for a period ending at least five years after the end of the performance evaluation.

The manufacturer shall take all the measures necessary for the manufacturing process to ensure that the products manufactured conform to the documentation mentioned in the first paragraph.

Annex IX

Criteria for the designation of notified bodies

1. The notified body, its director and the assessment and verification staff shall not be the designer, manufacturer, supplier, installer or user of the devices which they inspect, nor the authorised representative of any of these persons. They may not be directly involved in the design, construction, marketing or maintenance of the devices, nor represent the parties engaged in these activities. This in no way precludes the possibility of exchanges of technical information between the manufacturer and the body.

2. The notified body and its staff must carry out the assessment and verification operations with the highest degree of professional integrity and the requisite competence in the field of medical devices and must be free from all pressures and inducements, particularly financial, which might influence their judgment or the results of the inspection, especially from persons or groups of persons with an interest in the results of the verifications.

Should the notified body subcontract specific tasks connected with the establishment and verification of the facts, it must first ensure that the subcontractor meets the provisions of the Directive. The notified body shall keep at the disposal of the national authorities the relevant documents assessing the subcontractor's qualifications and the work carried out by the subcontractor under this Directive.

3. The notified body must be able to carry out all the tasks assigned to such bodies by one of Annexes 3 to 7 and for which it has been notified, whether these tasks are carried out by the body itself or on its responsibility. In particular, it must have the necessary staff and possess the facilities needed to perform properly the technical and administrative tasks entailed in assessment and verification. It must also have access to the equipment necessary for the verification required.

4. The notified body must have:

– sound vocational training covering all the assessment and verification operations for which the body has been designated,
– satisfactory knowledge of the rules on the inspections which they carry out and adequate experience of such inspections,
– the ability required to draw up the certificates, records and reports to demonstrate that the inspections have been carried out.

5. The impartiality of the inspection staff must be guaranteed. Their remuneration must not depend on the number of inspections carried out, nor on the results of the inspections.

6. The body must take out civil liability insurance, unless liability is assumed by the State under domestic legislation or the Member State itself carries out the inspections directly.

7. The staff of the inspection body are bound to observe professional secrecy with regard to all information gained in the course of their duties (except *vis-à-vis* the competent administrative authorities of the State in which their activities are carried out) under this Directive or any provision of national law putting it into effect.

Annex X

CE marking of conformity

The CE conformity marking shall consist of the initials "CE" taking the following form:

– If the CE marking is reduced or enlarged the proportions given in the above graduated drawing must be respected.
– The various components of the CE marking must have substantially the same vertical dimensions, which may not be less than 5 mm. This minimum dimension may be waived for small-scale devices.

COUNCIL DIRECTIVE 93/15/EEC

of 5 April 1993

on the harmonisation of the provisions relating to the placing on the market and supervision of explosives for civil uses

THE COUNCIL OF THE EUROPEAN COMMUNITIES,

Having regard to the Treaty establishing the European Economic Community, and in particular Article 100a thereof,

Having regard to the proposal from the Commission,[1]

In cooperation with the European Parliament,[2]

Having regard to the opinion of the Economic and Social Committee,[3]

Whereas Article 8a of the Treaty provides that the internal market must be established not later than 31 December 1992; whereas the internal market is to comprise an area without internal frontiers in which the free movement of goods, persons, services and capital is ensured in accordance with the provisions of the Treaty;

Whereas Article 100a (3) of the Treaty provides that the Commission, in its proposals concerning safety, will take as a base a high level of protection;

Whereas the free movement of goods presupposes that certain basic conditions are fulfilled; whereas, in particular, the free movement of explosives presupposes harmonisation of laws on the placing of explosives on the market;

Whereas explosives for civil use are covered by detailed national regulations, mainly in respect of safety and security requirements; whereas such national regulations provide, in particular, that marketing authorisations be granted only where explosives have satisfactorily undergone a series of tests;

Whereas harmonisation of provisions governing the placing of such explosives on the market presupposes that divergent national rules will be harmonised in order to ensure the free movement of these products without lowering optimum levels of safety and security;

Whereas this Directive defines only the essential requirements which must be met by explosives conformity tests; whereas, in order to facilitate the process of demonstrating compliance with the essential requirements, it would be very useful to process standards harmonised at European level concerning, *inter alia*, methods for testing exoplosives; whereas such standards do not exist at present;

Whereas standards harmonised at European level are drawn up by private bodies and must retain their status as non-mandatory text; whereas, in this connection, the European Committee for Standardisation (CEN) has been recognised as one of the two bodies competent to adopt harmonised standards in accordance with the general guidelines for cooperation between the Commission and CEN and European Committee for Electrotechnical Standardisation (CENELEC), ratified on 13 November 1984; whereas, for the purposes of this Directive, "harmonised standard" means a text setting out technical specifications adopted by CEN under a mandate conferred by the Commission, in accordance with Council Direc-

---

[1] OJ No C 121, 13. 5. 1992, p. 19.
[2] OJ No C 305, 23. 11. 1992, p. 128; and OJ No C 115, 26. 4. 1993.
[3] OJ No C 313, 30. 11. 1992, p. 13.

tive 83/189/EEC of 28 March 1983 laying down a procedure for the provision of information in the field of technical standards and regulations[4] and in keeping with the abovementioned general guidelines;

Wheras the Council, in its Decision 90/683/EEC of 13 December 1990 concerning the modules for the various phases of the conformity assessment procedures which are intended to be used in the technical harmonisation directives,[5] introduced harmonised means of applying procedures for conformity assessment; whereas the application of these modules to explosives will make it possible to determine the responsibility of manufacturers and of bodies responsible for applying procedures for conformity assessment by taking account of the nature of the explosives concerned;

Whereas, as regards safety, the rules concerning the transport of explosives are covered by international conventions and agreements; whereas, at international level, there are United Nations recommendations on the transport of dangerous goods (including explosives), the scope of which extends beyond the Community framework; whereas, in consequence, this Directive does not concern the transport rules;

Whereas, pyrotechnical articles require appropriate measures to ensure the protection of consumers and the safety of the public; whereas an additional directive is planned in this field;

Whereas the definition of the products covered by this Directive should be based on the definition of such products as set out in the abovementioned recommendations;

Whereas the scope of this Directive covers ammunition, but only as regards the rules governing controls on transfers and the associated arrangements; whereas, since ammunition is transferred under conditions similar to those under which arms are transferred, transfers of ammunition should be governed by provisions similar to those applicable to arms, as set out in Directive 91/447/EEC of 18 June 1991 on control of the acquisition and possession of weapons;[6]

Whereas the health and safety of workers producing or using explosives must also be protected; whereas an additional directive is in the course of preparation covering, *inter alia*, the health and safety of workers engaged in activities relating to the manufacture, storage and use of explosives;

Whereas it is appropriate in the case of a serious threat to, or attack on, public safety by reason of the illicit possession or use of explosives or ammunition falling within this Directive, to allow Member States to derogate, under certain conditions, from the provision of this Directive with regard to transfer;

Whereas, finally, it is essential to establish administrative cooperation mechanisms; whereas it is appropriate in this connection for the component authorities to base their approach on Council Regulation (EEC) No 1468/81 of 19 May 1981 on mutual assistance between the administrative authorities of the Member States and cooperation between the latter and the Commission to ensure the correct application of the law on customs or agricultural matters;[7]

Whereas this Directive does not affect the power of Member States to take measures with a view to preventing illegal trade in explosives and ammunition,

HAS ADOPTED THIS DIRECTIVE:

---

[4] OJ No L 109, 26. 4. 1983, p. 8. Directive as last amended by Commission Decision 90/230/EEC (OJ No L 128, 18. 5. 1990, p. 15).

[5] OJ No L 380, 31. 12. 1990, p. 13.

[6] OJ No L 256, 13. 9. 1991.

[7] OJ No L 144, 2. 6. 1981, p. 1. Regulation as amended by Regulation (EEC) No 945/87 (OJ No L 90, 2. 4. 1987, p. 4.)

Chapter I
General provisions

Article 1

1. This Directive shall apply to explosives as defined in paragraph 2.
2. "Explosives" shall mean the materials and articles considered to be such in the United Nations recommendations on the transport of dangerous goods and falling within Class 1 of those recommendations.
3. This Directive shall not apply to:
- explosives, including ammunition, intended for use, in accordance with national law, by the armed forces or the police,
- pyrotechnical articles,
- ammunition, except as provided in Articles 10, 11, 12, 13, 17, 18 and 19.

4. For the purpose of this Directive:
- "United Nations recommendations" shall mean the recommendations laid down by the United Nations Committee of Experts on the Transport of Dangerous Goods, as published in the UN (Orange Book) and as amended by the date when this Directive is adopted,
- "safety" shall mean the prevention of accidents and, where prevention fails, the containment of their effects,
- "security" shall mean the prevention of use contrary to law and order,
- "dealer" shall mean any natural or legal person whose occupation consists wholly or partly in the manufacture, trade, exchange, hiring out, repair or conversion of fire arms and ammunition,
- "approval" shall mean the decision taken to allow envisaged transfers of explosives within the Community,
- "undertaking in the explosives sector" shall mean any natural or legal person possessing a licence or authorisation which entitles him to engage in the manufacture, storage, use, transfer or trade in explosives,
- "placing on the market" shall mean any first disposal against payment or free of charge of explosives covered by this Directive with a view to their distribution and/or use on the Community market,
- "transfer" shall mean any physical movement of explosives within Community territory apart from movements within one and the same site.

5. This Directive shall not prevent Member States from designating certain substances not covered by this Directive as explosives under national law or regulations.

Chapter II
Harmonisation of laws relating to explosives

Article 2

1. Member States may not prohibit, restrict or hinder the placing on the market of explosives which fall within the scope of this Directive and which satisfy the requirements of this Directive.
2. Member States shall take the necessary measures to ensure that explosives falling within the scope of this Directive may be placed on the market only if they comply with all the provisions of this Directive, are provided with the CE marking described in Article 7 and their conformity has assessed in accordance with the procedures referred to in Annex II.
3. Where explosives falling within the scope of this Directive are subject to other Directives which cover other aspects and prescribe the fixing of the CE marking, this marking shall indicate that the abovementioned products are also presumed to conform to the provisions of these other directives which apply to them.

Article 3

Explosives falling within the scope of this Directive must comply with the essential safety requirements set out in Annex I which apply to them.

Article 4

1. Member States shall consider explosives falling within the scope of this

Directive which comply with the relevant national standards transposing the harmonised standards the references of which have been published in the *Official Journal of the European Communities* to be in conformity with the essential safety requirements referred to in Article 3. Member States shall publish the references of the national standards transposing those harmonised standards.

2. The Commission will give specific details of the work conducted in the field of harmonised standards in the framework of the report submitted to the European Parliament and the Council on the application of Directive 83/189/EEC and provided for in Article 11 (2) of that Directive.

Article 5

Where a Member State or the Commission considers that the harmonised standards referred to in Article 4 do not fully satisfy the essential requirements referred to in Article 3, the Commission or the Member State concerned shall bring the matter before the Standing Committee set up by Directive 83/189/EEC, giving its reasons. The Committee shall deliver its opinion without delay.

In the light of the Committee's opinion the Commission shall inform the Member States of the measures to be taken regarding the standards and the publication referred to in Article 4.

Article 6

1. The procedures for the attestation of the conformity of explosives shall be either:

(a) EC type examination (Module B) referred to in Annex 11 (1), and, at the choice of the manufacturer, either:
   - the type conformity (Module C) referred to in Annex II (2),
   - or the production quality assurance procedure (Module D) referred to in Annex II (3),
   - or the product quality assurance procedure (Module E) referred to in Annex II (4),
   - or the product verification (Module F) referred to in Annex II (5); or

(b) the unit verification (Module G) referred to in Annex II (6).

2. Member States shall inform the Commission and the other Member States of the bodies which they have appointed to carry out the procedures for assessing the conformity referred to above together with the specific tasks which these bodies have been appointed to carry out and the identification numbers assigned to them beforehand by the Commission.

The Commission shall publish in the *Official Journal of the European Communities* a list of the notified bodies and their identification numbers and the tasks for which they have been notified. The Commission shall ensure that this list is kept up to date.

Member States shall apply the minimum criteria set out in Annex III for the assessment of bodies of which the Commission is to be notified. Bodies which meet the assessment criteria laid down by the relevant harmonised standards shall be presumed to satisfy the relevant minimum criteria.

A Member State which has notified the Commission of a given body shall withdraw the notification if it discovers that that body no longer meets the criteria referred to in the second subparagraph. It shall immediately inform the other Member States and the Commission accordingly.

Article 7

1. The CE marking of conformity shall be affixed in such a way as to be visible, easily legible and indelible on the explosives themselves or, if this is not possible, on an identification plate attached thereto or, in the last resort, if the first two methods cannot be used, on the packaging. The identification plate must be so designed as to make its reuse impossible.

The model to be used for the CE marking shall be that reproduced in Annex IV.

2. It shall be prohibited to affix on explosives any mark or inscription

which may confuse third persons as to the meaning and style of writing of the CE marking. Any other mark may be affixed on explosives provided the visibility and legibility of the CE marking is not impaired.

3. Without prejudice to the provisions of Article 8:

(a) where a Member State establishes that the CE marking has been unduly affixed, the manufacturer, his agent or, failing these, the person responsible for placing the product in question on the Community market shall be obliged to restore the product to conformity with regard to the provisions on marking and end the infringement under conditions imposed by the Member States;

(b) where non-compliance continues, the Member State must take all. appropriate measures to restrict or prohibit the placing on the market of the product in question or to ensure that it is withdrawn from the market in accordance with the procedures laid down in Article 8.

Article 8

1. Where a Member State establishes that an explosive bearing CE conformity marking and being used for its intended purpose may compromise safety, it shall take all interim measures necessary to withdraw the explosive from the market or prohibit its being placed on the market or its freedom of movement.

The Member State shall immediately inform the Commission of such measures, indicating the reasons for its measures and, in particular, whether non-conformity is the result of:

- non-compliance with essential requirements,
- incorrect applications of standards, or
- a shortcoming in the standards.

2. The Commission shall consult the parties concerned as soon as possible. Where the Commission establishes, after consultation, that the measures are justified, it shall immediately inform the Member State which took the initiative, as well as the other Member States. Where the Commission establishes, after consultation, that the measures are unjustified, it shall immediately inform the Member State which took the decision.

In the particular case where the measures referred to in paragraph 1 are based on a shortcoming in the standards, the Commission shall first consult the parties concerned and then within two months bring the matter before the Standing Committee set up by Directive 83/189/EEC if the Member State which took the measures intends to maintain them and initiates the procedures referred to in Article 5.

3. Where CE conformity marking is borne by an explosive which does not comply with the requirements, the component Member State shall take appropriate measures in respect of the person who affixed the marking and shall inform the Commission and the other Member States.

Chapter III
Provisions governing the supervision of transfers in the community

Article 9

1. Explosives covered by this Directive may be transferred only in accordance with the following paragraphs.

2. Controls performed pursuant to Community law or national law in the event of transfers of the explosives governed by this Article shall no longer be performed as internal frontier controls but solely as part of the normal control procedures applied in a non-discriminatory fashion throughout the territory of the Community.

3. Approval to transfer explosives shall be obtained by the consignee from the recipient competent authority. The competent authority shall verify that the consignee is legally authorised to acquire explosives and that he is in possession of the necessary licences or authorisations. The person respon-

sible for the transfer must notify the competent authorities of the transit Member State or Member States of movement of explosives through this or these States, whose approval shall be required.

4. Where a Member State considers that there is a problem regarding the verification of the entitlement to acquire explosives referred to in paragraph 3, that Member State shall forward the available information on the subject to the Commission which will put the matter before the Committee provided for in Article 12 without delay.

5. Where the recipient competent authority approves a transfer, it shall issue to the consignee a document which includes all the information referred to in paragraph 7. Such a document must accompany the explosives until they arrive at their stated destination. It must be produced at the request of the relevant competent authorities. A copy of this document shall be retained by the consignee who shall present it for examination by the recipient competent authority, at the latter's request.

6. Where the competent authority of a Member State considers that special security requirements such as those referred to in paragraph 5 are unnecessary, explosives can be transferred on their territory or part thereof without prior provision of information within the meaning of paragraph 7. The recipient competent authority shall then grant an approval for a fixed period and liable to suspension or withdrawal at any time on the basis of a reasoned justification. The document referred to in paragraph 5, which must accompany the explosives until they arrive at their destination, shall refer solely to the abovementioned approval.

7. Where transfers of explosives must be specially supervised in order to comply with special security requirements in the territory or part of the territory of a Member State, prior to the transfer the following information shall be provided by the consignee to the recipient competent authority:

- the names and addresses of the operators concerned; this information must be detailed enough to enable the operators to be contacted and confirmation to be obtained that the persons in question are legally entitled to receive the consignment,
- the number and quantity of the explosives being transferred,
- a full description of the explosive in question and of the means of identification, including the United Nations identification number,
- where the explosives are to be placed on the market, information on compliance with conditions for placing on the market.
- the means of transfer and the itinerary,
- the expected dates of departure and arrival,
- where necessary, the precise points of entry to and exit from Member States.

Recipient competent authorities shall examine the conditions under which the transfer may take place, with particular regard to the special security requirements. If the special security requirements are satisfied, approval for the transfer shall be granted. In the event of transit through the territory of other Member States, those States shall likewise examine and approve, in the same conditions, the particulars concerning the transfer.

8. Without prejudice to the normal checks which the Member State of departure shall carry out in its territory, at the request of the competent authorities concerned, the consignees and the operators concerned in the explosives sector shall forward to the authorities of the Member State of departure and to those of the Member State of transit all relevant information they possess concerning the transfer of explosives.

9. No supplier may transfer explosives unless the consignee has obtained the necessary authorisation for the transfer in accordance with the provisions of paragraph 3, 5, 6 and 7.

Article 10

1. Ammunition may be transferred from one Member State to another only in accordance with the procedure laid down in the following paragraphs. These provisions shall also apply to transfers of ammunition under mail-order sales.

2. Where ammunition is to be transferred to another Member State the person concerned shall, before any dispatch, communicate to the Member State in which that ammunition is located:

- the names and addresses of the person selling or transferring the ammunition, of the person purchasing or acquiring the ammunition and, where appropriate, of the owner,
- the address to which the ammunition is to be consigned or transported,
- the quantity of ammunition to be consigned or transported,
- data making it possible to identify the ammunition and also an indication that the ammunition has undergone a check in accordance with the Convention of 1 July 1969 on the Reciprocal Recognition of Proofmarks on Small Arms,
- the means of transfer,
- the date of departure and the estimated date of arrival.

The information referred to in the last two indents need not be supplied in the event of a transfer between dealers. The Member State shall examine the conditions under which the transfer is to be carried out, in particular with regard to security. Where the Member State authorises such a transfer it shall issue a licence incorporating all the particulars referred to in the first subparagraph. That licence shall accompany the ammunition until it reaches its destination; it shall be produced whenever so required by the competent authorities of the Member States.

3. Each Member State may grant dealers the right to effect transfers of ammunition from its territory to a dealer established in another Member State without the prior authorisation referred to in paragraph 2. To that end it shall issue an authorisation valid for three years which may at any time be suspended or cancelled by reasoned decision. A document referring to that authorisation must accompany the ammunition until it reaches its destination. It must be produced whenever so required by the competent authorities of the Member States.

Before effecting the transfer, the dealer shall communicate to the authorities of the Member State from which the transfer is to be effected all the particulars listed in the first subparagraph of paragraph 2.

4. Each Member State shall supply the other Member States with a list of the ammunition the transfer of which to its territory may be authorised without its prior consent.

Such lists of ammunition shall be communicated to dealers who have obtained approval of transferring ammunition without prior authorisation under the procedure laid down in paragraph 3.

5. Each Member State shall communicate all useful information at its disposal concerning definitive transfers of ammunition to the Member State, to the territory of which such a transfer has been effected.

All information that Member States receive by way of the procedures laid down in this Article shall be communicated, not later than the time of the relevant transfers, to the Member States of destination and, where appropriate, not later than the time of transfer to the Member States of transit.

Article 11

By derogation from Article 9 (3), (5), (6) and (7), and from Article 10, a Member State, in case of grave threats to, or attacks upon, public security through the illicit possession or use of explosives or ammunition covered by the Directive, may take all necessary measures concerning transfers of explosives or ammunition in order to prevent such illicit possesison or use.

These measures shall respect the principle of proportionality. They must constitute neither a means of arbitrary discrimination nor a veiled restriction in trade between Member States.

Each Member State which adopts such measures shall notify the Commission of them forthwith; the Commission shall inform the other Member States thereof.

Chapter IV
Other provisions

Article 12

1. Member States shall set up information exchange networks for the implementation of Articles 9 and 10. They shall notify the other Member States and the Commission of the national authorities responsible for forwarding or receiving information and for applying the procedures referred to in the said Articles 9 and 10.

2. For the purpose of implementing this Directive, the provisions of Regulation (EEC) No 1468/81, in particular those relating to confidentiality, shall apply *mutatis mutandis*.

Article 13

1. The Commission shall be assisted by a committee composed of the representatives of the Member States and chaired by the representative of the Commission.

The committee shall examine any matter concerning the application of this Directive raised by its chairman either on his own initiative or at the request of a representative of a Member State.

2. The representative of the Commission shall submit to the committee a draft of the measures to be taken. The committee shall deliver its opinion on the draft within a time limit which the chairman may lay down according to the urgency of the matter. The opinion shall be delivered by the majority laid down in Article 148 (2) of the Treaty in the case of decisions which the Council is required to adopt on a proposal from the Commission. The votes of the representatives of the Member States within the committee shall be weighted in the manner set out in that Article. The chairman shall not vote.

The Commission shall adopt measures which shall apply immediately. However, if these measures are not in accordance with the opinion of the committee, they shall be communicated by the Commission to the Council forthwith. In that event the Commission shall defer applications of the measures which it has decided for three months from the date of communication.

The Council, acting by a qualified majority, may take a different decision within the period referred to in the second subparagraph.

3. The procedure laid down in paragraph 2 shall be followed in particular to take account of any future amendments to the United Nations recommendations.

Article 14

Member States shall keep at the disposal of the other Member States and the Commission updated information concerning undertakings in the explosives sector possessing licences or authorisations as referred to in Article 1 (4).

Member States shall ascertain whether such undertakings possess a system for keeping track of explosives such that those holding explosives can be identified at any time. The conditions for the application of this subparagraph shall be adopted in accordance with the committee procedure provided for in Article 13.

Undertakings in the explosives sector shall keep such records of their transactions as are necessary to fulfil the obligations set out in this Article.

The documents referred to in this Article must be kept for at least three years after the end of the calendar year in which the recorded transaction took place, even if the undertaking has ceased trading. They must be immediately available for inspection at the request of the competent authorities.

Article 15

Member States shall ensure that explosives are properly marked.

Article 16

When a Member State issues a licence or authorisation for the purpose of allowing an explosives manufacturing activity to be exercised, it shall check in particular that the persons responsible are capable of complying with the technical commitments they assume.

Chapter V
Final provisions

Article 17

Each Member State shall determine the penalties to be applied for infringement of the provisions adopted in implementation of this Directive. The penalties shall be sufficient to promote compliance with those provisions.

Article 18

Each Member State shall adopt, in the context of its national law, the necessary measures to enable the competent authorities to seize any product coming within the scope of this Directive if there is sufficient evidence that that product will be illicitly acquired, used or dealt in.

Article 19

1. Member States shall bring into force the provisions necessary to comply with Articles 9, 10, 11, 12, 13 and 14 before 30 September 1993.
2. Member States shall adopt and publish before 30 June 1994 the laws, regulations and administrative provisions necessary to comply with the provisions other than those mentioned in paragraph 1. They shall forthwith inform the Commission thereof.

They shall apply these provisions as from 1 January 1995.

3. When Member States adopt the provisions referred to in paragraphs 1 and 2 those provisions shall contain a reference to this Directive or shall be accompanied by such reference on the occasion of their official publication. The methods of making such a reference shall be determined by the Member States.
4. However, during the period up to 31 December 2002, Member States shall allow the placing on the markets of explosives complying with the national regulations in force in their territory before 31 December 1994.
5. Member States shall communicate to the Commission the provisions of national law which they adopt in the field governed by this Directive.

Article 20

This Directive is addressed to the Member States.

Done at Luxembourg, 5 April 1993.

For the Council
The President
J. TRØJBORG

Annex I
Essential safety requirements

I. General requirements

1. Each explosive must be designed, manufactured and supplied in such a way as to present a minimal risk to the safety of human life and health, and to prevent damage to property and the environment under normal, foreseeable conditions, in particular as

regards the safety and standard practices including until such time as it is used.

2. Each explosive must attain the performance characteristics specified by the manufacturer in order to ensure maximum safety and reliability.

3. Each explosive must be designed and manufactured in such a way that when appropriate techniques are employed it can be disposed of in a manner which minimises effects on the environment.

II. Special requirements

1. As a minimum, the following information and properties – where appropriate – must be considered. Each explosive should be tested under realistic conditions. If this is not possible in a laboratory, the tests should be carried out in the conditions in which the explosive is to be used.

(a) Construction and characteristic properties, including chemical composition, degree of blending and, where appropriate, dimensions and grain size distribution.

(b) The physical and chemical stability of the explosive in all environmental conditions to which it may be exposed.

(c) Sensitiveness to impact and friction.

(d) Compatibilty of all components as regards their physical and chemical stability.

(e) The chemical purity of the explosive.

(f) Resistance of the explosive against influence of water where it is intended to be used in humid or wet conditions and where its safety or reliability may be adversely affected by water.

(g) Resistance to low and high temperatures, where the explosive is intended to be kept or used at such temperatures and its safety or reliability may be adversely affected by cooling or heating of a component or of the explosive as a whole.

(h) The suitability of the explosive for use in hazardous environments (e.g. environment endangered by firedamp, hot masses, etc.) if it is intended to be used under such conditions.

(i) Safety features intended to prevent untimely or inadvertent initiation or ignition.

(j) The correct loading and functioning of the explosive when used for its intended purpose.

(k) Suitable instructions and, where necessary, markings in respect of safe handling, storage, use and disposal in the official language or languages of the recipient Member State.

(l) The ability of the explosive, its covering or other component to withstand deterioration during storage until the "use by" date specified by the manufacturer.

(m) Specification of all devices and accessories needed for reliable and safe functioning of the explosive.

The various groups of explosives must at least also comply with the following requirements:

A. *Blasting explosives:*

(a) The proposed method of initiation must ensure safe, reliable and complete detonation or deflagration as appropriate, of the blasting explosive. In the particular case of black powder, it is the capacity as regards deflagration which shall be checked.

(b) Blasting explosives in cartridge form must transmit the detonation safely and reliably from one end of the train of cartridges to the other.

(c) The gases produced by blasting explosives intended for underground use may contain carbon monoxide, nitrous gases, other gases, vapours or airborn solid residues only in quantities which do not impair health under normal operating conditions.

B. *Detonating cords, safety fuses, igniter cords and shock tubes*

(a) The covering of detonating cords, safety fuses and igniter cords must be of adequate mechanical strength and adequately protect the explosive filling when exposed to normal mechanical stress.

(b) The parameters for the burning times of safety fuses must be indicated and must be reliably met.

(c) Detonating cords must be capable of being reliably initiated, be of sufficient initiation capability and comply with requirements as regards storage even in particular climatic conditions.

C. *Detonators (including delay detonators) and relays*

(a) Detonators must reliably initiate the detonation of the blasting explosives which are intended to be used with them under all foreseeable conditions of use.

(b) Relays must be capable of being reliably initiated.

(c) The initiation capability must not be adversely affected by humidity.

(d) The delay times of delay detonators must be sufficiently uniform to ensure that the probability of overlapping of the delay times of adjacent time stops is insignificant.

(e) The electrical characteristics of electric detonators must be indicated on the packaging (e.g. no-fire current, resistance, etc.).

(f) The wires of electric detonators must be of sufficient insulation and mechanical strength including the solidity of the link to the detonator, taking account of their intended use.

D. *Propellants and rocket propellants*

(a) These materials must not detonate when used for their intended purpose.

(b) Propellants where necessary (e.g. those based on nitrocellulose) must be stabilised against decomposition.

(c) Solid rocket propellants, when in compressed or cast form, must not contain any unintentional fissures or gas bubbles which dangerously affect their functioning.

Annex II

1. MODULE B: EC type-examination

1. This module describes that part of the procedure by which a notified body ascertains and attests that an example, representative of the production envisaged, meets the relevant provisions of the Directive.

2. The application for EC type-examination is lodged by the manufacturer or his authorised representatives established within the Community with a notified body of his choice.

The application must include:

- the name and address of the manufacturer and, if the application is lodged with any other notified representative, the name and address in addition,
- a written declaration that the same application has not been lodged with any other notified body,
- the technical documents, as described in Section 3.

The applicant must place at the disposal of the notified body an example representative of the production envisaged, hereinafter called "type". The notified body may request further examples if needed for carrying out the test programme.

3. The technical documents must enable the conformity of the appliance with the requirements of the Directive to be assessed. They must, as far as is relevant for such assessment, cover the design, manufacture and operation of the appliance and contain as far as is relevant for assessment:

- a general type-description,
- conceptual design and manufacturing drawings and diagrams of components, sub-assemblies, circuits, etc.,
- descriptions and explanations

necessary for the understanding of the drawings and diagrams and the operation of the product,

– a list of the standards referred to in Article 4, applied in full or in part, and descriptions of the solutions adopted to meet the essential requirements of the Directive where the standards referred to in Article 5 have not been applied,
– results of design calculations made, examinations carried out, etc.,
– test reports.

4. The notified body must:

4.1. examine the technical documents, verify that the type has been manufactured in conformity with those documents and identify the elements which have been designed in accordance with the relevant provisions of the standards referred to in Article 4 as well as the components which have been designed without applying the relevant provisions of those standards;

4.2. perform or have performed the appropriate examinations and necessary tests to check whether, where the standards referred to in Article 4 have not been applied, the solutions adopted by the manufacturer meet the essential requirements of the Directive;

4.3. perform or have performed the appropriate examinations and necessary tests to check whether, where the manufacturer has chosen to apply the relevant standards, these have actually been applied:

4.4. agree with the applicant the location where the examinations and necessary tests are to be carried out.

5. Where the type meets the relevant provisions of this Directive, the notified body issues an EC type-examination certificate to the applicant. The certificate contains the name and address of the manufacturer, the conclusion of the examination and necessary data for identification of the approved type.

A list of the relevant parts of the technical documents is annexed to the certificate and a copy kept by the notified body.

If the manufacturer or his authorised representative established in the Community is refused a type certificate, the notified body must provide detailed reasons for such refusal.

Provision must be made for an appeals procedure.

6. The applicant informs the notified body that holds the technical documents concerning the EC type-examination certificate of all modifications to the approved appliance which must receive additional approval where such changes may affect the conformity with the essential requirements or the prescribed conditions for use of the product. This additional approval is given in the form of an addition to the original EC type-examination certificate.

7. Each notified body must communicate to the other notified bodies the relevant information concerning the EC type-examination certificates and additions issued and withdrawn.

8. The other notified bodies may receive copies of the EC type-examination certificates and/or their additions. The Annexes to the certificates must be kept at the disposal of the other notified bodies.

9. The manufacturer or his authorised representative established within the Community must keep with the technical documents copies of EC type-examination certificates and their additions for a period of at least 10 years after the last date of manufacture of the product concerned.

Where neither the manufacturer nor his authorised representative is established within the Community, the obligation to keep the technical documents available is the responsibility of the person who places the product on the Community market.

2. MODULE C: Conformity to type

1. The module describes that part of the procedure whereby the manufacture or his authorised representative established within the Community ensures and declares that the explosives concerned are in conformity with the type as described in the EC type-examination certificate and satisfy the

requirements of this Directive that apply to them. The manufacturer must affix the CE mark to each explosive and draw up a written declaration of conformity.

2. The manufacturer must take all measures necessary to ensure that the manufacturing process assures the conformity of the manufactured product with the type as described in the EC type-examination certificate with the essential safety requirements of the Directive.

3. The manufacturer or his authorised representative must keep a copy of the declaration of conformity for a period of at least 10 years after the last date of manufacture of the product concerned.

Where neither the manufacturer nor his authorised representative is established within the Community, the obligation to keep the technical documents available is the responsibility of the person who places the product on the Community market.

4. A notified body chosen by the manufacturer must perform or have performed examinations of the product at random intervals. A suitable sample of the finished products, taken on the spot by the notified body, is examined and appropriate tests, defined in the applicable standard or standards referred to in Article 4 or equivalent tests are carried out to check the conformity of the products with the requirements of the corresponding Directive. In the event of one or more samples of the products examined not conforming, the notified body must take the appropriate measures.

Under the responsibility of the notified body the manufacturer shall affix the identification symbol of that body during the manufacturing process.

3. MODULE D: Production quality assurance

1. This module describes the procedure whereby the manufacturer who satisfies the obligations of Section 2 ensures and declares that the explosives concerned are in conformity with the type as described in the EC type-examination certificate and satisfy the requirements of this Directive. The manufacturer affixes the CE mark to each explosive and draws up a written declaration of conformity. The CE mark is accompanied by the identification symbol of the notified body responsible for the checks referred to in Section 4.

2. The manufacturer must operate an approved quality system for production, final product inspection and testing as specified in Section 3. He is subject to the checks referred to in Section 4.

3. *Quality system*

3.1. The manufacturer lodges an application for assessment of his quality system with a notified body of his choice, for the explosives concerned.

The application must include:

- all relevant information for the explosive category envisaged,
- the documents concerning the quality system,
- the technical documents pertaining to the approved type and a copy of the EC type-examination certificate.

3.2. The quality system must ensure conformity of explosives with the type as described in the EC type-examination certificate and with the requirements of this Directive that apply to them.

All the elements, requirements and provisions adopted by the manufacturer must be documented in a systematic and orderly manner in the form of written policies, procedures and instructions. The quality system documents must permit a consistent interpretation of the quality programmes, plans, manuals and quality records.

It must contain in particular an adequate description of:

- the quality objectives and the organisational structure, responsibilities and powers of the management with regard to the quality of the explosives,
- the manufacturing, quality control and quality assurance techniques, processes and systematic actions that will be used,
- the examinations and tests that will

be carried out before, during and after manufacture, and the frequency with which they will be carried out,
- the quality records, such as inspection reports and test data, calibration data, qualification reports of the personal concerned, etc.,
- the means of monitoring the achievements of the required quality of explosive and the effective operation of the quality system.

3.3. The notified body must assess the quality system to determine whether it satisfies the requirements referred to in 3.2. It must presume conformity with those requirements in respect of quality systems that implement the relevant harmonised standard. The auditing team must have at least one member with experience of assessing the relevant product technology. The assessment procedure includes an inspection visit to the manufacturer's premises.

The decision is notified to the manufacturer. The notification must contain the conclusions of the examination and the duly substantiated assessment decision.

3.4. The manufacturer must undertake to fulfil the obligations arising out of the quality system as approved and maintain it at an adequate and efficient level.

The manufacturer or his authorised representative must keep the notified body that has approved the quality system informed of any proposed change in the quality system.

The notified body must assess the changes proposed and decide whether the altered quality system will still satisfy the requirements referred to in 3.2 or whether reassessment is required.

It must notify the manufacturer of its decision. The notification must contain the conclusions of the examination and the substantiated assessment decision.

4. *Monitoring under the responsibility of the notified body*

4.1. The purpose of monitoring is to make sure that the manufacturer duly fulfils the obligations arising out of the approved quality system.

4.2. The manufacturer must allow the notified body access for inspection purposes to the manufacturing, inspection, testing and storage premises and provide it with all necessary information, in particular:
- the quality system documents,
- the quality records, such as inspection reports and test data, calibration data, qualification reports of the personnel concerned, etc.

4.3. The notified body must periodically carry out audits to make sure that the manufacturer maintains and applies the quality system and provides an audit report to the manufacturer.

4.4. Additionally the notified body may pay unannounced visits to the manufacturer. During such visits the notified body may carry out tests or have them carried out to verify that the quality system is functioning correctly; if necessary, the notified body must provide the manufacturer with a visit report and, if a test has taken place, with a test report.

5. The manufacturer must, for a period of at least 10 years after the last date of manufacture of the product, keep at the disposal of the national authorities:
- the document referred to in the second indent of 3.1,
- the updating referred to in second paragraph of 3.4,
- the decision and reports from the notified body which are referred to in the final paragraph of 3.4, and in 4.3 and 4.4.

6. Each notified body must give the other notified bodies the relevant information concerning the quality system approvals issued and withdrawn.

4. MODULE E: Production quality assurance

1. This module describes the procedure whereby the manufacturer who satisfies the obligations of Section 2 ensures and declares that the explosives are in conformity with the type as described in the EC type-examination

certificate. The manufacturer must affix the CE mark to each explosive and draw up a written declaration of conformity. The CE mark must be accompanied by the identification symbol of the notified body responsible for the checks referred to in Section 4.

2. The manufacturer must operate an approved quality system for final explosive inspection and testing as specified in Section 3. He must be subject to the checks referred to in Section 4.

3. *Quality system*

3.1. The manufacturer lodges an application with a notified body of his choice for the assessment of the quality system for his explosives.

The application must include:

- all relevant information for the explosive category envisaged,
- the quality system's documentation,
- the technical documents pertaining to the approved type and a copy of the EC type-examination certificate.

3.2. Under the quality system, each explosive is examined and appropriate tests as defined in the relevant standard(s) referred to in Article 4 or equivalent tests are carried out in order to verify its conformity with the relevant requirements of the Directive. All the elements, requirements and provisions adopted by the manufacturer must be documented in a systematic and orderly manner in the form of written policies, procedures and instructions. This quality system documentation must enable the quality programmes, plans, manuals and records to be interpreted in a uniform manner.

It must in particular contain an adequate description of:

- the quality objectives and the organisational structure, responsibilities and powers of the management with regard to product quality,
- the examination and tests that will be carried out after manufacture,
- the means of monitoring the effective operation of the quality system,
- quality records, such as inspection reports and test data, calibration data, qualification reports of the personnel concerned, etc.

3.3. The notified body must assess the quality system to determine whether it satisfies the requirements referred to in 3.2. It must presume conformity with these requirements in respect of quality systems that implement the relevant harmonised standard.

The auditing team must have at least one member with experience of assessing the relevant product technology. The assessment procedure must include an inspection visit to the manufacturer's premises.

The manufacturer must be notified of the decision. The notification must contain the conclusions of the examination and the substantiated assessment decision.

3.4. The manufacturer must undertake to fulfil the obligations arising out of the quality system as approved and maintain it at an adequate and efficient level.

The maufacturer or his authorised representative must keep the notified body which has approved the quality system informed of any proposed change in the quality system.

The notified body must assess the changes proposed and decide whether the altered quality system will still satisfy the requirements referred to in 3.2 or whether a reassessment is required.

It must notify the manufacturer of its decision. The notification must contain the conclusions of the examination and the substantiated assessment decision.

4.1. The purpose of monitoring is to make sure that the manufacturer duly fulfils the obligations arising out of the approved quality system.

4.2. The manufacturer must allow the notified body access for inspection purposes to the inspection, testing and storage premises and provide it with all necessary information, in particular:

- the quality system documentation,
- the technical documents,
- the quality records, such as inspection reports and test data, calib-

ration data, qualification reports of the personnel concerned, etc.

4.3. The notified body must periodically carry out audits to ensure that the manufacturer maintains and applies the quality system and must provide an audit report to the manufacturer.

4.4. Additionally, the notified body may pay unannounced visits to the manufacturer. During such visits the notified body may carry out tests or have them carried out to verify that the quality system is functioning correctly; if necessary, the notified body must provide the manufacturer with a visit report and, if a test has been carried out, with a test report.

5. The manufacturer must for a period of at least 10 years after the last date of manufacture of the product keep at the disposal of the national authorities:

- the documents referred to in the second indent of 3.1,
- the changes referred to in the second paragraph of 3.4,
- the decisions and reports from the notified body which are referred to in the final paragraph of 3.4, and in 4.3 and 4.4.

6. Each notified body must forward to the other notified bodies the relevant information concerning the quality system approvals issued and withdrawn.

5. MODULE F: Product verification

1. This module describes the procedure whereby a manufacturer or his authorised representative established within the Community checks and attests that the explosives subject to the provisions of 3 are in conformity with the type as described in the EC type-examination certificates and satisfy the relevant requirements of the Directive.

2. The manufacturer shall take all measures necessary in order that the manufacturing process ensures conformity of the explosives with the type as described in the EC type-examination certificate and with the requirements of the Directive that apply to them. He shall affix the CE mark to each explosive and shall draw up a declaration of conformity.

3. The notified body shall carry out the appropriate examinations and tests in order to check the conformity of the explosive with the relevant requirements of the Directive by examination and testing of every explosive as specified in 4.

The manufacturer or his authorised representative shall keep a copy of the declaration of conformity for a period ending at least 10 years after the last explosive has been manufactured.

4. *Verification by examination and testing of every explosive*

4.1. All explosives shall be individually examined and appropriate tests are set out in the relevant standard(s) referred to in Article 4 or equivalent tests shall be carried out in order to verify their conformity with the relevant type and requirements of the Directive.

4.2. The notified body shall affix or cause to be affixed, its identification symbol to each approved explosive and draw up a written certificate of conformity relating to the tests carried out.

4.3. The manufacturer or his authorised representative shall ensure that he is able to supply the notified body's certificates of conformity on request.

6. MODULE G: Unit verification

1. This module describes the procedure whereby the manufacturer ensures and declares that the explosive which has been issued with the certificate referred to in Section 2 conforms to the relevant requirements of the Directive. The manufacturer must affix the CA mark to the explosive and draw up a declaration of conformity.

2. The notified body must examine the explosive and carry out the appropriate tests as set out in the relevant standard(s) referred to in Article 4, or equivalent tests, to ensure its conformity with the relevant requirements of the Directive.

The notified body must affix, or cause to be affixed, its identification symbol on the approved explosive and draw up a certificate of conformity concerning the tests carried out.

3. The aim of the technical documents is to enable conformity with the requirements of the Directive to be assessed and the design, manufacture and operation of the explosive to be understood.

The documents must contain, in so far as is necessary for the assessment:

- a general description of the type,
- conceptual design and manufacturing drawings and schemes of components, sub-assemblies, circuits, etc.
- descriptions and explanations necessary for the understanding of the said drawings and schemes and the operation of the explosive or protection system,
- a list of the standards referred to in Article 4, applied in full or in part, and descriptions of the solutions adopted to meet the essential requirements of the Directive where the standards referred to in Article 4 have not been applied.
- results of design calculations made, examinations carried out, etc.,
- test reports.

Annex III

Minimum criteria to be taken into account by Member States for the notification of bodies

1. The body, its director and the staff responsible for carrying out the verification tests shall not be the designer, manufacturer, supplier or installer of explosives which they inspect, nor the authorised representative of any of these parties. They shall not become either involved directly or as authorised representatives in the design, construction, marketing or maintenance of such explosives. This does not preclude the possibility of exchanges of technical information between the manufacturer and the body.

2. The body and its staff shall carry out the verification tests with the highest degree of professional integrity and technical competence and shall be free from all pressures and inducements, particularly financial, which might influence their judgement or the results of the inspection, especially from persons or groups of persons with an interest in the result of verification.

3. The body shall have at its disposal the necessary staff and possess the necessary facilities to enable it to perform properly the administrative and technical tasks connected with verification; it shall also have access to the equipment required for special verification.

4. The staff responsible for inspection shall have:

- sound technical and professional training,
- satisfactory knowledge of the requirements of the tests they carry out and adequate experience of such tests,
- the ability to draw up the certificates, records and reports required to authenticate the performance of the tests.

5. The impartiality of inspection staff shall be guaranteed. Their remuneration shall not depend on the number of tests carried out or on the results of such tests.

6. The body shall take out civil liability insurance unless its liability is assumed by the State in accordance with national law, or the Member State itself is directly responsible for the tests.

7. The staff of the body shall be bound

to observe professional secrecy with regard to all information gained in carrying out its tasks (except *vis-à-vis* the competent administrative authorities of the State in which its activities are carried out) under this Directive or any provision of national law giving effect to it.

Annex IV
Conformity markings

The CE conformity marking shall consist of the initials "CE" taking the following form:

If the marking is reduced or enlarged the proportions given in the above graduated drawing must be respected.

DIRECTIVE 94/9/EC of the European Parliament and the Council

of 23 March 1994

on the approximation of the laws of the Member States concerning equipment and protective systems intended for use in potentially explosive atmospheres

THE EUROPEAN PARLIAMENT AND THE COUNCIL OF THE EUROPEAN UNION,

Having regard to the Treaty establishing the European Economic Community, and in particular Article 100a thereof,

Having regard to the proposal from the Commission,[1]

Having regard to the opinion of the Economic and Social Committee,[2]

Acting in accordance with the procedure referred to in Article 189b of the Treaty establishing the European Community,

Whereas it is the duty of Member States to protect, on their territory, the safety and health of persons and, where appropriate, domestic animals and property and, in particular, that of workers, especially against the hazards resulting from the use of equipment and systems providing protection against potentially explosive atmospheres;

Whereas mandatory provisions within the Member States determine the level of safety to be achieved by protective equipment and systems intended for use in potentially explosive atmospheres; whereas these are generally electrical and non-electrical specifications having an effect on the design and structure of equipment which can be used in potentially explosive atmospheres;

Whereas the requirements to be met by such equipment differ from one Member State to another in respect of their extent and differing inspection procedures; whereas these differences are, therefore, likely to raise barriers to trade within the Community;

Whereas harmonisation of national legislation is the only way in which to remove these barriers to free trade; whereas this objective cannot be satisfactorily achieved by the individual Member States; whereas this Directive merely lays down the requirements vital to freedom of movement for the equipment to which it applies;

Whereas the regulations intended to remove technical barriers to trade are required to follow the new approach provided for in the Council resolution of 7 May 1985,[3] which requires a definition of the essential requirements regarding safety and other requirements of society without reducing existing, justified levels of protection

[1] OJ No C 46, 20. 2. 1992, p. 19.
[2] OJ No C 106, 27. 4. 1992, p. 9.

[3] OJ No C 136, 4. 6. 1985, p. 1.

within the Member States; whereas that resolution provides that a very large number of products be covered by a single Directive in order to avoid frequent amendments and the proliferation of Directives;

Whereas the existing Directives on the approximation of the laws of the Member States to electrical equipment for use in potentially explosive atmospheres have made positive steps towards protection against explosions via measures linked with the structure of the equipment at issue and which have helped to remove barriers to trade in this area; whereas, in parallel, a revision and expansion of the existing Directives is necessary since, more particularly, in an overall context, action must be taken to guard against the potential hazards arising from such equipment. This implies in particular that measures intended to guarantee effective protection of users and third parties must already be contemplated at the design and manufacturing states;

Whereas the form taken by the hazard, the protection measures and the test methods are often very similar, if not identical, for both mining and surface equipment; whereas it is, therefore, absolutely necessary to cover by a single Directive protective equipment and systems falling within both groups;

Whereas the two groups of equipment referred to above are used in a large number of commercial and industrial sectors and possess considerable economic significance;

Whereas compliance with the basic safety and health requirements is essential in order to ensure the safety of protective equipment and systems; whereas, in particular, the additional requirements are intended to take account of existing or potential hazards; whereas protective equipment and systems will, therefore, embody at least one of those requirements where this is necessary for their proper functioning or is to apply to their intended use; whereas the notion of intended use is of prime importance for the explosion-proofing of protective equipment and systems; whereas it is essential that manufacturers supply full information; whereas specific, clear marking of said equipment, stating its use in a potentially explosive atmosphere, is also necessary;

Whereas the intention is to prepare a Directive on operations in potentially explosive atmospheres which is based on Article 118a; whereas that additional Directive will, in particular, aim at explosion hazards which derive from a given use and/or types and methods of installation;

Where compliance with essential health and safety requirements is imperative if the safety of equipment is to be ensured; whereas judgment will have to be exercised in the implementation of those requirements in order to take account of both the technology obtaining at the time of manufacture and overriding technical and economic requirements;

Whereas, therefore, this Directive sets out essential requirements only; whereas, in order to facilitate the task of proving compliance with the essential requirements, harmonised European standards are necessary, more especially with regard to the non-electrical aspects of protection against explosions – standards relating to the design, manufacture and testing of equipment, compliance with which enables a product to be presumed to meet such essential requirements; whereas harmonised European standards are drawn up by private bodies and must retain their non-mandatory status; whereas, for this purpose, the European Committee for Standardisation (CEN) and the European Committee for Electrotechnical Standardisation (CENELEC) are recognised as the bodies competent to adopt harmonised standards which follow the general guidelines for cooperation between the Commission and those two bodies, signed on 13

November 1984; whereas, for the purposes of this Directive, a harmonised standard is a technical specification (European Standard or harmonisation document) adopted by one or other of those bodies, or by both, at the prompting of the Commission pursuant to Council Directive 83/189/EEC of the 28 March 1983 providing for a procedure governing the provision of information on technical standards and regulations[4] and pursuant to the general guidelines referred to above;

Whereas the legislative framework should be improved in order to ensure that employers and workers make an effective and appropriate contribution towards the standardisation process; whereas this should be completed by the time this Directive is implemented;

Whereas, in view of the nature of the risks involved in the use of equipment in potentially explosive atmospheres it is necessary to establish procedures applying to the assessment of compliance with the basic requirements of the Directives; whereas these procedures must be devised in the light of the level of risk which may be inherent in equipment and/or against which systems must protect the immediate environment; whereas, therefore, each category of equipment conformity must be supplemented by an adequate procedure or a choice between several equivalent procedures; whereas the procedures adopted comply fully with Council Decision 93/465/EEC of 22 July 1993 concerning the modules for the various phases of the conformity assessment procedures which are intended to be used in the technical harmonisation Directives;[5]

Whereas the Council has provided for the affixing of the CE marking by either the manufacturer or his authorised representative within the Community; whereas that marking means that the product complies with all the basic requirements and assessment procedures provided for by the Community law applying to that product;

Whereas it is appropriate that the Member States, as provided for by Article 100a of the Treaty, may take temporary measures to limit or prohibit the placing on the market and the use of equipment and protective systems in cases where they present a particular risk to the safety of persons and, where appropriate, domestic animals or property, provided that the measures are subject to a Community control procedure;

Whereas the recipients of any decision taken as part of this Directive must be aware of the reasons behind that decision and the means of appeal open to them;

Whereas, on 18 December 1985, the Council adopted a framework Directive on electrical equipment for use in potentially explosive atmospheres (76/117/EEC)[6] and, on 15 February 1982, a Directive concerning electrical equipment for use in potentially explosive atmospheres in mines susceptible to fire damp (82/130/EEC);[7] whereas, from the outset of harmonisation work, the conversion into total harmonisation of the optional and partial harmonisation on which these Directives are based had been contemplated; whereas this Directive fully covers the scope of the abovementioned Directives and whereas, therefore, these Directives must be repealed;

Whereas the internal market incorporates an area without internal frontiers within which the free movement of goods, persons, services and capital is assured;

---

[4] OJ No L 109, 26. 4. 1983, p. 8. Directive as last amended by Directive 88/182/EEC (OJ No L 81, 26. 3. 1988, p. 75).

[5] OJ No L 220, 30. 8. 1993, p. 23.

[6] OJ No L 24, 31. 1. 1976, p. 45. Directive at last amended by Directive 90/487/EEC (OJ No L 270, 2. 10. 1990, p. 23).

[7] OJ No L 59, 2. 3. 1982, p. 10.

Whereas it is necessary to provide for a transitional arrangement enabling equipment manufactured in compliance with the national regulations in force at the date of adoption of this Directive to be marketed and placed in service,

HAS ADOPTED THIS DIRECTIVE:

Chapter I
Scope, placing on the market and freedom of movement

Article 1

1. This Directive applies to equipment and protective systems intended for use in potentially explosive atmospheres.
2. Safety devices, controlling devices and regulating devices intended for use outside potentially explosive atmospheres but required for or contributing to the safe functioning of equipment and protective systems with respect to the risks of explosion are also covered by the scope of this Directive.
3. For the purpose of this Directive, the following definitions shall apply:

Equipment and protective systems intended for use in potentially explosive atmospheres

(a) "Equipment" means machines, apparatus, fixed or mobile devices, control components and instrumentation thereof and detection or prevention systems which, separately or jointly, are intended for the generation, transfer, storage, measurement, control and conversion of energy for the processing of material and which are capable of causing an explosion through their own potential sources of ignition.
(b) "Protective systems" means design units which are intended to halt incipient explosions immediately and/or to limit the effective range of explosion flames and explosion pressures. Protective systems may be integrated into equipment or separately placed on the market for use as autonomous systems.
(c) "Components" means any item essential to the safe functioning of equipment and protective systems but with no autonomous function.

Explosive atmospheres

Mixture with air, under atmospheric conditions, of flammable substances in the form of gases, vapours, mists or dusts in which, after ignition has occurred, combustion spreads to the entire unburned mixture.

Potentially explosive atmosphere

An atmosphere which could become explosive due to local and operational conditions.

Equipment groups and categories

Equipment group I applied to equipment intended for use in underground parts of mines, and to those parts of surface installations of such mines, liable to be endangered by firedamp and/or combustible dust.

Equipment group II applies to equipment intended for use in other places liable to be endangered by explosive atmospheres.

The categories of equipment defining the required levels of protection are described in Annex I.

Equipment and protective systems may be designed for a particular explosive atmosphere. In this case, they must be marked accordingly.

Intended use

The use of equipment, protective systems, and devices referred to in Article 1 (2) in accordance with the equipment group and category and with all the information supplied by the manufacturer which is required for the safe functioning of equipment, protective systems and devices.
4. The following are excluded from the scope of this Directive:
- medical devices intended for use in a medical environment,
- equipment and protective systems

where the explosion hazard results exclusively from the presence of explosive substances or unstable chemical substances,
- equipment intended for use in domestic and non-commercial environments where potentially explosive atmospheres may only rarely be created, solely as a result of the accidental leakage of fuel gas,
- personal protective equipment covered by Directive 89/686/EEC,[8]
- seagoing vessels and mobile offshore units together with equipment on board such vessels or units,
- means of transport, i.e. vehicles and their trailers intended solely for transporting passengers by air or by road, rail or water networks, as well as means of transport in so far as such means are designed for transporting goods by air, by public road or rail networks or by water. Vehicles intended for use in a potentially explosive atmosphere shall not be excluded,
- the equipment covered by Article 223 (1) (b) of the Treaty.

Article 2

1. Member States shall take all appropriate measures to ensure that the equipment, protective systems and devices referred to in Article 1 (2) to which this Directive applies may be placed on the market and put into service only if, when properly installed and maintained and used for their intended purpose, they do not endanger the health and safety of persons and, where appropriate, domestic animals or property.
2. The provisions of this Directive shall not affect Member States' entitlement to lay down, in due observance of the provisions of the Treaty, such requirements as they may deem necessary to ensure that persons and, in particular, workers are protected when using the equipment, protective systems, and devices referred to in Article 1 (2) in question provided that this does not mean that such equipment, protective systems, or devices are modified in a way not specified in the Directive.
3. At trade fairs, exhibitions, demonstrations, etc., Member States shall not prevent the showing of equipment, protective systems, or the devices referred to in Article 1 (2) which do not conform to the provisions of this Directive, provided that a visible sign clearly indicates that such equipment, protective systems, and devices referred to in Article 1 (2) do not conform and that they are not for sale until they have been brought into conformity by the manufacturer or his authorised representative established in the Community. During demonstrations, adequate safety measures shall be taken to ensure the protection of persons.

Article 3

Equipment, protective systems, and the devices referred to in Article 1 (2) to which this Directive applies must meet the essential health and safety requirements set out in Annex II which apply to them, account being taken of their intended use.

Article 4

1. Member States shall not prohibit, restrict or impede the placing on the market and putting into service in their territory of equipment, protective systems, or devices referred to in Article 1 (2) which comply with this Directive.
2. Member States shall not prohibit, restrict or impede the placing on the market of components which, accompanied by a certificate of conformity as referred to in Article 8 (3), are intended to be incorporated into equipment or protective systems within the meaning of this Directive.

Article 5

1. Member States shall regard as conforming to all the provisions of this Directive, including the relevant con-

[8] OJ No L 399, 30. 12. 1989, p. 18.

formity assessment procedures laid down in chapter II:

- equipment, protective systems, and devices referred to in Article 1 (2) accompanied by the EC declaration of conformity referred to in Annex X and bearing the CE marking provided for in Article 10,
- the components referred to in Article 4 (2) accompanied by the certificate of conformity referred to in Article 8 (3).

In the absence of harmonised standards, Member States shall take any steps which they deem necessary to bring to the attention of the parties concerned the existing national technical standards and specifications regarded as important or relevant to the proper implementation of the essential health and safety requirements in Annex II.

2. Where a national standard transposing a harmonised standard, the reference for which has been published in the *Official Journal of the European Communities*, covers one or more of the essential health and safety requirements, the equipment, protective system, device referred to in Article 1 (2), or the component referred to in Article 4 (2), constructed in accordance with that standard shall be presumed to comply with the relevant essential health and safety requirements.

Member States shall publish the references of national standards transposing harmonised standards.

Member States shall ensure that appropriate measures are taken to enable the social partners to influence the process of preparing and monitoring the harmonised standards at national level.

Article 6

1. Where a Member State or the Commission considers that the harmonised standards referred to in Article 5 (2) do not entirely satisfy the relevant essential health and safety requirements referred to in Article 3, the Commission of the Member State concerned shall bring the matter before the Committee set up under directive 82/189/EEC, hereinafter referred to as 'the Committee', giving reasons thereof. The Committee shall deliver an opinion without delay.

Upon receipt of the Committee's opinion, the Commission shall inform the Member States whether or not it is necessary to withdraw those standards from the published information referred to in Article 5 (2).

2. The Commission may adopt the appropriate measure with a view to ensuring the practical application in a uniform manner of this Directive in accordance with the procedure laid down in paragraph 3.

3. The Commission shall be assisted by a Standing Committee, consisting of representatives appointed by the Member States and chaired by a representative of the Commission.

The Standing Committee shall draw up its own rules of procedure.

The representative of the Commission shall submit to the Committee a draft of the measures to be taken. The Committee shall deliver its opinion on the draft, within a time limit which the chairman may lay down according to the urgency of the matter, if necessary by taking a vote.

The opinion shall be recorded in the minutes, in addition, each Member State shall have the right to ask to have its position recorded in the minutes.

The Commission shall take the utmost account of the opinion delivered by the committee. It shall inform the committee of the manner in which its opinion has been taken into account.

4. The Standing Committee may furthermore examine any question relating to the application of this Directive and raised by its chairman either on the latter's initiative, or at the request of a Member State.

Article 7

1. Where a Member State ascertains that equipment, protective systems or devices referred to in Article 1 (2) bearing the CE conformity marking and used in accordance with their

intended use are liable to endanger the safety of persons and, where appropriate, domestic animals or property, it shall take all appropriate measures to withdraw such equipment or protective systems from the market, to prohibit the placing on the market, putting into service or use thereof, or to restrict free movement thereof.

The Member State shall immediately inform the Commission of any such measures, indicating the reasons for its decision and, in particular, whether non-conformity is due to:

(a) failure to satisfy the essential requirements referred to in Article 3;
(b) incorrect application of the standards referred to in Article 5 (2);
(c) shortcomings in the standards referred to in Article 5 (2).

2. The Commission shall enter into consultation with the parties concerned without delay. Where the Commission considers, after this consultation, that the measure is justified, it shall immediately so inform the Member State which took the initiative and the other Member States. Where the Commission considers, after this consultation, that the action is unjustified, it shall immediately so inform the Member State which took the initiative and the manufacturer or his authorised representative established within the Community. Where the decision referred to in paragraph 1 is based on a shortcoming in the standards and where the Member State at the origin of the decision maintains its position, the Commission shall immediately inform the Committee in order to initiate the procedures referred to in Article 6 (1).

3. Where equipment or a protective system which does not comply bears the CE conformity marking, the competent Member State shall take appropriate action against the person(s) having affixed the marking and shall so inform the Commission and the other Member States.

4. The Commission shall ensure that the Member States are kept informed of the progress and outcome of this procedure.

Chapter II
Conformity assessment procedures

Article 8

1. The procedures for assessing the conformity of equipment, including where necessary the devices referred to in Article 1 (2), shall be as follows:

(a) *equipment-group I and II, equipment-category M 1 and 1*

 The manufacturer or his authorised representative established in the Community must, in order to affix the CE marking, follow the CE type-examination procedure (referred to in Annex III), in conjunction with:
 - the procedure relating to production quality assurance (referred to in Annex IV),

 or
 - the procedure relating to product verification (referred to in Annex V);

(b) *Equipment-group I and II, equipment-category M 2 and 2*
 (i) in the case of internal combustion engines and electrical equipment in these groups and categories, the manufacturer or his authorised representative established in the Community shall, in order to affix the CE mark, follow the EC-type examination procedure (referred to in Annex III), in conunction with:
 —the procedure relating to conformity to type referred to in Annex VI, or
 —the procedure relating to product quality assurance referred to in Annex VII;

 (ii) in the case of other equipment in these groups and categories, the manufacturer or his authorised representative established in the Community must, in order to affix the CE mark, follow the procedure relating

to internal control of production (referred to in Annex VIII)
and
communicate the dossier provided for in Annex VIII, paragraph 3, to a notified body, which shall acknowledge receipt of it as soon as possible and shall retain it.

(c) *equipment-group II, equipment-category 3*
The manufacturer or his authorised representative established in the Community must, in order to affix the CE marking, follow the procedure relating to internal control of production referred to in Annex VIII;

(d) *equipment-groups I and II*
In adddition to the procedures referred to in paragraph 1(a), (b) and (c), the manufacturer or his authorised representative established in the Community may also, in order to affix the CE marking, follow the procedure relating to CE unit verification (referred to in Annex IX).

2. The provision of 1(a) or 1(d) above shall be used for conformity assessment of autonomous protective systems.

3. The procedures referred to in paragraph 1 shall be applied in respect of components as referred to in Article 4 (2), with the exception of the affixing of the CE marking. A certificate shall be issued by the manufacturer or his authorised representative established in the Community, declaring the conformity of the components with the provisions of this Directive which apply to them and stating their characteristics and how they must be incorporated into equipment or protective systems to assist compliance with the essential requirements applicable to finished equipment or protective systems.

4. In addition, the manufacturer or his authorised representative established in the Community may, in order to affix the CE marking, follow the procedure relating to internal control of production (referred to in Annex VIII) with regard to the safety aspects referred to in point 1.2.7 of Annex II.

5. Notwithstanding the previous paragraphs, the competent authorities may, on a duly justified request, authorise the placing of the Member State concerned of the equipment, protective systems and individual devices referred to in Article 1 (2) in respect of which the procedures referred to in the previous paragraphs have not been applied and the use of which is in the interests of protection.

6. Documents and correspondence relating to the procedures referred to in the abovementioned paragraphs shall be drawn up in one of the official languages of the Member States in which those procedures are being applied or in a language accepted by the notified body.

7. (a) Where the equipment and protective systems are subject to other Community Directives covering other aspects which also provide for the affixing of the CE marking referred to in Article 10, that marking shall indicate that the equipment and protective systems are also presumed to conform with the provisions of those other Directives.

(b) However, where one or more of those Directives allow the manufacturer, during a transitional period, to choose which arrangements to apply, the CE marking shall indicate conformity only with the Directives applied by the manufacturer. In this case, particulars of the said Directives, as published in the *Official Journal of the European Communities*, must be given in the documents, notices or instructions required by the Directives and accompanying the equipment and protective systems.

Article 9

1. Member States shall notify the Commission and the other Member States

of the bodies which they have appointed to carry out the procedures referred to in Article 8, together with the specific tasks which these bodies have been appointed to carry out and the identification numbers assigned to them beforehand by the Commission.

The Commission shall publish in the *Official Journal of the European Communities* a list of the notified bodies, with their identification numbers and the tasks for which they have been notified. The Commission shall ensure that this list is kept up to date.

2. Member States shall apply the criteria laid down in Annex XI in assessing the bodies to be indicated in such notification. Bodies meeting the assessment criteria laid down in the relative harmonised standards shall be presumed to fulfil those criteria.

3. A Member State which has approved a body must withdraw its notification if it finds that the body no longer meets the criteria referred to in Annex XI. It shall immediately inform the Commission and the other Member States accordingly.

Chapter III
CE conformity marking

Article 10

1. The CE conformity marking shall consist of the initials "CE". The form of the marking to be used is shown in Annex X. The CE marking shall be followed by the identification number of the notified body where such body is involved in the production control stage.

2. The CE marking shall be affixed distinctly, visibly, legibly and indelibly to equipment and protective systems, supplementary to the provisions of point 1.0.5. of Annex II.

3. The affixing of markings on the equipment or protective systems which are likely to deceive third parties as to the meaning and form of the CE marking shall be prohibited. Any other marking may be affixed to the equipment or protective systems, provided that the visibility and legibility of the CE marking is not thereby reduced.

Article 11

Without prejudice to Article 7:

(a) where a Member State establishes that the CE marking has been incorrectly affixed, the manufacturer or his authorised representative established within the Community shall be obliged to make the product conform as regards the provisions concerning the CE marking and to end the infringement under the conditions imposed by the Member State;

(b) in the event of continuing non-conformity, the Member State must take all appropriate measures to restrict or prohibit the placing on the market of the product in question or to ensure that it is withdrawn from the market in accordance with the procedures laid down in Article 7.

Chapter IV
Final provisions

Article 12

Any decisison taken pursuant to this Directive which restricts or prohibits the placing on the market and/or the putting into service or requires the withdrawal from the market of equipment, a protective system, or a device referred to in Article 1 (2) shall state the exact grounds on which it is based. Such a decision shall be notified forthwith to the party concerned, who shall at the same time be informed of the legal remedies available to him under the laws in force in the Member State concerned and of the time limits to which such remedies are subject.

Article 13

Member States shall ensure that all the parties involved in the application of the Directive are bound to observe confidentiality in respect of all information obtained in the performance of carrying out their tasks. This does not affect the obligations of the Member States and of the notified bodies regarding reciprocal information and the dissemination of warnings.

Article 14

1. Directive 76/117/EEC, Directive 79/196/EEC[9] and Directive 82/130/EEC shall be repealed as from 1 July 2003.

2. EC certificates of conformity to the harmonised standards obtained in accordance with the procedures laid down in the Directives referred to in paragraph 1 shall continue to be valid until 30 June 2003 unless they expire before that date. Their validity shall continue to be limited to the harmonised standards indicated in the aforementioned Directives.

3. Member States shall take the necessary action to ensure that the notified bodies which are responsible pursuant to Article 8 (1) to (4) for the assessment of the conformity of electrical equipment placed on the market before 1 July 2003 take account of the results of tests and verifications already carried out under the Directives referred to in paragraph 1.

Article 15

1. Member States shall adopt and publish the laws, regulations and administrative provisions necessary to comply with this Directive before 1 September 1995. They shall forthwith inform the Commission thereof.

The Member States shall apply these measures with effect from 1 March 1996.

When Member States adopt the measures referred to in the first subparagraph, they shall contain a reference to this Directive or shall be accompanied by such reference at the time of their official publication. The methods of making such reference shall be laid down by Member States.

2. However, Member States shall allow the placing on the market and the putting into service of equipment and protective systems conforming with the national regulations in force in their territory at the date of adoption of this Directive for the period until 30 June 2003.

Article 16

This Directive is addressed to the Member States.

Done at Brussels, 23 March 1994.

For the European Parliament	*For the Council*
The President	*The President*
E. KLEPSCH	TH. PANGALOS

[9] OJ No L 43, 20. 2. 1979, p. 20. Directive as last amended by Directive 90/487/EEC (OJ No L 270. 2. 10. 1990, p. 23).

Annex I

Criteria determining the classification of equipment-groups into categories

1. Equipment-group I

(a) Category M 1 comprises equipment designed and, where necessary, equipped with additional special means of protection to be capable of functioning in conformity with the operational parameters established by the manufacturer and ensuring a very high level of protection.

Equipment in this category is intended for use in underground parts of mines as well as those parts of surface installations of such mines endangered by firedamp and/or combustible dust.

Equipment in this category is required to remain functional, even in the event of rare incidents relating to equipment, with an explosive atmosphere present, and is characterised by means of protection such that:

– either, in the event of failure of

one means of protection, at least an independent second means provides the requisite level of protection,
- or the requisite level of protection is assured in the event of two faults occurring independently of each other.

Equipment in this category must comply with the supplementary requirements referred to in Annex II, 2.0.1.

(b) Category M 2 comprises equipment designed to be capable of functioning in conformity with the operational parameters established by the manufacturer and ensuring a high level of protection.

Equipment in this category is intended for use in underground parts of mines as well as those parts of surface installations of such mines likely to be endangered by firedamp and/or combustible dust.

This equipment is intended to be de-energised in the event of an explosive atmosphere.

This means of protection relating to equipment in this category assure the requisite level of protection during normal operation and also in the case of more severe operating conditions, in particular those arising from rough handling and changing environmental conditions.

Equipment in this category must comply with the supplementary requirements referred to in Annex II, 2.0.2.

2. Equipment-group II

(a) Category 1 comprises equipment to be capable of functioning in conformity with the operational parameters established by the manufacturer and ensuring a very high level of protection.

Equipment in this category is intended for use in areas in which explosive atmospheres caused by mixtures of air and gases, vapours or mists or by air/dust mixtures are present continuously, for long periods or frequently.

Equipment in this category must ensure the requisite level of protection, even in the event of rare incidents relating to equipment, and is characterised by means of protection such that:
- either, in the event of failure of one means of protection, at least an independent second means provides the requisite level of protection,
- or the requisite level of protection is assured in the event of two faults occurring independently of each other.

Equipment in this category must comply with the supplementary requirements referred to in Annex II, 2.1.

(b) Category 2 comprises equipment designed to be capable of functioning in conformity with the operational parameters established by the manufacturer and of ensuring a high level of protection.

Equipment in this category is intended for use in areas in which explosive atmospheres caused by gases, vapours, mists or air/dust mixtures are likely to occur.

The means of protection relating to equipment in this category ensure the requisite level of protection, even in the event of frequently occurring disturbances or equipment faults which normally have to be taken into account.

Equipment in this category must comply with the supplementary requirements referred to in Annex II, 2.2.

(c) Category 3 comprises equipment designed to be capable of functioning in conformity with the operating parameters established by the manufacturer and ensuring a normal level of protection.

Equipment in this category is intended for use in areas in which explosive atmospheres caused by gases, vapours, mists, or air/dust mixtures are unlikely to occur or, if they do occur, are likely to do so only infrequently and for a short period only.

Equipment in this category ensures the requisite level of protection during normal operation.

Equipment in this category must comply with the supplementary requirements referred to in Annex II, 2.3.

Annex II

Essential health and safety requirements relating to the design and construction of equipment and protective systems intended for use in potentially explosive atmospheres

Preliminary observations

A. Technological knowledge, which can change rapidly, must be taken into account as far as possible and be utilised immediately.

B. For the devices referrred to in Article 1 (2), the essential requirements shall apply only in so far as they are necessary for the safe and reliable functioning and operation of those devices with respect to the risk of explosion.

1. COMMON REQUIREMENTS FOR EQUIPMENT AND PROTECTIVE SYSTEMS

1.0. **General requirements**

1.0.1. *Principles of integrated explosion safety*

Equipment and protective systems intended for use in potentially explosive atmospheres must be designed from the point of view of integrated explosion safety.

In this connection, the manufacturer must take measures:

- above all, if possible, to prevent the formation of explosive atmospheres which may be produced or released by equipment and by protective systems themselves,
- to prevent the ignition of explosive atmospheres, taking into account the nature of every electrical and non-electrical source of ignition,
- should an explosion nevertheless occur which could directly or indirectly endanger persons and, as the case may be, domestic animals or property, to halt it immediately and/or to limit the range of explosion flames and explosion pressures to a sufficient level of safety.

1.0.2. Equipment and protective systems must be designed and manufactured after due analysis of possible operating faults in order as far as possible to preclude dangerous situations.

Any misuse which can reasonably be anticipated must be taken into account.

1.0.3. *Special checking and maintenance conditions*

Equipment and protective systems subject to special checking and maintenance conditions must be designed and constructed with such conditions in mind.

1.0.4. *Surrounding area conditions*

Equipment and protective systems must be so designed and constructed as to be capable of coping with actual or foreseeable surrounding area conditions.

1.0.5. *Marking*

All equipment and protective systems must be marked legibly and indelibly with the following minimum particulars;

- name and address of the manufacturer,
- CE marking (see Annex X, point A),
- designation of series or type,
- serial number, if any,
- year of construction,
- the specific marking of explosion protection ⟨εₓ⟩ followed by the symbol of the equipment group and category,
- for equipment-group II, the letter "G" (concerning explosive atmospheres caused by gases, vapours or mists),
  and/or
  the letter "D" (concerning explosive atmospheres caused by dust).

Furthermore, where necessary, they must also be marked with all information essential to their safe use.

1.0.6. *Instructions*

(a) All equipment and protective systems must be accompanied by instructions, including at least the following particulars:
  - a recapitulation of the information with which the equipment or protective system is marked, except for the serial number (see 1.0.5.), together with any appropriate additional information to facilitate maintenance (e.g. address of the importer, repairer, etc.);
  - instructions for safe:
    - putting into service,
    - use,
    - assembling and dismantling,
    - maintenance (servicing and emergency repair),
    - installation,
    - adjustment;
  - where necessary, an indication of the danger areas in front of pressure-relief devices;
  - where necessary, training instructions;
  - details which allow a decision to be taken beyond any doubt as to whether an item of equipment in a specific category or a protective sytstem can be used in the intended area under the expected operating conditions;
  - electrical and pressure parameters, maximum surface temperatures and other limit values;
  - where necessary, special conditions of use, including particulars of possible misuse which experience has shown might occur;
  - where necessary, the essential characteristics of tools which may be fitted to the equipment or protective system.

(b) The instructions must be drawn up in one of the Community languages by the manufacturer or his authorised representative established in the Community.

On being put into service, all equipment and protective systems must be accompanied by a translation of the instructions in the language or languages of the country in which the equipment or protective system is to be used and by the instructions in the original language.

This translation must be made by either the manufacturer or his authorised representative established in the Community or the person introducing the equipment or protective system into the language area in question.

By way of derogation from this requirement, the maintenance instructions for use by the specialist personnel employed by the manufacturer or his authorised representative established in the Community may be drawn up in a single Community language understood by that personnnel.

(c) The instructions must contain the drawings and diagrams necessary for the putting into service, maintenance, inspection, checking of correct operation and, where appropriate, repair of the equipment or protective system, together with all useful instructions, in particular with regard to safety.

(d) Literature describing the equipment or protective system must not contradict the instructions with regard to safety aspects.

1.1 **Selection of materials**

1.1.1. The materials used for the construction of equipment and protective systems must not trigger off the explosion, taking into account foreseeable operational stresses.

1.1.2. Within the limits of the operating conditions laid down by the manufacturer, it must not be possible for a reaction to take place between the materials used and the constituents of the potentially explosive atmosphere which could impair explosion protection.

1.1.3. Materials must be so selected that predictable changes in their characteristics and their compatibility in combination with other materials will not lead to a reduction in the protection afforded; in particular, due account must be taken of the material's corrosion and wear resistance, electrical conductivity, impact strength, ageing resistance and the effects of temperature variations.

1.2. **Design and construction**

1.2.1. Equipment and protective systems must be designed and constructed with due regard to technological knowledge of explosion protection so that they can be safely operated throughout their foreseeable lifetime.

1.2.2. Components to be incorporated into or used as replacements in equipment and protective systems must be so designed and constructed that they function safely for their intended purpose of explosion protection when they are installed in accordance with the manufacturer's instructions.

1.2.3. *Enclosed structures and prevention of leaks*

Equipment which may release flammable gases or dusts must wherever possible employ enclosed structures only.

If equipment contains openings or non-tight joints, these must as far as possible be designed in such a way that developing gases or dusts cannot give rise to explosive atmospheres outside the equipment.

Points where materials are introduced or drawn off must, as far as possible, be designed and equipped so as to limit escapes of flammable materials during filling or draining.

1.2.4. *Dust deposits*

Equipment and protective systems, which are intended to be used in areas exposed to dust must be so designed that deposit dust on their surfaces is not ignited.

In general, dust deposits must be limited where possible. Equipment and protective systems must be easily cleanable.

The surface temperatures of equipment parts must be kept well below the glow temperature of the deposit dust.

The thickness of deposit dust must be taken into consideration and, if appropriate, means must be taken to limit the temperature in order to prevent a heat build up.

1.2.5. *Additional means of protection*

Equipment and protective systems which may be exposed to certain types of external stresses must be equipped, where necessary, with additional means of protection.

Equipment must withstand relevant stresses, without adverse effect on explosion protection.

1.2.6. *Safe opening*

If equipment and protective systems are in a housing or a locked container forming part of the explosion protection itself, it must be possible to open such housing or container only with a special tool or by means of appropriate protection measures.

1.2.7. *Protection against other hazards*

Equipment and protective systems must be so designed and manufactured as to:

(a) avoid physical injury or other harm which might be caused by direct or indirect contact;

(b) assure that surface temperatures

of accessible parts or radiation which would cause a danger, are not produced;

(c) eliminate non-electrical dangers which are revealed by experience;

(d) assure that foreseeable conditions of overload shall not give rise to dangerous situations.

Where, for equipment and protective systems, the risks referred to in this paragraph are wholly or party covered by other Community Directives, this Directive shall not apply or shall cease to apply in the case of such equipment and protective systems and of such risks upon application of those specific Directives.

1.2.8. *Overloading of equipment*

Dangerous overloading of equipment must be prevented at the design stage by means of integrated measurement, regulation and control devices, such as over-current cut-off switches, temperature limiters, differential pressure switches, flowmeters, time-lag relays, overspeed monitors and/or similar types of monitoring devices.

1.2.9. *Flameproof enclosure systems*

If parts which can ignite an explosive atmosphere are placed in an enclosure, measures must be taken to ensure that the enclosure withstands the pressure developed during an internal explosion of an explosive mixture and prevents the transmission of the explosion to the explosive atmosphere surrounding the enclosure.

1.3. **Potential ignition sources**

1.3.1. *Hazards arising from different ignition sources*

Potential ignition sources such as sparks, flames, electric arcs, high surface temperatures, acoustic energy, optical radiation, electromagnetic waves and other ignition sources must not occur.

1.3.2. *Hazards arising from static electricity*

Electrostatic charges capable of resulting in dangerous discharges must be prevented by means of appropriate measures.

1.3.3. *Hazards arising from stray electric and leakage currents*

Stray electric and leakage currents in conductive equipment parts which could result in, for example, the occurrence of dangerous corrosion, overheating of surfaces or sparks capable of provoking an ignition must be prevented.

1.3.4. *Hazards arising from overheating*

Overheating caused by friction or impacts occurring, for example, between materials and parts in contact with each other while rotating or through the intrusion of foreign bodies must, as far as possible, be prevented at the design stage.

1.3.5. *Hazards arising from pressure compensation operations*

Equipment and protective systems must be so designed or fitted with integrated measuring, control and regulation devices that pressure compensations arising from them do not generate shock waves or compressions which may cause ignition.

1.4. **Hazards arising from external effects**

1.4.1. Equipment and protective systems must be so designed and constructed as to be capable of performing their intended function in full safety, even in changing environmental conditions and in the presence of extraneous voltages, humidity, vibrations, contamination and other external effects, taking into account the limits of the operating conditions established by the manufacturer.

1.4.2. Equipment parts used must be appropriate to the intended mechanical and thermal stresses and capable of withstanding attack by existing or foreseeable aggressive substances.

1.5. **Requirements in respect of safety-related devices**

1.5.1. Safety devices must function independently of any measurement of control devices required for operation.

As far as possible, failure of a safety device must be detected sufficiently rapidly by appropriate technical means to ensure that there is only very little likelihood that dangerous situations will occur.

For electrical circuits the fail-safe principle is to be applied in general.

Safety-related switching must in general directly actuate the relevant control devices without intermediate software command.

1.5.2. In the event of a safety device failure, equipment and/or protective systems shall, wherever possible, be secured.

1.5.3. Emergency stop controls of safety devices must, as far as possible, be fitted with restart lockouts. A new start command may take effect on normal operation only after the restart lockouts have been intentionally reset.

1.5.4. *Control and display units*

Where control and display units are used, they must be designed in accordance with ergonomic principles in order to achieve the highest possible level of operating safety with regard to the risk of explosion.

1.5.5. *Requirements in respect of devices with a measuring function for explosion protection*

In so far as they relate to equipment used in explosive atmospheres, devices with a measuring function must be designed and constructed so that they can cope with foreseeable operating requirements and special conditions of use.

1.5.6. Where necessary, it must be possible to check the reading accuracy and serviceability of devices with a measuring function.

1.5.7. The design of devices with a measuring function must incorporate a safety factor which ensures that the alarm threshold lies far enough outside the explosion and/or ignition limits of the atmosphere to be registered, taking into account, in particular, the operating conditions of the installation and possible aberrations in the measuring system.

1.5.8. *Risks arising from software*

In the design of software-controlled equipment, protective systems and safety devices, special account must be taken of the risks arising from faults in the programme.

1.6. **Integration of safety requirements relating to the system**

1.6.1. Manual override must be possible in order to shut down the equipment and protective systems incorporated within automatic processes which deviate from the intended operating conditions, provided that this does not compromise safety.

1.6.2. When the emergency shutdown system is actuated, accumulated energy must be dispersed as quickly and as safely as possible or isolated so that it no longer constitutes a hazard.

This does not apply to electrochemically-stored energy.

1.6.3. *Hazards arising from power failure*

Where equipment and protective systems can give rise to a spread of additional risks in the event of a power failure, it must be possible to maintain them in a safe state of operation independently of the rest of the installation.

1.6.4. *Hazards arising from connections*

Equipment and protective systems must be fitted with suitable cable and conduit entries.

When equipment and protective systems are intended for use in combination with other equipment and protective systems, the interface must be safe.

1.6.5. *Placing of warning devices as parts of equipment*

Where equipment or protective systems are fitted with detection or alarm devices for monitoring the occurrence of explosive atmospheres, the necessary instructions must be provided to enable them to be provided at the appropriate places.

2. SUPPLEMENTARY REQUIREMENTS IN RESPECT OF EQUIPMENT

2.0 Requirements appalicable to equipment in category M of equipment-group I

2.0.1. *Requirements applicable to equipment in category M 1 of equipment-group I*

2.0.1.1. Equipment must be so designed and constructed that sources of ignition do not become active, even in the event of rare incidents relating to equipment.

Equipment must be equipped with means of protection such that:

- either, in the event of failure of one means of protection, at least an independent second means provide the requisite level of protection,
- or, the requisite level of protection is ensured in the event of two faults occurring independently of each other.

  Where necessary, this equipment must be equipped with additional special means of protection.

It must remain functional with an explosive atmosphere present.

2.0.1.2. Where necessary, equipment must be so constructed that no dust can penetrate it.

2.0.1.3. The surface temperatures of equipment parts must be kept clearly below the ignition temperature of the foreseeable air/dust mixtures in order to prevent the ignition of suspended dust.

2.0.1.4. Equipment must be so designed that the opening of equipment parts which may be sources of ignition is possible only under non-active or intrinsically safe conditions. Where it is not possible to render equipment non-active, the manufacturer must affix a warning label to the opening part of the equipment.

If necessary, equipment must be fitted with appropriate additional interlocking systems.

2.0.2. *Requirements applicable to equipment in category M 2 of equipment-group I*

2.0.2.1. Equipment must be equipped with means of protection ensuring that sources of ignition do not become active during normal operation, even under more severe operating conditions, in particular those arising from rough handling and changing environmental conditions.

The equipment is intended to be de-energised in the event of an explosive atmosphere.

2.0.2.2. Equipment must be so designed that the opening of equipment parts which may be sources of ignition is possible only under non-active conditions or via appropriate interlocking systems. Where it is not possible to render equipment non-active, the manufacturer must affix a warning label to the opening part of the equipment.

2.0.2.3. The requirements regarding explosion hazards arising from dust applicable to category M 1 must be applied.

2.1. Requirements applicable to equipment in category 1 of equipment-group II

2.1.1. *Explosive atmospheres caused by gases, vapours or hazes*

2.1.1.1. Equipment must be so designed and constructed that sources of ignition do not become active, even in event of rare incidents relating to equipment.

It must be equipped with means of protection such that:

- either, in the event of failure of one means of protection, at least an independent second means provides the requisite level of protection,

– or, the requisite level of protection is ensured in the event of two faults occurring independently of each other.

2.1.1.2. For equipment with surfaces which may heat up, measures must be taken to ensure that the stated maximum surface temperatures are not exceeded even in the most unfavourable circumstances.

Temperature rises caused by heat build-ups and chemical reactions must also be taken into account.

2.1.1.3. Equipment must be so designed that the opening of equipment parts which might be sources of ignition is possible only under non-active or intrinsically safe conditions. Where it is not possible to render equipment non-active, the manufacturer must affix a warning label to the opening part of the equipment.

If necessary, equipment must be fitted with appropriate additional interlocking systems.

2.1.2. *Explosive atmospheres caused by air/dust mixtures*

2.1.2.1. Equipment must be so designed and constructed that ignition air/dust mixtures does not occur even in the event of rare incidents relating to equipment.

It must be equipped with means of protection such that

– either, in the event of failure of one means of protection, at least an independent second means provides the requisite level of protection,
– or, the requisite level of protection is ensured in the event of two faults occurring independently of each other.

2.1.2.2. Where necessary, equipment must be so designed that dust can enter or escape from the equipment only at specifically designated points.

This requirement must also be met by cable entries and connecting pieces.

2.1.2.3. The surface temperatures of equipment parts must be kept well below the ignition temperature of the foreseeable air/dust mixtures in order to prevent the ignition of suspended dust.

2.1.2.4. With regard to the safe opening of equipment parts, requirement 2.1.1.3 applies.

2.2. **Requirements for category 2 of equipment-group II**

2.2.1. *Explosive atmosphere caused by gases, vapours or mists*

2.2.1.1. Equipment must be so designed as to prevent ignition sources arising, even in the event of frequently occurring disturbances or equipment operating faults, which normally have to be taken into account.

2.2.1.2. Equipment parts must be so designed and constructed that their stated surface temperatures are not exceeded, even in the case of risks arising from abnormal situations anticipated by the manufacturer.

2.2.1.3. Equipment must be so designed that the opening of equipment parts which might be sources of ignition is possible only under non active conditions or via appropriate interlocking systems. Where it is not possible to render equipment non-active, the manufacturer must affix a warning label to the opening part of the equipment.

2.2.2. *Explosive atmospheres caused by air/dust mixtures*

2.2.2.1. Equipment must be designed and constructed so that ignition of air/dust mixtures is prevented, even in the event of frequently occurring disturbances or equipment operating faults which normally have to be taken into account.

2.2.2.2. With regard to surface temperatures, requirement 2.1.2.3 applies.

2.2.2.3. With regard to protection against dust, requirement 2.1.2.2 applies.

2.2.2.4. With regard to the safe opening of equipment parts, requirement 2.2.1.3 applies.

2.3. **Requirements applicable to equipment in category 3 of equipment-group II**

2.3.1. *Explosive atmospheres caused by gases, vapours or mists*

2.3.1.1. Equipment must be so designed and constructed as to prevent foreseeable ignition sources which can occur during normal operation.
2.3.1.2. Surface temperatures must not exceed the stated maximum surface temperatures under intended operating conditions. Higher temperatures in exceptional circumstances may be allowed only if the manufacturer adopts special additional protective measures.

2.3.2. *Explosive atmospheres caused by air/dust mixtures*

2.3.2.1. Equipment must be so designed and constructed that air/dust mixtures cannot be ignited by foreseeable ignition sources likely to exist during normal operation.
2.3.2.2. With regard to surface temperatures, requirement 2.1.2.3. applies.
2.3.2.3. Equipment, including cable entries and connecting pieces, must be so constructed that, taking into account the size of its particles, dust can neither develop explosive mixtures with air nor form dangerous accumulations inside the equipment.

3. SUPPLEMENTARY REQUIREMENTS IN RESPECT OF PROTECTIVE SYSTEMS

3.0. **General requirements**

3.0.1. Protective systems must be dimensioned in such a way as to reduce the effects of an explosion to a sufficient level of safety.
3.0.2. Protective systems must be designed and capable of being positional in such a way that explosions are prevented from spreading through dangerous chain reactions or flashover and incipient explosions do not become detonations.
3.0.3. In the event of a power failure, protective systems must retain their capacity to function for a period sufficient to avoid a dangerous situation.
3.0.4. Protective systems must not fail due to outside interference.

3.1. **Planning and design**

3.1.1. *Characteristics of materials*

With regard to the characteristics of materials, the maximum pressure and temperature to be taken into consideration at the planning stage are the expected pressure during an explosion occurring under extreme operating conditions and the anticipated heating effect of the flame.
3.1.2. Protective systems designed to resist or contain explosions must be capable of withstanding the shock wave produced without losing system integrity.
3.1.3. Accessories connected to protective systems must be capable of withstanding the expected maximum explosion pressure without losing their capacity to function.
3.1.4. The reactions caused by pressure in peripheral equipment and connected pipe-work must be taken into consideration in the planning and design of protective systems.

3.1.5. *Pressure-relief systems*

If it is likely that stress on protective systems will exceed their structural strength, provision must be made in the design for suitable pressure-relief devices which do not endanger persons in the vicinity.

3.1.6. *Explosion suppression systems*

Explosion suppression systems must be so planned and designed that they react to an incipient explosion at the earliest possible stage in the event of an incident and counteract it to best effect, which due regard to the maximum rate of pressure increase and the maximum explosion pressure.

3.1.7. *Explosion decoupling systems*

Decoupling systems intended to disconnect specific equipment as swiftly as possible in the event of incipient explosions by means of appropriate devices must be planned and designed so as to remain proof against the transmission of internal ignition and to retain their mechanical strength under operating conditions.

3.1.8. Protective systems must be capable of being integrated into a circuit with a suitable alarm threshold so that, if necessary, there is cessation of product feed and output and shutdown of equipment parts which can no longer function safely.

Annex III

Module EC-type examination

1. This module describes that part of the procedure by which a notified body ascertains and attests that a specimen representative of the production envisaged meets the relevant applicable provisions of the Directive.

2. The application for the EC-type examination shall be lodged by the manufacturer or his authorised representative established within the Community with a notified body of his choice.

The application shall include:

- the name and address of the manufacturer and, if the application is lodged by the authorised representative, his name and address in addition;
- a written declaration that the same application has not been lodged with any other notified body;
- the technical documentation, as described in point 3.

The applicant shall place at the disposal of the notified body a specimen representative of the production envisaged and hereinafter called "type". The notified body may request further specimens if needed for carrying out the test programme.

3. The technical documentation shall enable the conformity of the product with the requirements of the Directive to be assessed. It shall, to the extent necessary for such assessment, cover the design, manufacture and operation of the product and shall to that extent contain:

- a general type-description;
- design and manufacturing drawings and layouts of components, sub-assemblies, circuits, etc.;
- descriptions and explanations necessary for the understanding of said drawings and layouts and the operation of the product;
- a list of the standards referred to in Article 5, applied in full or in part, and descriptions of the solutions adopted to meet the essential requirements of the Directive where the standards referred to in Article 5 have not been applied;
- results of design circulations made, examinations carried out, etc.;
- test reports.

4. The notified body shall:

4.1. examine the technical documentation, verify that the type has been manufactured in conformity with the technical documentation and identify the elements which have been designed in accordance with the relevant provisions of the standards referred to in Article 5, as well as the components which have been designed without applying the relevant provisions of those standards;

4.2. perform or have performed the appropriate examinations and necessary tests to check whether the solutions adopted by the manufacturer meet the essential requirements of the Directive where the standards referred to in Article 5 have not been applied;

4.3. perform or have performed the appropriate examinations and necessary tests to check whether these have actually been applied, where the manufacturer has chosen to apply the relevant standards;

4.4. agree with the applicant the location where the examinations and necessary tests shall be carried out.

5. Where the type meets the provisions of the Directive, the notified body shall issue an EC-type-examination certificate to the applicant. The certificate shall contain the name and address of the manufacturer, conclusions of the examination and the necessary data for identification of the approved type.

A list of the relevant parts of the technical documentation shall be annexed to the certificate and a copy kept by the notified body.

If the manufacturer or his authorised representative established in the Community is denied a type certificate, the notified body shall provide detailed reasons for such denial.

Provision shall be made for an appeals procedure.

6. The applicant shall inform the notified body which holds the technical documentation concerning the EC-type-examination certificate for all modifications to the approved equipment or protective system which must receive further approval where such changes may effect conformity with the essential requirements or with the prescribed conditions for use of the product. This further approval is given in the form of an addition to the original EC-type-examination certificate.

7. Each notified body shall communicate to the other notified bodies the relevant information concerning the EC-type-examination certificates and additions issued and withdrawn.

8. The other notified bodies may receive copies of the EC-type-examination certificates and/or their additions. The annexes to the certificates shall be kept at the disposal of the other notified bodies.

9. The manufacturer or his authorised representative established in the Community shall keep with the technical documentation copies of EC-type-examination certificates and their additions for a period ending at least 10 years after the last equipment or protective system was manufactured.

Where neither the manufacturer nor his authorised representative is established within the Community, the obligation to keep the technical documentation available shall be the responsibility of the person who places the product on the Community market.

Annex IV

Module: production quality assurance

1. This module describes the procedure whereby the manufacturer who satisfies the obligations of point 2 ensures and declares that the products concerned are in conformity with the type as described in the EC-type-examination certificate and satisfy the requirements of the Directive which apply to them. The manufacturer, or his authorised representative established in the Community, shall affix the CE marking to each piece of equipment and draw up a written declaration of conformity. The CE marking shall be accompanied by the identification number of the notified body responsible for EC monitoring, as specified in Section 4.

2. The manufacturer shall operate an approved quality system for production, final equipment inspection and testing as specified in Section 3 and shall be subject to monitoring as specified in Section 4.

3. **Quality system**

3.1. The manufacturer shall lodge an application for assessment of his quality system with a notified body of his choice, for the equipment concerned.

The application shall include:

- all relevant information for the product category envisaged;
- the documentation concerning the quality system;
- technical documentation on the approved type and a copy of the EC-type-examination certificate.

3.2. The quality system shall ensure compliance of the equipment with the type as described in the EC-type-examination certificate and with the requirements of the Directive which apply to them.

All the elements, requirements and provisions adopted by the manufacturer shall be documented in a systematic and orderly manner in the form of written policies, procedures and instructions. The quality system documentation must permit a consistent interpretation of quality programmes, plans, manuals and records.

It shall contain, in particular, an adequate description of

- the quality objectives and the organisational structure, responsibilities and powers of the management with regard to equipment quality;
- the manufacturing, quality control and quality assurance techniques, processes and systematic actions which will be used;
- the examinations and tests which will be carried out before, during and after manufacture and the frequency with which they will be carried out;
- the quality records, such as inspection reports and test data, calibration data, reports on the qualifications of the personnel concerned, etc.;
- the means to monitor the achievement of the required equipment quality and the effective operation of the quality system.

3.3. The notified body shall assess the quality system to determine whether it satisfies the requirements referred to in Section 3.2. It shall presume conformity with these requirements in respect of quality systems which implement the relevant harmonised standard. The auditing team shall have at least one member with experience of evaluation in the equipment technology concerned. The evaluation procedure shall include an inspection visit to the manufacturer's premises. The decision shall be notified to the manufacturer. The notification shall contain the conclusions of the examination and the reasoned assessment decision.

3.4. The manufacturer shall undertake to fulfil the obligations arising out of the quality system as approved and to uphold the system so that it remains adequate and efficient.

The manufacturer or his authorised representative shall inform the notified body which has approved the quality system of any intended updating of the quality system.

The notified body shall evaluate the modifications proposed and decide whether the amended quality system will still satisfy the requirements referred to in Section 3.2 or whether a re-assessment is required.

It shall notify its decision to the manufacturer. The notification shall contain the conclusions of the examination and the reasoned assessment decision.

4. **Surveillance under the responsibility of the notified body**

4.1. The purpose of surveillance is to make sure that the manufacturer duly fulfils the obligations arising out of the approved quality system.

4.2. The manufacturer shall, for inspection purposes, allow the notified body access to the manufacture, inspection, testing and storage premises and shall provide it with all necessary information, in particular

- the quality system documentation
- the quality records, such as inspection reports and text data, calibration data, reports on the qualifications of the personnel concerned, etc.

4.3. The notified body shall periodically carry out audits to ensure that the manufacturer maintains and applies the quality system and shall provide an audit report to the manufacturer.

4.4. Furthermore, the notified body may pay unexpected visits to the manufacturer. During such visits, the notified body may carry out tests, or arrange for tests to be carried out, to check that the quality system is functioning correctly, if necessary. The notified body shall provide the manufacturer with a visit report and, if a test has taken place, with a test report.

5. The manufacturer shall, for a period ending at least 10 years after the last piece of equipment was manufactured, keep at the disposal of the national authorities:

- the documentation referred to in the second indent of Section 3.1;
- the updating referred to in the second paragraph of Section 3.4;
- the decisions and reports from the notified body which are referred to in Section 3.4, last paragraph, Section 4.3 and Section 4.4.

6. Each notified body shall appraise the other notified bodies of the relevant information concerning the quality system approvals issued and withdrawn.

Annex V

Module: product verification

1. This module describes the procedure whereby a manufacturer or his authorised representative established within the Community checks and attests that the equipment subject to the provisions of point 3 are in conformity with the type as described in the EC-type-examination certificate and satisfy the relevant requirements of the Directive.

2. The manufacturer shall take all measures necessary to ensure that the manufacturing process guarantees conformity of the equipment with the type as described in the EC-type-examination certificate and with the requirements of the Directive which apply to them. The manufacturer or his authorised representative established in the Community shall affix the CE marking to each piece of equipment and shall draw up a declaration of conformity.

3. The notified body shall carry out the appropriate examinations and tests in order to check the conformity of the equipment, protective system or device referred to in Article 1 (2), with the relevant requirements of the Directive, by examining and testing every product as specified in Section 4.

The manufacturer or his authorised representative shall keep a copy of the declaration of conformity for a period ending at least 10 years after the last piece of equipment was manufactured.

4. **Verification by examination and testing of each piece of equipment.**

4.1. All equipment shall be individually examined and appropriate tests as set out in the relevant standard(s) referred to in Article 5 or equipment tests shall be carried out in order to verify their conformity with the type as described in the EC-type-examination certificate and the relevant requirements of the Directive.

4.2. The notified body shall affix or have affixed its identification number to each approved item of equipment and shall draw up a written certificate of conformity relating to the tests carried out.

4.3. The manufacturer or his authorised representative shall ensure that he is able to supply the notified body's certificates of conformity on request.

Annex VI
Module: conformity to type

1. This module describes that part of the procedure whereby the manufacturer or his authorised representative established within the Community ensures and declares that the equipment in question is in conformity with the type as described in the EC-type-examination certificate and satisfy the requirements of the Directive applicable to them. The manufacturer or his authorised representative established within the Commumnity shall affix the CE marking to each piece of equipment and draw up a written declaration of conformity.

21. The manufacturer shall take all measures necessary to ensure that the manufacturing process assures compliance of the manufactured equipment or protective systems with the type as described in the EC-type-examination certificate and with the relevant requirements of the Directive.

3. The manufacturer or his authorised representative shall keep a copy of the declaration of conformity for a period ending at least 10 years after the last piece of equipment was manufactured. Where neither the manufacturer nor his authorised representative is established within the Community, the obligation to keep the technical documentation available shall be the responsibility of the person who places the equipment or protective system on the Community market.

For each piece of equipment manufactured, tests relating to the anti-explosive protection aspects of the product shall be carried out by the manufacturer or on his behalf. The tests shall be carried out under the responsibility of a notified body, chosen by the manufacturer.

On the responsibility of the notified body, the manufacturer shall affix the former's identification number during the manufacturing process.

Annex VII
Module: product quality assurance

1. This module describes the procedure whereby the manufacturer who satisfies the obligations of Section 2 ensures and declares that the equipment is in conformity with the type as described in the EC-type-examination certificate. The manufacturer or his authorised representative established within the Community shall affix the CE marking to each product and draw up a written declaration of conformity. The CE marking shall be accompanied by the identification number of the notified body responsible for surveillance as specified in Section 4.

2. The manufacturer shall operate an approved quality system for the final inspection and testing of equipment as specified in Section 3 below and shall be subject to surveillance as specified in Section 4 below.

3. **Quality system**

3.1. The manufacturer shall lodge an application for assessment of his quality system for the equipment and protective systems, with a notified body of his choice.

The application shall include:

- all relevant information for the product category envisaged;
- documentation on the quality system;
- technical documentation on the approved type and a copy of the EC-

type-examination certificate.

3.2. Under the quality system, each piece of equipment shall be exam- nined and appropriate tests as set out in the relevant standard(s) referred to in Article 5 or equivalent tests shall be carried out in order to ensure its conformity with the relevant requirements of the Directive. All the elements, requirements and provisions adopted by the manufacturer shall be documented in a systematic and orderly manner in the form of written policies, procedures and instruments. This quality system documentation must permit a consistent interpretation of the quality programmes, plans, manuals and records.

It shall contain, in particular, an adequate description of:

- the quality objectives and the organisational structure, responsibilities and powers of the management with regard to product quality;
- the examinations and tests which will be carried out after manufacture;
- the means to monitor the effective operation of the quality system;
- quality records, such as inspection reports and test data, calibration data, reports on the qualifications of the personnel concerned, etc.

3.3. The notified body shall assess the quality system to determine whether it satisfies the requirements referred to in Section 3.2. It shall presume conformity with these requirements in respect of quality systems which implement the relevant harmonised standard.

The auditing team shall have at least one member experienced as an assessor in the product tehchnology concerned. The assessment procedure shall include an assessment visit to the manufacturer's premises.

The decision shall be notified to the manufacturer. The notification shall contain the conclusions of the examination and the reasoned assessment decision.

3.4. The manufacturer shall undertake to discharge the obligations arising from the quality system as approved and to maintain it in an appropriate and efficient manner.

The manufacturer or his authorised representative shall inform the notified body which has approved the quality system of any intended updating of the quality system.

The notified body shall evaluate the modifications proposed and decide whether the modified quality system will still satisfy the requirements referred to in Section 3.2 or whether a re-assessment is required.

It shall notify its decision to the manufacturer. The notification shall contain the conclusions of the examination and the reasoned assessment decision.

4. **Surveillance under the responsibility of the notified body**

4.1. The purpose of surveilllance is to ensure that the manufacturer duly fulfils the obligations arising out of the approved quality system.

4.2. The manufacturer shall for inspection purposes allow the notified body access to the inspection, testing and storage premises and shall provide it with all necessary information, in particular:

- quality system documentation;
- technical documentation;
- quality records, such as inspection reports and test data, calibration data, reports on the qualifications of the personnel concerned, etc.

4.3. The notified body shall periodically carry out audits to ensure that the manufacturer maintains and applies the quality system and shall provide an audit report to the manufacturer.

4.4. Furthermore, the notified body may pay unexpected visits to the manufacturer. At the time of such visits, the notified body may carry out tests or arrange for tests to be carried out in order to check the proper functioning of the quality system, where necessary, it shall provide the manufacturer with a visit report and, if a test has been carried out, with a test report.

5. The manufacturer shall, for a period ending at least 10 years after the last piece of equipment was manufac-

tured, keep at the disposal of the national authorities:
- the documentation referred to in the third indent of Section 3.1;
- the updating referred to in the second paragraph of Section 3.4;
- the decisions and reports from the notified body which are referred to in Section 3.4, last paragraph, Section 4.3 and Section 4.4.

6. Each notified body shall forward to the other notified bodies the relevant information concerning the quality system approvals issued and withdrawn.

Annex VIII
Module: internal control of production

1. This module describes the procedure whereby the manufacturer or his authorised representative established within the Community, who carries out the obligations laid down in Section 2, ensures and declares that the equipment satisfy the requirements of the Directive applicable to it. The manufacturer or his authorised representative established within the Community shall affix the CE marking to each piece of equipment and draw up a written declaration of conformity.

2. The manufacturer shall establish the technical documentation described in Section 3 and he or his authorised representative established within the Community shall keep it at the disposal of the relevant national authorities for inspection purposes for a period ending at least 10 years after the last piece of equipment was manufactured.

Where neither the manufacturer nor his authorised representative is established within the Community, the obligation to keep the technical documentation available shall be the responsibility of the person who places the equipment on the Community market.

3. Technical documentation shall enable the conformity of the equipment with the relevant requirements of the Directive to be assessed. It shall, to the extent necessary for such assessment, cover the design, manufacture and operation of the product. It shall contain:
- a general description of the equipment,
- conceptual design and manufacturing drawings and schemes of components, sub-assemblies, circuits, etc.,
- description and explanations necessary for the understanding of said drawings and schemes and the operation of the equipment,
- a list of the standards applied in full or in part, and descriptions of the solutions adopted to meet the safety aspects of the Directive where the standards have not been applied,
- results of design calculations made, examinations carried out, etc.,
- test reports.

4. The manufacturer or his authorised representative shall keep a copy of the declaration of conformity with the technical documentation.

5. The manufacturer shall take all measures necessary to ensure that the manufacturing process guarantees compliance of the manufactured equipment with the technical documentation referred to in Section 2 and with the requirements of the Directive applicable to such equipment.

Annex IX
Module: unit verification

1. This module describes the procedure whereby the manufacturer ensures and declares that the equipment or protective system which has been issued with the certificate referred to in Section 2 conforms to the requirements of the Directive which are applicable to it. The manufacturer or his authorised representative in the Community shall affix the CE marking to the equipment or protective system and draw up a declaration of conformity.

2. The notified body shall examine the individual equipment or protective system and carry out the appropriate tests as set out in the relevant standard(s) referred to in Article 5, or equivalent tests, to ensure its conformity with the relevant requirements of the Directive.

3. The aim of the technical documentation is to enable conformity with the requirements of the Directive to be assessed and the design, manufacture and operation of the equipment or protective system to be understood.

The documentation shall contain:
- a general description of the product;
- conceptual design and manufacturing drawings and layouts of components, sub-assemblies, circuits, etc.;
- descriptions and explanations necesary for the understanding of said drawings and layouts and the operation of the equipment or protective system;
- a list of the standards referred to in Article 5, applied in full or in part, and descriptions of the solutions adopted to meet the essential requirements of the Directive where the standards referred to in Article 5 have not been applied;
- results of design calculations made, examinations carried out, etc.;
- test reports.

Annex X

A. **CE marking**

The CE coinformity marking shall consist of the initials "CE" taking the following form:

If the marking is reduced or enlarged, the proportions given in the above graduated drawing must be respected.

The various components of the CE marking must have substantially the same vertical dimensions, which may not be less than 5 mm.

This minimum dimension may be waived for small-scale eqipment, protective sysetms or devices referred to n Article 1 (2).

B. **Cotent of the EC declaration of conformity**

The EC declaration of conformity must contain the following elements:
- the name or identification mark and the address of the manufacturer or his authorised representative established within the Community;
- a description of the equipment, protective system, or device referred to in Article 1 (2);
- all relevant provisions fulfilled by

the equipment, protective system, or device referred to in Article 1 (2);
- where appropriate, the name, identification number and address of the modified body and the number of the EC-type-examination certificate;
- where appropriate, reference to the harmonised standards;
- where appropriate, references to other Community Directives which have been applied;
- identification of the signatory who has been empowered to enter into commitments on behalf of the manufacturer or his authorised representative established within the community.

Annex XI

Minimum criteria to be taken into account by Member States for the notification of bodies

1. The body, its director and the staff responsible for carrying out the verification tests shall not be the designer, manufacturer, supplier or installer of equipment, protective systems, or devices referred to in Article 1 (2) which they inspect, nor the authorised representative of any of these parties. They shall become involved neither directly nor as authorised representatives in the design, construction, marketing or maintenance of the equipment, protective systems or devices referred to in Article 1 (2) in question. This does not preclude the possibility of exchanges of technical information between the manufacturer and the body.

2. The body and its inspection staff shall carry out the verification tests with the highest degree of professional integrity and technical competence and shall be free from all pressures and inducements, particularly financial, which may influence their judgement or the results of the inspection, especially from persons or groups of persons with an interest in the result of verifications.

3. The body shall have at its disposal the necessary staff and possess the necessary facilities to enable it to perform properly the administrative and technical tasks connected with verification; it shall also have access to the equipment required for special verification.

4. The staff responsible for inspection shall have:
- sound technical and professional training;
- satisfactory knowledge of the requirements of the tests which they carry out and adequate experience of such tests;
- the ability to draw up the certificates, records and reports required to authenticate the performance of the tests.

5. The impartiality of inspection staff shall be guaranteed. Their remuneration shall not depend on the number of tests carried out or on the results of such tests.

6. The body shall take out liability insurance unless its liability is assumed by the State in accordance with national law or the Member State itself is directly responsible for the tests.

7. The staff of the body shall be bound to observe professional secrecy with regard to all information gained in carrying out its tasks (except *vis-à-vis* the component administrative authorities of the State in whcih its activities are carried out) under this Directive or any provision of national law giving effect to it.

DIRECTIVE 94/25/EC OF THE EUROPEAN PARLIAMENT AND OF THE COUNCIL

of 16 June 1994

on the approximation of the laws, regulations and administrative provisions of the Member States relating to recreational craft

THE EUROPEAN PARLIAMENT AND THE COUNCIL OF THE EUROPEAN UNION,

Having regard to the Treaty establishing the European Community, and in particular Article 100a thereof,

Having regard to the proposal from the Commission,[1]

Having regard to the opinion of the Economic and Social Committee,[2]

Acting in accordance with the procedure laid down in Article 189b of the Treaty,[3]

Whereas the internal market is to comprise an area without internal frontiers in which the free movement of goods, persons, services and capital is ensured;

Whereas the laws, regulations and administrative provisions in force in the various Member States relating to the safety characteristics of recreational craft differ in scope and content; whereas such disparities are liable to create barriers to trade and unequal conditions of competition within the internal market;

Whereas harmonisation of national legislation is the only way in which to remove these barriers to free trade; whereas this objective cannot be satisfactorily achieved by the individual Member States; whereas this Directive merely lays down the requirements vital to freedom of movement for recreational craft;

Whereas this Directive should cover only recreational craft of a minimum length of 2.5 m and a maximum length of 24 m, derived from the ISO standards;

Whereas the removal of technical barriers in the field of recreational craft and their components, to the extent that they cannot be removed by mutual recognition of equivalence among all the Member States, should follow the new approach set out in the Council resolution of 7 May 1985[4] which calls for the definition of essential requirements on safety and other aspects which are important for the general well-being; whereas paragraph 3 of Article 100a provides that, in its proposals, concerning health, safety, environmental protection and consumer protection, the Commission will take as a base a high level of protection; whereas the essential requirements constitute the criteria with which recreational craft, partly completed craft and their components when separate and when installed must comply;

Whereas, therefore, this Directive sets out essential requirements only; whereas, in order to facilitate the task of proving compliance with the

---

[1] OJ No C 123, 15. 5. 1992, p. 7.

[2] OJ No C 313, 30. 11. 1992, p. 38.

[3] Opinion of the European Parliament of 16 November 1992 (OJ No C 337, 21. 12. 1992, p. 17); Council common position of 16 December 1993 (OJ No 137, 19. 5. 1994, p. 1); Decision of the European Parliament of 9 March 1994 (OJ No C 91, 28. 3. 1994).

[4] OJ No C 136, 4. 5. 1985, p. 1.

essential requirements, harmonised European standards are necessary for recreational craft and their components as referred to in Annex II; whereas harmonised European standards are drawn up by private bodies and must retain their non-mandatory status; whereas, for this purpose, the European Committee for Standardisation (CEN) and the European Committee for Electrotechnical Standardisation (CENELEC) are recognised as the bodies competent to adopt harmonised standards which follow the general guidelines for cooperation between the Commission and those two bodies, signed on 13 November 1984; whereas, for the purposes of this Directive, a harmonised standard is a technical specification (European Standard or harmonisation document) adopted by one or other of those bodies, or by both, at the prompting of the Commission pursuant to Council Directive 83/189/EEC of 28 March 1983 laying down in the field of technical standards and regulatsions[5] and pursuant to the general guidelines referred to above;

Whereas, the view of the nature of the risks involved in the use of recreational craft and their components, it is necessary to establish procedures applying to the assessment of compliance with the essential requirements of the Directive; whereas these procedures must be devised in the light of the level of risk which may be inherent in recreational craft and their components; whereas, therefore, each category of conformity must be supplemented by an appropriate procedure or a choice between several equivalent procedures; whereas the procedures adopted comply with Council Decision 93/465/EEC of 22 July 1993 concerning the modules for the various phases of the conformity assessment procedures and the rules for the affixing and use of the CE conformity marking which are intended to be used in the technical harmonisation Directives;[6]

Whereas the Council has provided for the affixing of the CE marking by either the manufacturer or his authorised representative within the Community; whereas that mark means that the recreational craft and components comply with all the essential requirements and assessment procedures provided for by the Community law applying to the product;

Whereas it is appropriate that the Member States, as provided for by Article 100a (5) of the Treaty, may take provisional measures to limit or prohibit the placing on the market and the use of recreational craft or constituent products thereof in cases where they present a particular risk to the safety of persons and, where appropriate, domestic animals or property, provided that the measures are subject to a Community control procedure;

Whereas the recipients of any decision taken as part of this Directive must be aware of the reasons behind that decision and the means of appeal open to them;

Whereas it is necessary to provide for a transitional arrangement enabling recreational craft and their components manufactured in compliance with the national regulations in force at the date of adoption of this Directive to be marketed and placed in service;

Whereas this Directive does not contain any provisions directed towards limiting the use of the recreational craft after it has been put into service;

Whereas the construction of recreational craft may have an impact on the environment to the extent that the craft may discharge polluting substances; whereas it is therefore necessary to include provisions on the protection of the environment in the

[5] OJ No L 109, 26. 4. 1983, p. 8. Directive as last amended by Directive 88/182/EEC (OJ No L 81, 26. 3. 1988, p. 75).

[6] OJ No L 220, 30. 8. 1993, p. 23.

Directive, in so far as those provisions concern the construction of recreational craft from the point of view of its direct impact on the environment;

Whereas the provisions of this Directive should not affect Member States' entitlement to lay down, in accordance with the Treaty, such requirements as they may deem necessary concerning navigation on certain waters for the purpose of protection of the environment, the fabric of waterways and ensuring safety on waterways, provided that this does not mean that the recreational craft is modified in a way not specified in this Directive.

HAVE ADOPTED THIS DIRECTIVE:

Chapter I

Article 1

1. This Directive shall apply to recreational craft, partly completed boats and components referred to in Annex II when separate and when installed.
2. "Recreational craft" shall mean any boat of any type, regardless of the means of propulsion, from 2.5 to 24 m hull length, measured according to the appropriate harmonised standards intended for sports and leisure purposes. The fact that the same boat could be used for charter or for recreational boating training shall not prevent it being covered by this Directive when it is placed on the market for recreational purposes.
3. The following shall be excluded from the scope of this Directive:
(a) craft intended solely for racing, including rowing racing boatss and training rowing boats labelled as such by the manufacturer;
(b) canoes and kayaks, gondolas and pedalos;
(c) sailing surfboards;
(d) powered surfboards, personal watercraft and other similar powered craft;
(e) original, and individual replicas of, historical craft designed before 1950, built predominantly with the original materials and labelled as such by the manufacturer;
(f) experimental craft, provided that they are not subsequently placed on the Community market;
(g) craft built for own use, provided that they are not subsequently placed on the Community market during a period of five years;
(h) craft specifically intended to be crewed and to carry passengers for commercial purposes, without prejudice to paragraph 2, in particular those defined in Directive 82/716/EEC of 4 October 1982 laying down technical requirements for inland waterway vessels,[7] regardless of the number of passengers;
(i) submersibles;
(j) air cushion vehicles;
(k) hydrofoils.

Article 2

Placing on the market and putting into service

1. Member States shall take all necessary measures to ensure that the products referred to in Article 1 (1) may be placed on the market and put into service for use in accordance with their intended purpose only if they do not endanger the safety and health of persons, property or the environment when correctly constructted and maintained.
2. The provisions of this Directive shall not prevent Member States from adopting, in compliance with the Treaty, provisions concerning navigation on certain waters for the purpose of protection of the environment, the fabric of waterways, and ensuring safety of waterways, providing that this does not require modification to craft conforming to this Directive.

Article 3

Essential requirements

The product referred to in Article 1

[7] OJ No L 301, 28. 10. 1982, p. 1.

(1) shall meet the essential safety, health, environmental protection and consumer protection requirements set out in Annex I.

Article 4

Free movement of the products referred to in Aticle 1 (1)

1. Member States shall not prohibit, restrict or impede the placing on the market and putting into service in their territory of the products referred to in Article 1 (1) bearing the CE marking referred to in Annex IV, which indicates their conformity with all provisions of this Directive, including the conformity procedures set out in Chapter II.

2. Member States shall not prohibit, restrict or impede the placing on the market of partly-completed boats where the builder or his authorised representative established in the Community or the person responsible for the placing on the market declares, in accordance with Annex IIIA, that they are intended to be completed by others.

3. Member States shall not prohibit, restrict or impede the placing on the market and putting into service of components referred to in Annex II and bearing the CE marking referred to in Annex IV which indicates their conformity with the relevant essential requirements where these components are intended to be incorporated into recreational craft, in accordance with the declaration, referred to in Annex IIIB, of the manufacturer, his authorised representative established in the Community or, in the case of imports from a third country, of any person who places those components on the Community market.

4. At trade fairs, exhibitions, demonstrations, etc., Member States shall not create any obstacles to the showing of the products referred to in Article 1 (1) which do not comply with this Directive, provided that a visible sign clearly indicates that such products may not be marketed or put into service until they have been made to comply.

5. Where the products referred to in Article 1 (1) are subject to other Directives concerning other aspects and which also provide for the affixing of the CE marking, the latter shall indicate that such products also fulfil the provisions of those other Directives. However, should one or more of those Directives allow the manufacturer, during a transitional period, to choose which arrangement to apply, the CE marking shall indicate that the product fulfils the provisions only of those Directives applied by the manufacturer. In this case, the particulars of those Directives, as published in the *Official Journal of the European Communities*, must be given in the documents, notices or instructions required by those Directives and accompanying such products.

Article 5

Member States shall presume compliance with the essential requirements referred to in Article 3 or products referred to in Article 1 (1) which meet the relevant national standards adopted pursuant to the harmonised standards the references of which have been published in the *Official Journal of the European Communities*; Member States shall publish the references of such national standards.

Article 6

1. Where a Member State or the Commission is of the opinion that the harmonised standards referred to in Article 5 do not fully meet the essential requirements referred to in Article 3, the Commission or the member State shall notify the committee set up under Directive 83/189/ECE, setting out its reasons. The committee shall deliver an urgent opinion.

In the light of the opinion of the committee, the Commission shall inform Member States if the standards concerned should be withdrawn from the publications referred to in Article 5.

2. The Commission may adopt any appropriate measure with a view to ensuring that this Directive is applied

practically in a uniform manner in pursuance of paragraph 3.
3. The Commission shall be assisted by a Standing Committee composed of representatives appointed by the Member States and chaired by a representative from the Commission.

The Standing Committee shall draw up its own rules of procedure.

The representative of the Commission shall submit to the Committee a draft of the measures to be taken. The Committee shall deliver its opinion on the draft within a time limit which the chairman may lay down according to the urgency of the matter, if necessary by taking a vote.

The opinion shall be recorded in the minutes; in addition, each Member State shall have the right to ask to have its position recorded in the minutes.

The Commission shall take the utmost account of the opinion delivered by the Committee. It shall inform the Committee of the manner in which its opinion has been taken into account.
4. The Standing Committee may, in addition, examine any question concerning the application of this Directive and raised by its chairman, either at the chairman's initiative or at the request of a Member State.

Article 7

Safeguard clause

1. Where a Member State ascertains that recreational craft or components referred to in Annex IV when correctly constructed, installed, maintained and used in accordance with their intended purpose may endanger the safety and health of persons, property or the environmment, it shall take all appropriate interim measures to withdraw them from the market or prohibit or restrict their being placed on the market or put into service.

The Member State shall immediataely inform the Commission of any such measures, indicating the reasons for its decision, in particular where non-conformity is the result of:

(a) failure to comply with the essential requirements referred to in Article 3;

(b) incorrect application of the standards referred to in Article 5, in so far as it is claimed that those standards have been applied;

(c) shortcomings in the standards referred to in Article 5 themselves.

2. The Commission shall enter into consultation with the parties concerned as soon as possible. Where, after such consultation, the Commission finds that:

– the measures are justified, it shall immediately so inform the Member State which took the initiative and the other Member States; where the decision referred to in paragraph 1 is attributed to shortcomings in the standards, the Commission shall, after consulting the parties concerned, bring the matter before the Committee referred to in Article 6 (1) within two months, if the Member State which has taken the decision intends to maintain it, and shall initiate the procedure referred to in Article 6 (1).

– the measures are unjustified, it shall immediately so inform the Member State which took the initiative and the manufacturer or his authorised representative established in the Community.

3. Where a non-complying component referred to in Annex II or craft bears the CE marking, the appropriate measures shall be taken by the Member State which has authority over whomsoever affixed the marking; that Member State shall inform the Commission and the other Member States thereof.

4. The Commission shall ensure that the Member States are kept informed of the progress and outcome of this procedure.

Chapter II
Conformity assessment

Article 8

Before producing and placing on the market products referred to in Article

1 (1), the manufacturer or his authorised representative established in the Community shall apply the following procedures for boat design categories A, B, C and D as referred to in Section 1 of Annex I.

1. *For categories A and B:*
- for boats of less than 12 m hull length: the internal production control plus tests (module Aa) referred to in Annex VI,
- for boats from 12 m to 24 m hull length: the EC type-examination (module B) referred to in Annex VII supplemented by module C (type conformity) referred to in Annex VIII, or any of the following modules: B+D, or B+F, or G or H.

2. *For category C:*

(a) for boats from 2.5 m to 12 m hull length:
- where the harmonised standards relating to sections 3.2 and 3.3 of Annex I are complied with: the internal production control (module A), referrred to in Annex V,
- where the harmonised standards relating Sections 3.2 and 3.3 of Annex I are not complied with: the internal production control plus tests (module Aa) referred to in Annex VI.

(b) for boats from 12 m to 24 m hull length: the EC type-examination (module B) referred to in Annex VII followed by module C (type conformity) referred to in Annex VIII, or any of the following modules: B+D, or B+F, or G or H.

3. *For category D:*

For boats from 2.5 m to 24 m hull length: the internal production control (module A) referred to in Annex V.

4. For components referred to in Annex II: any of the following modules: B+C, or B+D, or B+F, or G or H.

Article 9

Notified bodies

1. Member States shall notify the Commission and other Member States of the bodies which they have appointed to carry out the tasks pertaining to the conformity assessment procedures referred to in Article 8, together with the specific tasks which these bodies have been appointed to carry out and the identification numbers asssigned to them beforehand by the Commission.

The Commission shall publish a list of the notified bodies, together with the identification numbers it has allocated to them and the tasks for which they have been notified, in the *Official Journal of the European Communities*. It shall ensure that the list is kept up to date.

2. Member States shall apply the criteria laid down in Annex XIV in assessing the bodies to be indicated in such notification. Bodies meeting the assessment criteria laid down in the relevant harmonised standards shall be presumed to fulfil those criteria.

3. A Member State shall withdraw its approval from such a body if it is established that the latter no longer satisfies the criteria referred to in Annex XIV. It shall inform the Commission and the other Member States of its action forthwith.

Chapter III
CE marking

Article 10

1. Recreational craft and components as referred to in Annex II which are regarded as meeting the essential requirements referred to in Article 3 must bear the CE marking of conformity when they are placed on the market.

2. The CE marking of conformity, as shown in Annex IV, must appear in a visible, legible and indelible form on the recreational craft as in point 2.2 of Annex I and on components as referred to in Annex II and/or on their packaging.

The CE marking shall be accompanied by the identification number of the notified body responsible for implementation of the procedures set out in Annexes VI, IX, X, XI and XII.

3. The affixing of markings or inscrip-

tions on the craft which are likely to mislead third parties with regard to the meaning or the form of the CE marking shall be prohibited. Any other markings may be affixed to the recreational craft and components as referred to in Annex II and/or on their packaging, provided that the visibilty and legibility of the CE marking is not thereby reduced.

4. Without prejudice in Article 7:

(a) where a Member State establishes that the CE marking has been affixed wrongly, the manufacturer or his authorised representative established in the Community shall be obliged to end the infringement under conditions laid down by the Member State;

(b) where non-compliance continues, the Member State shall take all appropriate measures to restrict or prohibit the placing on the market, in accordance with the procedure laid down in Article 7.

Chapter IV
Final provisions

Article 11

Detailed grounds shall be given for any decision taken pursuant to this Directive leading to a restriction on the marketing and putting into service of products rerferred to in Article 1 (1). The party concerned shall be informed of the decision as soon as possible together with the means of redress available under the laws in force in the Member State concerned and the periods within which appeals must be lodged.

Article 12

The Commission shall take the necessary measures to ensure that data affecting all pertinent decisions concerning the management of this Directive are made available.

Article 13

1. Member States shall adopt and publish the laws, regulations and administrative provisions necessary to comply with this Directive not later than 16 December 1995. They shall immediately inform the Commission thereof.

Member States shall apply these provisions from 16 June 1996.

The Standing Committee referred to in Article 6 (3) may assume its tasks from the date of the entry into force of this directive. Member States may take the measures referred to in Article 9 of such date.

When Member States adopt the provisions referred to in the first subparagraph, these shall contain a reference to this directive or shall be accompanied by such a reference at the time of their official publication. The procedure for such reference shall be adapted by Member States.

2. Member States shall communicate to the Commission the text of the provisions of national law which they adopt in the field governed by this Directive.

3. Member States shall accept the placings on the market and putting into service of products referred to in Article 1 (1) which comply with the rules in force in their territory on the date of adoption of this Directive during a period of four years from that date.

Article 14

This Directive shall enter into force on the day of its publication in the *Official Journal of the European Communities*.

Article 15

This Directive is addressed to the Member States.

Done at Brussels, 16 June 1994.

For the European Parliament	*For the Council*
The President	*The President*
E. KLEPSCH	Y. PAPANTONIOU

Annex I

Essential safety requirements for the design and construction of recreational craft

1. BOAT DESIGN CATEGORIES

Design category	Wind force (Beaufort scale)	Significant wave height (H $\frac{1}{3}$, metres)
A – "Ocean"	exceeding 8	exceeding 4
B – "Offshore"	up to, and including, 8	up to, and including, 4
C – "Inshore"	up to, and including, 6	up to, and including, 2
D – "Sheltered waters"	up to, and including, 4	up to, and including, 0.5

Definitions:

A. OCEAN: Designed for extended voyages where conditions may exceed wind force 8 (Beaufort scale) and significant wave heights of 4 m and above, and vessels largely self-sufficient.

B. OFFSHORE: Designed for offshore voyages where conditions up to, and including, wind force 8 and significant wave heights up to, and including, 4 m may be experienced.

C. INSHORE: Designed for voyages in coastal waters, large bays, estuaries, lakes and rivers where conditions up to, and including, wind force 6 and significant wave heights up to, and including, 2 m may be experienced.

D. SHELTERED WATERS: Designed for voyages on small lakes, rivers, and canals where conditions up to, and including, wind force 4 and significant wave heights up to, and including, 0.5 m may be experienced.

Boats in each Category must be designed and constructed to withstand these parameters in respect of stability, buoyancy, and other relevant essential requirements listed in Annex I, and to have good handling characteristics.

2. GENERAL REQUIREMENTS

Recreational craft and components as referred to in Annex II shall comply with the essential requirements in so far as they apply to them.

2.1. **Hull identification**

Each craft shall be marked with a hull identification number including the following information:

- manufacturer's code,
- country of manufacture,
- unique serial number,
- year of production,
- model year.

The relevant harmonised standard gives details of these requirements.

2.2. **Builder's plate**

Each craft shall carry a permanently affixed plate mounted separately from the boat hull identification number, containing the following information:

- manufacturer's name,
- CE marking (see Annex IV),
- boat design category according to section 1,
- manufacturer's maximum recommended load according to section 3.6,
- number of persons recommended by the manufacturer for which the boat was designed to carry when under way.

2.3. **Protection from falling overboard and means of reboarding**

Depending on the design category, craft shall be designed to minimise the risks of falling overboard and to facilitate reboarding.

2.4. **Visibility from the main steering position**
For motor boats, the main steering position shall give the operator, under normal conditions of use (speed and load), good all-round visibility.

2.5. **Owner's manual**
Each craft shall be provided with an owner's manual in the official Community language or languages which may be determined by the Member State in which it is marketed in accordance with the Treaty. This manual should draw particular attention to risks of fire and flooding and shall contain the information listed in sections 2.2, 3.6 and 4 as well as the unladen weight of the craft in kilograms.

3. INTEGRITY AND STRUCTURAL REQUIREMENTS

3.1. **Structure**
The choice and combination of materials and its construction shall ensure that the craft is strong enough in all respects. Special attention shall be paid to the design category according to section 1, and the manufacturer's maximum recommended load in accordance with section 3.6.

3.2. **Stability and freeboard**
The craft shall have sufficient stability and freeboard considering its design category according to section 1 and the manufacturer's maximum recommended load according to section 3.6.

3.3. **Buoyancy and flotation**
The craft shall be constructed to ensure that it has buoyancy characteristics appropriate to its design category according to section 1.1, and the manufacturer's maximum recommended load according to section 3.6. All habitable multihull craft shall be so designed as to have sufficient buoyancy to remain afloat in the inverted position.

Boats of less than six metres in length that are susceptible to swamping when used in their design category shall be provided with appropriate means of flotation in the swamped condition.

3.4. **Openings in hull, deck and superstructure**
Openings in hull, deck(s) and superstructure shall not impair the structural integrity of the craft or its weathertight integrity when closed.

Windows, portlights, doors and hatchcovers shall withstand the water pressure likely to be encountered in their specific position, as well as point-loads applied by the weight of persons moving on deck.

Through hull fittings designed to allow water passage into the hull or out of the hull, below the waterline corresponding to the manufacturer's maximum recommended load according to section 3.6, shall be fitted with shutoff means which shall be readily accessible.

3.5. **Flooding**
All craft shall be designed so as to minimise the risk of sinking.

Particular attention should be paid where appropriate to:

- cockpits and wells, which should be self-draining or have other means of keeping water out of the boat interior,
- ventilation fittings,
- removal of water by pumps or other means.

3.6. **Manufacturer's maximum recommended load**
The manufacturer's maximum recommended load (fuel, water, provisions, miscellaneous equipment and people (in kilograms)) for which the boat was designed, as marked on the builder's plate, shall be determined according to the design category (section 1), stability and freeboard (section 3.2) and buoyancy and flotation (section 3.3).

3.7. **Liferaft stowage**
All craft of categories A and B, and craft of categories C and D longer than six metres shall be provided with one or more stowage points for a liferaft (liferafts) large enough to hold the number of persons the boat was designed to carry as recommended by the manufacturer. This (these) stowage point(s) shall be readily accessible at all times.

3.8. **Escape**
All habitable multihull craft over 12 metres long shall be provided with viable means of escape in the event of inversion.

All habitable craft shall be provided with viable means of escape in the event of fire.

3.9. **Anchoring, mooring and towing**
All craft, taking into account their design category and their characteristics shall be fitted with one or more strong points or other means capable of safely accepting anchoring, mooring and towing loads.

4. HANDLING CHARACTERISTICS
The manufacturer shall ensure that the handling characteristics of the craft are satisfactory with the most powerful engine for which the boat is designed and constructed. For all recreational marine engines, the maximum rated engine power shall be declared in the owner's manual in accordance with the harmonised standard.

5. INSTALLATION REQUIREMENTS
5.1. **Engines ands engine spaces**
5.1.1. *Inboard engine*
All inboard mounted engines shall be placed within an enclosure separated from living quarters and installed so as to minimise the risk of fires or spread the fires as well as hazards from toxic fumes, heat, noise or vibrations in the living quarters.

Engine parts and accessories that require frequent inspection and/or servicing shall be readily accessible.

The insulating materials inside engine spaces shall be non-combustible.

5.1.2. *Ventilation*
The engine compartment shall be ventilated. The dangerous ingress of water into the engine compartment through all inlets must be prevented.

5.1.3. *Exposed parts*
Unless the engine is protected by a cover or its own enclosure, exposed moving or hot parts of the engine that could cause personal injury shall be effectively shielded.

5.1.4. *Outboard engines starting*
All boats with outboard engines shall have a device to prevent starting the engine in gear, except:

(a) when the engine produces less than 500 newtons (N) of static thrust;
(b) when the engine has a throttle limiting device to limit thrust to 500 N at the time of starting the engine.

5.2. **Fuel system**
5.2.1. *General*
The filling, storage, venting and fuel-supply arrangements and installations shall be designed and installed so as to minimiuse the risk of fire and explosion.

5.2.2. *Fuel tanks*
Fuel tanks, lines and hoses shall be secured and separated or protected from any source of significant heat. The material the tanks are made of and their method of construction shall be according to their capacity and the type of fuel. All tank spaces shall be ventilated.

Liquid fuel with a flash point below 55°C shall be kept in tanks which do not form part of the hull and are:

(a) insulated from the engine compartment and from all other sources of ignition;
(b) separated from living quarters.

Liquid fuel with a flash point equal to or above 55°C may be kept in tanks that are integral with the hull.

5.3. **Electrical system**
Electrical systems shall be designed and installed so as to ensure proper operation of the craft under normal conditions of use and shall be such as to minimise risk of fire and electric shock.

Attention shall be paid to the provision of overload and short-circuit protection of all circuits, except engine starting circuits, supplied from batteries.

Ventilation shall be provided to prevent the accumulation of gases which might be emitted from batteries. Batteries shall be firmly secured and protected from ingress of water.

5.4. **Steering system**

5.4.1. *General*

Steering systems shall be designed, constructed and installed in order to allow the transmission of steering loads under foreseeable operating conditions.

5.4.2. **Emergency arrangements**

Sailboat and single-engined inboard powered motor boats with remote-controlled rudder steering systems shall be provided with emergency means of steering the craft at reduced speed.

5.5. **Gas system**

Gas systems for domestic use shall be of the vapour-withdrawal type and shall be designed and installed so as to avoid leaks and the risk of explosion and be capable of being tested for leaks. Materials and components shall be suitable for the specific gas used to withstand the stresses and exposures found in the marine environment.

Each appliance shall be equipped with a flame failure device effective on all burners. Each gas-consuming appliance must be supplied by a separate branch of the distribution system, and each appliance must be controlled by a separate closing device. Adequate ventilation must be provided to prevent hazards from leaks and products of combustion.

All craft with a permanently installed gas system shall be fitted with an enclosure to contain all gas cylinders. The enclosure shall be separated from the living quarters, accessible only from the outside and ventilated to the outside so that any escaping gas drains overboard. Any permanent gas system shall be tested after installation.

5.6. **Fire protection**

5.6.1. *General*

The type of equipment installed and the layout of the craft shall take account of the risk and spread of fire. Special attention shall be paid to the surroundings of open flame devices, hot areas or engines and auxiliary machines, oil and fuel overflows, uncovered oil and fuel pipes and avoiding electrical wiring above hot areas of machines.

5.6.2. *Fire-fighting equipment*

Craft shall be supplied with fire-fighting equipment appropriate to the fire hazard. Petrol engine enclosures shall be protected by a fire extinguishing system that avoids the need to open the enclosure in the event of fire. Where fitted, portable fire extinguishers shall be easily accessible and one shall be so positioned that it can easily be reached from the main steering position of the craft.

5.7. **Navigation lights**

Where navigation lights are fitted, they shall comply with the 1972 Colreg or CEVNI regulations, as appropriate.

5.8. **Discharge prevention**

Craft can be constructed so as to prevent the accidental discharge of pollutants (oil, fuel, etc.) overboard.

Craft fitted with toilets shall have either:

(a) holding tanks; or
(b) provision to fit holding tanks on a temporary basis in areas of use where the discharge of human waste is restricted.

In addition, any through-the-hull pipes for human waste shall be fitted with valves which are capable of being sealed shut.

Annex II
Components

1. Ignition-protected equipment for inboard and stern drive engines.
2. Start-in-gear protection devices for outboard engines.
3. Steering wheels, steering mechanisms and cable assemblies.
4. Fuel tanks and fuel hoses.
5. Prefabricated hatches and portlights.

Annex III
Declaration by the builder or his authorised representative established in the Community or the person responsible for placing on the market

(Article 4 (2) and (3))

(a) The declaration by the builder or his authorised representative estalished in the Community referred to in Article 4 (2) (partly completed craft) shall contain the following:
   - the name and address of the builder,
   - the name and address of the representative of the builder established in the Community or, if appropriate, of the person responsible for the placing on the market,
   - a description of the partly completed craft,
   - a statement that the partly completed craft is intended to be completed by others and that it complies with the essential requirements that apply at this stage of construction.

(b) The declaration by the builder, his authorised representative established in the Community or the person responsible for placing on the market referred to in Article 4 (3) (components) shall contain the following:
   - the name and address of the builder,
   - the name and address of the representative of the builder established in the Community or, if appropriate, of the person responsible for the placing on the market,
   - a description of the component,
   - a statement that the component complies with the relevant essential requirements.

Annex IV
CE marking

The CE conformity marking must consist of the initials "CE" taking the following form:

If the marking is reduced or enlarged, the proportions given in the above graduated drawing must be respected.

The various elements of the CE marking must have about the same vertical dimensions, which shall not be less than 5 mm.

The CE marking is followed by the identification number of the notified body, if it intervenes in the control of production, as well as by the last two figures of the year that the CE marking is affixed.

Annex V
Internal production control

(module A)

1. The manufacturer or his authorised representative established within the Community, who carries out the obligations laid down in point 2, ensures and declares that the products concerned satisfy the requirements of the Directive that apply to them. The manufacturer or his authorised representative established within the Community shall affix the CE marking to each product and draw up a written declaration of conformity (see Annex XV).

2. The manufacturer shall establish the technical documentation described in paragraph 3 and he or his authorised representative established within the Community shall keep it for a period ending at least 10 years after the last product has been manufactured at the disposal of the relevant national authorities for inspection purposes.

Where neither the manufacturer nor his authorised representative is established within the Community, the obligation to keep the technical documentation available shall be the responsibility of the person who places the product on the Community market.

3. Technical documentation shall enable the conformity of the products with the requirements of the Directive to be assessed. It shall, as far as relevant for such assessment, cover the design, manufacture and operation of the product (see Annex XIII).

4. The manufacturer or his authorised representative shall keep a copy of the declaration of conformity with the technical documentation.

5. The manufacturer shall take all measures necessary in order that the manufacturing process shall ensure compliance of the manufactured products with the technical documentation referred to in point 2 and with the requirements of the Directive that apply to them.

Annex VI
Internal production control plus tests

(module Aa, option 1)

This module consists of module A, as referred to in Annex V, plus the following supplementary requirements:

On one of several boats representing the production of the manufacturer one or more of the following tests, equivalent calculation or control shall be carried out by the manufacturer or on his behalf:

- test of stability according to point 3.2 of the Essential Requirements,
- test of buoyancy characteristics according to point 3.3 of the Essential Requirements.

Provisions common to both variations

These tests or calculations or control shall be carried out on the responsibility of a notified body chosen by the manufacturer. On the responsibility of the notified body, the manufacturer shall affix the former's distinguishing number during the manufacturing process.

Annex VII
EC type-examination

(module B)

1. A notified body ascertains and attests that a specimen, representative of the production envisaged, meets the provisions of the Directive that apply to it.

2. The application for the EC type-examination shall be lodged by the manufacturer or his authorised representative established within the Community with a notified body of his choice.

The application shall include:

- the name and address of the manufacturer and, if the application is lodged by the authorised representative, his name and address in addition,
- a written declaration that the same application has not been lodged with any other notified body,
- the technical documentation, as described in point 3.

The applicant shall place at the disposal of the notified body a specimen, representative of the production envisaged and hereinafter called "type".[1]

The notified body may request further specimens if needed for carrying out the test programme.

3. The technical documentation shall enable the conformity of the product with the requirements of the Directive to be assessed. It shall, as far as relevant for such assessment, cover the design, manufacture and functioning of the product (see Annex XIII).

4. The notified body shall:

4.1. examine the technical documentation, verify that the type has been manufactured in conformity with the technical documentation and identify the elements which have been designed in accordance with the relevant provisions of the standards referred to in Article 5, as well as the components which have been designed

[1] A type may cover several versions of the product provided that the differences between the versions do not affect the level of safety and the other requirements concerning the performance of the product.

without applying the relevant provisions of those standards;
4.2. perform or have performed the appropriate examinations and necessary tests to check whether, where the standards referred to in Article 5 have not been applied, the solutions adopted by the manufacturer meets the Essential Requirements of the Directive;
4.3. perform or have performed the appropriate examinations and necessary tests to check whether, where the manufacturer has chosen to apply the relevant standards, these have actually been applied;
4.4. agree with the applicant the location where the examination and necessary tests shall be carried out.
5. Where the type meets the provisions of the Directive, the notified body shall issue an EC type-examination certificate to the applicant. The certificate shall contain the name and address of the manufacturer, conclusions of the examination, conditions for its validity and the necessary data for identification of the approved type.

A list of the relevant parts of the technical documentation shall be annexed to the certificate and a copy kept by the notified body.

If the manufacturer is denied a type certification, the notified body shall provide detailed reasons for such denial.
6. The applicant shall inform the notified body that holds the technical documentation concerning the EC type-examination certificate of all modifications to the approved product which must receive additional approval where such changes may affect the conformity with the essential requirements or the prescribed conditions for use of the product. This additional approval is given in the form of an addition to the original EC type-examination certificate.
7. Each notified body shall communicate to the other notified bodies the relevant information concerning the EC type-examination certificates and additions issued and withdrawn.
8. The other notified bodies may receive copies of the EC type-examination certificates and/or their additions. The annexes to the certificates shall be kept at the disposal of the other notified bodies.
9. The manufacturer or his authorised representative shall keep with the technical documentation copies of EC type-examination certificates and their additions for a period ending at least 10 years after the last product has been manufactured.

Where neither the manufacturer nor his authorised representative is established within the Community, the obligation to keep the technical documentation available shall be the responsibility of the person who places the product on the Community market.

Annex VIII

Conformity to type

(module C)

1. The manufacturer or his authorised representative established within the Community ensures and declares that the products concerned are in conformity with the type as described in the EC type-examination certificate and satisfy the requirements of the Directive that applies to them. The manufacturer shall affix the CE marking to each product and draw up a written declaration of conformity (see Annex XV).
2. The manufacturer shall take all measures necessary to ensure that the manufacturing process assures compliance of the manufactured products with the type as described in the EC type-examination certificate and with

the requirements of the Directive that apply to them.
3. The manufacturer or his authorised representative shall keep a copy of the declaration of conformity for a period ending at least 10 years after the last product has been manufactured.

Where neither the manufacturer nor his authorised representative is established within the Community, the obligation to keep the technical documentation available shall be the responsibility of the person who places the product on the Community market (see Annex XIII).

Annex IX

Production quality assurance

(module D)

1. The manufacturer who satisfies the obligations of point 2 ensures and declares that the products concerned are in conformity with the type as described in the EC type-examination certificate and satisfy the requirements of the Directive that apply to them. The manufacturer or his authorised representative established within the Community shall affix the CE marking to each product and draw up a written declaration of conformity (see Annex XV). The CE marking shall be accompanied by the distinguishing number of the notified body responsible for the monitoring as specified in point 4.
2. The manufacturer shall operate an approved quality system for production, final product inspection and testing as specified in paragraph 3 and shall be subject to monitoring as specified in point 4.
3. **Quality system**
3.1. The manufacturer shall lodge an application for assessment of his quality system with a notified body of his choice, for the products concerned.

The application shall include:
- all relevant informataion for the product category envisaged,
- the documentation concerning the quality system,
- where appropriate, the technical documentation of the approved type (see Annex XIII) and a copy of the EC type-examination certificate.

3.2. The quality system shall ensure compliance of the products with the type as described in the EC type-examination certificate and with the requirements of the Directive that apply to them.

All the elements, requirements and provisions adopted by the manufacturer shall be documented in a systematic and orderly manner in the form of written policies, procedures and instructions. The quality system documentation must permit a consistent interpretation of the quality programmes, plan, manuals and records.

It shall contain in particular an adequate description of:
- the quality objectives and the organisational structure, responsibilities and powers of the management with regard to product quality,
- the manufacturing, quality control and quality assurance techniques, processes and systematic actions that will be used.
- the examinations and tests that will be carried out before, during and after manufacture, and the frequency with which they will be carried out,
- the quality records, such as inspection reports and test data, calibration data, qualification reports of the personnel concerned, etc.,
- the means to monitor the achievement of the required product quality and the effective operation of the quality system.

3.3. The notified body shall assess the quality system to determine whether it satisfies the requirements referred to in point 3.2. It shall presume conformity with these requirements in respect of quality systems that implement the relevant harmonised standard.

The auditing team shall have at least one member with experience of evaluation in the product technology concerned. The evaluation procedure shall include an inspection visit to the manufacturer's premises.

The decision shall be notified to the manufacturer. The notification shall contain the conclusions of the examination and the reasoned assessment decision.

3.4. The manufacturer shall undertake to fulfil the obligations arising out of the quality system as approved and to uphold it so that it remains adequate and efficient.

The manufacturer or his authorised representative shall keep the notified body that has approved the quality system informed of any intended updating of the quality system.

The notified body shall evaluate the modifications proposed and decide whether the amended quality system will still satisfy the requirements referred to in paragraph 3.2 or whether a reassessment is required.

It shall notify its decisions to the manufacturer. The notification shall contain the conclusions of the examination and the reasoned assessment decision.

4. **Surveillance under the responsibility of the notified body**

4.1. The purpose of surveillance is to make sure that the manufacturer duly fulfils the obligations arising out of the approved quality system.

4.2. The manufacturer shall allow the notified body entrance for inspection purposes to the locations of manufacture, inspection and testing, and storage and shall provide it with all necessary information, in particular:

- the quality system documentation,
- the quality records, such as inspection reports and test data, calibration data, qualification reports of the personnel concerned, etc.

4.3. The notified body shall periodically carry out audits to make sure that the manufacturer maintains and applies the quality system and shall provide an audit report to the manufacturer.

4.4. Additionally the notified body may pay unexpected visits to the manufacturer. During such visits the notified body may carry out, or cause to be carried out, tests to verify that the quality system is functioning correctly, if necessary. The notified body shall provide the manufacturer with a visit report and, if a test has taken place, with a test report.

5. The manufacturer shall, for a period ending at least 10 years after the last product has been manufactured, keep at the disposal of the national authorities:

- the documentation referred to in the second indent of the second subparagraph of point 3.1,
- the updating referred to in the second subparagraph of point 3.4,
- the decision and reports from the notified body which are referred to in the final subparagraph of point 3.4, point 4.3 and 4.4.

6. Each notified body shall give the other notified bodies the relevant information concerning the quality system approvals issued and withdrawn.

Annex X

Product verification

(module F)

1. This module describes the procedure whereby a manufacturer or his authorised representative established within the Community checks and

attests that the products subject to the provisions of point 3 are in conformity with the type as described in the EC type-examination certificate and satisfy the requiremnents of the Directive that apply to them.

2. The manufacturer shall take all measures necessary in order that the manufacturing process ensures conformity of the products with the type as described in the EC type-examination certificate and with the requirement of the Directive that apply to them. The manufacturer or his authorised representative established within the Community shall affix the CE marking to each product and shall draw up a declaration of conformity (see Annex XV).

3. The notified body shall carry out the appropriate examinations and tests in order to check the conformity of the product with the requirements of the Directive either by examination and testing of every product as specified in point 4 or by examination and testing of products on a statistical basis, as specified in point 5, at the choice of the manufacturer.

3a. The manufacturer or his authorised representative shall keep a copy of the declaration of conformity for a period ending at least 10 years after the last product has been manufactured.

4. **Verification by examination and testing of every product**

4.1. All products shall be individually examined and appropriate tests as set out in the relevant standard(s) referred to in Article 5 or equivalent tests shall be carried out in order to verify their conformity with the type as described in the EC type-examination certificate and the requirements of the Directive that apply to them.

4.2. The notified body shall affix, or cause to be affixed, its distinguishing number to each approved product and draw up a written certificate of conformity relating to the tests carried out.

4.3. The manufacturer or his authorised representative shall ensure that he is able to supply the notified body's certificates of conformity on request.

5. **Statistical verification**

5.1. The manufacturer shall present his products in the form of homogeneous lots and shall take all measures necessary in order that the manufacturing process ensures the homogeneity of each lot produced.

5.2. All products shall be available for verification in the form of homogeneous lots. A random sample shall be drawn from each lot. Products in a sample shall be individually examined and appropriate tests as set out in the relevant standard(s) referred to in Article 5, or equivalent tests, shall be carried out to ensure their conformity with the requirements of the Directive which apply to them and to determine whether the lot is accepted or rejected.

5.3. The statistical procedure shall use the following elements:

- the statistical method to be applied,
- the sampling plan with its operational characteristics.

5.4. In the case of accepted lots, the notified body shall affix, or cause to be affixed, its distinguishing number to each product and shall draw up a written certificate of conformity relating to the tests carried out. All products in the lot may be put on the market except those products from the sample which were found not to be in conformity.

If a lot is rejected, the notified body or the competent authority shall take appropriate measures to prevent the putting on the market of that lot. In the event of frequent rejection of lots the notified body may suspend the statistical verification.

The manufacturer may, under the responsibility of the notified body, affix the latter's distinguishing number during the manufacturing process.

5.5. The manufacturer or his authorised representative shall ensure that he is able to supply the notified body's certificates of conformity on request.

Annex XI
Unit verification

(module G)

1. This module describes the procedure whereby the manufacturer ensures and declares that the product concerned, which has been issued with the certificate referred to in point 2, conforms to the requirements of the Directive that apply to it. The manufacturer or his authorised representative established within the Community shall affix the CE marking to the product and draw up a declaration of conformity (see Annex XV).
2. The notified body shall examine the individual product and carry out the appropriate tests as set out in the relevant standard(s) referred to in Article 5, or equivalent tests, to ensure its conformity with the relevant requirements of the Directive.

The notified body shall affix, or cause to be affixed, its distinguishing number on the approved product and shall draw up a certificate of conformity concerning the tests carried out.
3. The aim of the technical documentation is to enable conformity with the requirement of the Directive to be assessed and the design, manufacture and operation of the product to be understood (see Annex XIII).

Annex XII
Full quality assurance

(module H)

1. This module describes the procedure whereby the manufacturer who satisfies the obligations of paragraph 2 ensures and declares that the products concerned satisfy the requirements of the Directive that apply to them. The manufacturer or his authorised representative established within the Community shall affix the CE marking to each product and draw up a written declaration of conformity (see Annex XV). The CE marking shall be accompanied by the distinguishing number of the notified body responsible for the surveillance as specified in point 4.
2. The manufacturer shall operate an approved quality system for design, manufacture and final production inspection and testing as specified in point 3 and shall be subject to surveillance as specified in point 4.
3. **Quality system**
3.1. The manufacturer shall lodge an application for assessment of his quality system with a notified body.

The application shall include:
- all relevant information for the product category envisaged,
- the quality system's documentation.

3.2. The quality system shall ensure compliance of the products with the requirements of the Directive that apply to them.

All the elements, requirements and provisions adopted by the manufacturer shall be documented in a systematic and orderly manner in the form of written policies, procedures and instructions. This quality system documentation shall ensure a common understanding of the quality policies and procedures such as quality programmes, plans, manuals and records.

It shall contain in particular an adequate description of:
- the quality objectives and the organisational structure, responsibilities

and powers of the management with regard to design and product quality,
- the technical design specifications, including standards, that will be applied and, where the standards referred to in Article 5 will not be applied in full, the means that will be used to ensure that the essential requirements of the Directive that apply to the products will be met.
- the design control and design verification techniques, processes and systematic actions that will be used when designing the products pertaining to the product category covered,
- the corresponding manufacturing, quality control and quality assurance techniques, processes and systematic actions that will be used,
- the examinations and tests that will be carried out before, during and after manufacture, and the frequency with which they will be carried out,
- the quality records, such as inspection reports and test data, calibration data, qualification reports of the personnel concerned, etc.,
- the means to monitor the achievement of the required design and product quality and the effective operation of the quality system.

3.3. The notified body shall assess the quality system to determine whether it satisfies the requirements referred to in point 3.2. It shall presume compliance with these requirements in respect of quality systems that implement the relevant harmonised standard (EN 29001).

The auditing team shall have at least one member experienced as an assessor in the product technology concerned. The evaluation procedure shall include an assessment visit to the manufacturer's premises.

The decision shall be notified to the manufacturer. The notification shall contain the conclusions of the examination and the reasoned assessment decision.

3.4. The manufacturer shall undertake to fulfil the obligations arising out of the quality system as approved and to uphold it so that it remains adequate and efficient.

The manufacturer or his authorised representative shall keep the notified body that has approved the quality system informed of any intended updating of the quality system.

The notified body shall evaluate the modifications proposed and decide whether the amended quality system will still satisfy the requirements referred to in paragraph 3.2 or whether a reassessment is required.

It shall notify its decision to the manufacturer. The notification shall contain the conclusions of the examination and the reasoned assessment decision.

4. **EC surveillance under the responsibility of the notified body**

4.1. The purpose of surveillance is to make sure that the manufacturer duly fulfils the obligations arising out of the approved quality system.

4.2. The manufacturer shall allow the notified body entrance for inspection purposes to the locations of design, manufacture, inspection and testing, and storage, and shall provide it with all necessary information, in particular:
- the quality system documentation,
- the quality records as foreseen by the design part of the quality system, such as results of analyses, calculations, tests, etc.,
- the quality records as foreseen by the manufacturing part of the quality system, such as inspection reports and test data, calibration data, qualification reports of the personnel concerned, etc.

4.3. The notified body shall periodically carry out audits to make sure that the manufacturer maintains and applies the quality system and shall provide an audit report to the manufacturer.

4.4. Additionally the notified body may pay unexpected visits to the manufacturer. At the time of such visits, the notified body may carry out tests or have them carried out in order to check the proper functioning of the quality system where necessary; it shall provide the manufacturer with a visit report and, if a test has been carried out, with a test report.

5. The manufacturer shall, for a period ending at least 10 years after the last product has been manufactured, keep at the disposal of the national authorities:

- the documentation referred to in the second indent of the second subparagraph of point 3.1,
- the updating referred to in the second subparagraph of point 3.4,
- the decisions and reports from the notified body which are referred to in the final subparagraph of point 3.4, point 4.3 and point 4.4.

6. Each notified body shall forward to the other notified bodies the relevant information concerning the quality system approvals issued and withdrawn.

Annex XIII

Technical documentation supplied by the manufacturer

The technical documentation referred to in Annexes V, VII, VIII, IX and XI must comprise all relevant data or means used by the manufacturer to ensure that components or craft comply with the essential requirements relating to them.

The technical documentation shall enable understanding of the design, manufacture and operation of the product, and shall enable assessment of conformity with the requirements of this Directive.

The documentation shall contain so far as relevant for assessment:

- a general description of the type,
- conceptual design and manufacturing drawings and schemes of components, sub-assemblies, circuits, etc.,
- descriptions and explanations necessary for the understanding of said drawings and schemes and the operation of the product,
- a list of the standards referred to in Article 5, applied in full or in part, and descriptions of the solutions adopted to fulfil the essential requirements when the standards referred to in Article 5 have not been applied,
- results of design calculations made, examinations carried out, etc.,
- test reports, or calculations namely on stability according to point 3.2 of the Essential Requirements and on buoyancy according to point 3.3 of the Essential Requirements.

Annex XIV

Minimum criteria to be taken into account by Member States for the notification of bodies

1. The body, its director and the staff responsible for carrying out the verification tests shall not be the designer, manufacturer, supplier or installer of boat or components which they inspect, nor the authorised representative of any of these parties. They shall not become either involved directly or as authorised representatives in the design, construction, marketing or maintenance of the said products. This does not preclude the possibility of

exchanges of technical information between the manufacturer and the body.

2. The body and its staff shall carry out the verification tests with the highest degree of professional integrity and technical competence and shall be free from all pressures and inducements, particularly financial, which might influence their judgment or the result of the inspection, especially from persons or groups of persons with an interest in the result of verifications.

3. The body shall have at its disposal the necessary staff and possess the necessary facilities to enable it to perform promptly the administrative and technical tasks connected with verification; it shall also have access to the equipment required for special verification.

4. The staff responsible for inspection shall have:
- sound technical and professional training,
- satisfactory knowledge of the requirements of the tests they carry out and adequate experience of such tests,
- the ability to draw up the certificates, records and reports required to authenticate the performance of the tests.

5. The impartiality of inspection staff shall be guaranteed. Their remuneration shall not depend on the number of tests carried out or on the results of such tests.

6. The body shall take out liability insurance unless its liability is assumed by the State in accordance with national law, or the Member State itself is directly responsible for the tests.

7. The staff of the body shall be bound to observe professional secrecy with regard to all information gained in carrying out its tasks (except *vis-à-vis* the competent administrative authorities of the State in which its activities are carried out) under the Directive or any provision of national law giving effect to it.

Annex XV

Written declaration of conformity

1. The written declaration of conformity to the provisions of the Directive must always accompany:
- the recreational craft and be joined to the owner's manual (Annex I, point 2.5),
- the components as referred to in Annex II.

2. The written declaration of conformity shall include the following:[1]
- name and address of the manufacturer or his authorised representative established in the Community,[2]
- description of the recreational craft[3] or of the component,[3]
- references to the relevant harmonised standards used, or references to the specifications in relation to which conformity is declared.
- where appropriate, reference to the EC type-examination certificate issued by a notified body,
- where appropriate, the name and address of the notified body,
- identification of the person empowered to sign on behalf of the manufacturer or his authorised representative established within the Community.

[1] and be drawn up in the language(s) as foreseen under point 2.5 of Annex I.

[2] business name and full address; authorised representative must also give the business name and address of the manufacturer.

[3] description of the product make, type, serial number, where appropriate.

Index